Mathematical and Statistical Applications in Food Engineering

Editors

Surajbhan Sevda

Department of Biosciences and Bioengineering
Indian Institute of Technology Guwahati
Guwahati-781039, India

Department of Biotechnology
National Institute of Technology Warangal
Warangal-506004, India

Anoop Singh

Department of Scientific and Industrial Research (DSIR)
Ministry of Science and Technology
Government of India, Technology Bhawan
New Delhi-110016, India

CRC Press
Taylor & Francis Group
Boca Raton London New York

CRC Press is an imprint of the
Taylor & Francis Group, an **informa** business

A SCIENCE PUBLISHERS BOOK

CRC Press
Taylor & Francis Group
6000 Broken Sound Parkway NW, Suite 300
Boca Raton, FL 33487-2742

Version Date: 20191219

International Standard Book Number-13: 978-1-138-34767-0 (Hardback)

Visit the Taylor & Francis Web site at
http://www.taylorandfrancis.com

and the CRC Press Web site at
http://www.crcpress.com

Foreword

Mathematical and Statistical techniques have a very broad applications in a vast number of fields including food technology. The food industry is growing very fast with increasing food demand and facing challenge to supply safe, nutritious and palatable products. The development of new food products/processes is complex, expensive and multistage risky process. The utilization and applications of mathematical and statistical modelling have increased in food science and technology to overcome the risks involved in the developmental processes of various food products.

Several Food Engineering and Food Science students face problems in understanding the mathematical application, especially in their research. This is the latest and updated book that having 25 different chapters with details of various new mathematical and statistical applications that will be used in food engineering to optimize process parameter.

This book explained the use of the design of the experiment, full factorial design, cluster analysis, multi-way statistical methods, partial least squares regression, principal component regression, response surface methodology, CFD simulation, 3 paradigm method, neural networks in experiments in food engineering and technology. The use of correlation, association, and regression to analyse food processes and products are also included. All these above methods are explained with their basic principle and used case study so that Food engineering researchers can understand these methods well and can use these in the domain of food engineering.

I congratulate the editors, Dr Surajbhan Sevda and Dr Anoop Singh, for seeing the need for such a book and producing it. This is a well-edited book and its material will be helpful for food engineering and food process students and food industry professionals and at a level appropriate to their backgrounds. This will also enhance student's maths skills with respect to the process parameter optimization and food quality measurement.

Overall, the information provided in this book is highly scientific, up-dated and would be beneficial for the researchers and practitioners equally; this will also be useful for those entering into this area. I strongly recommend this excellent book to food engineering scientists, Food process engineers and research students who are interested in Food Engineering in general and Food Process optimization in particular.

Poonam Singh Nigam

PhD CBiol PGCUT FRSB SFHEA FCHERP FBRSI FIFIBiop FAMSc DYEd DNSc
Programme Director Biotechnology Research,
School of Biomedical Sciences
Ulster University
Northern Ireland, UK

Food is a very complex matrix having different components such as carbohydrates, proteins, fats, vitamins, minerals besides moisture. This makes Food Processing also very complex and difficult proposition for more than one reason. Unlike inert chemical material, food cannot be subjected to harsh processing conditions because food is meant for human consumption. Hence apart from sensory attributes of the processed foods, care should be taken to minimize the loss in color, texture, nutrients, etc.

As working woman is sign of times, premium on time is increasing. Accordingly, the demand for processed foods as convenience foods is increasing. So the challenge for food processing engineer is to give the consumer the processed food which is very close to raw material in sensory quality and nutrition.

To achieve this, the food needs to be processed under optimal conditions. It requires optimization of process parameters of different unit operations. Another complexity in food processing, unlike that of chemical processing, the physical (density, viscosity, porosity, texture, etc.), chemical (protein denaturation, enzyme inactivation, etc.) and thermal (thermal conductivity, thermal diffusivity, etc.) properties of food change during processing, making it difficult to apply mathematical models for prior prediction of performance output.

In spite of these challenges, in order to arrive at the best possible process conditions, it is essential to employ different forms of modelling. The present book is an important contribution in this direction. It is aimed at bridging the gap between statistical science and engineering science and also at ensuring that food process engineers and food technologists are better equipped to serve the food processing industry at large.

The book presents application of various mathematical and statistical methods for food processing by renowned researchers from different countries. It covers application of the following in general for food processing: Optimization techniques, Role of design of experiments in process optimization, Application of correlation, regression and cluster analysis, Application of partial least square and principal component regression methods and Surface response methodology.

It also covers mathematical modelling for specific unit operations/processes such as high pressure processing, drying, baking, microwave heating, microwave drying, deep fat frying. In addition, it presents application of neural networks and computational fluid dynamics simulation for food processing.

This book will be of immense use to the Food Technologists and Food Process Engineers working in design and development of new products/process besides those involved in production. This is mainly because it offers a choice of suitable statistical techniques/methods which can be helpful in prior prediction of sensory quality and also in addressing variations/problems during large scale production before releasing the product into the market.

This book is invaluable for Researchers, Academicians and Students.

(KSMS Raghavarao)
FNAE, FNAAS, FASc
Director
CSIR-Central Food Technological Research Institute (CSIR-CFTRI),
Mysuru, India

Preface

Mathematical/statistical modelling is a prerequisite operation in food engineering as it is a complex process due to the presence of various scalable throughputs, rheological properties of food molecules, different process and environmental parameters, quality and stability concerns. The complexity lies in raw and processed food commodities as variability, which impacts the composition and their processability in further operations. The influencing parameters also impacted the other parameters due to mutual dependencies. Modelling facilitates the easy understanding of the proposed system by taking care of interaction and square terms along with the individual effects of the process. The application of mathematical and statistical tools is an essential component of a food-engineering program in both teaching and research. The importance of reaping the benefits of mathematical techniques has been used in the process analysis, design and optimization from an empirical to a scientific and model-based approach. This book is a cornucopia of information on mathematical and statistical methods that can be applied in food engineering. The use of these techniques is also included as a case study, which will make things easier for the researcher in terms of the development of alternative processes and their optimization in food engineering and technology.

The book on "Mathematical and Statistical Applications in Food Engineering' provides state-of-the-art information on the different mathematical and statistical methods application in food engineering. We have put together a host of highly relevant topics, ranging from the importance of the artificial neural network, design of experiments, full factorial design, correlation, association, regression, cluster analysis, multiway statistical methods for food processes, product development for food engineering and technology. Various applications such as multivariate statistical analysis, partial least squares regression, principal component regression, CFD simulations, reaction engineering approach are discussed in detail for optimizing different food processes, sensory evaluation, baking behaviour dough biscuits and analysis for food safety and quality assurance. Modelling facilitates the easy understanding of the proposed system by taking care of interaction and square terms along with the individual effects of the process. These aspects have been dealt with by their peers. Mathematical and statistical application in food engineering is a very powerful tool that can be used in different areas in order to reduce the number of experiments needed to decide which factors influence dependent variables. All these have been achieved in the book by describing the specialty processes and pioneering works. The editors have brought together a pool of expertise to present the state-of-art information, which has presented an in-depth analysis of the knowledge on various aspects.

This book provides a complete solution for all the latest mathematical methods used in food engineering and fermentation technology. This book provides the basic knowledge of how statistical methods can be used to solve the typical problem associated with food engineering and fermentation technology. Combining theory with a practical, hands-on approach, this book covers the key aspects of food engineering. A complement to Mathematical and statistical applications in food engineering. Presenting cutting-edge information, this book is an essential reference for the fundamental concepts associated with food engineering today.

With contributions from a broad range of leading professors and scientists from all over the globe, this book focuses on new areas of mathematical and statistical methods for food engineering to help meet the increasing food demand of the rapidly growing populations of the world. This is a friendly book on the mathematical method for food engineering, it will help researchers, and students to overcome this apathy, appreciate the beauty of analytical tools, sharpen their mathematical skills through examples, and exercise problems. This book offers an accessible guide to applying statistical and mathematical technologies in the food-engineering field whilst also addressing the statistical methods and theoretical foundations. Using clear examples and case studies by way of practical illustration, the book is more than just a theoretical guide for non-statisticians is, and may, therefore, be used by scientists, students and food industry professionals at different levels and with varying degrees of statistical skill. This book also provides all different mathematical methods used in food engineering, which will be helpful to the researchers in this area.

Guwahati, India   **Surajbhan Sevda**

New Delhi, India   **Anoop Singh**

Contents

CHAPTER **1**

Role of Mathematical and Statistical Modelling in Food Engineering

Surajbhan Sevda,[1,]* *Vijay Kumar Garlapati*[2] *and Anoop Singh*[3]

1. Importance of Mathematical Modelling in Food Engineering

The food engineering domain is considered to be a complex process due to the presence of various scalable throughputs, rheological properties of food molecules, different processes and environmental parameters, quality and stability concerns. Moreover, the influencing parameters impact the other parameters through the mutual dependencies. The complexity also lies in the raw and processed food commodities in the form of variability, which impacts the composition and their processability in further operations. Various food engineering problems, such as heat and mass transfer operations, are heterogeneous and variable. Therefore, it is not possible to handle them with only basic disciplines, such as mathematics, physics and science (Barnabé et al., 2018).

The heterogeneity and variability encountered in the food engineering domain can be made manageable by utilizing the modelling approaches which put forth the non-linear relationship of the different process parameters on the performance of the process or on the final yield of the food product (Shenoy et al., 2015). Mathematical modelling plays a vital role in understanding the food process related to thermal (heat and mass transfer) and non-thermal processes (high-pressure processing, pulsed electric field) and in assessing the possible microbial and biochemical changes during the shelf-life of the food products (Farid, 2010). The modelling approaches also tackle the scale-up associated problem and are useful in simulations, since they take care of impossible process parameters, such as high temperatures (Dutta, 2016).

The machine-learning/simulation-based optimization approaches help in attaining/predicting the optimum output by considering the individual, square and interaction parameters of the process. The machine-learning based optimization approaches produce a huge set of data for processing with the intuitive knowledge of the nature-based phenomenon. Moreover, the modelling and optimization

[1] Department of Biosciences and Bioengineering, Indian Institute of Technology Guwahati, Guwahati, India.
[2] Department of Biotechnology and Bioinformatics, Jaypee University of Information Technology, Wakhanaghat-173234, Himachal Pradesh, India.
[3] Government of India, Ministry of Science and Technology, Department of Scientific and Industrial Research (DSIR), Technology Bhawan, New Mehrauli Road, New Delhi-110016, India.
* Corresponding author: sevdasuraj@gmail.com

approach also helps in attaining a deductive conclusion with a smaller set of experiments with the positive expert-knowledge based methodologies (Perrot et al., 2016). Computational Fluid Mechanics (CFM) also play an essential role in the food engineering domain, with various applications in process operations, such as drying, freezing, cooking, sterilization and baking, by providing food engineers with new insights into the probable performance of the equipment during the time-course of quality or safety (Dutta, 2016). Artificial Neural Networks (ANN) are also one of the machine learning tools with the potential for use in the food industry to tackle the food characteristics and understand the structure-process interaction during the production step of food products (Huang et al., 2007).

2. Experiment Design and Data Analysis

The most important step in food engineering is the experiment design. In all the related work, such as enzyme production and other food related products, there is a need to design the experiment before beginning in the laboratory or at pilot scale. All of the dependent and independent variables need to be defined before the experiment, and based on the optimization methods, an accurate mathematical method should be chosen. Figure 1 explains the various basic terminologies used in experiment design.

Figure 2 shows the three basic principles used for the design of experiment (DOE). The three basic principles for DOE are 'replication', 'randomization' and 'control of experimental error'.

Using randomization eliminates the systematic bias and aids in the distribution of all the designed experiments. In order to make a completely randomized design (CRD), the entire experiment must be divided into a number of experimental units, say x. In this design, the total number of replications for different treatments are fixed in advance. The replication procedure helps in repeating the experiment at the centre stage and it also estimates the experimental error. The experiments are repeated twice or thrice and mean values are also estimated. The control of experimental error will in turn increase the efficiency.

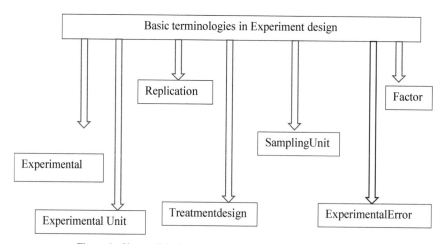

Figure 1: Shows all the basic terminologies in the experiment design.

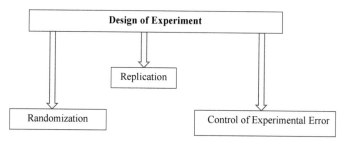

Figure 2: Shows the basic principles used for the design of experiment.

If the sample size is large, the CRD experiments are not homogenous and there are n treatments to be compared, then it may be possible to: # Blocks are build and this procedure is called as randomized block design (RBD). The experimental material is divided into rows and columns in the Latin square method (LSD). In the LSD, each experiment occurred only once in the each row and column.

3. Typical Problems in Application of Mathematical Modelling in Food Engineering

The typical problems encountered while using the modelling and optimization approaches in a food engineering discipline are mainly due to the usage of low-level manual-coded languages, such as FORTRAN or C, for implementation. This is highly resource-consuming and error-prone. Furthermore, it is impossible to utilize the approaches to plant-wide simulation by taking care of multiple interacting units. The FORTRAN or C-based modelling approaches also lack robustness and efficiency in dealing with the multitask problems associated with the food engineering domain (Banga et al., 2003).

Despite the well-proven modelling and optimization approaches in the food industry for the typical processes, the utilization of the adopted approaches for other food engineering operations is limited due to the following factors:

- The developed food process model's applicability is not straightforward (Bimbenet et al., 2007).
- Uncertainties encountered due to the raw material properties, process parameters and unreliability associated with the process conditions (Ioannou et al., 2006).
- The developed process models are applicable to tiny sectors of the food engineering domain (Perrot et al., 2006).
- The proposed process models are not able to integrate with the other developed models (Bimbenet et al., 2007).

Moreover, due to the dynamic nature of food processes, the developed food process models are very complicated in terms of the structural understanding, nature of simulation and optimized control. The complexity will be greater while developing a process model and taking care of relationships between composition, structure, properties, quality and shelf-life of the food, resulting in a very challenging task (Fito et al., 2007).

4. Conclusion

Mathematical/statistical modelling facilitates a real understanding of the non-linear problems associated with food engineering operations with the aid of statistical heuristics and helps in the development of validated non-linear regression models. This approach helps in attaining the process optimization with well-planned and less experimental DOE. The developed mathematical models help in straightforward extrapolation towards the scale-up studies. In conclusion, there is a need for the industry and academia to work together in order to further the progression in modelling approaches by handling the typical problems encountered in the execution of modelling approaches in food engineering.

References

Banga, J.R., Balsa-Canto, E., Moles, C.G. and Alonso, A.A. 2003. Improving food processing using modern optimization methods. Trends in Food Science & Technology 14: 131–144.

Barnabé, M., Blanc, N., Chabin, T., Delenne, J.Y., Duri, A., Frank, X. et al. 2018. Multiscale modelling for bioresources and bioproducts. Innovative Food Science & Emerging Technologies 46: 41–53.

Bimbenet, J.J., Schubert, H. and Trystram, G. 2007. Advances in research in food process engineering as presented at ICEF9. Journal of Food Engineering 78: 390–404.

Datta, A.K. 2016. Toward computer-aided food engineering: Mechanistic frameworks for evolution of product, quality and safety during processing. Journal of Food Engineering 176: 9–27.

Farid, M.M. 2010. Mathematical Modelling of Food Processing, published by CRC Press, Taylor & Francis Group, 2010, ISBN 978-1-4200-5351-7.

Fito, P., LeMaguer, M., Betoret, N. and Fito, P.J. 2007. Advanced food process engineering to model real foods and processes: The SAFES methodology. Journal of Food Engineering 83: 390–404.

Huang, Y., Kangas, L.J. and Rasco, B.A. 2007. Applications of artificial neural networks (ANNs) in food science. Crit Rev Food SciNutr 47(2): 113–26.

Ioannou, I., Mauris, G., Trystram, G. and Perrot, N. 2006. Back-propagation of imprecision in a cheese ripening fuzzy model based on human sensory evaluations. Fuzzy Sets and Systems 157: 1179–1187.

Perrot, N., Ioannou, I., Allais, I., Curt, C., Hossenlopp, J. and Trystram, G. 2006. Fuzzy concepts applied to food product quality control: A review. Fuzzy Sets and Systems 157: 1145–1154.

Perrot, N., De Vries, H., Lutton, E., Van Mil, H.G., Donner, M. et al. 2016. Some remarks on computational approaches towards sustainable complex agrifood systems. Trends in Food Science & Technology 48: 88–101.

Shenoy, P., Viau, M., Tammel, K., Innings, F., Fitzpatrick, J. and Ahrné, L. 2015. Effect of powder densities, particle size and shape on mixture quality of binary food powder mixtures. Powder Technology 272: 165–172.

CHAPTER 2

Evolutionary Optimization Techniques as Effective Tools for Process Modelling in Food Processing

Lakshmishri Roy,[1,*] *Debabrata Bera*[2] and *Vijay Kumar Garlapati*[3]

1. Introduction

Most food processing firms are making persistent efforts to maximize their returns and minimize their process costs to compete in the existing market scenario. Consequently, these industries need to opt for advanced alternative technologies for improving, monitoring, optimizing and controlling process parameters like nutrients, moisture content, temperatures, etc. (Rodríguez-Fernández et al., 2007). Processing operations in these industries are conducted in a dynamic, unpredictable environment, subject to a large number of constraints, i.e., quality of the final product, financial, environmental, safety aspects, etc. Therefore, extracting an optimal solution from a large set of options for a food processing problem is an arduous task. Hence, a useful model-based optimization tool is essential to accomplish it. An exhaustive evaluation of the cons of the existing tools has been summarized below:

1.1 Limitations of mathematical optimization techniques

Specific characteristics of the food processing operations, like those mentioned below, make it difficult for application of mathematics-based optimization tools:

- Most of the processes are conducted in a batch or semi-batch mode. Hence, the models employed need to be dynamic, non-linear models with discrete events.

- Many process variables of these studies (temperature, pH, concentration, etc.), are more often spatially distributed and coupled with transport phenomena, thus making it difficult for mathematical models using only partial differential equations.

[1] Dept. of Food Technology, Techno India, Kolkata, West Bengal-700091, India.
[2] Dept. of Food and Biochemical Engineering, Jadavpur University, Kolkata, West Bengal-700032, India
[3] Dept. of Biotechnology & Bioinformatics, Jaypee University of Information Technology (JUIT), Waknaghat, Himachal Pradesh-173234, India.
* Corresponding author: lakshmi1371@gmail.com

- Complicated nonlinear constraints issued from safety and quality aspects associated with food processing operations cannot be effectively represented in mathematical optimization models.
- Also, the food processes more often involve coupled time-dependent transport phenomena, making it even more difficult.

Thus, optimization of such processes requires an alternative physics-based model capable of being used in a systemic search approach in conjunction with explicit and implicit constraints.

1.2 Empirical equation-based models

Operational barriers limit extensive use of statistical, empirical equation-based models (Garlapati and Roy, 2017; Chauhan and Garlapati, 2014; Sharma et al., 2016) for optimization in food process engineering operations. Most of the simulators consequentially developed using these tools trail the traditional path of employing low-level languages. These tools are both highly resource-consuming and error-prone, thereby making them non-applicable for plant-wide simulation.

1.3 Challenges in extensive utilization of tools and simulators in food industries

Modern-day simulators can increase productivity much more effectively in comparison to the traditional modelling approach. These high-level modeling systems are advantageous in terms of (i) better and ease of maintainability, (ii) flexibility in facilitating effective communication between co-workers and partners, (iii) ease in development, reusability, etc.

- In spite of their advantages, many of these models lack robust and efficient optimization solvers and, hence, preclude a more widespread use for optimization studies in the food industry.
- Another type of barrier arises from human-resources and knowledge issues:
  In most food processing industries, the managerial and technical human resources are often not familiar with these simulation and optimization tools. Even competent people with the relevant technological know-how are skeptical in applying these tools for food industries as these processing operations are incredibly complex.
- The food industries are need dynamic models that mimic their processes because, for a long time, there have been a lack of tailor-made modeling and optimization software tools. These may be like the tools developed by de Prada (2001) for the sugar industry.
- Most of the real-world problems of food processing operations have multiple, often competing for objectives as the raw materials involved are complex and of wide variations, making it difficult.
- Food processing operations more often encompass multiple suboptimal and equivalent solutions, thus posing a major challenge in developing an optimization model for them adequately.

Non-convex problems of these industries are solvable using conventional global optimization methodologies but, for issues of non-identifiability, the complexibility to be dealt with persists. Additionally, the desired process performances of these operations encompass variables and constraints that attribute to the economic impact on efficiency product quality and safety. A class of linear search algorithms, i.e., Evolutionary Algorithms (EAs), are seemingly vital tools for challenges that make things difficult in existing search and optimization situations. These algorithms have gained popularity in recent times because of their ease in the way of handling multiple objective problems, irrespective of the multi-objective optimization problems being constrained or unconstrained (Karaboga, 2004; Saputelli et al., 2004). Thus, evolutionary optimization tools may be successfully applied to these food processing industries.

2. Evolutionary Algorithms/Optimization Tools

These computational biological-inspired optimization algorithms, based on natural evolution and selection principles, are popularly used for solving non-differentiable, intermittent and multimodal optimization problems.

Salient aspects of EAs include

- They operate on a population of potential solutions and yield effective and improved results using evolutionary-like operations that work on the principle of survival of the fittest (selection, reproduction and mutation) (Ronen et al., 2002).
- These optimization tools can generate Pareto optimal solutions for complex processes with many objective functions and constraints and, hence, can be used for optimization processes (Garlapati et al., 2017; Garlapati and Banerjee, 2013; Garlapati et al., 2011).
- Evolutionary algorithms (EAs) are the ultimate tool to overcome limitations (Price, 1999; Boillereaux et al., 2003; Mariani et al., 2008) of a situation lacking problem-solving technique because of multiple local minima due to unidentified process parameters.

EAs differ from the traditional methods in the following aspects

- These algorithms work with coded versions of the parameter set and do not operate with the parameters themselves directly.
- Optimal search is made from a population of points and not a single point.
- Objective functions are used, not derivatives or other ancillary information.
- Probabilistic transition rules are applied instead of deterministic rules.

3. Basic Operational Characteristics of Evolutionary Algorithms

An evolutionary algorithm is a biologically inspired, generic, population-based optimization algorithm. Its mechanism includes:

- **Reproduction/procreation:** The process of producing new "offspring" from their "parents".
- **Mutation:** Alteration in the order of the process being considered (e.g., organism, production or business process, code).
- **Recombination:** A process of exchange of information between two processes yielding a new combination of processes (e.g., operations in a workflow process).
- **Selection:** A method by which traits become either more or less common in a population as a function of the influence of traits concerning the intended goal (e.g., increased production efficiency in a production process). Selection is a key evolution mechanism. Probable solutions of the optimization problem for which an evolutionary algorithm is employed to arrive at, are viewed as entities in a population. A fitness function is used to assess its suitability as a solution. **A fitness function** is an objective function that is used to summarize how close a given solution is to fulfilling the optimization goals. All the stated operators are applied several times in the process and, hence, the term "evolutionary".

The evolutionary process thus involves

- ➤ **Generation** of the initial population (i.e., first-generation) of individuals randomly.
- ➤ **Evaluation of the fitness** of each entity of the population based on the optimization criteria given.
- ➤ **Repetition of the fitness evaluation** on this generation till its termination, wherein the termination criteria can be time limit, etc.
- ➤ **Selection of the best-fit individuals,** i.e., parents for subsequent reproduction.

> ➢ **Breeding of** new individuals through crossover (for bringing in variation from one generation to the other) and mutation (for varying the programming from one generation to the next) operations to yield offspring from the best fit individuals.
> ➢ **Evaluation of** the new individuals fitness.
> ➢ **Replacement of least-fit population** with new individuals.
> ➢ This sequence of the evolutionary process is repeated until an individual fulfilling the fitness criteria within the given parameters is obtained.

4. Types of Evolutionary Algorithms

Evolutionary algorithms are robust global optimal solutions that help in overcoming the limitations of traditional methods. The various evolutionary optimization techniques available include: Genetic algorithm (GA), differential evolution (DE), particle swarm optimization (PSO), artificial neural networks (ANNs), fuzzy logic (FL), and ant colony optimization (ACO) (Bhattacharya et al., 2011; Adeyemo, 2011; Sarker and Ray, 2009; Kennedy and Eberhart, 1995).

4.1 Genetic Algorithm (GA)

GAs, as depicted in Fig. 1 below, are optimization algorithms that mimic natural evolution (Holland 1975, 1973; Mohebbi et al., 2008; Babu and Munawar, 2007). They have been employed to obtain near-optimum solutions for a large number of situations (Gen and Cheng, 1996). One limitation of GAs is the long processing time required for the near-optimum solution to evolve.

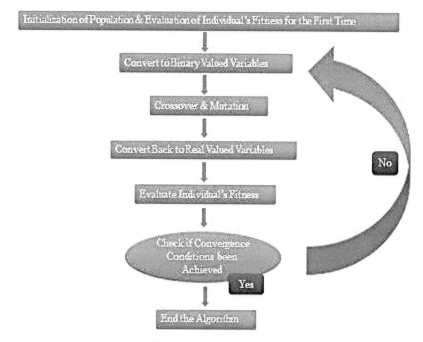

Figure 1: Flow chart of GA.

4.2 Differential Evolution (DE)

DE algorithm is a stochastic, population-based optimization method like GA; optimization functions with real variables and multiple local optima (Storn and Price, 1997; Pierreval et al., 2003) can be effectively optimized with this algorithm. A mutation is the primary search mechanism (Godfrey and Babu, 2004) for these search optimization tools. DE is self-adaptive (Karaboga, 2004). These algorithms have many advantages (Abbass et al., 2001; Strens and Moore, 2002). DE exhibits more convergence speed than genetic algorithms (Abbass et al., 2001; Strens and Moore, 2002; Karaboga, 2004). Its process flowchart has been depicted in Fig. 2 below.

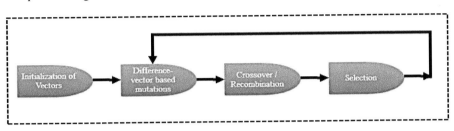

Figure 2: Sequence of events in DE.

4.3 Fuzzy Modeling (FM)

It is a robust method, encompassing scientific and heuristic modelling approaches. It mimics human control logic wherein they utilize the data and expert knowledge. Its input data may be an imprecise, descriptive language as a human operator (Huang et al., 2010). Fuzzy systems have been extensively applied to solve different problems. The present trend is towards enhancing their effectivity by employing soft-computing methods, such as fuzzy genetic systems.

4.4 Particle Swarm Optimization (PSO)

These are metaheuristic algorithms. PSO's mimic the social behavior of flocks of birds and schools of fish. Its process flowchart is depicted by Fig. 3.

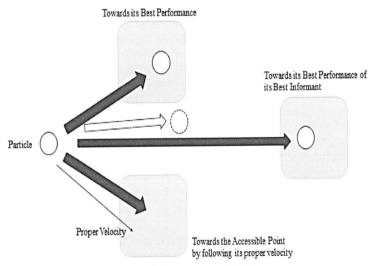

Figure 3: Process events in PSO.

5. Overview of Application of EA's in Food Processing Industries

Food processing industries involve a large number of unit operations, each **governed by a series of dynamic conditions** that include mass, heat and momentum transfer operations. Likewise, **market-driven parameters,** like cost, demand and consumer acceptability factors and **the regulatory norms**

Table 1: Comparative summary of evolutionary algorithms/optimization techniques.

Algorithm	Pros	Cons
Genetic algorithms	• Alter information (crossover or mutation) • Effective to solve continuous process issues	• Lack retention • Early convergence • Meager local search ability • Effective and impressive computational effort • Challenging to translate a problem in the form of a chromosome
Particle swarm optimization	• Possesses memory • Convenient for execution as it employs a simple operator • Promising to resolve continuous problems	• Early convergence • Poor local search ability
Ant colony optimization	• Retains the information • Yields good solutions rapidly • Effective in solving discrete and varied types of problems	• Untimely convergence • Ineffective local search ability • Yields changes in probability distribution with iterations • Ineffective in cracking the continuous problems

on the quality parameters of the product, all **dictate the decision of fixing the operable strategy** for a product. Thus, modeling and optimization of these processes are highly challenging with the development of models governed by laws of mass, energy, etc., and capable of predicting the physicochemical, quality properties and safety aspects of the products. Kinetic models reflect the **change in relevant state variables** with time and position when the food sample is subjected to **different processing conditions** (Tijskens et al., 2001; Wang and Sun, 2003). The **Shelf life of products** impact, shortage and surplus of goods which in turn may impact income for the manufacturing units and, hence, this aspect also needs to be considered. Therefore, it may be summarized that optimization techniques are essential tools in food processing operations, used to enhance the economic values of processing and for the marketing of food.

Currently, there lies a pressing need to ensure admirable product quality. Consequentially, the food industries are focusing more attention on improving their processing operations (e.g., Effective methods for drying, wetting, heating, cooling and freezing of foods are necessary (Doganis et al., 2006). Thus, it is becoming imperative to implement advanced optimization tools like EAs and the related techniques thereof, in the complex operations of modern food processing industries. EAs, like Differential evolution (DE) algorithms, have been successfully applied to solve several optimization problems of chemical and biological processes (Liu and Wang, 2010; Cheng and Ramaswamy, 2002; Chiou and Wang, 2001; Lu and Wang, 2001) while other similar EA tools have been used for the fuzzy-decision making problems of fuel ethanol production (Wang and Cheng, 1999), fermentation process (Wang and Cheng, 1999) and other engineering issues (Garlapati et al., 2010; Garlapati and Banerjee, 2010a,b; Babu, 2004, 2007; Angira and Babu, 2006; Babu and Angira, 2002; Babu and Jehan, 2003; Sarimveis and Bafas, 2003). These studies concluded that these techniques are less time consuming than the existing techniques and can adequately estimate the optimal parameters. Summarized below is the current status of application of these tools in various operations.

5.1 Role of EA's in food-based fermentation

Fermentative processes are dynamic and involve a large number of process variables (e.g., media parameters and process parameters like aeration rate, temperature, duration of incubation, etc.). These processes are governed by mass transfer, heat transfer principles, kinetic models and operational constraints. Traditional optimization techniques for resolving the multiple intended objectives of these operations are mostly non-lucrative. Evolutionary algorithms (EAs) are preferred alternative methods for monitoring the state variables of these dynamic fermentative operations (Soons et al., 2008). Artificial neural networks (ANNs) and genetic algorithm (GA) that mimic different aspects of biological information processing for data modelling and media optimization have proven to be effective for optimization problems of these sectors (Baishan et al., 2003). Impressive results have been obtained from ANN-GA for simultaneous maximization of biomass and conversion of product, e.g., pentafluoroacetophenon with *Synechococcus* PCC 7942 (Franco-Lara et al., 2006), and fermentative production of xylitol from *Candida mogii* (Desai et al., 2006; Baishan et al., 2003), exopolysaccharides production by *Lactobacillus plantarum* isolated from the fermented *Eleusine coracan*. In the latter application, Plackett Burman (PB) was applied to identify the three most influential media components, ANN for modeling the non-linear relationship between the operating variables and the intended objectives, finally the ANN model was used as an input for the optimization through GA. The optimization of hydantoinase production from *Agrobacterium radiobacter*, lipase production from a mixed culture and glucansucrase production from *Leuconostoc dextranicum* NRRL B-1146 was performed with ANN–GA model using RSM-based data by Nagata and Chu, 2003; Haider et al., 2008; Singh et al., 2008, respectively. Kovarova-Kovar et al., 2000 demonstrated optimization using hybrid algorithms in the fed-batch process for riboflavin production. GAs with their multi-objective problem-solving capabilities have been applied in synthesis and optimization of non-ideal distillation systems (Fraga and Senos, 1996), computer-aided molecular design (Shunmugam et al., 2000), optimal design of xylitol synthesis reactor (Baishan et al., 2003), estimation of parameters in trickle bed reactors (González-Sáiz et al., 2008), on-line optimization of culture temperature for yeast fermentation (Yüzgeç et al., 2009) and optimal ethanol production (Guo et al., 2010; Rivera et al., 2006).

5.2 Evolutionary optimization for extrusion-based processes

RSM integrated GA-based optimization was reported to be effective in predicting the optimal process conditions with the intended quality of an extruded fish product. The conditions yielded a product with more desirable features than those obtained by specific condition optimization. Process variables taken for consideration included: Screw speed, feed moisture content, expansion ratio, water solubility index, barrel temperature, bulk density, hardness, and fish contents for single-screw extrusion cooking of a fish and rice flour blend (Shankar and Bandyopadhyay, 2004).

5.3 EA's application in the dairy industry

Preparation of different types of milk powder, i.e., whole milk powder, spray dried milk powder, skimmed milk powder, etc., in the dairy industry involves operations with multi-process parameters affecting the final product quality. Independent parameters required may include screw speed, process temperature, milk powder feed rate to the drier, addition rate of additive, etc. (Koc et al., 2007). The depended parameters include free fat content, lactose crystallinity, particle size and colour, while the constraint was desired power consumption. The intended multi-objectives include maximization of free fat content, the crystallinity of lactose and minimization of particle size (Koe et al., 2007). Fuzzy logic was also used in the real-time control of a spray–drying of whole milk powder processing. The algorithm used controlled the process at the desired power consumption and yielded entire milk products with the desired attributes (Queiroz and Nebra, 2001).

5.4 Application of EA's in oil processing

Neural network-based genetic algorithm optimization tools have been employed for multi-objective estimations during oil processing. Experimental data from vegetable oil hydrogenation process plant was used to develop the model and the intended objectives included minimization of isomer and maximization of cis –oleic acid (Izadifar and Jahromi, 2007).

5.5 EA tools for food product quality evaluation

Quality parameters of the final food product play an important role in the consumer acceptability and approval by the food safety standard norms. Quality parameters may vary for the type of food products and the processing conditions also affect these parameters. Artificial neural networks have been applied in predicting the selected intended quality parameters with variations in process conditions during different unit operations for varied types of food products, e.g., extruded products (Linko et al., 1992), rheological dough properties in bakery operations (Ruan et al., 1997), meat quality (Yan et al., 1998), bakery products (Cho and Kim, 1998) and post-harvest processed products (Morimoto et al., 1997a,b). The impact of thawing conditions on Thermal properties of gelatin was determined using artificial neural networks (Boillereaux et al., 2003). Mittal and Zhang, 2000 developed a feed-forward neural network to predict the freezing and thawing time of food products with simple regular shapes. The results demonstrated that the developed ANN-GA-based models were useful for the estimation of parameters that were usually considered for foods of varied structural, morphological configurations and compositions.

5.6 Utilization of evolutionary optimization tools in drying operations

Performance of a drying process in the food industry is assessed from improvement in manufacturing quality and reduction in energy consumption. Optimization techniques, when applied to drying operations, are intended for reduction of drying time and occasionally the process cost. Nonlinear predictive control genetic algorithm and the like have been developed and reported (Yüzgeç et al., 2006, 2009; Na et al., 2002; Potocnik and Grabec, 2002; Mankar et al., 2002; Quirijns et al., 2000). The intended objective was final product quality enhancement, minimal energy consumption during drying and reduced process cost by developing a control procedure for the drying process.

In recent times, ANNs have been receiving more considerable attention in modelling the drying operations (Chen et al., 2000; Kaminski et al., 1998; Sreekanth et al., 1998), food rheology (Ruan et al., 1995) and thermal processing (Sablani et al., 1997a,b). Structural identifiability analysis of model methods for improvement in the efficacy and robustness of the model parameter has been proposed and demonstrated in many reports (Rodríguez-Fernández et al., 2007; Movagharnejad and Nikzad, 2007). ANN model was reported to be more accurate than empirical correlations in describing the drying behavior of tomato.

The optimization of multiproduct batch plants design issues for protein production using fuzzy multi-objective algorithm concepts was demonstrated by Dietz et al., 2008. The model developed provided an up-and-coming framework that could take imprecision into account during the new product development stage and finally in making the decision. Kiranoudis and Markatos, 2000 considered the multi-objective design of food dryers using a static mathematical model for simultaneous minimization of economic measure and the colour deviation of the final product.

6. Case Study of EA in Thermal Processing Operation

Thermal processing is an active food preservation strategy, for inactivation of microbial spores that are a public health concern or of microbial species responsible for spoilage of foods in containers. These operations are conducted at temperatures well above the ambient boiling point of water. For these purposes, pressurized steam retorts/autoclaves are operated at conditions not detrimental to food quality (Simpson et al., 2003; Holdsworth and Simpson, 2007; Abakarov et al., 2009). For obtaining the extended shelf life of products and their intended safety, cost-efficient treatments are widely preferred in industries. The imperatives, like health value, drive these thermal treatment operations and the economic aspects of sustainable food supply, they also minimize food-borne illness and food waste and retain or enhance nutritive quality to ensure affordability. Thermal treatments are mostly capable of catering to the imperatives mentioned above. There is an increasing concern regarding the harmful effects of these thermal treatments, i.e., compromised nutritive quality, deteriorating effects on essential nutrients and unique bioactive phytochemicals.

6.1 Basic objective

Thermal treatment optimization is a **dynamic process** where the **intended objective** is to determine the **optimal heating temperature-time combination** that effectively **maximizes the final nutrient retention** of a per packaged conduction heated food. The operable **constraint** involved is that the heating temperature should be effective **to impart microbiological lethality**.

Thus, there are two contradictory demands: To obtain the **desired minimum lethality**, all the sections of the food must be subjected to a high enough temperature for sufficient duration. However, the same exposure is likely to destroy nutrients; therefore, it is desired to **minimize that undesirable effect**.

6.2 Challenges

- Thermal destruction of microbes is conventionally proven to follow a first-order semi-logarithmic rate. Consequentially, it is not likely that, theoretically, a sterile product can't be produced with certainty even after exposure of the food product for long process time.
- If the intended product is to be rendered utterly void of microorganisms, then the thermal treatment is likely to yield a product which is unwholesome or inferior in quality. Thus, commercial sterility or shelf stability of the products is the most preferred or sought-after objective by the processing authorities of industries.
- Thermal destruction issue is dynamic and, hence, it implicates dynamic optimization techniques to determine optimal operating policies. The EA techniques are more effective than the traditional (constant temperature) processes. The optimal strategies can enhance the quality of the final product, and/or reduce the processing time to yield the desired quality level. Reduced-order models have generated cost-effective simulations (Banga et al., 2003). With these "accelerated" models, the dynamic optimization issue can be performed in just a few seconds. Since EA tools are capable of minimizing the complexity of a process model which seems promising for food process operations with a new avenue for real-time optimization and control.
- Treatments engaged in thermal destruction of microbes encompass a multi-objective optimization problem where the reduction of total process time and significant retention of several nutrients and quality factors need to be deliberated simultaneously (Fryer and Robbins, 2005). To this effect, Sendín et al., 2006, 2010 proposed and applied a novel multicriteria optimization method to the thermal processing of foods.

6.3 Background and principle

Designing an effective thermal processing strategy makes it imperative to have an extensive understanding of process methods, the heating behavior of the product and its impact on a target microorganism. Thus, the dependable factors for gauging the severity of any thermal process must be known and they include:

- **The physical characteristics of the food product.**
- **The type and thermal resistance of the target microorganisms.**
- **The changes in intrinsic properties of the food which affect the survival of microbes in thermal processes.**

6.4 Basic design premise and concerns of thermal processes

For design optimization studies, the user should be reminiscent of the following:

- The **heat resistance of microorganisms** for each specific product formulation and composition. Thermal inactivation kinetics of microorganisms is essential and may be obtained from a survivor curve, i.e., a logarithmic plot of the number of microorganisms surviving a given heat treatment at a given temperature against the heating time. Thermal inactivation generally follows a first-order reaction. Two key parameters (D and z values) are then determined by the survivor and resistance curves, respectively.

- **The heating rate of the specific product:** This is essential for mathematical modeling of experimental data, which aids in understanding the impact of process parameters, on relevant pathogens. The effective heating rate of the product is accomplished from a detailed analysis of product and system parameters affecting the heating behavior of the product.
- **The conditions for which such models apply** and
- Their **limitations** since food matrices are complex and can influence microbial resistances in different ways.

Design concerns include

- Simple, Robust, flexible models operable for process deviation analysis and ensuring appropriate levels of public safety are becoming popular. A single "fit-all-data" model is not effective for explaining or describing the complex behavior of microbes when subjected to external agents (such as temperature, salt, pH, etc.), and their interactions.
- Varied time-temperature combinations, processing methods, systems or techniques may yield the desired lethality. However, these variations are also likely to impact the quality of the end product to different extents. Therefore, minimal changes to the desired sensory and organoleptic attributes of food products are always intended through process optimization routines and thereby determine system appropriateness using kinetic data for the most heat-sensitive nutrient.
- The time-temperature history of a product undergoing thermal treatment will depend on several factors that include but are not limited to: (i) the processing system (conventional, static or agitating retorts, etc.), (ii) the heating medium (steam, water immersion, etc.), (iii) product characteristics including consistency, solid/liquid ratio, and thermophysical properties, (iv) product initial and heating medium temperatures, and (v) container type, shape and size.

6.5 Guidelines for applying evolutionary algorithms for thermal process optimization studies

Step 1: Identification of process parameters affecting the process under consideration.

Step 2: Collection of data from experiments conducted based on chosen Design of experiments (DOE). Selection of DOE is from amongst Plackett Burman, Factorial designs, Central composite design, etc., to have representations from the significant combinations of the process parameters (Kumari et al., 2013; Chauhan et al., 2013; Mahapatra et al., 2009).

Step 3: Developing the thermal process schedule, i.e., model development using the experimental data from the heat penetration and kinetic data (z and Freq values) by the conventional methods. Formula methods have been currently developed and employed to impart flexibility to establish times to achieve the desired cumulative lethality. Incidentally, these formula methods have limited implementation in optimization studies and automatic control systems as they are incapable of defining dynamic functions during the entire processing. Artificial neural networks (ANN) are effective to computerize mathematical aspects of thermal process calculations. These models mitigate the need for a large storage space while computerizing.

Step 4: Application of optimization tool to the developed model.
The main factors/selected objective functions taken into consideration during the optimization of thermal processing include **final product quality and safety, consumption of energy** and **total processing time**. The diversity of the sighted processing objectives imposes different optimal conditions to sterilization/thermal processing. Hence, an algorithm capable of considering

various objective functions is preferred for the determination of the optimal thermal processing of food. These software packages are intended to ascertain the optimum variable temperature profiles concerning the intended objective functions, geometric options and constraints chosen by the processor/user. The packages should facilitate/automate **the calculation of heat transfer coefficient** for irregular or regular geometries, various shapes and heating conditions, **predict moisture loss, shrinkage, yield loss, internal temperatures** and **lethality for different sizes of product**. These capabilities would make them a useful tool in taking processing decisions. These algorithm packages should be proficient at simulating food safety by combining a physics-based model of food processes, with the microbial kinetics and chemical transformations to provide a microbial count and/or nutrient and/or undesired chemicals amounts at any time and in any location in the food during processing.

Step 5: Mathematical formulation of the Problem statement for thermal sterilization of foods.

In canning operations, the problems mentioned below may be addressed using modeling and optimization techniques:

- Estimation of a retort function, where the final quality retention or surface quality retention is maximized, while the final process lethality is held to a specified minimum.
- Determination of a retort function, such that the final process time is minimized subject to the same lethality requirement, while the quality retention does not fall beneath some specified minimum.
- Fixation of a retort function, where the cooked value is minimized, while the final process lethality is held to a specified minimum.
- Fixation of a retort function, such that the final process time is minimized subject to the same lethality requirement, while the quality retention is not below some specified minimum, and the energy consumption is not above a specified maximum; minimum and maximum values are computed at constant retort temperature profiles.
- The thermal process optimization problem is, thus, posed as a multi-objective optimization problem.

In each of these cases:

- **The lethality constraint** is specified as: (i)

$$F_{0(t)} = \int_0^{tf} 10^{\frac{T - Teff}{zf}} \, dt$$

where $F_{0(t)}$ is the final required lethality which is calculated using: $\int_0^{tf} 10^{\frac{T - Teff}{zf}} \, dt$

where T is the temperature at the critical point or cold spot, normally the geometric center of the container (in the case of conduction-heated canned foods), Teff is the reference temperature and Zf the thermal resistance of the microorganisms.

- **The quality retention constraint** is specified as:

$$C_{av}(t_f) = \frac{1}{Vt} \int_0^{Vt} \exp[-\frac{ln10}{Drefc} \int_0^{tf} 10^{\frac{T - Treff}{Zc}} dt] dVt$$

where Cv is the desired volume-average final quality retention value and is calculated using the equation given above, where Treff is the reference temperature, z c and D are kinetics of the degradation of nutrients.

- **The surface retention is given by**

$$S(t_f) = \exp[-\frac{ln10}{Dfefs} \int_0^{tf} 10^{\frac{T - Trefs}{Zs}} dt]$$

where Trefs is the reference temperature, z and D are kinetics of the degradation of nutrients, and the surface retention constraint can be specified as S(t), which is the desired final surface retention value.

- Also, a common relationship for estimating quality losses is the "Cook or Cvalue", which is calculated using .... $C(t_f) = \int_0^{tf} \frac{T - Trefv}{Zq} dt$

where z and Trefv represent the z-value and reference temperature for the most heat-labile component. The z-value for cooking degradation within the given range corresponds to sensory attributes, texture softening, and color changes. The z-value of 33.1°C and Tref equal to 100°C are often used to compute a cook value to describe the overall quality loss. The cook value constraint is specified as Cd, where Cd is the desired minimum final cook value.

These expressions are standard model equations and, for consideration of new process parameters, model equations may be obtained using regression analysis or artificial neural network algorithms. The obtained models are then subjected to numerous iterations within the set range of the individual parameters and subject to the constraint of the individual problem statements to attain the set levels of the objective functions. The iterations are performed using evolutionary optimization techniques until the deviations between the set and predicted values are minimal enough to achieve the optimal conditions, which are further validated by performing the processes at those conditions.

7. Advantages of EA's

Evolutionary computation techniques only require an evaluation of the objective function and not an exhaustive mathematical requirement on the optimization problem. There are zero order methods capable of handling nonlinear problems and dependent on discrete, mixed or continuous spaces, irrespective of whether they are unconstrained or constrained using operators that are global in scope.

- Evolutionary algorithms are a potential source of breakthroughs for most of the food industrial engineering processes that include challenging, unstructured, real-life problems to be modeled as they include unfamiliar factors ranging from risk factors to aesthetics. They have the potential to provide many near-optimal solutions at the end of an optimization run which facilitates selection of the best solution later, based on criteria that were either incoherent from the expert or poorly modeled. The efficiency of EA's can be enhanced because of their flexibility and comparative ease of being hybridized with domain-dependent heuristics.

- These optimizers are global optimization methods that can be scaled up to higher-dimensional problems. EAs are robust concerning noisy evaluation functions, and can effectively handle evaluation functions which do not yield a sensible result in a given period.

- The algorithms are incredibly flexible and, hence, can be moderated, changed and customized to fit the problem at hand. They are applicable in many complex problem-solving applications, unlike classical search and optimization techniques.

- EAs are inspired by natural evolution and, hence, conceptually flexible and straightforward.

- EAs use prior information and, thus, outperform the methods which utilize the prior information minimally and with restricted search space.

- EA is representation independent, i.e., applies to constrained or unconstrained sets and to sets whether discrete or continuous, unlike most of the numeric techniques.

- Evaluation in Evolutionary optimization processes is performed as a parallel operation and only operations of the selection process are serially processed.

- Evolutionary algorithms that develop adaptability to yield a solution in changing environment are robust, unlike the traditional optimization tools which vary according to variations in the surrounding environment.

- EAs are capable of solving problems without any human intervention, hence, handy tools. However, these tools do not perform satisfactorily for automating problem-solving routines.

8. Disadvantages of EA's

- Evolutionary algorithms do not always assure an optimal solution to a definite problem within the anticipated time. There lies a great need for tuning of parameters by trial-and-error, thereby necessitating lots of computational resources.

- The performance of evolutionary search methods in the optimization of food engineering problems is highly impacted as the majority of these are constrained problems.

- The confirmatory conclusion of the best suited evolutionary algorithm for a given problem remains unanswered. The standard values provide good performance, but, interestingly, a variation in configurations tends to yield better results. The adverse configuration may lead to premature convergence, generating local optima and not the global optima.

9. Conclusion

This chapter provides an overview of the use of computational-based optimization algorithms in major real-world applications, intending to find global optimum solutions for food processing industry issues. Multi-objective optimization problems of modern food processing operations, whether constrained or unconstrained, have been resolved using new hybrid optimizers. Process treatments affect product quality, safety and marketing. Hence, the use of new techniques for the optimization of food treatment processes becomes vital. Consequentially, basic research on the modeling, simulation, designing and evaluation of parameters affecting different food processes is vital. It is suggestive that EA's are likely to have a positive impact on solving real-world issues/challenges in the food processing industries shortly. Also, it is noteworthy that, though these EA's are extensively applicable in many areas, these too come with marginal success in performance. Hence, the current efforts are focused on the application of some parallel algorithms along with Evolutionary Algorithms, that is, to hybridize two or more algorithms or to improve the existing algorithms.

References

Abakarov, A., Sushkov, Y., Almonacid, S. and Simpson, R. 2009. Thermal processing optimization through a modified adaptive random search. Journal of Food Engineering 93: 200–209.

Adeyemo, J.A. 2011. Reservoir operation using multi-objective evolutionary algorithms—a review. Asian Journal of Science and Research 4: 16–27.

Angira, R. and Babu, B.V. 2006. Optimization of process synthesis and design problems: A modified differential evolution approach. Chemical Engineering Science 61: 4707–4721.

Babu, B.V. and Angira, R. 2002. A differential evolution approach for global optimization of MINLP problems. In Proceedings of Fourth Asia-Pacific Conference on Simulated Evolution and Learning (SEAL'02), November 18–22, 2002, Singapore 1033(2): 880–884.

Babu, B.V. and Jehan, M.M. 2003. Differential evolution for multi-objective optimization. In Proceedings of IEEE Congress on Evolutionary Computation, Dec. 8–12, 2696–2703. Canberra: IEEE Press.

Babu, B.V. 2004. Process Plant Simulation. New Delhi: Oxford University Press.

Babu, B.V. 2007. Improved differential evolution for single- and multiobjective optimization: MDE, MODE, NSDE, and MNSDE. *In:* K. Deb, P. Chakroborty, N.G.R. Iyengar and S.K. Gupta (eds.). Advances in Computational Optimization and its Applications, 1–7. Hyderabad: Universities Press.

Babu, B.V. and Munawar, S.A. 2007. Differential evolution strategies for optimal design of shell-and-tube heat exchangers. Chemical Engineering Science 62: 3720–3739.

Baishan, F., Hongwen, C., Xiaolan, X., Ning, W. and Zongding, H. 2003. Using genetic algorithms coupling neural networks in a study of xylitol production: Medium optimisation. Process Biochemistry 38: 979–985.

Banga, J.R., Balsa-Canto, E., Moles, C.G. and Alonso, A.A. 2003. Improving food processing using modern optimization methods. Trends in Food Science and Technology 14(4): 131–144.

Bhattacharya, S.S., Garlapati, V.K. and Banerjee, R. 2011. Optimization of laccase production using response surface methodology coupled with differential evolution. New Biotechnology 28(1): 31–39.

Boillereaux, L., Cadet, D. and Le Bail, A. 2003. Thermal properties estimation during thawing via real-time neural network learning. Journal of Food Engineering 57(1): 17–23.

Chauhan, M., Chauhan, R.S. and Garlapati, V.K. 2013. Modelling and optimization Studies on a novel lipase production by staphylococcus arlettae through submerged fermentation. Enzyme Research 2013: 1–8.

Chauhan, M. and Garlapati, V.K. 2014. Modelling embedded optimization strategy for formulation of bacterial lipase based bio-detergent. Industrial & Engineering Chemistry Research 53(2): 514–520.

Chen, C.R., Ramaswamy, H.S. and Alli, I. 2000. Neural network-based optimization of quality of osmo-convective dried blueberries. In Proceedings of FOOD SIM' 2000, First International Conference on Simulation in Food and Bio Industries 33–35. Nantes: Food Sim Press.

Chen, C.R. and Ramaswamy, H.S. 2002. Modeling and optimization of variable retort temperature (VRT) thermal processing using coupled neural networks and genetic algorithms. Journal of Food Engineering 53: 209–220.

Chiou, J.P. and Wang, F.S. 2001. Estimation of monod parameters by hybrid differential evolution. Bioprocess and Biosystems Engineering 24: 109–113.

Desai, K.M., Akolkar, S.K., Badhe, Y.P., Tambe, S.S. and Lele, S.S. 2006. Optimization of fermentation media for exopolysaccharide production from Lactobacillus plantarum using artificial intelligence-based techniques. Process Biochemistry 41: 1842–1848.

Dietz, A., Aguilar-Lasserre, A., Azzaro-Pantel, C., Pibouleau, L. and Domenech, S. 2008. A fuzzy multiobjective algorithm for multiproduct batch plant: Application to protein production. Computers & Chemical Engineering 32(1-2): 292–306.

Doganis, P., Alexandridis, A., Patrinos, P. and Sarimveis, H. 2006. Time series sales forecasting for short shelf-life food products based on artificial neural networks and evolutionary computing. Journal of Food Engineering 75: 196–204.

Fraga, E.S. and Senos, M.T.R. 1996. Synthesis and optimization of a nonideal distillation system using a parallel genetic algorithm. Computers & Chemical Engineering 20: 79–84.

Franco-Lara, E., Link, H. and Weuster-Botz, D. 2006. Evaluation of artificial neural networks for modeling and optimization of medium composition with a genetic algorithm. Process Biochemistry 41: 2200–2206.

Fryer, P. and Robbins, P. 2005. Heat transfer in food processing: ensuring product quality and safety. Applied Thermal Engineering 25: 2499–2510.

Garlapati, V.K., Vundavilli, P.R. and Banerjee, R. 2010. Evaluation of lipase production by Genetic algorithm and Particle swarm optimization and their comparative study. Applied Biochemistry and Biotechnology 162: 1350–1361.

Garlapati, V.K. and Banerjee, R. 2010a. Optimization of lipase production using Differential evolution. Biotechnology and Bioprocess Engineering 15(2): 254–260.

Garlapati, V.K. and Banerjee, R. 2010b. Evolutionary and swarm intelligence based approaches for optimization of lipase extraction from fermented broth. Engineering in Life Sciences 10(3): 1–9.

Garlapati, V.K., Vundavilli, P.R. and Banerjee, R. 2011. Integration of RSM model for optimization of immobilized lipase mediated solvent-free synthesis of flavour ester by genetic algorithm. In Proceedings of IEEE First International Conference on Image Information Processing (ICIIP) 1–4. JUIT Waknaghat, HP, India: IEEE press.

Garlapati, V.K. and Banerjee, R. 2013. Enhanced lipase recovery through RSM integrated differential evolutionary approach from the fermented biomass. Brazilian Archives of Biology and Technology 56(5): 699–709.

Garlapati, V.K., Vundavilli, P.R. and Banerjee, R. 2017. Optimization of flavour ester production through artificial bee colony algorithm: ABC optimization approach for flavour ester production. In Proceedings of IEEE Fourth International Conference on Image Information Processing (ICIIP) 2017, 1–4. JUIT Waknaghat, HP, India: IEEE press.

Garlapati, V.K. and Roy, L.S. 2017. Utilization of response surface methodology for modeling and optimization of tablet compression process. Journal of Young Pharmacists 9(3): 417–421.

Gen, M. and Cheng, R. 1996. Genetic Algorithms and Engineering Design. New York: Wiley.

Godfrey, C.O. and Babu, B.V. 2004. New Optimization Techniques in Engineering. Heidelberg: Springer-Verlag.

González-Sáiz, J.M., Pizarro, C. and Garrido-Vidal, D. 2008. Modelling gas liquid and liquid-gas transfers in vinegar production by genetic algorithms. Journal of Food Engineering 87(1): 136–147.

Guo, Y., Xu, J., Zhang, Y., Xu, H., Yuan, Z. and Li, D. 2010. Medium optimization for ethanol production with Clostridium autoethanogenum with carbon monoxide as sole carbon source. Bioresource Technology 101: 8784–8789.

Haider, M.A., Pakshirajan, K., Singh, A. and Chaudhry, S. 2008. Artificial neural network and genetic algorithm approach to optimize media constituents for enhancing lipase production by a soil microorganism. Applied Biochemistry and Biotechnology 144: 225–235.

Holdsworth, S.D. and Simpson, R. 2007. Thermal Processing of Packaged Foods. New York: Springer.

Holland, J.H. 1973. Genetic algorithms and the optimal allocation of trials. SIAM Journal of Computation 2(2): 88–105.

Holland, J.H. 1975. Adaptation in Natural and Artificial Systems. Ann Arber: University of Michigan Press.

Huang, Y., Lan, Y., Thomson, S.J., Fang, A., Hoffmann, W.C. and Lacey, R.E. 2010. Development of soft computing and applications in agricultural and biological engineering. Computers and Electronics in Agriculture 71(2): 107–127.

Izadifar, M. and Jahromi, M.Z. 2007. Application of genetic algorithm for optimization of vegetable oil hydrogenation process. Journal of Food Engineering 78: 1–8.

Kaminski, W.P., Strumillo, P. and Romczak, E. 1998. Neurocomputing approaches to modeling of drying process dynamics. Drying Technology 16(6): 967–992.

Karaboga, D. 2004. A simple and global optimization algorithm for engineering problems: differential evolution algorithm. Turk. Journal of Electrical Engineering 12(1): 53–60.

Kennedy, J. and Eberhart, R. 1995. Particle swarm optimization. In Proceedings of IEEE International Conference on Neural Networks 1942–1948. Perth: IEEE Press.

Kiranoudis, C.T. and Markatos, N.C. 2000. Pareto design of conveyor-belt dryers. Journal of Food Engineering 46: 145–155.

Koc, A.B., Heinemann, P.H. and Ziegler, G.R. 2007. Optimization of whole milk powder processing variables with neural networks and genetic algorithms. Food and Bioproducts Processing 85(4): 336–343.

Kovarova-Kovar, K., Gehlen, S., Kunze, A., Keller, T., Von Daniken, R., Kolb, M. and VanLoon, A.P.G.M. 2000. Application of model-predictive control based on artificial neural networks to optimize the fed-batch process for riboflavin production. Journal of Biotechnology 79: 39–52.

Kumari, A., Garlapati, V.K., Mahapatra, P. and Banerjee, R. 2013. Modeling, simulation and kinetic studies of solvent-free biosynthesis of Benzyl acetate. Journal of Chemistry 2013: 1–9.

Linko, P., Uemura, K. and Eerikainen, T. 1992. Application of neural network models in fuzzy extrusion control. Transactions in Chemical Engineering 70(3): 131–137.

Liu, P.K. and Wang, F.S. 2010. Hybrid differential evolution including geometric mean mutation for optimization of biochemical systems. Journal of Taiwan Institute of Chemical Engineers 41: 65–72.

Mahapatra, P., Kumari, A., Garlapati, V.K., Banerjee, R. and Nag, A. 2009. Enzymatic synthesis of fruit flavor esters by immobilized lipase from Rhizopus oligosporus optimized with response surface methodology. Journal of Molecular Catalysis B: Enzymatic 60: 57–63.

Mankar, R.B., Saraf, D.N. and Gupta, S.K. 2002. On-line optimizing control of bulk polymerization of methyl methacrylate: Some experimental results for heater failure. Journal of Applied Polymer Science 85: 2350–2360.

Mariani, V.C., Barbosa de Lima, A.G. and Coelho, L.S. 2008. Apparent thermal diffusivity estimation of the banana during drying using inverse method. Journal of Food Engineering 85(4): 569–579.

Mittal, G.S. and Zhang, J. 2000. Prediction of freezing time for food products using a neural network. Food Research International 33: 557–562.

Mohebbi, M., Barouei, J., Akbarzadeht, M.R., Rowhanimanesh, A.R., Habibinajafi, M.B. and Yavarmanesh, M. 2008. Modeling and optimization of viscosity in enzyme-modified cheese by fuzzy logic and genetic algorithm. Computers and Electronics in Agriculture 62: 260–265.

Morimoto, T., Baerdemaeker, J.D. and Hashimoto, Y. 1997a. An intelligent approach for optimal control of fruit-storage process using neural networks and genetic algorithms. Computers and Electronics in Agriculture 18: 205–224.

Morimoto, T., Purwanto, W., Suzuki, J. and Hashimoto, Y. 1997b. Optimization of heat treatment for fruit during storage using neural networks and genetic algorithms. Computers and Electronics in Agriculture 19: 87–101.

Movagharnejad, K. and Nikzad, M. 2007. Modeling of tomato drying using artificial neural network. Computers and Electronics in Agriculture 59(1-2): 78–85.

Na, J.G., Chang, Y.K., Chung, B.H. and Lim, H.C. 2002. Adaptive optimization of fed-batch culture of yeast by using genetic algorithms. Bioprocess Engineering 24: 299–308.

Nagata, Y. and Chu, K.H. 2003. Optimization of a fermentation medium using neural networks and genetic algorithms. Biotechnology Letters 25: 1837–1842.

Potocnik, P. and Grabec, I. 2002. Nonlinear model predictive control of a cutting process. Neurocomputing 43: 107–126.

Price, K. 1999. New Ideas in Optimization. San Francisco: McGraw-Hill Publishing Company.

Queiroz, M.R. and Nebra, S.A. 2001. Theoretical and experimental analysis of the drying kinetics of bananas. Journal of Food Engineering 47: 127–132.

Quirijns, E.J., van Willigenburg, L.G., van Boxtel, A.J.B. and van Straten, G. 2000. The significance of modeling spatial distributions of quality in optimal control of drying processes. JA Benelux Quartly Automatic Control 41: 56–64.

Ronen, M., Shabtai, Y. and Guterman, H. 2002. Optimization of feeding profile for a fed-batch bioreactor by an evolutionary algorithm. Journal of Biotechnology 97: 253–263.

Ruan, R., Almaer, S. and Zhang, J. 1995. Prediction of dough rheological properties using neural networks. Cereal Chemistry 72(3): 308–311.

Ruan, R., Almaer, S., Zou, C. and Chen, P.L. 1997. Spectrum analyses of mixing power curves for neural network prediction of dough rheological properties. Transactions in ASAE 40(3): 677–681.

Sablani, S.S., Ramaswamy, H.S., Sreekanth, S. and Prasher, S.O. 1997a. Neural network modeling of heat transfer to liquid particle mixtures in cans subjected to end-over-end processing. Food Research International 30(2): 105–116.

Sablani, S.S., Ramaswamy, H.S., Sreekanth, S. and Prasher, S.O. 1997b. A neural network approach for thermal processing application. Journal of Food Processing and Preservation 19(4): 283–301.

Sarimveis, H. and Bafas, G. 2003. Fuzzy model predictive control of nonlinear processes using genetic algorithms. Fuzzy Sets and Systems 139: 59–80.

Sarker, R. and Ray, T. 2009. An improved evolutionary algorithm for solving multi-objective crop planning models. Computers and Electronics in Agriculture 68: 191–199.

Sendín, J.O.H., Otero-Muras, I., Alonso, A.A. and Banga, J.R. 2006. Improved optimization methods for the multiobjective design of bioprocesses. Industrial Engineering & Chemistry Research 45: 8594–8603.

Sendín, J.O.H., Alonso, A.A. and Banga, J.R. 2010. Efficient and robust multiobjective optimization of food processing: A novel approach with application to thermal sterilization. Journal of Food Engineering 98(3): 317–324.

Shankar, T.J. and Bandyopadhyay, S. 2004. Optimization of extrusion process variables using a genetic algorithm. Food and Bioproducts Processing 82(2): 143–150.

Sharma, D., Garlapati, V.K. and Goel, G. 2016. Bioprocessing of wheat bran for the production of lignocellulolytic enzyme cocktail by Cotylidia pannosa under submerged conditions. Bioengineered 7(2): 88–97.

Shunmugam, M.S., Reddy, S.V.B. and Narendran, A.A. 2000. Selection of optimal conditions in multi-pass face-milling using a genetic algorithm. International Journal of Machine Tools and Manufacture 40: 401–414.

Simpson, R., Almonacid, S. and Teixeira, A. 2003. Optimization criteria for batch retort battery design and operation in food canning-plants. Journal of Food Process Engineering 25(6): 515–538.

Singh, A., Majumder, A. and Goyal, A. 2008. Artificial intelligence based optimization of exocellular glucansucrase production from Leuconostoc dextranicum NRRL B-1146. Bioresource Technology 99: 8201–8206.

Soons, Z., Streefland, M., Van straten, G. and Van boxtel, A.J.B. 2008. Assessment of near infrared and "software sensor" for biomass monitoring and control. Chemometrics and Intelligent Laboratory Systems 94: 166–174.

Sreekanth, S., Ramaswamy, H.S. and Sablani, S. 1998. Prediction of psychrometric parameters using neural networks. Drying Technology 16(3-5): 825–837.

Storn, R. and Price, K. 1997. Differential evolution—a simple and efficient heuristic for global optimization over continuous spaces. Journal of Global Optimization 11: 341–359.

Strens, M. and Moore, A. 2002. Policy search using paired comparisons. Journal of Machine Learning Research 3: 921–950.

Tijskens, L.M.M., Hertog, M. and Nicolai, B.M. 2001. Food Process Modeling. Cambridge/Boca Raton: Woodhead Pub. Lim./CRC Press LLC.

Wang, F.S. and Cheng, M.W. 1999. Simultaneous optimization of feeding rate and operation parameters for fed-batch fermentation processes. Biotechnology Progress 15(5): 949–952.

Wang, L.J. and Sun, D.W. 2003. Recent developments in numerical modeling of heating and cooling processes in the food industry—A review. Trends in Food Science and Technology 14(10): 408–423.

Yüzgeç, U., Becerikli, Y. and Turker, M. 2006. Nonlinear predictive control of a drying process using genetic algorithms. ISA Transactions 45(4): 589–602.

Yüzgeç, U., Turker, M. and Hocalar, A. 2009. On-line evolutionary optimization of an industrial fed-batch yeast fermentation process. ISA Transactions 48: 79–92.

CHAPTER 3

Optimization of Food Processes Using Mixture Experiments

Some Applications

Daniel Granato,[1,*] *Verônica Calado*[2,*] *and Edmilson Rodrigues Pinto*[3,*,#]

1. Introduction

Design of experiments (DOE) is a very powerful tool that can be used in different areas in order to reduce the number of experiments needed to decide which factors influence dependent variables. Herein, we will discuss a specific type of design, Mixture Design, that is very important any time we need to use and optimize formulations in order to obtain the best product in many industries, such as food, cosmetics, pharmaceuticals and so on.

2. Theoretical and Methodological Aspects

A mixture experiment involves mixing proportions of two or more components to make different compositions of an end product. In mixture experiments, the main aim is to determine the proportions of the mixture components which lead to desirable results with respect to some property of interest. Mixture component proportions x_i are subject to the constraints

$$0 \leq x_i \leq 1 \qquad i = 1, 2, \ldots, q \qquad \text{and} \qquad \sum_{i=1}^{q} x_i = 1, \tag{3.1}$$

where q is the number of components involved in the mixture experiment. Consequently, the design space or experimental region is a $(q - 1)$-dimensional simplex. An n-dimensional simplex experimental region is

[1] Department of Food Engineering, State University of Ponta Grossa - Brazil.
[2] School of Chemistry Federal University of Rio de Janeiro - Brazil.
[3] Faculty of Mathematics Federal University of Uberlândia - Brazil.
* Corresponding authors: dgranato@uepg.br; calado@ufrj.br; edmilson.pinto@ufu.br
# The author thanks FAPEMIG for financial support.

an *n*-dimensional polytope determined by the mixture restrictions. For example, a 1-dimensional simplex is a line segment; a 2-dimensional simplex is a triangle (including its interior); a 3-dimensional simplex is a tetrahedron (see Boyd and Van-denberghe 2004 for a more comprehensive treatment). Figure 1 shows the 2-dimensional simplex experimental region for a mixture with three components.

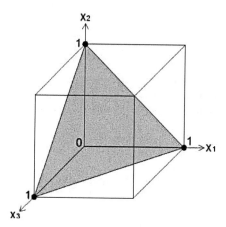

Figure 1: 2-dimensional simplex experimental region for a mixture with three components.

The experimental region is part of a *n*-dimensional simplex if there are further conditions on the proportions such as $L_i \leq x_i \leq U_i$ for $i = 1, 2, \ldots, q - 1$, with the proportion x_q taking values which make up the mixture. L_i and U_i are, respectively, the lower and upper limits of mixture. Figure 4 shows an experimental region for a constrained mixture experiment with three components.

When the existence of lower and/or upper limits does not affect the shape of the region of the mixture, that is, it is still a simplex region, it is possible to define a new set of components that take on the values 0 to 1 over the feasible region. These new components are called pseudocomponents (see Cornell, 2002 for more details). Myers and Montgomery, 2002 recommend the use of pseudocomponents to reduce multicollinearity among the predictors.

2.1 Models and standard designs

According to Khuri, 2005, consider the linear model

$$y = X\beta + \epsilon,$$ (3.2)

where $y^t = (y_1, \ldots y_n)$ is a response vector, X is a known design matrix of order $n \times p$ and rank $p \leq n$, β is an unknown parameter vector, and $\epsilon^t = (\epsilon_1, \ldots \epsilon_n)$ is a random error vector with a zero mean and a variance-covariance matrix $\sigma^2 I$, where $\sigma^2 = Var(\epsilon_i)$ for $i = 1, \ldots, n$ and I is the identity matrix. The response value at any point $x^t = (x_1, \ldots, x_q)$ in a region of interest $\mathcal{X}$ is given by $y(x) = \zeta(x, \beta) + \epsilon$, where $\zeta(x, \beta) = f^t(x)\beta$ is a polynomial model of a certain degree in the control variables $x_1, \ldots, x_q$. The vector $f(x)$ is of the same form as a row of X, but is evaluated at the point x.

Models commonly used in experiments with mixture are the so-called Scheffé's models. Scheffé 1958 proposed canonical polynomial models for mixture experiments, considering a reparametrization, based on the mixture constraint, of the polynomial model. The Scheffé's first-order polynomial model is given by

$$\zeta(x, \beta) = \sum_{i=1}^{q} \beta_i x_i,$$ (3.3)

which does not explicitly contain the constant term. The Scheffé's second-order model (quadratic model) is given by

$$\zeta(\boldsymbol{x}, \boldsymbol{\beta}) = Q(\boldsymbol{x}, \boldsymbol{\beta}) = \sum_{i=1}^{q} \beta_i x_i + \sum_{i=1}^{q-1} \sum_{j=i+1}^{q} \beta_{ij} x_i x_j. \tag{3.4}$$

For higher order models, the reparametrization does not lead to simple models; for example, for the third order model (cubic model), the Scheffé's canonical polynomial model is given by

$$\zeta(\boldsymbol{x}, \boldsymbol{\beta}) = Q(\boldsymbol{x}, \boldsymbol{\beta}) + \sum_{i=1}^{q-1} \sum_{j=i+1}^{q} \beta_{i-j} x_i x_j (x_i - x_j) + \sum_{i=1}^{q-2} \sum_{j=i+1}^{q-1} \sum_{k=j+1}^{q} \beta_{ijk} x_i x_j x_k. \tag{3.5}$$

For the estimation of model parameters, Scheffé proposed the simplex-lattice design (q, m), which is characterized by symmetrical arrangements of points within the experimental region. The number of parameters in the canonical polynomial model is exactly equal to the number of points chosen within the experimental region. The simplex-lattice design (q, m) consists of $\dfrac{(q+m-1)!}{m!(q-1)!}$ experimental points. Each component of the mixture takes $m + 1$ equally spaced values between 0 and 1, that is, $x_i = 0, \dfrac{1}{m}, \dfrac{2}{m}, \dots, \dfrac{m}{m} = 1$ for $i = 1, \dots, q$. All mixtures involving these proportions are used in experimental design. For example, a simplex-lattice design $(3, 2)$ consists of $\dfrac{(3+2-1)!}{2!(3-1)!} = 6$ experimental points. Each mixture component x_i can take $2 + 1 = 3$ possible values, that is, $x_i = 0, \dfrac{1}{2}, 1$; which results in the following experimental points $(1, 0, 0), (0, 1, 0), (0, 0, 1), (1/2, 1/2, 0), (0, 1/2, 1/2), (1/2, 0, 1/2)$. The simplex-lattice design $(3, 2)$ is shown in Fig. 2.

Scheffé, 1963 defined a special case of a canonical polynomial model given by

$$\zeta(\boldsymbol{x}, \boldsymbol{\beta}) = \sum_{i=1}^{q} \beta_i x_i + \sum_{i=1}^{q-1} \sum_{j=i+1}^{q} \beta_{ij} x_i x_j + \dots + \beta_{1, \dots, q} x_1 \dots x_q. \tag{3.6}$$

Again, for parameter estimation of the canonical polynomial model, Scheffé proposed an experimental design called simplex-centroid, which consists of $2^q - 1$ experimental points. The points of the simplex-centroid experimental design are formed by q permutations of pure mixtures $(1, 0, \dots, 0), \dots, (0, \dots, 0, 1)$, $\dfrac{q!}{2!(q-2)!}$ permutations of binary mixtures $(1/2, 1/2, 0, \dots, 0), \dots, (0, \dots, 0, 1/2, 1/2)$, and the global centroid, the mixture of the q components $(1/q, 1/q, \dots, 1/q)$. For example, for $q = 3$, the simplex-centroid consists of seven points, given by: $(1, 0, 0), (0, 1, 0), (0, 0, 1), (1/2, 1/2, 0), (1/2, 0, 1/2), (0, 1/2, 1/2)$ and $(1/3, 1/3, 1/3)$. The simplex-centroid for $q = 3$ is shown in Fig. 3.

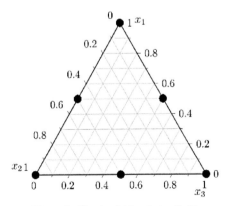

Figure 2: Simplex-lattice design (3, 2).

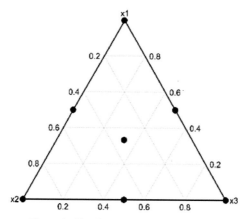

Figure 3: Simplex-centroid design for $q = 3$.

Alternatively to the Scheffé's canonical models, we can also use models of mixture with slack variable. Due to the mixture restrictions (3.1), the proportions of the components are not independent; thus, knowing the proportions of the first $q - 1$ components we can determine the proportion of the remaining component. In this way, the mixture models with a slack variable can be obtained from the Scheffé's canonical models by replacing the slack variable chosen, say x_l, by $1 - \sum_{i \neq l}^{q} x_i$. The purpose of this process is to produce mixture models that depend on $q - 1$ independent variables. For example, by defining x_q as the slack variable, the mixture model with slack variable obtained from Scheffé's quadratic model (3.4) is given by

$$\psi(x, \alpha) = \alpha_0 + \sum_{i=1}^{q-1} \alpha_i x_i + \sum_{i=1}^{q-1} \alpha_{ii} x_i^2 + \sum_{i=1}^{q-2} \sum_{j=i+1}^{q-1} \alpha_{ij} x_i x_j. \tag{3.7}$$

Note that, by comparing the coefficients of models (3.4) and (3.7), we have the following equivalences $\alpha_0 = \beta_q$, $\alpha_i = \beta_i - \beta_q + \beta_{iq}$, $\alpha_{ii} = -\beta_{iq}$ and $\alpha_{ij} = \beta_{ij} - (\beta_{iq} + \beta_{jq})$.

The use of models with slack variable as well as the choice of which component should be designated as a slack variable is not clearly presented in the literature. A discussion about the pros and cons of the use of slack variable is given by Cornell, 2000. Comparative studies between models of mixtures with a slack variable and Scheffé's models are given in Cornell, 2002; Cornell and Gorman, 2003 and Khuri, 2005.

2.2 Mixture experiments with process variables

Problems of mixture experiments with process variables arise when, in the mixture experiment, the property of interest is a function of the proportions of the ingredients and of other factors that do not form any portion of the mixture, as heat or time, for example. These factors, which depend on process conditions, are called process variables. In this way, the response variable depends not only on the proportion of the mixture components but also on the process variables.

Usually, in problems of mixture experiments including process variables, the goal is to determine the proportions of the mixture components along with situations of process conditions, in order to satisfy some property or characteristic of interest involving the variable response.

Thus, if in addition to the q mixture components $x^t = (x_1, \ldots, x_q)$ there are r process variables $z^t = (z_1, \ldots, z_r)$, we can consider additive models like $\eta(x, z, \beta, \gamma) = \zeta(x, \beta) + \vartheta(z, \gamma)$ or complete cross product models of the type $\eta(x, z, \beta, \gamma) = \zeta(x, \beta)\vartheta(z, \gamma)$ or combinations of these, such as $\eta(x, z, \beta, \gamma) = \zeta(x, \beta) + v(x, z, \beta, \gamma)$, where $\vartheta(z, \gamma)$ represents the process variable model and $v(x, z, \beta, \gamma)$ comprises products of terms in $\zeta(x, \beta)$ and $\vartheta(z, \gamma)$.

In general, the methodology used to construct mixture designs involving process variables is a combination of two designs, one being a mixture design for the mixture components and the other being factorial or fractional factorial design for the process variables. For example, Fig. 5 in the bread-making problem, discussed in Subsection 3.1, shows a 3^2 full factorial design in two process variables, z_1 and z_2, crossed with a third order simplex-centroid design in three mixture components, x_1, x_2 and x_3.

A comprehensive presentation of several aspects of statistical modeling and experimental design involving mixture experiments with process variables is given by Cornell, 2002.

When some process variables are either uncontrollable or difficult to be controlled they should be treated as noise variable. Noise variables are considered random variables with a supposedly known probability distribution. Thus, the presence of noise variables in the model means that the variance can no longer be assumed to be constant. One way to control the variability present in the model is by modeling the variance.

Statistical modeling of variance in mixture experiments with noise variables has been considered in Steiner and Hamada, 1997, who proposed a combined mixture-process-noise variable model. They built and solved an optimization problem to minimize a quadratic loss function, taking into account both the mean and variance of response. Another approach to modeling the variance is due to Goldfarb et al., 2003 using the delta method, which employs a first-order Taylor series approximation of the regression model at a vector of noise variables. The delta method is a well-known technique, based on Taylor series expansions, for finding approximations to the mean and variance of functions of random variables.

The delta method, considering the first-order Taylor approximation, is given as follows. Let $Z_1, \ldots,$ Z_r be random variables with means $\mu_1, \ldots, \mu_r$ and variance-covariance matrix Σ_z. Define $\mathbf{Z}^t = (Z_1, \ldots, Z_r)$ and $\boldsymbol{\mu}^t = (\mu_1, \ldots, \mu_r)$. For our problem of mixture experiments, we also consider x as the vector of mixture components. Suppose there is a differentiable function $\eta(x, \mathbf{Z}, \beta, \gamma)$ for which we want an approximate estimate of mean and variance. Consider $\eta'(x, \mu, \beta, \gamma) = \frac{\partial}{\partial z} \eta(x, z, \beta, \gamma)\big|_{z=\mu}$. The first-order Taylor series expansion of η about μ is

$$\eta(\boldsymbol{x}, z, \beta, \gamma) = \eta(\boldsymbol{x}, \mu, \beta, \gamma) + (z - \mu)^t \eta'(\boldsymbol{x}, \mu, \beta, \gamma) + \mathcal{R}, \tag{3.8}$$

where $\mathcal{R}$ is the remainder of the approximation and will be ignored. Thus, as $\eta(x, \mathbf{Z}, \beta, \gamma)$ is a random variable, because it is a function of $\mathbf{Z}$ that is a random variable, we can obtain

$$E\left[\eta(\boldsymbol{x}, \mathbf{Z}, \beta, \gamma)\right] \approx \eta(\boldsymbol{x}, \mu, \beta, \gamma) \tag{3.9}$$

and

$$Var\left[\eta(\boldsymbol{x}, \mathbf{Z}, \beta, \gamma)\right] \approx [\eta'(\boldsymbol{x}, \mu, \beta, \gamma)]^t \Sigma_z \eta'(\boldsymbol{x}, \mu, \beta, \gamma). \tag{3.10}$$

For a comprehensive treatment about the delta method, see Casella and Berger, 2002, p. 240.

Piepel 2004 makes a bibliographical survey on mixture experiments over a period of 50 years, ranging from 1955 to 2004. A bibliography of papers on mixture designs with process variables as well as various other types of mixture experiments is given by Piepel and Cornell, 2001.

2.3 Optimal experimental designs

Optimal experimental design is a powerful and flexible tool for generating efficient experimental designs, enabling a reduction in the number of experimental trials and, thus, reducing costs and saving time and money. However, the optimal experimental design depends on the model considered and if the model is not suitable the design will not be appropriate.

The construction of optimal designs for mixture experiments is of paramount importance, especially when the traditional designs used for standard situations can not be applied, for example, when we have mixture experiments with restriction, where the experimental region is a part of the simplex region.

Consider the linear model given in (3.2), where the response variable $y(x)$ is associated with an experimental unit x. Suppose we had k experimental units where n responses should be observed. The set of vectors $\{x_i, i = 1, \ldots, k\}$ represents a possible choice of k experimental units, in which n responses will be observed. The experimental design can be represented by $\xi = \{(x_1, w_1), \ldots, (x_k, w_k)\}$, with $w_j = \frac{r_j}{n}$, where r_j is the number of replicates of the jth experimental unit, for $j = 1, \ldots, k$. The experimental units $x_1, \ldots, x_k$ are the support points of the experiment and $w_1, \ldots, w_k$ are the weights associated with the support points. Thus, for the experimental design ξ, the weighted least squares estimator for β is given by $\hat{\beta} = (X^t W X)^{-1} X^t W y$, where the diagonal matrix $W = n \, Diag\{w_1, \ldots, w_k\}$. From literature, it is known that the variance-covariance matrix of $\hat{\beta}$ is given by $Cov(\hat{\beta}) = \sigma^2 (X^t W X)^{-1}$, which is the inverse of Fisher's information. In the theory of optimal experimental design, the information of Fisher is of fundamental importance. Define the experimental matrix as $M(\xi) = X^t W X$. Note that, for the model (3.2), $M(\xi)$ depends only on ξ; thus, optimal designs can be obtained from it according to some criterion of optimality. Two of the main optimality criteria are presented as follows.

D-optimality is used to minimize the covariance of $\hat{\beta}$. There is a known result that the ellipsoid of confidence for $\hat{\beta}$ is proportional to $|M(\xi)|^{-1/2}$, where $|\;|$ represents the determinant of a matrix. Thus, we can define a D-optimal design as follows. An experiment ξ^* is D-optimal if $\xi^* = \max_{\xi} \log |M(\xi)|$.

G-optimality is related to the predictions that can be obtained for the expected value of the response variable. For a given point x, the value predicted by the model is $\hat{y}(x) = f^t(x)\hat{\beta}$, with variance given by $Var[\hat{y}(x)] = \sigma^2 f^t(x)[M(\xi)]^{-1} f(x)$. An experiment ξ^* is G-optimal if $\xi^* = \min_{\xi_n} \max_{x \in \mathcal{X}} f^t(x)M^{-1}(\xi)f(x)$.

The equivalence between the two criteria mentioned above is shown by the celebrated General Theorem of Equivalence of Kiefer and Wolfowitz, 1960. For a good introduction to the theory of optimal experimental designs, with practical situations, see Atkinson et al., 2007.

Kiefer, 1961 proves that the simplex designs used by Scheffé for the first and second order polynomial models are D-optimal. Atkinson et al., 2007 discuss the D-optimality of the experimental design simplex-centroid for third order models and verify that the efficiency of simplex-centroid design in relation to D-optimal design is close to 1. Donev, 1989 establishes conditions for D-optimality of designs with both mixture and qualitative factors.

As an illustration of an optimal design in a constrained mixture experiment, consider the following example taken from Atkinson et al. 2007. Consider three mixture variables x_1, x_2 and x_3 and suppose that the model is the second-order canonical polynomial given by: $\zeta(x, \beta) = \beta_1 x_1 + \beta_2 x_2 + \beta_3 x_3 + \beta_{12} x_1 x_2 + \beta_{13} x_1 x_3 + \beta_{23} x_2 x_3$. Also consider, in addition to the usual mixture constraints, the following restrictions: $0.2 \leq x_1 \leq 0.7$, $0.1 \leq x_2 \leq 0.6$ and $0.1 \leq x_3 \leq 0.6$. The D-optimal design is $\xi = \{(x_1, w_1), \ldots, (x_{10}, w_{10})\}$, with $x_1 = (0.7, 0.1, 0.2)$, $x_2 = (0.2, 0.6, 0.2)$, $x_3 = (0.7, 0.2, 0.1)$, $x_4 = (0.2, 0.2, 0.6)$, $x_5 = (0.3, 0.6, 0.1)$, $x_6 = (0.3, 0.1, 0.6)$, $x_7 = (0.2, 0.4, 0.4)$, $x_8 = (0.5, 0.4, 0.1)$, $x_9 = (0.5, 0.1, 0.4)$, $x_{10} = (0.4, 0.3, 0.3)$ and $w_i = 0.1$ for $i = 1, \ldots, 6$, $w_7 = w_8 = w_9 = 0.11$ and $w_{10} = 0.07$. The experimental region and the points of D-optimal design are shown in Fig. 4.

Lawson and Willden, 2016 discuss optimal design and analysis of mixture experiments using software R (R Development Core Team, 2013). They illustrate the use of a recent package in R.

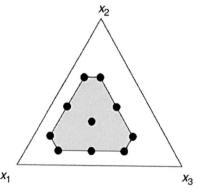

Figure 4: Experimental region and D-optimal design for a constrained mixture experiment.

3. Applications

3.1 Statistical modeling of variability in mixture experiments with noise variables—an application in a bread-making problem

The bread-making problem, originally presented by Faergestad and Naes, 1997, according to Naes et al., 1998, consisted of an experiment with three ingredients of mixture and two noise variables, and had as objective to investigate and to value the final quality of flour, composed by different mixtures of wheat flour, for production of bread. Also, according to Naes et al., 1998; Faergestad and Naes, 1997 considered three types of wheat flour: Two Norwegian, Tjalve (x_1) and Folke (x_2) and one American, Hard Red Spring (x_3), that were considered as control variables, and two types of process variables: Mixing time (z_1) and the proofing (resting) time of the dough (z_2), considered as noise variables. The response variable was considered as the loaf volume after baking with target value of 530 ml. The flour blends were considered to be mixing ingredients with $x_1 + x_2 + x_3 = 1$ and with constraints $0.25 \leq x_1 \leq 1.0$; $0 \leq x_2 \leq 0.75$ and $0 \leq x_3 \leq 0.75$, where x_1, x_2 and x_3 are the proportions of Tjalve, Folke and Hard Red Spring flour, respectively. For the noise variables, three situations for the mixing time were considered: 5, 15 and 25 minutes and also three situations for proofing time: 35, 47.5, and 60 minutes.

A full 3^2 factorial design was used for the noise variables and the 10 runs corresponding to a simplex-lattice design were replicated at each of the nine combinations of the mixing and proofing times, so that the complete design involved 90 experimental runs, as shown in Fig. 5. We consider the noise variables coded as z_1 = (mixing time − 15)/10 and z_2 = (proofing time − 47.5)/12.5, therefore, the noise variables are coded as −1, 0, 1 according to their minimum, mean and maximum values.

The volumes recorded for the 10 flour types and the 9 combinations of the noise variables are reproduced in Table 1. Additional details of the way in which the experiment was conducted are given by Naes et al., 1998, and further description of the practical aspects of the study is provided by Faergestad and Naes, 1997 according to Naes et al., 1998.

Naes et al., 1998 carried out a study of the use of robust design methodology to the bread-making problem in order to investigate the underlying relationships between the response variable loaf volume and the mixture and noise variables, comparing three techniques for analysing the loaf volume, i.e., the mean square error, the analysis of variance and the regression approach, where all factors, the three mixtures components and the two noise variables, are modeled simultaneously. In the analysis carried out by them, the full crossed model for three mixture ingredients and two noise variables were taken as the starting model. However, a detailed argument was presented for reducing the number of parameters by removing some of the second and third-order mixture terms before performing the cross. Thus, the initial reduced model had 28 terms.

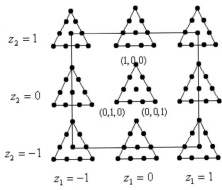

Figure 5: Full factorial 3^2 in two process variables crossed with a third order simplex-centroid design in three mixture components.

Table 1: Loaf volume for the 10 flour types and the 9 combinations of the noise variables.

					Noise Factors							
Design factors			z_1	-1	0	1	-1	0	1	-1	0	1
n^o	x_1	x_2	x_3	z_2	-1	-1	-1	0	0	0	1	1	1
1	0.25	0.75	0.00		378.89	396.67	392.22	445.56	452.22	487.78	457.22	500.56	472.78
2	0.50	0.50	0.00		388.89	423.33	416.11	460.00	488.89	475.78	472.78	478.00	506.11
3	0.75	0.25	0.00		426.11	483.33	389.44	474.44	514.44	462.78	506.67	591.67	522.22
4	1.00	0.00	0.00		386.11	459.11	423.33	458.33	506.11	514.44	545.56	522.22	551.11
5	0.25	0.50	0.25		417.78	437.22	444.56	484.44	490.00	495.00	497.78	531.11	577.78
6	0.50	0.25	0.25		389.44	447.22	415.00	490.89	528.89	507.78	517.78	567.22	538.33
7	0.75	0.00	0.25		448.33	459.44	455.56	436.00	535.00	552.22	507.44	578.89	590.00
8	0.25	0.25	0.50		413.89	485.56	462.22	483.89	529.44	540.00	565.00	598.89	580.56
9	0.50	0.00	0.50		415.56	514.44	437.78	493.89	583.33	578.89	524.44	694.44	640.00
10	0.25	0.00	0.75		432.78	498.33	517.22	474.44	568.33	579.44	541.11	638.89	638.89

The analysis presented here begins with the same initial reduced model, involving 28 terms, considered by Naes et al., 1998 and given in Eq. (3.11).

$$\eta(x, z, \beta, \gamma) = \beta_1^0 x_1 + \beta_2^0 x_2 + \beta_3^0 x_3 + \beta_{12}^0 x_1 x_2 + \beta_{13}^0 x_1 x_3 + \gamma_{12}^0 x_1 x_2 (x_1 - x_2) +$$
$$\gamma_{13}^0 x_1 x_3 (x_1 - x_3) + \{\beta_1^1 x_1 + \beta_2^1 x_2 + \beta_3^1 x_3 + \beta_{12}^1 x_1 x_2 + \beta_{13}^1 x_1 x_3 +$$
$$\gamma_{12}^1 x_1 x_2 (x_1 - x_2) + \gamma_{13}^1 x_1 x_3 (x_1 - x_3)\} z_1 + \{\beta_1^2 x_1 + \beta_2^2 x_2 +$$
$$\beta_3^2 x_3 + \beta_{12}^2 x_1 x_2 + \beta_{13}^2 x_1 x_3 + \gamma_{12}^2 x_1 x_2 (x_1 - x_2) +$$
$$\gamma_{13}^2 x_1 x_3 (x_1 - x_3)\} z_2 + \{\beta_1^{11} x_1 + \beta_2^{11} x_2 + \beta_3^{11} x_3 + \beta_{12}^{11} x_1 x_2 +$$
$$\beta_{13}^{11} x_1 x_3 + \gamma_{12}^{11} x_1 x_2 (x_1 - x_2) + \gamma_{13}^{11} x_1 x_3 (x_1 - x_3)\} z_1^2 \tag{3.11}$$

Initially, a linear regression model for the mean is fitted by Ordinary Least Squares (OLS) method. After applying the stepwise backward method, which consists of adding or eliminating variables according to their significance in the model, it was possible to eliminate the regressors that do not really influence the response variable and obtain a parsimonious model with 18 terms given in Table 2.

The goodness of fit of the model found is assessed using different types of diagnostic displays shown in Fig. 6. By observing the graphs, it is possible to conclude that there are no outliers and that the model fits well to the data. In addition, the Breusch–Pagan test applied to the data indicates homoscedasticity (with p-value $p = 0.0644$).

In order to facilitate our analysis, the model found will be written as

$$\hat{\eta}(x, z, \hat{\beta}, \hat{\gamma}) = c_1 + c_2 z_1 + c_3 z_2 + c_4 z_1^2, \tag{3.12}$$

where $c_1 = 484.62 x_1 + 474.88 x_2 + 436.38 x_3 + 468.31 x_1 x_3 + 375, 34 x_1 x_2 (x_1 - x_2) - 403.03 x_1 x_3 (x_1 - x_3)$, $c_2 = 16.77 x_1 + 51.88 x_3 - 144.55 x_1 x_2 (x_1 - x_2)$, $c_3 = 54.93 x_1 + 42.50 x_2 + 188.76 x_1 x_3 - 202.82 x_1 x_3 (x_1 - x_3)$ and $c_4 = -52.64 x_2 + 164.08 x_3 - 600.05 x_1 x_3 - 440.72 x_1 x_2 (x_1 - x_2) + 525.48 x_1 x_3 (x_1 - x_3)$. Note that, c_1, c_2, c_3 and c_4 are constant with respect to z.

As the process variables are considered noise variables, they influence the variability of the model. In this way, the statistical modeling of variance should be considered. Thus, supposing the noise variables are uncorrelated and that the noise variables and the model errors are uncorrelated with each other, the mean and variance of the model can be obtained by delta method. Consider $E(Z_i) = \mu_i$ and $Var(Z_i) = \sigma_i^2$, for $i = 1, 2$. Note that, since Z_1 and Z_2 are continuous variables, it could have been considered $Z_i \sim N(\mu_i, \sigma_i^2)$, for $i = 1, 2$, but this is not necessary since it is only necessary to know the mean and variance of the noise variables. From Eq. (3.12) it is possible to obtain $[\eta'(x, \mu, \beta, \gamma)]' = (c_2 + 2c_4 \mu_1, c_3)$. In this way, as $Y = \eta(x, Z, \beta, \gamma) + \epsilon$, where $\epsilon \sim N(0, \sigma^2)$, $\Sigma_z = Diag(\sigma_1^2, \sigma_2^2)$, using Eqs. (3.9) and (3.10), the expressions for mean and variance are approximated by

Table 2: Regression coefficients for the mean model by OLS method.

Terms	Estimate	Std. Error	t value	p-value
x_1	484.624	6.363	76.161	0.0000
x_2	474.875	13.369	35.521	0.0000
x_3	436.381	64.837	6.730	0.0000
$x_1 x_3$	468.313	164.234	2.851	0.0057
$x_1 x_2 (x_1 - x2)$	375.341	94.623	3.397	0.0002
$x_1 x_3 (x_1 - x_3)$	−403.031	199.679	−2.018	0.0473
$x_1 z_1$	16.768	5.452	3.076	0.0029
$x_3 z_1$	51.876	8.406	6.171	0.0000
$x_1 x_2 (x_1 - x_2)z_1$	−144.553	60.706	−2.381	0.0199
$x_1 z_2$	54.933	6.703	8.195	0.0000
$x_2 z_2$	42.504	8.470	5.018	0.0000
$x_1 x_3 z_2$	188.762	25.167	7.500	0.0000
$x_1 x_3 (x_1 - x_3)z_2$	−202.822	61.681	−3.288	0.0016
$x_2 z_1^2$	−52.644	14.972	−3.516	0.0008
$x_3 z_1^2$	164.077	79.249	2.070	0.0420
$x_1 x_3 z_1^2$	−600.046	199.173	−3.013	0.0036
$x_1 x_2 (x_1 - x_2)z_1^2$	−440.721	109.730	−4.016	0.0001
$x_1 x_3 (x_1 - x_3)z_1^2$	525.480	244.486	2.149	0.0349

$$E(Y) \approx c_1 + c_2\mu_1 + c_3\mu_2 + c_4\mu_1^2, \tag{3.13}$$

and

$$Var(Y) \approx (c_2 + 2c_4\mu_1)^2\, \sigma_1^2 + c_3^2\sigma_2^2 + \sigma^2. \tag{3.14}$$

If it is assumed that $E(\mathbf{Z}) = \mathbf{0}$, then $E(Y) = c_1$ and $Var(Y) = c_2^2\sigma_1^2 + c_3^2\sigma_2^2 + \sigma^2$.

The mean square error for the fitted model is used as an estimate of σ^2. For our model $\hat{\sigma}^2 = 58.36$.

Following Taguchi's idea for quality improvement, see Taguchi, 1986, after we found the equations for $E(Y)$ and $Var(Y)$ and since the target value for the response variable (loaf volume after baking) is 530 ml, the minimization problem shown in Eq. (3.15) must be solved.

$$M\, in\ Var(Y)$$

$$Subject\ to \quad \begin{cases} E(Y) = 530.0 \\ x_1 + x_2 + x_3 = 1.0 \\ 0.25 \leq x_1 \leq 1.0 \\ 0.0 \leq x_2 \leq 0.75 \\ 0.0 \leq x_3 \leq 0.75 \end{cases} \tag{3.15}$$

where $E(Y)$ and $Var(Y)$ are functions of x_1, x_2, x_3, μ_1, μ_2, σ_1^2 and σ_2^2, being that μ_1, μ_2, σ_1^2 and σ_2^2 are fixed and known. The expressions for $E(Y)$ and $Var(Y)$ are given in Eqs. (3.13) and (3.14), respectively.

The optimization problem was solved considering some scenarios involving the values of mean and variance for the random variables mixing time and proofing time. Table 3 shows the optimum combination for the mixture, its estimated variance and its coefficient of variation, given by $\sqrt{Var(Y)}/530$, for each scenario involving the noise variables. In Table 3, according to the encoding previously assumed for the noise variables, the mean and variance values of Z_1 and Z_2 were obtained by $\mu_1 = (\mu_m - 15)/10$, $\sigma_1^2 = \sigma_m^2/100$, $\mu_2 = (\mu_p - 47.5)/12.5$ and $\sigma_2^2 = \sigma_p^2/156.25$, where μ_m and σ_m^2 represent,

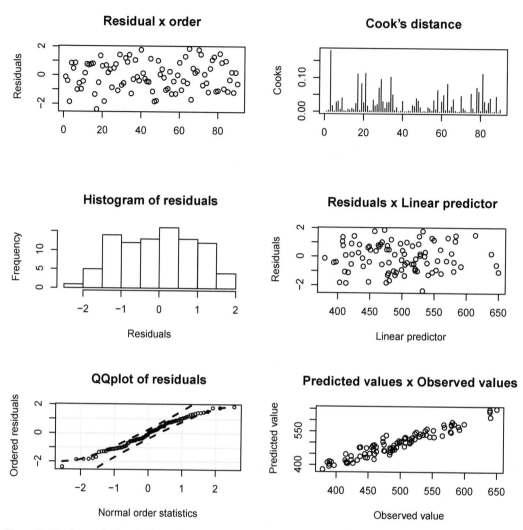

Figure 6: Six diagnostic plots for data of bread mixture. The upper left panel plots standardized residuals against the order of observations. Since no pattern is found in that plot, we can assume that the data are independent. The upper right panel plots the Cook's distances versus the order of observations. Since all values of Cook's distance do not exceed $F_{(0.5,18,72)} \approx 0.97$, there is no evidence of outliers. The middle left panel displays the residuals histogram, showing that the normality hypothesis should not be rejected. The middle right panel plots the standardized residuals versus the linear predictor. The plot indicates no evidence of heteroskedasticity or non-linearity. The lower left panel displays the Q-Q plot of residuals, reinforcing the hypothesis of residual normality. The lower right panel presents the predicted values versus observed values, showing that the model is well fitted to the data.

respectively, the mean and variance of the random variable mixing time and, similarly, μ_p and σ_p^2 represent, in this order, the mean and variance of the random variable proofing time.

From scenarios considered in Table 3, it is possible to observe that some optimum points were robust for some scenarios and that the coefficient of variation in all cases was less than 12%, indicating a small variability in the response. It is noteworthy that the scenarios were considered only in an illustrative way in order to show how the optimization problem could be driven. Scenarios could be appropriately chosen if there was some previous study on the behavior of the noise variables in the productive process. The decision about which mixture composition would be ideal, that is, robust to the noise factors, depends on the knowledge about the behavior of the noise variables and, since this information is not known, it is not possible to evaluate which scenarios should be considered for the choice of the optimal composition.

Table 3: Optimal values of mixture with their respective values of $Var(Y)$ and coefficient of variation for some scenarios involving the noise variables: Mixing time and proofing time.

| Uncoded noise variables | | Coded noise variables | | Optimum | $Var(Y)$ | Coef. of |
Mixing Time† (μ_m, σ_m^2)	Proofing Time‡ (μ_p, σ_p^2)	Z_1 (μ_1, σ_1^2)	Z_2 (μ_2, σ_2^2)	(x_1, x_2, x_3)	estimated	variation
(10.0, 6.25)	(47.50, 9.766)	(−0.50, 0.0625)	(0.00, 0.0625)	(0.25,0.10,0.65)	604.91	0.046
(12.5, 6.25)	(44.375, 9.766)	(−0.25, 0.0625)	(−0.25, 0.0625)	(0.25,0.10,0.65)	480.79	0.041
(12.5, 6.25)	(53.75, 9.766)	(−0.25, 0.0625)	(0.50, 0.0625)	(0.27,0.00,0.73)	643.55	0.048
(12. 5, 25.0)	(44.375, 39.063)	(−0.25, 0.25)	(−0.25, 0.25)	(0.25,0.10,0.65)	1748.08	0.078
(12.5, 56.25)	(44.375, 87.891)	(−0.25, 0.5625)	(−0.25, 0.5625)	(0.25,0.10,0.65)	3860.23	0.117
(15.0, 6.25)	(53.75, 9.766)	(0.00, 0.0625)	(0.50, 0.0625)	(0.61,0.36,0.03)	216.34	0.028
(15.0, 6.25)	(41.25, 9.766)	(0.00, 0.0625)	(−0.50, 0.0625)	(0.44,0.10,0.46)	396.81	0.038
(15.0, 6.25)	(47.50, 9.766)	(0.00, 0.0625)	(0.00, 0.0625)	(0.62,0.17,0.21)	260.33	0.030
(15.0, 25.0)	(47.50, 39.063)	(0.00, 0.250)	(0.00, 0.250)	(0.62,0.17,0.21)	866.23	0.056
(15.0, 56.25)	(47.50, 87.891)	(0.00, 0.5625)	(0.00, 0.5625)	(0.62,0.17,0.21)	1876.07	0.082
(20.0, 6.25)	(47.50, 9.766)	(0.50, 0.0625)	(0.00, 0.0625)	(0.25,0.30,0.45)	234.32	0.029
(20.0, 6.25)	(53.75, 9.766)	(0.50, 0.0625)	(0.50, 0.0625)	(0.26,0.43,0.31)	211.24	0.027
(20.0, 25.0)	(53.75, 39.063)	(0.50, 0.250)	(0.50, 0.250)	(0.26,0.43,0.31)	669.87	0.049
(20.0, 56.25)	(53.75, 87.891)	(0.50, 0.5625)	(0.50, 0.5625)	(0.26,0.43,0.31)	1434.25	0.071

† Mixing time is a random variable with mean μ_m and variance σ_m^2.

‡ Proofing time is a random variable with mean μ_p and variance σ_p^2.

3.2 Some other examples

Simplex-centroid designs are widely used for food development and for the assessment of different factors on processes. In this regard, Karnopp et al., 2017 optimized an organic yogurt manufactured with purple grape skin flour (GSF), oligofructose (OLI) and Bordeaux grape juice (PGJ). For this purpose, a simplex-centroid design containing 10 different formulations was used and data were subjected to response surface methodology (RSM). The regression models were significant ($p < 0.05$) and the determination coefficient for the proximal composition, instrumental texture profile, sensory analysis, and antioxidant activity data were higher than 0.740. Thus, authors concluded that the use of a simplex-centroid coupled with RSM was able to explain the effects of OLI, GSF, and PGJ on the nutritional, physicochemical, and functional properties of yogurt. Moreover, the residual analysis was applied and authors verified that the residuals followed the normal distribution. Aiming to optimize the yogurt formulation, the desirability function was employed with the aim of maximizing the fiber and ash contents, total phenolic content (TPC), and sensory acceptance. A formulation (based on 100 g) containing 1.7 g GSF and 8 g PGJ had over 5.5 g dietary fibers, 0.76 g ash, 28 mg TPC, and 57.8 mg ascorbic acid equivalent of antioxidant activity measured by the DPPH assay. This formulation also presented 79% of sensory acceptance. Overall PGJ and GSF were shown to be suitable alternatives for the yogurt sector.

Not only product development can be assessed by using DOE. In an interesting study, Bochi et al., 2014 assessed the effects of different factors on the extraction of bioactive polyphenols from *Dovyalis hebecarpa* pulp (Ceylon gooseberry). For this purpose, authors used a simplex-lattice design containing 14 experiments, whereby propapone, water, and ethanol and different concentrations. Total phenolic content and anthocyanins (TA) were the responses. The data analysis showed that the TPC ranged from 5.31 to 14.32 mg of gallic acid equivalent/g of freeze-dried sample and TA ranged from 0.25 to 13.8 mg of cyanidin-3-glucoside equivalent/g of freeze-dried sample. Overall, for TPC the highest extraction yield was obtained with ethanol as the main solvent in the mixture. A similar behavior was observed for TA, where the highest extraction yield was obtained in a mixture of solvents with higher concentration of water than acetone and ethanol. For the TA, the quadratic and cubic models presented the best modeling results ($R^2 = 0.91$) and no lack of fit was observed ($p = 0.24$), while for TPC, lower R^2 values were obtained for the quadratic and cubic models (0.68 and 0.69, respectively) with significant lack of fit ($p = 0.00008$). The regression equation, obtained with significant ($p < 0.05$) coefficients, to explain the effects of different solvents on the TA extraction was:

$$E(Y) = 0.533x_1 + 2.456x_2 + 2.911x_3 + 4.780x_1x_2 + 4.110x_1x_3 + 0.100, \tag{3.16}$$

where x_1 = acetone, x_2 = ethanol, and x_3 = water.

The models for TPC cannot be used for prediction purposes as they did not explain the experimental results at a large extent. When this situation occurs, RSM can be used in order to gain an idea of the effects of solvent ratios on the extraction of bioactive compounds. Other examples of simplex-lattice design and simplex-centroid design applied in various areas of food science, technology and engineering can be found in recent literature (Fauzi et al., 2013; Santana et al., 2016; Handa et al., 2016; Moreno-Vilet et al., 2017).

An interesting work was developed by Ouedrhiri et al., 2016, about an optimization of a bacterial effect using a mixture of *Origanum compactum* (Oc), *Origanum majorana* (Om) and *Thymus serpyllum* (Ts) essential oils. They used an augmented simplex-centroid design in order to find the optimum levels of these three oils in order to obtain a minimum inhibitory concentration (MIC). The response variables were bacterial effect against *B. subtilis* ATCC 3366, *S. aureus* ATCC 29213 and *E. coli* ATCC 25922. The three types of oils were statistically significant. The interaction terms are only significant for some of the bacteria. Some of these effects are antagonistic and some are synergistic. Figure 7 presents the 3D ternary plots showing the minimum region for each type of bacteria studied. These graphs were obtained by Statistica 13.1 (Statsoft), Statistica 2018. It is possible to notice that there is no minimum for *E. coli*. For 31%, 23% and 46% of oregano, marjoram and wild thyme oils, respectively, the authors found the minimum inhibitory concentration (MIC) for *B. subtilis*, 0.050% (v/v) and for *S. aureus*, 0.021% (v/v). In order to get a lower value of MIC for *E. coli*, about 75% of oregano oil and 25% of marjoram oil must be used.

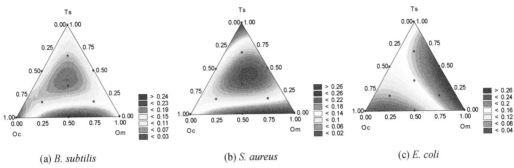

Figure 7: 3D ternary plots showing the minimum region for each type of bacteria studied.
(a) *B. subtilis*; (b) *S. aureus* and (c) *E. coli*.

Color version at the end of the book

Doriya and Kumar, 2018, worked with a simplex-centroid design, augmented by the interior points and global centroid, followed by a Box-Behnken design in order to optimize L-asparaginase production from newly isolated *Aspergillus* sp., in which the solid substrate was from agroindustrial waste. The authors used wheat bran, cotton seed oil cake and red gram husk to form 14 experiments with four replicates. The maximum L-asparaginase production was 9.47 *U/mL* for 6.5 g of cotton seed cake, 1.5 g of wheat bran and 2 g of red gram husk. The authors proposed a special quadratic model, with a determination coefficient (R_2) equal to 0.96. The maximum predicted by the model was 9.508 *U/mL*, representing am error of only 0.4%. After finding this best composition and using it, the authors applied a Box-Behnken design, using pH (6 to 8), moisture content (60 to 80%) and fermentation temperature (25°C to 35°C) as process variables. Now, the maximum L-asparaginase activity was 12.57 *U/mL* for pH = 8.0, temperature = 35°C and initial moisture content = 70% (*v/w*). The regression coefficient was 0.91 and the linear effects for all factors were statistically significant ($p \leq 0.05$), as well as the quadratic effect of pH and moisture content.

It is possible to point out a problem with the work developed by Doriya and Kumar, 2018. They optimize the L-asparaginase activity in two steps instead of only one. The authors did not say at which conditions (process variables values) they found the maximum L-asparaginase activity by using the mixture design. The answer can change considerably without considering the integration of both designs simultaneously like the bread-making problem.

4. Final Considerations

In this chapter, all the models considered had as response a continuous variable with Gaussian distribution. However, other types of response variables could also be considered, such as, for example, counting or proportions. The class of generalized linear models (see Mc-Cullagh and Nelder, 1989 for details) extends the class of regression models to a wider class of models, which allows models to be obtained for a wide range of situations, in particular, for those previously mentioned.

An interesting practical and didactic article addressing mixture experiments with process variables that consider counting data and Poisson distribution for response variable is given by Steiner et al., 2007. Optimal mixture experimental designs can also be obtained for such models (see Atkinson et al., 2007 for details).

It is worth emphasizing that the simplex-lattice is used more when we want to test binary combinations between minimum and maximum concentrations. The simplex-centroid is used more to test both combinations: Binary and ternary of the three components ($q = 3$). Thus, the optimization is more precise for process and product development.

In summary, mixture designs should be used when we want to find the best composition for a product. However, we usually also need to involve process variables, such as temperature, time, pressure, and so on. Thus, the problem becomes complex because we need to use two different designs of experiments: Mixture inside factorial, central composite or Box-Behnken design or vice-versa. Some examples were discussed herein and attention must be taken. Sometimes, someone optimizes one factor each time, although this is not right as the process variables can interact with each other. This leads to an illusory best condition. Thus, all factors need to be analyzed at the same time.

Acknowledgements

The authors thank Dr. Leandro Alves Pereira (Federal University of Uberlândia - Faculty of Mathematics) for his help in obtaining the regression model and the diagnostic plots for the bread-making problem discussed in Subsection 3.1.

References

Atkinson, A.C., Donev, A.N. and Tobias, R.D. 2007. Optimum Experimental Designs, with SAS. Oxford University Press, New York.

Bochi, V.C., Barcia, M.T., Rodrigues, D., Speroni, C.S., Giusti, M.M. and Godoy, M.T. 2014. Polyphenol extraction optimisation from Ceylon gooseberry (Dovyalis hebecarpa) pulp. Food Chemistry 164: 347–354.

Boyd, S. and Vandenberghe, L. 2004. Convex Optimization. Cambridge University Press, Cambridge.

Casella, G. and Berger, R.L. 2002. Statistical Inference. 2nd ed. Duxbury - Thomson Learning - New York.

Cornell, J.A. 2000. Fitting a slack-variable model to mixture data: Some questions raised. Journal of Quality Technology 32: 133–147.

Cornell, J.A. 2002. Experiments with Mixtures: Designs, Models, and the Analysis of Mixture Data. 3rd ed. Wiley, New York.

Cornell, J.A. and Gorman, J.W. 2003. Two new mixture models: Living with collinearity but removing its influence. Journal of Quality Technology 35: 78–88.

Donev, A.N. 1989. Design of experiments with both mixture and qualitative factors. Journal of the Royal Statistical Society B 50: 297–302.

Doriya, K. and Kumar, D.S. 2018. Optimization of solid substrate mixture and process parameters for the production of L-asparaginase and scale-up using tray bioreactor. Biocatalysis and Agricultural Biotechnology 13: 244–250.

Faergestad, E.M. and Naes, T. 1997. Evaluation of baking quality of wheat flours: I: small scale straight dough baking test of heart bread with variable mixing time and proofing time. In: Report MATFORSK, As, Norway.

Fauzi, S.H.M., Rashid, N.A. and Omar, Z. 2013. Effects of chemical interesterification on the physicochemical, microstructural and thermal properties of palm stearin, palm kernel oil and soybean oil blends. Food Chemistry 137: 8–17.

Goldfarb, H., Borror, C. and Montgomery, D. 2003. Mixture-process variable experiments with noise variables. Journal of Quality Technology 35: 393–405.

Handa, C.L., de Lima, F.S., Guelfi, M.F.G., Georgetti, S.R. and Ida, E.I. 2016. Multi-response optimisation of the extraction solvent system for phenolics and antioxidant activities from fermented soy flour using a simplex-centroid design. Food Chemistry 197: 175–184.

Karnopp, A.R., Oliveira, K.G., de Andrade, E.F., Postingher, B.M. and Granato, D. 2017. Optimization of an organic yogurt based on sensorial, nutritional, and functional per-spectives. Food Chemistry 233: 401–411.

Khuri, A.I. 2005. Slack variable models versus Scheffé's Mixture models. Journal of Applied Statistics 32: 887–908.

Kiefer, J. and Wolfowitz, J. 1960. The Equivalence of two extremum problems. Canadian Journal of Statistics 12: 363–366.

Kiefer, J. 1961. Optimum designs in regression problems II. Annals of Mathematical Statistics 32: 298–325.

Lawson, J. and Willden, C. 2016. Mixture experiments in R using mixexp. Journal of Statistical Software 72: 1–20.

McCullagh, P. and Nelder, J.A. 1989. Generalized Linear Models. 2nd ed., Chapman and Hall, London.

Moreno-Vilet, L., Bostyn, S., Flores-Montaño, J.L. and Camacho-Ruiz, R.M. 2017. Size-exclusion chromatography (HPLC-SEC) technique optimization by simplex method to estimate molecular weight distribution of agave fructans. Food Chemistry 237: 833–840.

Myers, R.H. and Montgomery, D.C. 2002. Response Surface Methodology: Process and Product Optimization Using Designed Experiments. 2nd ed., John Wiley & Sons, New York.

Naes, T., Faergestad, E.M. and Cornell, J.A. 1998. A comparison of methods for analyzing data from a three component mixture experiment in the presence of variation created by two process variables. Chenometrics and Intelligence Laboratory Systems 41: 221–235.

Ouedrhiri, W., Balouiri, M., Bouhdid, S., Moja, S., Chahdi, F.O., Taleb, M. et al. 2016. Mixture design of Origanum compactum, Origanum majorana and Thymus serpyllum essential oils: Optimization of their antibacterial effect. Industrial Crops and Products 89: 1–9.

Piepel, G.F. and Cornell, J.A. 2001. A Catalogue of Mixture Experiment Examples. BN-SA-3298, Battelle, Pacific Northwest Laboratories. Richland, WA.

Piepel, G.F. 2004. 50 Years of Mixture Experiment Research: 1955–2004, Singapore. pp. 283–327. *In*: A.I. Khuri (ed.). Response Surface Methodology and Related Topics. World Scientific, Singapore.

R Development Core Team, 2019. R: A language and environment for statistical computing, reference index version 3.5.3. R Foundation for Statistical Computing, Viena, Austria, ISBN 3-900051-07-0, URL http://www.R-project.org.

Santana, A.A., Cano-Higuita, D.M., de Oliveira, R.A. and Telis, V.R.N. 2016. Influence of different combinations of wall materials on the microencapsulation of jussara pulp (Euterpe edulis) by spray drying. Food Chemistry 212: 1–9.

Scheffé, H. 1958. Experiments with mixture. Journal of the Royal Statistical Society B 20: 344–366.

Scheffé, H. 1963. Simplex-centroid designs for experiments with mixtures. Journal of the Royal Statistical Society B 25: 235–263.

Statistica, 2018. 3307 Hillview Avenue, Palo Alto, CA 94304, USA.

Steiner, S.H. and Hamada, M. 1997. Making mixtures robust to noise and mixing measurement errors. Journal of Quality Technology 29: 441–450.

Steiner, S.H., Hamada, M., White, B.J.G., Kutsyy, V., Mosesova, S. and Salloum, G. 2007. A bubble mixture experiment project for use in an advanced design of experiments class. Journal of Statistics Education 15: 1–20.

Taguchi, G. 1986. Introduction to Quality Engineering. Unipub/Kraus International Publications, White Plains, New York.

CHAPTER **4**

Microorganisms and Food Products in Food Processing Using Full Factorial Design

Davor Valinger,* Jasna Gajdoš Kljusurić, Danijela Bursać Kovačević, Predrag Putnik and
Anet Režek Jambrak*

1. Introduction

Food processing is of great importance when obtaining products with specific organoleptic, physical and chemical properties, and is the most important step to assure food safety. The main concern of food processes and food product development is to meet standards in order to obtain safe, accepted by consumers and highest possible quality product. Microorganisms can be easily destroyed (inactivated) by using standard heat processes, like pasteurisation, sterilisation, ultrahigh temperature (UHT) processing, high temperature short time (HTST) and other short and efficient processes. On the other hand, there is a rapidly growing area of non-thermal food processing. The research efforts are well-known in UV processing and gamma processing, and, in the last 15 years, emerging in high-pressure processing (cold pressurisation), pulsed electric field, pulsed light, colds plasma, electronic beam and ultrasound processing. The main task here is to inactivate microbes using lower operating temperatures (30–60ºC), shorter treatment time and lower carbon emission. Therefore, there is emphasis on sustainable preservation techniques while assuring food safety. The agri-food sector is faced with sustainability issues with great concerns for complex quantitative and qualitative mathematical factors, legislations and the analysis of large datasets (Perrot et al., 2016). Mathematical modelling, optimisation and virtualisation are very important in small- and large-scale processing. It is an efficient tool to determine, implement, analyse and control the whole process so as to manufacture the desired food products. The main limitation is the fact that microorganisms are dynamic living entities and the food is a very complex matrix. Therefore, inactivation and behaviour of microorganisms in a complex matrix will have a large number of influencing factors regarding their destruction.

There are many research papers dealing with the modelling and predicting of safety and quality of thermal and non-thermal food processing. There is food safety microbial modelling with user-friendly models and non-user friendly matrices focusing on food-quality changes (Valdramidis et al., 2012). The integration of safety and quality quantitative approaches are the main tasks that need to be carefully planned and optimised in order to have the most efficient food processes. Sensory analysis is based on

Faculty of Food Technology and Biotechnology, University of Zagreb, Pierroti street 6, Zagreb, Croatia.
* Corresponding authors: dvalinger@pbf.hr; anet.rezek.jambrak@pbf.hr

classical experimental designs and linear multivariate analysis techniques. Putnik et al. (2017) researched the shelf-life of minimally-processed fresh-cut apples (Putnik et al., 2017d). They used mathematical models to calculate their shelf-life by application of product quality attributes such as: pH, soluble solids content (SSC), color parameters, sensory evaluation and microbial spoilage. In another study, researchers tried to assess through a fractional experimental design the environmental factors that could affect the survival of *L. monocytogenes* in food processing (Overney et al., 2017). These authors tried to prevent the persistence of this pathogen in conditions encountered in food processing plants.

There are many examples where mathematical models were applied in order to assure food safety. Therefore, the aim of this chapter is to present the principal component analysis (PCA) and other multivariate (chemometric) analyses in food processing. The focus will be on modelling and virtualisation for predicting microbial growth in food engineering. Due to the open debate on which design of experiments to use, e.g., full factorial or fractional factorial design (Plackett-Burman), central composite (full, half, quarter), Box-Behnken, or some other type, the main concern in selection is the purpose of experiments and available funding. The full or fractional factorial design is used for screening and separation of effects since it is based on the numbers of factors (k) which most commonly have 2 values (usually high/low, on/off). The number of experiments is arranged by 2^k and depends on the number of experimental factors, so the total number of runs in an experiment can reach large numbers. To save resources, researchers presently use one of the above-mentioned models that provide credible data. However, there are always advantages and limitation to each particular type of design. For instance, only full factorial design (FFD) can give information about all the interactions of process variables and the obtained results, on the other hand, this choice increases the number of experimental runs, which may be a burden on economic resources in production. That is the reason why current trends in using FFD in non-thermal food processing will be addressed, as well as future perspectives of its application.

2. Chemometrics in Food Processing

Current technological achievements in the analytical industry enabled researchers to generate large (chemical) data sets (Granato et al., 2018). One of the possible solutions, useful for data mining and removal of noise from the hidden associations, can be sought in the realm of chemometrics. This word commonly describes various multivariate statistical analyses able to derive useful conclusions from such data with numerous independent and dependent variables (Varmuza and Filzmoser, 2009). A common scenario in research includes situation(s) where researchers have large number of independent variables and have to evaluate them against single outcome (dependent) variable, usually without application of multivariate statistics. The main reason for multivariate statistics is the necessity to decrease the possibility of false positive results during hypothesis testing in a study, or, to put it bluntly, to prevent reporting findings that are not true (Dumancas et al., 2015).

Multivariate statistical approaches employed for chemometric assessment range from multivariate analysis of variance (MANOVA), various factor analysis (e.g., principal components analysis, PCA), different mathematical modelling, artificial neural networks, discriminant analysis, and others (Dziurkowska and Wesolowski, 2015). However, probably the most popular approaches in chemometrics are PCA and partial least squares (PLS) analysis (Granato et al., 2018). In food science, chemometrics is adaptable to various situations, but it is commonly employed for food authenticity, functionality, bioactivity, processing and safety (Skov et al., 2014). The three main approaches in chemometrics are explorative analysis, classification and calibration (Granato et al., 2018). Selection of the right choice depends of the type of experimental design and problem that needs to be tackled.

One example is a study that evaluated the colour of pomegranate juice and the corresponding content of anthocyanins processed with cold atmospheric gas phase plasma by MANOVA and regression analysis (Bursać Kovačević et al., 2016). Here, the influence of three independent variables, with four different levels, on 12 dependent variables (six anthocyanins and six colour parameters) were evaluated. This ensured that each of the comparisons per dependent variable had type I error equal to 0.05. Optimal parameters obtained by modelling with the highest anthocyanin content were t = 3 min, V = 5 cm³, with gas flow of 0.75 dm³/min. Plasma treated juices had higher change of colour as opposed to untreated samples, where colour decreased with increase of plasma gas flow, although it remained constant with alteration in treatment time and volume. In conclusion, this study showed positive influence of plasma processing to anthocyanin stability and colour of juices.

The PCA is an established type of factor analysis (FA) with the concept of forming a smaller number of factors from a larger number of variables (called items). It is able to explain the highest possible quantity of variance in the entire dataset. Subsequently, such factors can be used to reveal the patterns within the set of studied variables (Wold et al., 1987). The general idea in PCA is to start with a large number of variables, then reduce them with the guidance of statistics. The PCA is particularly useful for forming various indexes (e.g., aroma index, nutritive index, authenticity score, bioavailability index, etc.). Prerequisites for PCA include the Kaiser-Meyer-Olkin (KMO) and Bartlett's test of sphericity to gauge the suitability of the dataset for factor analysis/PCA (Tabachnick and Fidell, 2007). KMO implies a proportion of variance that might be due to underlying factors, while Bartlett's test of sphericity indicates usefulness of PCA for the dataset. KMO suggested minimum is 0.6 or more (some sources evaluate this threshold to 0.7), and Bartlett's test should be significant at $p \leq 0.05$. High KMO indicates usefulness of PCA analysis, while significance below 0.05 for Bartlett's test of sphericity indicates related variables (e.g., that the correlation matrix is an identity matrix), hence, being suitable for pattern detection in PCA. Factors are formed based on the eigenvalues and explained variance, where the first factor should have the greatest variance and the highest eigenvalue. For interpretation of the PCA, it is important to report factor loadings, factor eigenvalues, and the amount of explained variance (Cukelj et al., 2016).

Chemometrics in food processing is useful for determining the influence of the production procedure on changes in the chemical composition of foods. Commonly, this is referred to thermal processing, and often to a negative influence on the nutritive value (van Boekel et al., 2010). Fresh produce (Putnik et al., 2017a; Poojary et al., 2017; Putnik et al., 2017b), their products (Koubaa et al., 2018; Gabrić et al., 2017; Putnik et al., 2017c; Barba et al., 2017), and important byproducts (Bursać Kovačević et al., Dragović-Uzelac et al., Putnik et al., 2017e; Putnik et al., 2018) lose their nutritive value due to thermal processing. Similarly, thermal processing can damage the production of the biologically active compounds in extracts from various plants that can serve as raw materials for functional food production (Vinceković et al., 2017; Kovačević et al., 2018). In many of the above-mentioned examples, mathematical modelling provided the optimal parameters, usually done by some form of multivariate regression analysis outlining them also as very useful tools for chemometrics.

3. Methods and Principle of Modelling

Planning any kind of production in the modern age with large number of complex parameters and without using some kind of modelling is likely unthinkable. Mathematical modelling involves a solution of two problems—constructing a mathematical model and its investigation (Niemark, 2003). As Niemark explained, problems are very different and require various knowledge, habits and intuition, but in spite of this, they are closely interconnected. When a model is constructed, it should account for difficulties and possibilities for each situation, while testing of a model can reveals a necessity for its correcting.

Before attempting to generate a model, the first step is to think carefully about the main purpose of that model. For instance, what questions do we want to address with it? The answers to the main questions help the researcher(s) to define the system, select the system boundary and identify relevant system variables (Imboden and Pfenninger, 2013). Some of the basic questions that need to be addressed are: (i) What is needed to know about the process? (ii) What data are available? (iii) What type of data will be obtained (categorical or numerical)? (iv) What are the boundaries? (v) What is the required complexity of a model in a terms of required steps? (vi) What kind of prediction is needed? (vii) Can it be validated and verified? For instance, these principles and Surface Enhanced Raman Spectroscopy (SERS) were applied in the research where model should predict histamine in fish by silver colloid SERS substrates (Jančí et al., 2017).

Variables in modelling can be of different types, representing type of material, temperature, concentration, viscosity, etc. It is important to distinguish which variables are inputs or outputs, which are dependant or independent, to identify their relationship (based on chemical or physical law, previous experience, etc.) and to predict their assumed influence. A good example presents the study of Jurinjak Tušek and co-workers (2016) about kinetics and thermodynamics of the solid-liquid extraction process *vs.* total polyphenols, antioxidants and extraction yield from *Asteraceae* plants.

Sometimes it is possible to generate a "simplified" model with the "black box" principle which dates back to 1941 (Cauer et al., 2000) and was improved by Bunge in 1962 (Bunge, 1963). Here, only input and output variables are observed without knowing how the whole system is constructed or how it works. In order to increase the stability and prediction of any model, the more we know about the system, the more variables are available and the whole process can be explained by their interactions.

There are four very common approaches to mathematical modelling (Magdić, 2011): (i) Empirical Modelling, which includes the testing of empirical data. (ii) Simulation modelling, which refers to the testing of particular scenarios based on a set of predetermined rules. These rules define the evolution of a particular process. (iii) Deterministic modelling, which tests equations or sets of equations for predicting the events or system. (iv) Stochastic, which is similar to deterministic modelling, but also accounts for the randomness and likelihood that an event will happen.

Due to variety of mathematical models, one of the main division is to classify them as linear and non linear models. According to Dym (2004), the imperative concept for mathematical models is the linearity. They provide a proportional connection between the parameters that are monitored (input) and results in term of output parameters. The two terms regarding linear models, which are widely used, are general linear model (GLM) and generalized linear model (GLIM or GLM). The difference between these two is mainly based on the means of solving certain mathematical problems, GLM includes different linear regressions, e.g., analysis of variance (ANOVA), and the analysis of covariance (ANCOVA) (PennState, 2018).

A persistent problem for describing complex systems is that linearity is not sufficient to describe them accurately, so nonlinear models need to be applied (Graybill and Iyer, 1994). For nonlinear cases, least-squares techniques are usually used, where models have nonlinearity in variables, but the parameters could be made linear via a transformation. In the case that parameters could not be transformed, least squares technique is extended to an estimation procedure known as nonlinear least squares. There are numerous articles with linear and nonlinear models in the food industry, including microbiology testing in terms of safety, milling of foods for different particle size distributions, different kinds of extraction procedures, drying (e.g., lyophilisation), testing of physical and chemical properties, etc. Although nonlinear models have wider versatility, their main drawback is their sensitivity to outliers, which can have negative influences on the results.

According to Marion (Marion, 2008) one of the other divisions of models is based on whether they predict the same outcome (deterministic models) or they predict several different outcomes (stochastic models). One example is the work of Flick et al. (2012), which was based on the cold chain food production, where authors tested the evolution of food products and their variability. For the equipment, authors used deterministic models and stochastic laws in order to get a certain randomness needed to solve the equations (generated by Monte-Carlo algorithm). Also, mathematical models can be static and dynamic. While static models describe a process at a certain time of the process duration, dynamic models are time dependent and give perspective of changes in the process over time.

Modelling as a part of chemometrics has been used in the past decades due to the progression of a user-friendly software with various algorithms (Nunes et al., 2015). The most common approach is response surface methodology (RSM), that is, linear regression approach which gives overall insight into how independent and dependent variables describe industrial process (Nwabueze, 2010) under appropriate DOE. It is used for screening purpose and to reduce the number of research factors, while it is only applicable to continuous variables. The RSM has a wide variety of uses for optimisation, ranging from physical and chemical characteristics of tested food products or added ingredients (Montgomery, 2009; Myers et al., 2009).

Most of the available software allows for the selection of different DOEs (e.g., full factorial design, fractional factorial design, saturated design, central composite design and mixture design). It is important to think of the final goal of the research and whether or not it is necessary to use interactions between variables (Granato et al., 2014). One of the most complex, is an FFD that is able to single-out effects of the independent variables and their interactions, while it can be applied to categorical and continuous variables. As mentioned in the introduction, this design is based on the numbers of factors (k) that are tested and the levels (values that are tested for each factor) that can be 2, 3, 4, etc., where the number of experiments grows rapidly (2^k, 3^k ...) and leads to a large number of runs that have to be performed. Numerous papers use FFD in the food industry, but most of them are based on certain process optimisation where this design is still one of the best options. The further use of FFD is explained in Section 6.

In some cases, when there is a low number of variables (factors) that are tested, the two-level designs are most commonly used. When the number of variables increases, the two-level designs are not applicable (Ferreira et al., 2017) and three-level designs or fractional factorial designs are used. During the last two decades, Box-Behnken has been one of the most used three-level designs since it requires less experiments than FFD and more than normal fractional factorial design, which enables more precise analysis of investigated factors interactions (Haaland, 1989). In order to lower the cost of the analysis, one could simply use fractional factorial design. Although the accuracy of the model will suffer, it is not a bad way to reduce the number of assays. As Siala et al. (2012) suggest, Plackett-Burman design is a very useful design for testing many factors, especially in the early stages of experimental phase. This design is often used on large sets of factors in order to find global effects using a low number of experiments. The appropriate use of particular DOE is explained elsewhere. If some process is already well known and documented, then there is no reason to use FFD for testing different products in terms of monitoring physical and chemical properties of different extracts, ultrasound, freeze-drying or high pressure influence on different products, etc. For a screening of new process or the product testing, one could use fractional design to get an insight into which variables will be most influential. However, in order to optimize the process, especially for microbial inactivation, one has to consider all the influences, because sometimes even the small difference (simpler *vs.* full factorial model at laboratory scale) can cause economic losses when scaled-up to industrial proportions, especially over a period of few years.

With the increase of different algorithms that are now implemented to a user-friendly software (Nunes et al., 2015), the importance of modelling is increasing rapidly. Software packages, like Statistica, Mathematica, MatLab, Design-experiment, and even Excel (originally developed as spreadsheet), now include options from basic modelling to the newest DOEs needed for process optimisation. Granato and Calado (2014) outlined that there are still researchers that do not use an experimental design in order to get more information and understanding of generated data. Nowadays, in order to save time and money, models are needed to reduce the numbers of experimental runs and identify interactions that are of crucial importance for production.

4. Modelling and Virtualization in Food Processing

Anticipation of the future has been mysterious and attractive to humans from the beginning of time, and likely fueled the development of modern mathematical equations (models) that are able to calculate outcomes of the complex events with certain probability. In food science, mathematical modelling found versatile applications, such as for food design (Musina et al., 2017), predicting the efficacy of extraction (Putnik et al., 2017e), microbial growth (Putnik et al., 2017f; Oladunjoye et al., 2017; Fujikawa and Sakha, 2014; Fujikawa, 2016), product quality (Putnik et al., 2017b; Putnik et al., 2017d) and processing (Putnik et al., 2017a; Silva et al., 2016; González-Tejedor et al., 2017).

Modelling and Simulation in food processing is used as a tool in designing and optimizing the operational parameters (Bursać Kovačević et al., 2017) and is an important base of product quality and safety control (Belščak Cvitanović et al., 2018) as well as a useful tool for developing innovative food processing technologies (Rumora et al., 2013). Development of models, i.e., modelling is often used to extrapolate some value of a property that is difficult to directly measure, and different tools are available to achieve the goal (Santos Guerreiro et al., 2013).

The first explained model in the literature was the model of the heat transfer during sterilization by Ball in the year 1923 (Ho et al., 2013). This was the milestone for food engineers to develop and apply mathematical models in food processes. This approach helped in the designing and optimising of new food processes for the improvement and understanding of physical phenomena that occur during manufacturing (Perrot et al., 2011; Ho et al., 2013).

Modelling approaches depend on the complexity of the observed process. They can range from simple relationships between variables that are often regression models, or just the observed relationship between variables, called statistical models (Datta, 2008). When variables vary as a function of time, ordinary differential equations (and partial different equations) can describe variable dependence on time and space (Perrot et al., 2011).

To solve the partial differential equations, the most common tools are numerical methods used in a software for heat transfer, deformation of food and material, fluid flow or diffusion. Some common types of software are W.R. Mathematica, COMSOL, Multiphysics, etc. The mentioned programs 5 offer different facilities that allow us to define complicated geometries (Ho et al., 2013). Those programs also enable us to observe different interactions between physical processes what is providing the Multiphysics capabilities.

According to Ho et al. (2013), multiscale models are basically a hierarchy of sub-models that are interconnected in order to describe the investigated material properties in different scales. The advantage of these models is to predict the macro system behaviour that is consistent with the fundamental structure of matter at different scales without the need for excessive computer resources (Steinhauser, 2017). In modelling and virtualization, it is relevant to define the set of important variables that are additionally analysed in order to investigate if reduction of data (observed variables) can be objectively verified and to identify variables that are most significant (Guerreiro et al., 2013; Olmedo et al., 2013; Kim et al., 2014).

In the data reduction multivariate tools like FAs are used, as in the paper of Budić-Leto and co-workers (2017) in the investigation of an aroma profile, based on the phenolic composition of Croatian dessert wine Prošek. The FA was used to identify important variables from the set of 66 observations (aroma components).

The term for coupling information technology (IT) and computer science is called virtualization (IBM, 2017). According to IBM, 2017 the main advantage here is the separation of specific physical hardware from software operations, thus enabling more flexible operations and better utilization of the physical hardware. Since all the industrial manufacturing processes and operation costs relate to economic limitations, with emphasis on energetic requirements, virtualization and modelling can find their place as useful tools in the food industry. The application of mathematical modelling, although with increasing popularity in literature, still lacks data for entire food processing lines at an industrial scale (European Commission, 2013). One of the biggest challenges, which is related to heat transfer and other transport phenomena, occurs during entire processes and can alter the structure of materials (Fagundes et al., 2013). The main goal is to develop a set of mathematical equations based on physical laws, which are mainly partial differential equations (PDEs). To simplify and avoid time-consuming experiments, such solutions require computer simulations of the process. Subsequently, mathematical models should be tested to check whether or not this process is stable under the required conditions.

In food manufacturing, it is common to replace a product or ingredient in certain production lines. Here, reaction kinetics and structural mechanical modelling may play a significant role in simulations and virtualizations (Belščak-Cvitanović et al., 2018), especially if industrial alterations can thermally denature nutrients, effect structural or sensory qualities, inactivate microorganisms or generate contaminants. The intrinsic multidisciplinary character required in food processing integrates various aspects of engineering that serve the same goal (Edogdu et al., 2017).

Erdogdu and co-workers (2017) summarized the basic steps for modelling complex food processing. They emphasized the following: (i) the processing problem has to be defined by computational description; (ii) the process modelling is governed by the physical laws with defining initial process conditions and status and boundaries; (iii) the aim of modelling should be the simplification of mathematical complexity with different algorithms while offering simulation and numerical solving. Software that are purposeful for modelling are Mathematica, LabView, MatLab, Comsol and others. At the end of modelling, maths equations should be evaluated by statistics with, e.g., goodness-of-fit tests, and by examining modelled data by comparing computational and experimental results.

Virtualisation by 3D software identifies the spectral variations generated from each sample while reducing the number of dimensions to identify the spectral features of greatest importance. Data from Fig. 1 is re-projected along the axes of greatest variation using the principal component analysis (PCA). Wolfram Research Mathematica v.10 (Wolfram Research, USA) was used for virtualization and plotting in 3D. Such virtualization allows plotting of three factors in three dimensions (3D PCA) for better differentiation between different particle size fractions (Valinger et al., 2017).

Based on the visual distribution of the particle sizes for different medical plant extracts, it is possible to categorise observed samples, and lay the foundation for finding optimal particle sizes for optimal extractions (e.g., for chamomile, nettle, dandelion, yarrow and broadleaf plantain samples).

Models and virtualization will save production resources and generate new data which will help to better understand the mechanisms behind a process (Faruh et al., 2017). Saguy (2016) emphasized that, regardless of the complexity and challenges of modelling and virtualization, they still offer opportunities for food engineering by innovation and social responsibility supportive of R&D in food processing industry. In the food production process, it is crucial to understand the main transfer mechanisms, energy transfer and its consumption. The main energy transfer processes utilized in the food processing industry are focused on inactivation of microbial-enzymatic activity as: Heating treatments (e.g., pasteurization, sterilization, baking, roasting, etc.), cooling, freezing, short or long storage, or for emerging technologies (radio frequency, infrared and ohmic heating). Therefore, modelling in food engineering continues to be described as a unique but complex approach, useful for manufacturing, anticipating consumer needs and expectations, preserving the health of the public, and maintaining sustainability in order to bind food safety and nutrition with socially responsible production (Steinhauser, 2017).

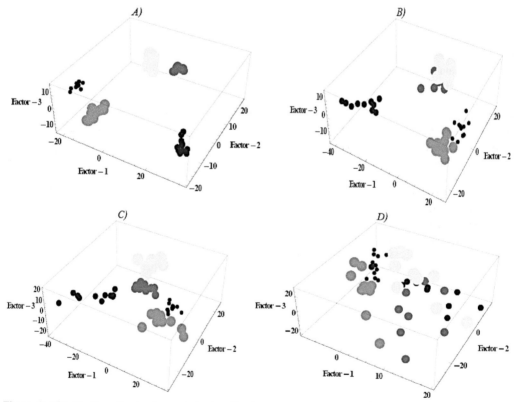

Figure 1: Virtualization of extraction optimization (for five different medical plants) using first derivatives of NIR spectra presented by first three factors for particle size fractions: < 100 μm: (A); 100–280 μm: (B); 280–450 μm: (C) and > 450 μm: (D) (Valinger et al., 2017).

5. Modelling and Prediction of Microbial Growth

Multivariate regression statistics offer an interesting possibility to develop models from raw experimental data. Such fundamental models are flexible, tailor-made, accurate and able to incorporate a larger number of relevant variables (Putnik et al., 2017f). On the other hand, they are very complex to construct and evaluate, and require carefully structured DOE and data standardization with minimal influence on the quality of information (Karnopp et al., 2017; Nunes et al., 2015). Alternatives to regression can be found in semi-fundamental Michaelis–Menten models that are a simplification of the complex biochemical processes (Fagundes et al., 2013), Arrhenius equations that are able to integrate effects of temperature on product quality (Tsironi et al., 2017), or with some other predictive kinetic models (Dermesonluoglu et al., 2016).

For instance, most of the models for predicting storage employ isothermal conditions and might need adjustments and validation for the non-isothermal cold food chains. Recently, modelling of shelf-life of minimally processed fresh-cut salad was reported for both non-isothermal and temperature conditions (Tsironi et al., 2017). Here, Baranyi Growth equation predicted the microbial growth, while the deterioration effect of temperature was evaluated by Arrhenius model.

A different study employed kinetic models for predicting the storage temperature of dandelion salad in terms of microbial growth and product quality (Dermesonluoglu et al., 2016). Here, the quality of the stored salad was assessed based on microbiology, texture and sensory evaluation, where non-isothermal dynamic conditions had the potential for mathematical prediction of storage. Another study aimed to study quality changes (sensory, microbial, nutritive) of untreated and mild heat-treated fruit and vegetable smoothies, using mathematical models to estimate the potential shelf-life of the products at different storage temperatures (5, 15 and 25°C). Linear regression was used to study the relationship between investigated variables. Obtained results revealed that the modelling methodology used in this study was useful for prediction, thus making the results reliable enough to describe the process relationships (González-Tejedor et al., 2017).

Freely available Anti-browning Apple Calculator - "C.A.P.P.A.B.L.E.©" (Pizent et al., 2015) is an online application developed solely from fundamental models. From the website that is intended as the production aid, it is possible to estimate microbial and non-microbial spoilage of various types of apples stored in modified atmospheres. The user is able to alter a large number of industrial factors and obtain projections for the microbial and non-microbial spoilage. Non-microbial spoilage is estimated from physiochemical changes on various apple cultivars and minimal processing with numerous anti-browning agents. Models account for browning, type and size of packaging, respiration rates, mass of the product and the content of modified atmosphere gases.

Calculations for microbial spoilage are based on length of time, apple variety, processing with anti-browning agent, use of advanced technology, physiochemical parameters, type of microorganism, numerous industrial packaging parameters, mass of product and content of the modified atmosphere. Details and explanations of the experimentations of the research are available elsewhere (Putnik et al., 2017a; Putnik et al., 2016). These tools can be useful for food technology professionals to control, optimize and improve industrial production. Similar models can be developed for different types of foods with their specific industrial parameters providing products with higher quality and safety (e.g., improving HACCP procedures), and manufacturing with better economic perspective.

6. Current Trends in Using Full Factorial Design in Non-thermal Food Processing

In summary, the FFD is currently applied when it is necessary to determine independent variables that have important influences on output values in the study. It is also possible to determine the range of levels of independent values and to visualize the effects of different factor combinations on the selected response values (output). When there are too many factors it is recommended to use the fractional factorial design, aiming at reducing the number of assays (Jambrak, 2011). DOE can be applied in order to save resources and to have faster insights into the observed influence on variable parameters (factors) on results (desired output parameters) (Yu et al., 2018). Even though the accuracy of the fractional factorial design is lower, less time and money are spent. FFD can be efficiently used in many R&D of food processes using different technologies. It can be used to optimize thermal processes (drying process, pasteurization conditions, inactivation of enzymes, etc.) in terms of increasing yield and reducing drying time and/or energy input. In extraction processes, FFD can be used in order to reduce the extraction time, solvent input and material costs. Also, FFD can be applied in non-thermal processing. Ultrasound, plasma, high pressure processing, high voltage electrical discharge, electron beam, pulsed light, pulsed electric field and others can be studied in order to have optimal output results for inactivation of microorganisms (Jambrak et al., 2016; Herceg et al., 2015; Režek Jambrak et al., 2017), extraction processes (Jovanović et al., 2017), waste-water treatment (Misra et al., 2015) and many others. In two studies, RSM was used to show results in relation to polynomic of optimization (Batur et al., 2010; Jambrak et al., 2010). Researchers studied the influence of ultrasound on physical and rheological properties of ice cream in model mixtures with FFD. As they described for the input variables, one can take into consideration amplitude of ultrasound, treatment time, power, pulse, temperature of process, volume of sample and others and as response variables which are the result of measurement one can take, for example, the amount of extracted bioactive component, different yields, aroma compounds, anthocyanins content, texture profile, activity of enzymes and others. They also stated that, in order to determine the importance of a certain factor (independent variable) and the number of levels (range of values) for those factors, one should certainly use full factorial design. If one is faced with large numbers of factors that need to be tested, and there is a shortage of materials and time/money available they recommend the use of fractional factorial design in order to lower the number of experiments and reduce the cost. In a study by Seyed Shahabadi and Reyhani (2014), they applied a two-level FFD analysis in order to optimize the ultrafiltration process for produced water treatment. In total, 19 experiments with eight duplicate experiments for fractional points and three replicates at the center point were performed in order to overview the main and interaction effects of three operating parameters (independent variables), temperature, transmembrane pressure and cross flow velocity on permeate flux, fouling resistance and total organic carbons rejection (response variables) using PAN350 membrane. They demonstrated that it would be a good potential choice for industrial applications (Seyed Shahabadi and Reyhani, 2014). In another study, Lohani (2018) used FFD for an experimental plan and analysis of results using statistical software. They researched extrusion in order to enhance the total phenolic content (TPC) and antioxidant activity (AA) of apple pomace (AP) prior to extrusion process (Lohani, 2018). Satari et al. (2017) used FFD for simultaneous pectin extraction and pretreatment (SPEP) of citrus waste (CW). The optimum conditions for obtaining the maximum SPEP yield (glucose + pectin (g)/raw material (g))*100) were pH 1.8, 80°C, and 2 h, which resulted in a yield of 58.7% (g/g CW). This data was optimised and the processes were considered to be satisfactory and saved the energy and time required (Satari et al., 2017). In a study by Vajić et al. (2015) a full three level factorial design was implemented for optimization of extraction parameters. They were trying to maximize the total phenolic (TP) yield from stinging nettle leaf. The optimal extraction conditions were 54% aqueous methanol and 38 min extraction time, while maximal theoretical TP yield was 8.9 mgGAE/g DW. They showed that ultrasound extraction has a better extraction capability, affecting yield and time of extraction (Vajić et al., 2015). In the study by Chinatangkul et al. (2017), the factors influencing the formation and properties of shellac (SHL) nanofibers loaded with an antimicrobial monolaurin (ML) were investigated. They also researched the main and interaction effects of formulation and process parameters, including SHL content, ML content, applied voltage and flow rate on the characteristic of nanofibers. They used 19 experiments based on FFD with three replicated center points. They showed that ML might be entrapped between the chains of SHL during the electrospinning process, exhibiting an excellent encapsulation (Chinatangkul et al., 2017). Jovanović et al. (2017) studied the optimization of the extraction parameters in order to improve the efficiency of polyphenol extraction from *T. serpyllum*. They used FFD and factors like particle size, solid-to-solvent ratio, solvent type and extraction time, by using maceration, heat- and ultrasound-assisted extraction (HAE and UAE). Output values were extraction efficiency expressed via total polyphenol content (TPC) and total flavonoid content (TFC). The obtained optimal conditions for achieving the best polyphenol yield were particle size of 0.3 mm, 1:30 solid-to-solvent ratio and 50% ethanol as an environmentally friendly extraction medium (Jovanović et al., 2017). Sinha et al. (2013) researched microwave-assisted extraction of yellow-red natural dye from seeds of Bixa orellana (Annatto). They researched the influence of process parameters pH, extraction time and amount of Annatto seeds on the extraction efficiency through a two level three factor (23) full factorial central composite design (CCD). They used RSM and artificial neural network (ANN) to develop predictive models for simulation and optimization of the dye extraction process and their results suggest that ANN has better prediction performance in comparison to RSM (Sinha et al., 2013). Tiwari et al. (2010) researched grape juice processing by ultrasound by changing processing factors. They used full factorial experimental design with regression modeling in order to investigate the main effects of amplitude level and treatment time on anthocyanins and color parameters. They found significant retention of anthocyanin content in grape juice and showed that sonication could be employed as a preservation technique for fruit juice processing where anthocyanin retention is desired (Tiwari et al., 2010).

7. Future Perspectives of Application

Herein is the overview of application of chemometrics, principal component and other multivariate analysis, characteristic of full factorial design in food processing. The future perspectives of modelling and virtualization in food processing, in particular prediction and modelling of microbial growth, are very important for the entire food chain. In the future, there is necessity to use statistical and mathematical software, especially for "big data", in order to tackle the challenges in food processing and food design. The future processing lies in shorter and efficient treatments, using less energy, raw materials and less experimental runs in research. Regarding the microbial inactivation, it is very difficult to manage and "predict" what will happen with microorganisms in different matrix. Microorganisms are living things and there are many unsolved questions regarding adaptation to different environments and particularly to their resistance. Therefore, all mentioned mathematical protocols provide good tools to overcome and/or to adapt models to specific tasks.

Acknowledgments

Authors would like to acknowledge the Croatian Science Foundation for their financing of the project titled: "High voltage discharges for green solvent extraction of bioactive compounds from Mediterranean herbs (IP-2016-06-1913)".

References

Anti-browning Apple Calculator – CAPPABLE. Faculty of Food Technology and Biotechnology, University in Zagreb. http://apple.pbf.hr. (Accessed May 1, 2016).

Barba, Francisco J., Predrag Putnik, Danijela Bursać Kovačević, Mahesha M. Poojary, Shahin Roohinejad, José M. Lorenzo et al. 2017. Impact of conventional and non-conventional processing on prickly pear (*Opuntia* spp.) and their derived products: From preservation of beverages to valorization of by-products. Trends in Food Science & Technology 67: 260–270. doi: 10.1016/j.tifs.2017.07.012.

Batur, V., Lelas, V., Jambrak, A.R., Herceg, Z. and Badanjak, M. 2010. Influence of High Power Ultrasound on Rheological and Foaming Properties of Model Ice-Cream Mixtures | Utjecaj Ultrazvuka Visoke Snage Na Reološka Svojstva I Svojstva Pjenjenja Modelnih Sladolednih Smjesa. Mljekarstvo 60(1): 10–18.

Bleoancă, Iulia, Klemen Saje, Liliana Mihalcea, Elena-Alexandra Oniciuc, Sonja Smole-Mozina, Anca Ioana Nicolau et al. 2016. Contribution of high pressure and thyme extract to control Listeria monocytogenes in fresh cheese—A hurdle approach. Innovative Food Science & Emerging Technologies 38(A): 7–14. doi: 10.1016/j.ifset.2016.09.002.

Brimacombe, M. 2014. High-dimensional data and linear models: A review. Open Access Medical Statistics 4: 17–27. https://doi.org/10.2147/OAMS.S56499.

Budić-Leto, Irena, Goran Zdunić, Jasenka Gajdoš-Kljusurić, Ana Mucalo and Urška Vrhovšek. 2017. Differentiation of Croatian dessert wine Prošek and dry wines based on the phenolic composition. Food Composition and Analysis 62: 211–216. doi:10.1016/j.jfca.2017.05.015.

Bunge, M. 1963. A general black-box theory. Philosophy of Science 30(4): 346–358. doi:10.1086/287954.

Bursać Kovačević, D., Putnik, P., Pedisić, S., Ježek, D., Karlović, S. and Dragović-Uzelac, V. 2015. High hydrostatic pressure extraction of flavonoids from freeze-dried red grape skin as winemaking by-product. Annals of Nutrition and Metabolism 67: 521–522.

Bursać Kovačević, Danijela, Predrag Putnik, Verica Dragović-Uzelac, Sandra Pedisić, Anet Režek Jambrak and Zoran Herceg. 2016. Effects of cold atmospheric gas phase plasma on anthocyanins and color in pomegranate juice. Food Chemistry 190: 317–323. doi: 10.1016/j.foodchem.2015.05.099.

Cauer, E., Mathis, W. and Pauli, R. 2000. Life and Work of Wilhelm Cauer (1900–1945). Proceedings of the Fourteenth International Symposium of Mathematical Theory of Networks and Systems (MTNS2000) (June, 2000), 19–23.

Chinatangkul, Nawinda, Chutima Limmatvapirat, Jurairat Nunthanid, Manee Luangtana-Anan, Pornsak Sriamornsak and Sontaya Limmatvapirat. 2017. Design and characterization of monolaurin loaded electrospun shellac nanofibers with antimicrobial activity. Asian Journal of Pharmaceutical Sciences, December. Elsevier. doi:10.1016/J.AJPS.2017.12.006.

Christensen, R. 1987. Plane Answers to Complex Questions: The Theory of Linear Models. New York: Springer-Verlag.

Cox, D.R. and Wermuth, N. 1996. Multivariate Dependencies. London: Chapman and Hall.

McCullagh, P. 2002. What is a statistical model? The Annals of Statistics. 30(5): 1225–1310. doi:10.1214/aos/1035844977.

Cukelj, Nikolina, Predrag Putnik, Dubravka Novotni, Saša Ajredini, Bojana Voucko and Duška Curic. 2016. Market potential of lignans and omega-3 functional cookies. British Food Journal 118(10): 2420–2433. doi: 10.1108/bfj-03-2016-0117.

Datta, A.K. 2008. Status of physics-based models in the design of food products, processes, and equipment. Comprehensive Reviews in Food Science and Food Safety 7(1): 121–129. doi: 10.1111/j.1541-4337.2007.00030.x.

Dermesonluoglu, Efimia, Kalliopi Fileri, Argiro Orfanoudaki, Maria Tsevdou, Theofania Tsironi and Petros Taoukis. 2016. Modelling the microbial spoilage and quality decay of pre-packed dandelion leaves as a function of temperature. Journal of Food Engineering 184: 21–30. doi: 10.1016/j.jfoodeng.2016.03.017.

Devore, J.L. and Berk, K.N. 2012. Modern Mathematical Statistics with Applications. New York: Springer-Verlag.

Dragović-Uzelac, V., Putnik, P., Zoran, Zorić, Ježek, D., Karlović, S. and Bursać Kovačević, D. Winery by-products: Anthocyanins recovery from red grape skin by high hydrostatic pressure extraction (HHPE). Annals of Nutrition and Metabolism 67(1): 522–523.

Dumancas, Gerard G., Sindhura Ramasahayam, Ghalib Bello, Jeff Hughes and Richard Kramer. 2015. Chemometric regression techniques as emerging, powerful tools in genetic association studies. TrAC Trends in Analytical Chemistry 74: 79–88. doi: 10.1016/j.trac.2015.05.007.

Dym, C. 2004. Principles of Mathematical Modeling 2nd Edition. Cambridge, Massachusetts: Academic Press.

Dziurkowska, Ewelina and Marek Wesolowski. 2015. Multivariate statistical analysis as a supplementary tool for interpretation of variations in salivary cortisol level in women with major depressive disorder. The Scientific World Journal 2015: 1–8. doi: 10.1155/2015/987435.

Erdogdu, F., Sarghini, F. and Marra, F. 2017. Mathematical Modeling for Virtualization in Food Processing. Food Engineering Reviews.

European Commission. 2013. 2013 EU industrial research and development (R&D) investment scoreboard. doi: 10.2791/25901.

Fagundes, Cristiane, Bruno Augusto Mattar Carciofi and Alcilene Rodrigues Monteiro. 2013. Estimate of respiration rate and physicochemical changes of fresh-cut apples stored under different temperatures. Food Science and Technology (Campinas) 33(1): 60–67. doi: 10.1590/s0101-20612013005000023.

Ferreira, S.L.C., Silva Junior, M.M., Felix, C.S.A., da Silva, D.L.F., Santos, A.S., Santos Neto, J.H. et al. 2017. Multivariate optimization techniques in food analysis—A review, Food Chemistry, doi:https://doi.org/10.1016/j.foodchem.2017.11.114.

Fick, D., Hoang, H.M., Alvarez, G. and Laguerre, O. 2012. Combined deterministic and stochastic approaches for modeling the evolution of food products along the cold chain. Part I: Methodology. International Journal of Refrigeration 35(4): 907–914. doi:10.1016/j.ijrefrig.2011.12.010.

Fujikawa, H. and Sakha, M.Z. 2014. Prediction of competitive microbial growth in mixed culture at dynamic temperature patterns. Biocontrol Science 19(3): 121–127. doi: 10.4265/bio.19.121.

Fujikawa, H. 2016. Prediction of competitive microbial growth. Biocontrol Science 21(4): 215–223. doi: 10.4265/bio.21.215.

Gabrić, Domagoj, Francisco Barba, Shahin Roohinejad, Seyed Mohammad Taghi Gharibzahedi, Milivoj Radojčin, Predrag Putnik et al. 2017. Pulsed electric fields as an alternative to thermal processing for preservation of nutritive and physicochemical properties of beverages: A review. Journal of Food Process Engineering. doi: 10.1111/jfpe.12638.

Gelman, A., Carlin, J.B., Stern, H. and Rubin, D.B. 1995. Bayesian Data Analysis. London: Chapman and Hall.

González-Tejedor, Gerardo A., Ginés Benito Martínez-Hernández, Alberto Garre, Jose A. Egea, Pablo S. Fernández et al. 2017. Quality changes and shelf-life prediction of a fresh fruit and vegetable purple smoothie. Food and Bioprocess Technology 10(10): 1892–1904. doi: 10.1007/s11947-017-1965-5.

Granato, D. and Araújo Calado, V.M. 2014. The use and importance of design of experiments (DOE) in process modelling in food science and technology. Mathematical and Statistical Methods in Food Science and Technology, First Edition. New York: John Wiley & Sons, Inc.

Granato, Daniel, Predrag Putnik, Danijela Bursać Kovačević, Jânio Sousa Santos, Verônica Calado, Ramon Silva Rocha et al. 2018. Trends in chemometrics: Food authentication, microbiology, and effects of processing. Comprehensive Reviews in Food Science and Food Safety.

Graybill, F.A. and Iyer, H.K. 1994. Regression analysis: Concepts and Applications. California: Duxbury Press.

Guerreiro, Joana Santos, Mário Barros, Paulo Fernandes, Preciosa Pires and Ronald Bardsley. 2013. Principal component analysis of proteolytic profiles as markers of authenticity of PDO cheeses. Food Chemistry 136(3-4): 1526–1532. doi: 10.1016/j.foodchem.2012.02.066.

Haaland, P.D. 1989. Statistical problem solving. In: Experimental Design in Biotechnology. New York: Marcel Dekker.

Herceg, Z., Juraga, E., Sobota-Šalamon, B. and Režek-Jambrak, A. 2012. Inactivation of mesophilic bacteria in milk by means of high intensity ultrasound using response surface methodology. Czech Journal of Food Sciences 30(2): 108–17.

Herceg, Zoran, Anet Režek Jambrak, Vesna Lelas and Selma Mededovic Thagard. 2012. The Effect of High Intensity Ultrasound Treatment on the Amount of Staphylococcus aureus and Escherichia coli in Milk 50(1): 46–52.

Herceg, Z., Jambrak, A.R., Vukušić, T., Stulić, V., Stanzer, D. and Milošević, S. 2015. The effect of high-power ultrasound and gas phase plasma treatment on Aspergillus spp. and Penicillium spp. count in pure culture. Journal of Applied Microbiology 118(1): 132–41. doi:10.1111/jam.12692.

IBM. 2007. Virtualization in education. White Paper. http://www-07.ibm.com/solutions/in/education/download/Virtualization%20in%20Education.pdf (accessed November 19, 2017).

Imboden, D.M. and Pfenninger, S. 2013 Introduction to Systems Analysis. Berlin Heidelberg: © Springer-Verlag.

Jánči, Tibor, Davor Valinger, Jasenka Gajdoš Kljusurić, Lara Mikac, Sanja Vidaček and Mile Ivanda. 2017. Determination of histamine in fish by Surface Enhanced Raman Spectroscopy using silver colloid SERS substrates. Food Chemistry 224: 48–54. doi:10.1016/j.foodchem.2016.12.032.

Jambrak, A.R., Vukušić, T. and Herceg, Z. 2016. State of the Art of the Use of Ultrasound in the Beverage Industry (II): Effects on Microorganisms. Applications of Ultrasound in the Beverage Industry.

Jambrak, Anet Režek, Doriane Lerda, Ranko Mirčeta, Marina Šimunek, Vesna Lelas, Farid Chemat et al. 2010. Journal of Food Processing & Technology. Food Processing & Technology. OMICS International. http://agris.fao.org/agris-search/search.do?recordID=US2016B00374.

Jambrak, Anet Režek. 2011. Experimental design and optimization of ultrasound treatment of food products. Journal of Food Processing & Technology 2(3). doi:10.4172/2157-7110.1000102e.

Jovanović, Aleksandra A., Verica B. Đorđević, Gordana M. Zdunić, Dejan S. Pljevljakušić, Katarina P. Šavikin, Dejan M. Gođevac et al. 2017. Optimization of the extraction process of polyphenols from Thymus serpyllum L. Herb using maceration, heat- and ultrasound-assisted techniques. Separation and Purification Technology 179: 369–80. doi:10.1016/j.seppur.2017.01.055.

Judge, George G., Griffiths, William E. and Hill, R. Carter. 2000. Undergraduate Econometrics, 2nd edition. Somerset, New Jersey, USA.: John Wiley & Sons Inc.

Juraga, Edita, Brankica Sobota Šalamon, Zoran Herceg and Anet Režek Jambrak. 2011. Application of High Intensity Ultrasound Treatment on Enterobacteriae Count in Milk 61(2): 125–34.

Jurinjak Tušek, Ana, Maja Benković, Ana Belščak-Cvitanović, Davor Valinger, Tamara Jurina and Jasenka Gajdoš Kljusurić. 2016. Kinetics and thermodynamics of the solid-liquid extraction process of total polyphenols, antioxidants and extraction yield from asteraceae plants. Industrial Crops and Products 91: 205–214. doi: 10.1016/j.indcrop.2016.07.015.

Karnopp, Ariadne Roberto, Katherine Guimarães Oliveira, Eriel Forville de Andrade, Bruna Mara Postingher and Daniel Granato. 2017. Optimization of an organic yogurt based on sensorial, nutritional, and functional perspectives. Food Chemistry 233: 401–411. doi: 10.1016/j.foodchem.2017.04.112.

Kim, Nam Sook, Ji Hyun Lee, Kyoung Moon Han, Ji Won Kim, Sooyeul Cho and Jinho Kim. 2014. Discrimination of commercial cheeses from fatty acid profiles and phytosterol contents obtained by GC and PCA. Food Chemistry 143: 40–47. doi: 10.1016/j.foodchem.2013.07.083.

Koubaa, M., Barba, F.J., Bursać Kovačević, D., Putnik, P., Santos, M.D., Queirós, R.P. et al. 2018. Pulsed Electric Field Processing of Fruit Juices. In Fruit Juices 437–449.

Kovačević, Danijela Bursać, Francisco J. Barba, Daniel Granato, Charis M. Galanakis, Zoran Herceg, Verica Dragović-Uzelac et al. 2018. Pressurized Hot Water Extraction (PHWE) for the green recovery of bioactive compounds and steviol glycosides from stevia rebaudiana bertoni leaves. Food Chemistry. doi: 10.1016/j.foodchem.2018.01.192.

Lehmann, Erich L. 2011. Fisher, Neyman, and the Creation of Classical Statistics, New York: Springer-Verlag.

Lohani, Umesh Chandra. 2018. Understanding the Impact of Non-Thermal Processing and CO_2 Assisted Extrusion on Antioxidant, Textural and Functional Properties of Corn, Sorghum and Apple Pomace Based Extrudates. Accessed February 8. http://openprairie.sdstate.edu/etd.

Magdić, D. 2011. Uvod u matematičko modelitanje. http://www.ptfos.unios.hr/joomla/modeli/images/files/prezentacije/ Uvod%20u%20matematicko%20modeliranje.pdf (accessed October 12, 2017).

Marion, G. 2008. An Introduction to Mathematical Modelling. Given 2008 by Daniel Lawson and Glenn Marion (March): 1–32. https://people.maths.bris.ac.uk/~madjl/course_text.pdf.

McCullagh, P. and Nelder, J.A. 1989. Generalized Linear Models, Second Edition. New York: Chapman and Hall.

MCCullagh, P. and Nelder, J.A. 1989. Generalized Linear Models, 2nd ed. London: Chapman and Hall.

Myers, R.H., Montgomery, D.C. and Anderson-Cook, C.M. 2009. Response Surface Methodology: Process and Product Optimization Using Designed Experiments, 3rd ed. New York: John Wiley & Sons, Inc.

Misra, N.N., Cullen, P.J., Barba, F.J., Lik Hii, C., Jaeger, H., Schmidt, J. et al. 2015. Emerging macroscopic pretreatment. Food Waste Recovery: Processing Technologies and Industrial Techniques. doi:10.1016/B978-0-12-800351-0.00009-2.

Mohammed, Is'haq A., Mercy T. Bankole, Ambali S. Abdulkareem, Stephen S. Ochigbo, Ayo S. Afolabi and Oladiran K. Abubakre. 2017. Full factorial design approach to carbon nanotubes synthesis by CVD method in argon environment. South African Journal of Chemical Engineering. doi:10.1016/j.sajce.2017.06.001.

Montgomery, D.C. 2009. Design and Analysis of Experiments, 5th ed. New York: John Wiley & Sons, Inc.

Musina, Olga, Predrag Putnik, Mohamed Koubaa, Francisco J. Barba, Ralf Greiner, Daniel Granato et al. 2017. Application of modern computer algebra systems in food formulations and development: A case study. Trends in Food Science & Technology 64: 48–59. doi: 10.1016/j.tifs.2017.03.011.

Neimark, J.I. 2003. Mathematical Models in Natural Science and Engineering. Berlin Heidelberg: © Springer-Verlag.

NIST/SEMATECH. e-Handbook of Statistical Methods. http://www.itl.nist.gov/div898/handbook/(accessed January 10, 2018).

Nunes, C.A., V. O., Alvarenga, A., de Souza Sant'Ana, J. S., Santos, D. and Granato, D. 2015. The use of statistical software in food science and technology: Advantages, limitations and misuses. Food Research International 75: 270–280. doi: 10.1016/j.foodres.2015.06.011.

Nwabueze, Titus U. 2010. Review article: Basic steps in adapting response surface methodology as mathematical modelling for bioprocess optimization in the food systems. International Journal of Food Science and Technology 45(9): 1768–1776. doi: 10.1111/j.1365-2621.2010.02256.x.

Oladunjoye, Adebola O., Stanley A. Oyewole, Singh, S. and Oluwatosin A. Ijabadeniyi. 2017. Prediction of Listeria monocytogenes ATCC 7644 growth on fresh-cut produce treated with bacteriophage and sucrose monolaurate by using artificial neural network. LWT - Food Science and Technology 76: 9–17. doi: 10.1016/j.lwt.2016.10.042.

Olmedo, Rubén H., Valeria Nepote and Nelson R. Grosso. 2013. Preservation of sensory and chemical properties in flavoured cheese prepared with cream cheese base using oregano and rosemary essential oils. LWT - Food Science and Technology 53(2): 409–417. doi: 10.1016/j.lwt.2013.04.007.

Overney, Anaïs, Joséphine Jacques-André-Coquin, Patricia Ng, Brigitte Carpentier, Laurent Guillier and Olivier Firmesse. 2017. Impact of Environmental Factors on the culturability and viability of listeria monocytogenes under conditions encountered in food processing plants. International Journal of Food Microbiology 244 (March). Elsevier: 74–81. doi:10.1016/J.IJFOODMICRO.2016.12.012.

Parente Eugenio, Himanshu Patel, Veronica Caldeo, Paolo Piraino and Paul L.H. McSweeney. 2012. RP-HPLC peptide profiling of cheese extracts: A study of sources of variation, repeatability and reproducibility. Food Chemistry 131(4): 1552–1560. doi:10.1015/j.foodchem.2011.10.003.

Passari, L.M.Z.G., Soares, P.K. and Bruns, R.E. 2011. Estatística aplicada à química: dez dúvidas comuns. Quimica Nova 34: 888–892. http://dx.doi.org/10.1590/S0100-40422011000500028.

Perrot, Nathalie, Hugo De Vries, Evelyne Lutton, Harald G.J. van Mil, Mechthild Donner, Alberto Tonda et al. 2016. Some remarks on computational approaches towards sustainable complex agri-food systems. Trends in Food Science & Technology 48 (February). Elsevier: 88–101. doi:10.1016/J.TIFS.2015.10.003.

Penn State Eberly College of Science. 2018. Analysis of Discrete Data. https://onlinecourses.science.psu.edu/stat504/ node/216 (accessed November 19, 2017).

Poojary, Mahesha M., Predrag Putnik, Danijela Bursać Kovačević, Francisco J. Barba, Jose Manuel Lorenzo, Daniel A. Dias et al. 2017. Stability and extraction of bioactive sulfur compounds from Allium genus processed by traditional and innovative technologies. Journal of Food Composition and Analysis 61: 28–39. doi: 10.1016/j.jfca.2017.04.007.

Pripp, Are Hugo. 2013. Statistics in Food Science and Nutrition. New York: Springer-Verlag.

Putnik, P., Bursać Kovačević, D., Herceg, K. and Levaj, B. 2016. Influence of respiration on predictive microbial growth of Aerobic mesophilic bacteria and Enterobacteriaceae in fresh-cut apples packaged under modified atmosphere. Journal of Food Safety 37(1): e12284. doi: 10.1111/jfs.12284.

Putnik, Predrag, Francisco J. Barba, Jose Manuel Lorenzo, Domagoj Gabrić, Avi Shpigelman, Giancarlo Cravotto et al. 2017a. An integrated approach to mandarin processing: Food safety and nutritional quality, consumer preference, and nutrient bioaccessibility. Comprehensive Reviews in Food Science and Food Safety 16(6): 1345–1358. doi: 10.1111/1541-4337.12310.

Putnik, Predrag, Francisco J. Barba, Ivana Španić, Zoran Zorić, Verica Dragović-Uzelac and Danijela Bursać Kovačević. 2017b. Green extraction approach for the recovery of polyphenols from Croatian olive leaves (Olea europea). Food and Bioproducts Processing 106: 19–28. doi: 10.1016/j.fbp.2017.08.004.

Putnik, Predrag, Danijela Bursać Kovačević, Korina Herceg, Ivan Pavkov, Zoran Zorić and Branka Levaj. 2017c. Effects of modified atmosphere, anti-browning treatments and ultrasound on the polyphenolic stability, antioxidant capacity and microbial growth in fresh-cut apples. Journal of Food Process Engineering 12539. doi: 10.1111/jfpe.12539.

Putnik, Predrag, Danijela Bursać Kovačević, Korina Herceg, Shahin Roohinejad, Ralf Greiner, Alaa El-Din A. Bekhit et al. 2017d. Modelling the shelf-life of minimally-processed fresh-cut apples packaged in a modified atmosphere using food quality parameters. Food Control 81 (November). Elsevier: 55–64. doi: 10.1016/j.foodcont.2017.05.026.

Putnik, Predrag, Danijela Bursać Kovačević, Anet Režek Jambrak, Francisco Barba, Giancarlo Cravotto, Arianna Binello et al. 2017e. Innovative "Green" and novel strategies for the extraction of bioactive added value compounds from citrus wastes—A review. Molecules 22(5): 680. doi: 10.3390/molecules22050680.

Putnik, Predrag, Shahin Roohinejad, Ralf Greiner, Daniel Granato, Alaa El-Din A. Bekhit and Danijela Bursać Kovačević. 2017f. Prediction and modeling of microbial growth in minimally processed fresh-cut apples packaged in a modified atmosphere: A review. Food Control 80: 411–419. doi: 10.1016/j.foodcont.2017.05.018.

Putnik, Predrag, Danijela Bursać Kovačević, Damir Ježek, Ivana Šustić, Zoran Zorić and Verica Dragović-Uzelac. 2018. High-pressure recovery of anthocyanins from grape skin pomace (Vitis vinifera cv. Teran) at moderate temperature. Journal of Food Processing and Preservation 42(1). doi: 10.1111/jfpp.13342.

Režek Jambrak, Anet, Marina Šimunek, Silva Evačić, Ksenija Markov, Goran Smoljanić and Jadranka Frece. 2017. Influence of high power ultrasound on selected moulds, yeasts and alicyclobacillus acidoterrestris in apple, cranberry and blueberry juice and nectar. Ultrasonics. doi:10.1016/j.ultras.2017.02.011.

Rumora Ivana, Irena Kobrehel Pintarić, Jasenka Gajdoš Kljusurić, Olivera Marić and Damir Karlović. 2013. Efficient use of modelling in new food-product design and development. Acta Alimentaria 42(4): 565–575. doi: 10.1556/AAlim.42.2013.4.11.

Sarghini Fabrizio, Francesco Marra and Ferruh Erdogdu. 2017. Mathematical modeling for virtualization in food processing. Food Engineering Review 9: 295–313. doi:10.1007/s12393-017-9161-y.

Satari, Behzad, Jonny Palhed, Keikhosro Karimi, Magnus Lundin, Mohammad J. Taherzadeh and Akram Zamani. 2017. Process optimization for citrus waste biorefinery via simultaneous pectin extraction and pretreatment. BioResources 12(1): 1706–22. doi:10.15376/biores.12.1.1706-1722.

Seyed Shahabadi, Seyed Mahdi and Amin Reyhani. 2014. Optimization of operating conditions in ultrafiltration process for produced water treatment via the full factorial design methodology. Separation and Purification Technology. doi:10.1016/j.seppur.2014.04.051.

Saguy, S. 2016. Challenges and opportunities in food engineering: Modeling complexity, virtualization, open innovation and social responsibility. Journal of Food Engineering 176: 2–8. doi:10.1016/j.jfoodeng.2015.07.012.

Siala, R., Frikha, F. and Mhamdi, S. 2012 Optimization of acid protease production by Aspergillus niger I1 on shrimp peptone using statistical experimental design. The Scientific World Journal 2012: 1–11. doi:10.1100/2012/564932.

Sinha, Keka, Shamik Chowdhury, Papita Das Saha and Siddhartha Datta. 2013. Modeling of Microwave-Assisted Extraction of Natural Dye from Seeds of Bixa Orellana (Annatto) Using Response Surface Methodology (RSM) and Artificial Neural Network (ANN). Industrial Crops and Products 41(1). Elsevier B.V.: 165–71. doi:10.1016/j.indcrop.2012.04.004.

Silva, Nathália Buss da, Daniel Angelo Longhi, Wiaslan Figueiredo Martins, Gláucia Maria Falcão de Aragão and Bruno Augusto Mattar Carciofi. 2016. Mathematical modeling of Lactobacillus Viridescens growth in vacuum packed sliced ham under non isothermal conditions. Procedia Food Science 7: 33–36. doi: 10.1016/j.profoo.2016.02.081.

Skov, Thomas, Anders H. Honoré, Henrik Max Jensen, Tormod Næs and Søren B. Engelsen. 2014. Chemometrics in foodomics: Handling data structures from multiple analytical platforms. TrAC Trends in Analytical Chemistry 60: 71–79. doi: 10.1016/j.trac.2014.05.004.

Steinhauser, M.O. 2017. Multiscale Modeling of Fluids and Solids—Theory and Applications. Berlin Heidelberg: Springer-Verlag.

Tabachnick, B.G. and Fidell, L.S. 2007. Principal components and factor analysis. In Using multivariate statistics (5th ed.), 607–615. Needham Heights, MA: Allyn & Bacon.

Tsironi, Theofania, Efimia Dermesonlouoglou, Marianna Giannoglou, Eleni Gogou, George Katsaros and Petros Taoukis. 2017. Shelf-life prediction models for ready-to-eat fresh cut salads: Testing in real cold chain. International Journal of Food Microbiology 240: 131–140. doi: 10.1016/j.ijfoodmicro.2016.09.032.

van Boekel, Martinus, Vincenzo Fogliano, Nicoletta Pellegrini et al. 2010. A review on the beneficial aspects of food processing. Molecular Nutrition & Food Research 54(9): 1215–1247. doi: 10.1002/mnfr.200900608.

Vajić, Una-Jovana, Jelica Grujić-Milanović, Jelena Živković, Katarina Šavikin, Dejan Gođevac, Zoran Miloradović et al. 2015. Optimization of extraction of stinging nettle leaf phenolic compounds using response surface methodology. Industrial Crops and Products 74: 912–17. doi:10.1016/j.indcrop.2015.06.032.

Valinger Davor, Maja Benković, Tamara Jurina, Ana Jurinjak Tušek, Ana Belščak-Cvitanović, Jasenka Gajdoš Kljusurić et al. 2017. Use of NIR spectroscopy and 3d principal component analysis for particle size control of dried medicinal plants. Journal on Processing and Energy in Agriculture 21(1): 17–22.

Varmuza, Kurt and Peter Filzmoser. 2009. Chemoinformatics - chemometrics - statistics. In Introduction to multivariate statistical analysis in Chemometrics, 1–26. Boca Raton, FL: CRC Press Taylor and Francis Group.

Valdramidis, V.P., Taoukis, P.S., Stoforos, N.G. and Van Impe, J.F.M. 2012. Modeling the kinetics of microbial and quality attributes of fluid food during novel thermal and non-thermal processes. In Novel Thermal and Non-Thermal Technologies for Fluid Foods, 433–71. Elsevier. doi:10.1016/B978-0-12-381470-8.00014-1.

Vinceković, Marko, Marko Viskić, Slaven Jurić et al. 2017. Innovative technologies for encapsulation of Mediterranean plants extracts. Trends in Food Science & Technology 69: 1–12. doi: 10.1016/j.tifs.2017.08.001.

Wold, Svante, Kim Esbensen and Paul Geladi. 1987. Principal component analysis. Chemometrics and Intelligent Laboratory Systems 2(1-3): 37–52. doi: 10.1016/0169-7439(87)80084-9.

Yu, Peigen, Mei Yin Low and Weibiao Zhou. 2018. Design of experiments and regression modelling in food flavour and sensory analysis: A review. Trends in Food Science & Technology 71 (January). Elsevier: 202–15. doi:10.1016/J.TIFS.2017.11.013.

CHAPTER 5

The Use of Correlation, Association and Regression Techniques for Analyzing Processes and Food Products

Jimy Oblitas,[1] *Miguel De-la-Torre,*[2] *Himer Avila-George*[2] and *Wilson Castro*[3]

1. Introduction

The food industry requires analytical methods in order to evaluate physicochemical, microbiological, bromatological, and sensory aspects of food products. These traditional methods, such as gravimetric, volumetric and colorimetry determinations, use techniques with limited sensitivity levels, little specificity and most of them measure information of only one parameter. The analysis of data retrieved with such methods, requires the application of univariate or bivariate statistical techniques, for descriptive or predictive purposes.

The need for new methods of evaluating different aspects of foods at the same time has led to the development of new approaches and sensors for this purpose. In this scene, food analysis methods are being modernized, applying technological developments such as multiparameter sensors and indirect measurements of the physical, chemical, physicochemical properties (Oliveri and Simonetti, 2016). Techniques such as near-infrared (NIR) spectroscopy, surface-enhanced Raman spectroscopy (SERS), medium infrared spectroscopy (MIR) and hyperspectral imaging (HSI) are being used more and more frequently (Ai et al., 2018; Alamprese et al., 2016; Azevedo et al., 2018; Casale et al., 2012; Grassi and Alamprese, 2018; Panagou et al., 2014; Vásquez et al., 2018).

These techniques constitute new and promissory analytic tools for food quality and safety throughout the food chain, mainly due to their specificity, sample preparation simplicity and multi-parameter data handling. However, statistical tools are required in order to examine whole information or reduce non-relevant variables of the acquired high amount of data (Fayez et al., 2018; López et al., 2015), giving space to new research fields in data treatments.

[1] Departamen to de Tecnología de Alimentos, Universidad de Lleida. Lleida 25198, Spain.
Email: joc5@alumnes.udl.cat
[2] Centro Universitario de los Valles, Universidad de Guadalajara. Ameca, Jalisco 46600, Mexico.
Emails: miguel.dgomora@academicos.udg.mx; himer.avila@academicos.udg.mx
[3] Facultad de Ingeniería, Universidad Privada del Norte. Cajamarca, Cajamarca 06002, Peru.
Email:wilson.castro@upn.edu.pe

In this panorama, the multivariate statistical analysis, capable of evaluating the interaction between the different parameters that the food scientist needs to assess, was developed (Eriksson et al., 2014). Another fact that increased the development of new methods was the increase of computing capacities, observed in the growing number of programs developed, and its advantages in minimizing error in the generated models (Mishra and Datta-Gupta, 2018).

In this chapter, the reader is introduced to the use of correlation, association and regression applied to the analysis of instrumental data in food analysis. The purpose of these multivariate statistics techniques is growing significantly as a consequence of the computer revolution, which facilitates the analysis of big data. Spectrophotometric data and other multisensor systems employed in modern food analysis techniques are receiving attention due to their ability to extract information for complex systems and reduce the cost and amount of time required for processing. Consequently, in this chapter, we present the description of data exploration techniques, such as Principal Component Analysis (PCA), Cluster Analysis and hierarchical cluster analysis (HCA), classification techniques, such as Linear discriminant analysis (LDA), and regression techniques, such as Partial least square regression (PLSR) and neural networks. Finally, in order to exemplify, a decision strategy for multivariate analysis is shown together with a detailed implementation of the applied techniques in Matlab 2016.

2. Multivariate Statistics Used in Food Analysis

Multivariate data analysis simultaneously studies more than two variables from some observations (Manly and Navarro, 2016); and it is currently considered to be an excellent option to obtain useful information when it is necessary to recognize patterns of a significant amount of data originated in the food industry, during either the quality or process controls.

Multiple cases of usage of multivariate analyses have been reported in the literature, e.g., the work of Callao and Ruisánchez (2018) mentions that it is common to use this type of analysis to authenticate food products. In the area of traceability, Bona et al. (2017) reported the use of NIR images to determine the origin of coffee beans, and Zhang et al. (2017) used an isotope ratio analysis to identify the source of shrimp. These techniques have also been successfully used to identify adulterated or contaminated products, e.g., honey, milk, olive oil and fruit juice, among others (Azevedo et al., 2017; Boggia et al., 2013; Chen et al., 2017; Georgouli et al., 2017). In these works, adequate results are reported both for selection and classification processes.

In order to revise the elements of multivariate statistics, the methods for data exploration, classification and regression are discussed in the following sections. In particular, artificial neural networks are revised as a method that can be applied for both regression and classification.

2.1 Pre-treatment

Before applying any data analysis, raw data is pre-processed, which makes it possible for the analysed variables to have a proportional influence on the results. This process is known as z-scores standardisation and shows how far the raw data are from the average $(\bar{X}_k)$ in terms of the standard deviation (σ_K), as shown in the next equation and sketched in Fig. 1.

$$Z_k = \frac{X_k - \bar{X}_k}{\sigma_k}$$

Likewise, other pre-treatments, e.g., normalization, smoothing, outlier's removal, and similar processes, are common preliminary steps for later analysis. But, in the same way, some researchers like Amodio et al. (2017) and Marquetti et al. (2016) have shown that different pre-treatments affect model accuracy. Due to this issue, modeling should begin with the testing of as many pre-treatments as possible.

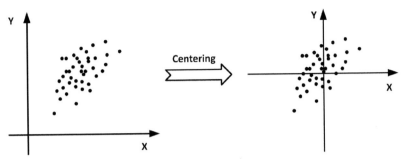

Figure 1: Data centering.

2.2 Data exploration

The exploration of the data uses techniques whose objective is to analyze the behavior of the observations relative to all the variables at the same time. Some of the primary methods for data exploration are Principal Components Analysis (PCA), Correspondence Analysis (CA), Multiple Correspondence Analysis (MCA) and Hierarchical Cluster Analysis (HCA). When the variables are categorical, CA or MCA are usually used. On the other hand, when the variables are quantitative, PCA is generally preferred (Husson et al., 2017). Among these methods, the most commonly used is PCA, being one of the most frequently used forms for the reduction of multidimensional data (Granato et al., 2017).

2.2.1 Principal component analysis

PCA consists of finding orthogonal transformations of the original variables in order to get a new set of uncorrelated variables, denominated Principal Components, which are obtained in descending order of importance. This concept could be better understood by looking at the scheme of the process in Fig. 2.

The PCA algorithm creates these new variables or components, sorting them according to their capacity in order to explain variance. In this sense, the first principal component (PC_1) corresponds to the variable with the largest variance, the second component (PC_2) with the second largest variance, and so on (Oliveri and Simonetti, 2016). Likewise, the follow equation is commonly used to calculate the PCA components.

$$Z = SP^T + E$$

where Z is the original data matrix of size $m \times n$; m is the number of n-dimensional samples, n corresponds to the number of variables; S is the score matrix; P is the loading orthogonal matrix; and E is the error matrix.

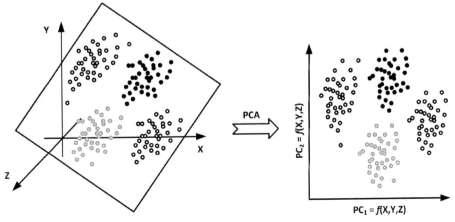

Figure 2: PCA process scheme.

In simple terms, PCA is a mathematical tool that aims to represent the present variation in the data set using a small number of factors and is applied to obtain the principal components $PC_j(j = 1,...)$ of the total data record after the pre-processing.

The use of PCA for feature compression is prevalent in the food industry. For example, De Araújo et al. (2017) used PCA analysis to eliminate the less significant input variables in the purification of baking yeast invertase. Results are shown in Fig. 3, which presents a bar diagram of the variance of the first eight principal components (PC). It can be observed that there is a variance concentration of 30.82% for PC_1. Furthermore, the cumulative variance reaches 90% with the sum of the first 5 PCs. This indicates that the yeast invertase data set used in this research can be represented in a multidimensional space, preferable with more than 5 PCs.

When PCA is used as an exploratory technique, the information of the first two or three PCs are the most important values. Its main limitation is when the first PCs do not contain enough information (Callao and Ruisánchez, 2018). According to a search in Scopus for PCA used in topics of *Agricultural and Biological Sciences*, there are 8,015 documents since 2013 which show the extensive use of this tool in this specific field. The countries that have published the most are China, followed by the USA (Scopus, 2018). In this field, the use of PCA for the classification of parameters in food studies is extensive. Table 1 shows a summary of some researches that employed PCA method to analyze food products.

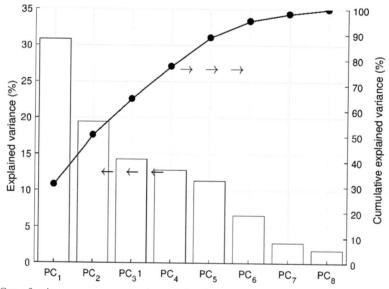

Figure 3: Case of variance per component and cumulative. Source: Based on the results of De Araújo et al. (2017).

2.2.2 Cluster analysis

Cluster analysis is the second most used method in the food industry. According to information obtained from Scopus, in the last five years, 1,433 manuscripts were reported in which this type of analysis was used within the *Agricultural and Biological Sciences* area (Scopus, 2018).

Hierarchical Cluster Analysis (HCA) is one of the most used cluster analysis methods. According to Lee and Yang (2009), this clustering method explores the organization of samples in groups and between groups that represent a hierarchy. Figure 4 shows two common representations employed for cluster analysis: A dendrogram and a nested cluster diagram.

In HCA, the initial number of clusters must be specified, then the observations between the clusters are iteratively reassigned until a predetermined point of convergence is reached. In general, clustering algorithms use the distance between data points in order to classify them (Mishra and Datta-Gupta, 2018). The most common measure is the Euclidean distance between vectors X_1 and X_2, which is given by

Table 1: Systematic analysis of PCA as a multivariate technique for data exploration.

Material	Objective	PCA use/result	Reference
Baking yeast extract	Predict Baker's yeast invertase content using the PEG/MgSO$_4$ Aqueous Two-Phase System (ATPS).	The PCA/LS-SVM method achieving R^2 values around 0.974.	De Araújo et al., 2017
Camel milk and goat milk	Detection and quantification of the contamination of camel milk with goat milk using NIR.	The PCA achieved a differentiation and complete separation between camel, cow and goat milk.	Mabood et al., 2017
Wine	Discrimination of grape varieties from red and white wines using NMR spectroscopy.	PCA/LDA technique obtained discrimination ratio over 82% in validation stage.	Fan et al., 2017
Liver Paste	Study the effect of salt reduction on liver paste.	Two PCs cover most of the data variation (91%).	Næs et al., 2014
Minced turkey meat	Determination of turkey meat quality, based on the texture.	PCA allowed the differentiation between two brands of products.	Probola and Zander, 2007
Whey powder	Evaluation of damage in whey and whey proteins caused by heat.	Three PCs were found, explaining 79.0% of the variance, and were the basis for the cluster analysis.	Gómez-Narváez et al. (2017)

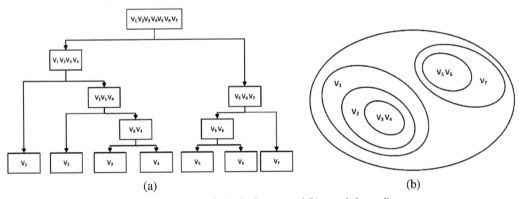

(a) (b)

Figure 4: Cluster analysis (a) dendrogram and (b) nested cluster diagram.

$$d(x_1, x_2) = \|X_1 - X_2\|$$

There are two principal approaches to solving the clustering problem with HCA: Agglomerative and divisive. In the first, a bottom-up approach is followed, starting from as many classes as objects to be classified and classes of similar objects are obtained successively. Conversely, in the divisive approach, a top-down strategy is followed. It starts from a single class formed by all the objects, and it is successively divided into sub-classes. Usually, the results from HCA are presented in the form of a dendrogram, which is a graphical representation in the form of a tree, summarizing the grouping process. In this sense, some works that used HCA in the food industry and their results are given in Table 2.

The principal weakness of HCA is that it does not provide the general relationship between all the samples, only among the nearby samples. Furthermore, it does not give any information about the relationship between the samples and the variables; in this sense, HCA can be considered to complement the PCA analysis.

Another widely employed algorithm for solving the clustering problems is known as *k*-means. Here, *k*-groups must be specified, as well as their initial centroids. Subsequently, an array of similarities is calculated between the *n* data points and the *k*-centroids; finally, each observation is assigned to the cluster that has the closest centroid (Mishra and Datta-Gupta, 2018). Groups are allocated by minimizing the sum of the squares of their distances within the cluster for the *k*-groups (see next equation).

Table 2: Systematic analysis of HCA as a multivariate technique for data exploration.

Sample	Objective	HCA use/result	Reference
Foods from different restaurants	Categorize foods according to their nutritional values.	The classification and identification of the foods was achieved according to their composition and place of sale.	Da Silva et al., 2006
Radishes	Select obstacle technologies (physical and chemical) for the inhibition of polyphenolxidase activity and color stability.	The HCA was able to choose the best methods capable of maintaining the characteristics of fresh type and color of radish slices.	Goyeneche et al., 2014
Rice cultivation	Classify different rice cultivars based on their starch processability indicators.	Twelve rice cultivars were analyzed and classified according to their hydration and pasting properties.	Lee et al., 2012
Tasters	Measure the gustatory function of individuals and their associations.	HCA, make possible to classify in more sensitive, semi-sensitive and less sensitive tasters.	Puputti et al., 2017

$$\arg\min_{\mathbf{C}} \sum_{g=1}^{k} \sum_{x \in C_g} || d(x, \bar{x}_g) ||^2$$

where $\mathbf{C} = \{C_1, C_2, ..., C_k\}$ corresponds to the set of n samples $\mathbf{x} = \{x_1, x_2, ..., x_n\}$ to be grouped into k clusters, and $\bar{x}_g$ is the mean of the samples in C_g.

2.3 Classification methods

Classification is another of the elements applied in multivariate statistics, which commonly involves finding a mathematical model capable of recognizing an object belonging to a class. After the classification model has been obtained, it is possible to predict to which class each new observation belongs. Classification is a task of data analysis that is commonly addressed in statistics by *discriminant analysis*.

A classification problem is related to cluster analysis in the fact that there are objects that need to be categorized in distinct groups. However, in classification, the number of groups and the membership of some training samples are known a priori. Classification and modeling techniques for classes can be grouped into three main categories: (a) Techniques based on distance as K-nearest neighbors (KNN); (b) Techniques based on probability as Linear discriminant analysis (LDA), Quadratic discriminant analysis (QDA) or Unequal class models (UNEQ) and (c) Techniques based on experience, such as Artificial Neuronal Network (ANN).[1] The representation of the decision boundaries generated by example 2-class classifiers are shown in Fig. 5.

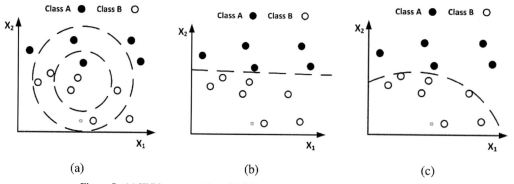

(a)   (b)   (c)

Figure 5: (a) KNN representation. (b) LDA representation. (c) QDA representation.

---

[1] Due to the wide use of ANNs in literature and practice, this technique is further detailed in a later section.

According to a review of the more popular classification techniques applied in food engineering, LDA is the most relevant according to the number and quality in its results. For this reason, there is an opportunity for research in the use of different classifiers in this field. In the next paragraphs the underlying concepts and application of LDA are detailed.

Data analysis methods used in food research typically start with a pre-treatment and then an exploratory data analysis, i.e., use unsupervised learning techniques to find groups of samples that share certain features. Finally, supervised learning techniques are employed in order to recognize patterns in raw food data. Linear discriminant analysis (LDA) is one of the most popular supervised methods for food analysis (Cai et al., 2018); it is a supervised technique used for the classification of patterns concerning the reduction of the dimensionality of the problem. Among its applications in food analysis, its use has been highlighted to determine product adulteration and authentication.

LDA is a probability-based technique, which estimates the multivariable probability density functions for each class. It begins with the estimation of location and dispersion parameters which are determined using experimental data; therefore, it is considered a probabilistic parametric technique. During the process, LDA assumes that the probability density distributions within each class are standard multivariate, with the same dispersion for all classes. For this reason, the same variance-covariance matrix is used to describe all probability distributions, and it is calculated as a combined variance-covariance matrix.

Numerous investigations employ multivariate analysis in conjunction to distinct technologies, like spectroscopy, to measure the relevant characteristics of food and improve food analysis. For instance, Reis et al. report the use of characteristics obtained from attenuated total reflection from Fourier transform infrared spectroscopy (ATR-FTIR), diffuse reflectance and data fusion to detect adulteration in various types of roasted coffee. They separate distinct coffee mixtures, achieving up to 100% sensitivity with 95% specificity in the test samples using data fusion from both sources (Reis et al., 2017). Data fusion from fluorescence and UV spectroscopy was also used for the verification of cocoa adulteration, obtaining 95% sensitivity on validation data (Dankowska, 2017).

Other authors report using LDA on features from hyperspectral imagery from a VIS-NIR hyperspectral camera (400–1000 nm) to detect adulteration in honey (Shafiee et al., 2016). Some techniques like the high-performance liquid chromatography with diode array detection (HPLC-DAD) and UV-vis spectroscopy have been used, coupled with LDA, for geographic classifications, as in the works of D'Archivio et al. (2016) and D'Archivio and Maggi (2017). These and other examples are summarized in Table 3, and evidence the wide applicability of LDA in the solution of distinct issues of food analysis.

2.4 Regression methods

Another type of analysis of particular interest in food engineering is related to the prediction of food composition. Regression methods are commonly employed for such a task, either with univariate or multivariate analysis, depending on the number of predictors and, eventually, on the response variables involved. Considering the order of the resulting relation modeling, the models can be linear or non-linear. Among the most known models, commonly used are the ordinary least squares (OLS), Fig. 6a, principal component regression (PCR), Fig. 6b, and partial least squares (PLS) regression.

In Fig. 6 the difference between the distinct regression methods is sketched, i.e., how the predictors are used. In OLS new variables—without treatment—named components are created prior to model creation. On the other hand, in PCR strategy, the principal components of the variables are used as the independent variables (regressors). Following the structure of the last section, the more relevant techniques used for multivariate regression are summarized.

Partial least square regression (PLSR) has been used by several researchers in Agricultural and Biological fields. As an example, a search in SCOPUS aiming to find the works that employed PLSR, showed 799 publications in the last 5 years (Scopus, 2018).

Table 3: Applications of LDA in the Food industry.

Sample	Objective	LCA use/result	Reference
Chestnuts	Evaluate the quality of healthy and moldy chestnuts using NIR.	The LDA model shown better accuracy than the PCA-DA model.	Huetal, 2017
Cereal bars	Classification of cereal bars using NIR spectroscopy.	Samples were classified into three different types (conventional, dietary and light); with better results than LDA/GA model.	Brito et al., 2013
Extra virgin olive oil	Detection of adulteration of extra virgin olive oil.	LDA showed a good performance in conjunction with the other multivariate tests (supported vector machine and KNN).	Georgouli et al., 2017
Saffron	Optimization of geographical volatile markers identification of saffron.	Geographical differentiation of samples using all aroma components was around 91% and below 73% using the major aroma components.	D'Archivio et al., 2018
Roasted coffee	Detection of adulteration in roasted coffee using diffuse reflectance, ATR-FTIR spectroscopy and data fusion.	ATR-FTIR capable of detecting adulteration in roasted and ground coffee, report up to 100% sensitivity with 95% specificity on test.	Reis et al., 2017
Cocoa butter	Detection of adulteration in cocoa butter using fluorescence, UV spectroscopy, and data fusion.	Capable of detecting adulteration up to 95% sensitivity on validation test set.	Dankowska, 2017
Honey	Detection of adulteration honey using hyperspectral imagery.	Reached classification accuracy of 90% on test, using LDA.	Shafiee et al., 2016

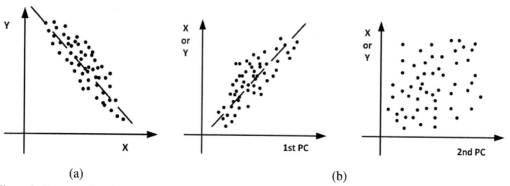

(a)   (b)

Figure 6: Two examples of regression methods, (a) ordinary least squares (OLS), and (b) principal component regression (PCR) representations.

PLSR establishes linear relationships between the input (X) and output (Y) variables using principal components instead of using the original data. The development of the PLSR models is carried out based in the following equations.

$$X = WQ^T + E_1 = \sum_{h_1}^{a} w_h q_h^T + E_1$$

$$Y = US^T + E_2 = \sum_{h_1}^{a} u_h s_h^T + E_2$$

where W and U are the score matrices of X and Y. Q and S represent the orthogonal loading matrices of X and Y, and E_1 and E_2 are residual matrices. The term a refers to the number of principal components (also called latent variables) that explain most of the variations in X and Y. In other words, the determination of the *a* value (usually done through cross-validation) is critical to the generalization capability of the PLS model. The correlation between *X* and *Y* is a linear relationship obtained from the linear regression coefficients.

As in the classification methods, multivariate regression requires the stages of model creation and validation (Oliveri and Simonetti, 2016). This regression technique is one of the most used in the food industry. The regression by PLSR is observed in multiple applications also associated with spectrophotometric methods, such as the one used to predict carotenoid content in carrots (Lawaetz et al., 2016), in the evaluation of cheese in its maturation phase (Vásquez et al., 2018), or of determination of amylose in rice (Sampaio et al., 2018).

2.5 Artificial neural networks

A typical artificial neural network (ANN) architecture known as multilayer perceptron (MLP) contains a series of layers, composed of neurons, processing units, and their connections (see Fig. 7).

An artificial neuron calculates the weighted sum of its inputs, then applies an activation function in order to obtain the signal that will be transmitted to the next neuron (Castro et al., 2017).

This technique does not require an explicit relationship between inputs and results; in this way, ANNs are preferable when inputs are qualitative and when the inputs or outputs cannot be represented in mathematical terms. However, this technique requires advanced numerical methods to solve the training problem in a system of highly interconnected processing elements.

During the learning phase of an ANN, similar to a biological neural network, arrangements are made in the connections between the neurons in order to reduce the variation of the target values (Delgado et al., 2015). The equation above shows the mathematical operation of a single neuron in ANN.

$$y_i = f\left(\sum_j w_{ij} u_j\right)$$

where w_{ij} represents the weight of the neuron *j* at the output of the neuron *i*, u_j is the output of the neuron *j*, and *f* is the activation function of the neuron *i*.

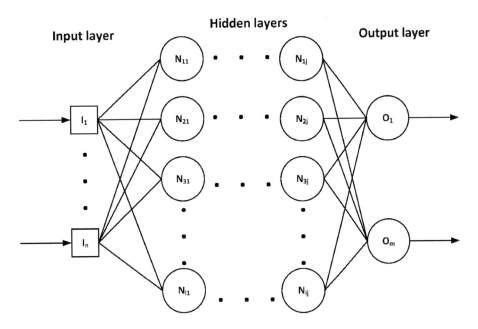

Figure 7: General structure of an ANN.

In the food industry, ANNs have been used in multiple applications that include growth models, control in the food industry, interpretation of spectroscopic data and prediction of physical, chemical, functional and sensory properties of various food products during the stages of processing and distribution. Özbalci et al. (2013) used ANN to generate a discriminant analysis of sugar content in honey; Sanchéz-González and Oblitas (2017) used the same technique to predict the shelf life of packaged cheese. Finally, Sanahuja et al. (2017) used ANNs and qSVMs to classify sensory attributes for its use in the food industry.

2.6 Multivariate analysis strategy

The multivariate methods described in this chapter are not always applied separately; in a real analysis of complex systems, the objectives of the study require a specific procedure sequence. In this way, Callao and Ruisánchez (2018) proposed a qualitative multivariate method of detecting food fraud that involves several steps, such as data collection, pretreatment, exploration techniques, classification techniques and method validation. A more general adaptation of this method is observed in Fig. 8, and, as can be seen, it is useful to plan an analysis strategy using multivariate techniques.

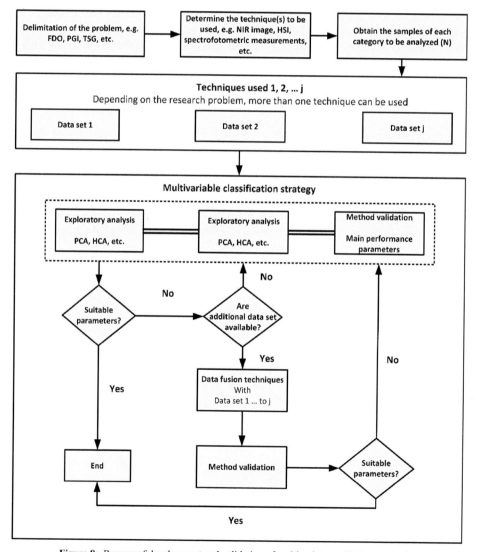

Figure 8: Process of development and validation of multivariate qualitative methods.

It is worth mentioning that each investigation should be treated as a particular topic. And the strategy to be used can be selected according to the particular objective and research hypothesis.

3. Application Example

In this section, an example where some of the techniques analyzed in the previous sections are applied is presented. The problem to be addressed is the classification of fresh fruits according to their level of maturity, see Fig. 9, which is usually a personal and tedious task. Consequently, there is a growing interest in the use of non-intrusive techniques, such as those based on computer vision and machine learning.

The data for this example was generated with the objective of evaluating the classification efficiency of Cape gooseberry (*Physalis peruviana* L.) based on color parameters $L*a*b*$. The values of each piece of fruit in the three-color spaces and their corresponding degrees of maturity were organized for use in the creation, testing and comparison of the classification models developed. The file with the data used in this example is attached to this book.

The analysis was divided into three parts: Data normalization, PCA and KNN as a classification technique; all routines were implemented in the script of Matlab 2016 to carry out the analysis.

1   2   3   4   5   6   7

Figure 9: Maturity levels in *Physalis peruviana* evaluated in this example.

> **Color version at the end of the book**

3.1 Data pre-treatment

One of the primary objectives of signal pretreatment is to improve the quality of signals by eliminating unwanted information (i.e., data not related to the chemical characterization of the system under investigation).

The script Listing 1, shown in Appendix A, performs the data normalization of samples; the resulting normalized data for each class (maturity level and variable) are shown in Fig. 10.

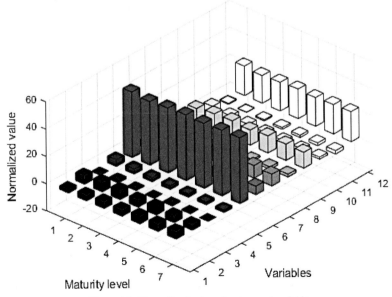

Figure 10: Normalized values for class and variable.

3.2 *Principal component analysis*

Once the data has been normalized, an analysis of the principal components is carried out, for which the Matlab code proposed Listing 2 can be seen in Appendix A.

The results are shown in Fig. 11, PC_1 explains 55.31% of the variance and when the cumulative variance is calculated using components 1 to 4 it is possible to explain up to 99.57% of the variance.

Next, it is necessary to calculate the components in order to evaluate the ability to discriminate between classes. Appendix A shows the Matlab code needed to carry out these tasks, see Listing 3. Also, Fig. 12 shows that the model based on these components could successfully separate the data.

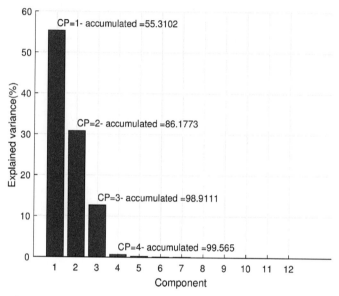

Figure 11: Explanation of variance per component.

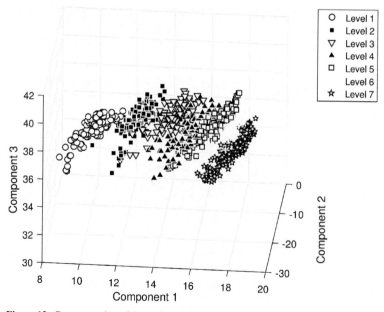

Figure 12: Representation of the analyzed data according to three principal components.

3.3 Classification method

For this example, the KNN technique was used. This technique represents a very intuitive approach to performing a classification based on the distances between samples in the multivariate space of the measured variables. The Matlab code proposed for this task is shown in Appendix A, see Listing 4.

Finally, the statistical analysis of the model is tested. For this purpose, several techniques can be applied but, in this case, the confusion matrix (CF) was selected due to its representativeness of performance for each class. This method evaluates the correspondence between rows (predicted - output class) and columns (true - objective class); this correspondence is placed in diagonal cells. Off-diagonal cells represent incorrectly classified observations. So, both the number of observations and the percentage of the total number of observations are commonly shown in each cell.

For our example, Fig. 13 shows the results of the confusion matrix when the model was evaluated and, as it is observed, the KNN classification model obtained 88.8% precision.

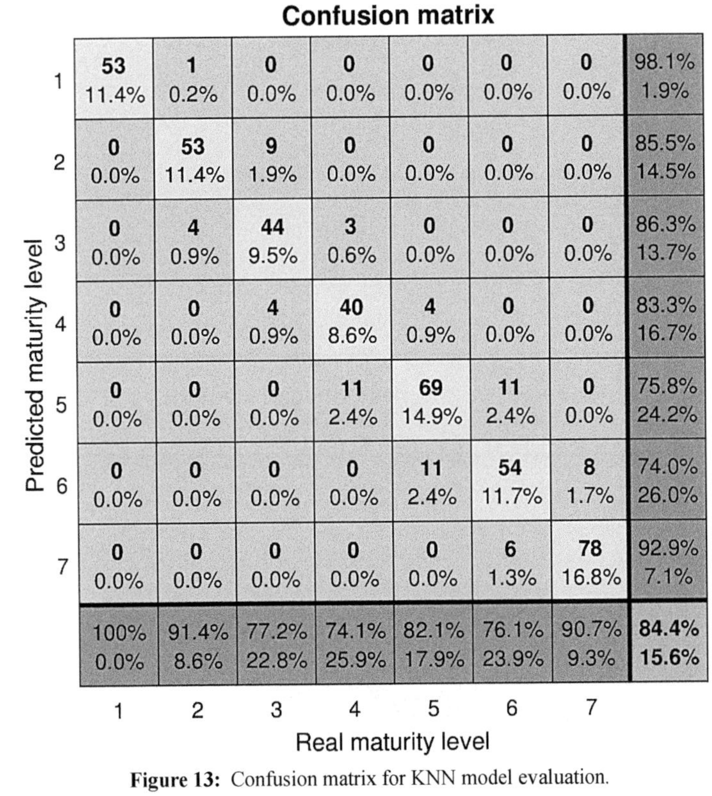

Confusion matrix

	1	2	3	4	5	6	7	
1	**53** / 11.4%	1 / 0.2%	0 / 0.0%	0 / 0.0%	0 / 0.0%	0 / 0.0%	0 / 0.0%	98.1% / 1.9%
2	0 / 0.0%	**53** / 11.4%	9 / 1.9%	0 / 0.0%	0 / 0.0%	0 / 0.0%	0 / 0.0%	85.5% / 14.5%
3	0 / 0.0%	4 / 0.9%	**44** / 9.5%	3 / 0.6%	0 / 0.0%	0 / 0.0%	0 / 0.0%	86.3% / 13.7%
4	0 / 0.0%	0 / 0.0%	4 / 0.9%	**40** / 8.6%	4 / 0.9%	0 / 0.0%	0 / 0.0%	83.3% / 16.7%
5	0 / 0.0%	0 / 0.0%	0 / 0.0%	11 / 2.4%	**69** / 14.9%	11 / 2.4%	0 / 0.0%	75.8% / 24.2%
6	0 / 0.0%	0 / 0.0%	0 / 0.0%	0 / 0.0%	11 / 2.4%	**54** / 11.7%	8 / 1.7%	74.0% / 26.0%
7	0 / 0.0%	0 / 0.0%	0 / 0.0%	0 / 0.0%	0 / 0.0%	6 / 1.3%	**78** / 16.8%	92.9% / 7.1%
	100% / 0.0%	91.4% / 8.6%	77.2% / 22.8%	74.1% / 25.9%	82.1% / 17.9%	76.1% / 23.9%	90.7% / 9.3%	**84.4% / 15.6%**

Predicted maturity level (vertical axis) — Real maturity level (horizontal axis: 1 2 3 4 5 6 7)

Figure 13: Confusion matrix for KNN model evaluation.

4. Conclusions

In this chapter, several research works that use multivariate analysis techniques to address problems of exploration, classification and regression have been shown. These techniques have been used increasingly frequently in research within the field of the food industry. Specific techniques, such as the PCA or the HCA, have been used with great success in several research works, demonstrating that these techniques can significantly improve data visualization, reduction, and understanding of the data structure and its dimensions, as well as for the recognition of patterns, data classification, and regression analysis.

We believe that the inclusion of these techniques in research work has increased because the results are easy to interpret and analyse; this is due to the advantages offered by multivariate statistics over classical methods, with which the complexity of the problem is reduced, thus reducing the time and speed of the analysis.

Finally, we consider that current analysis in food is continuously evolving and current results produce a significant amount of data. The research cases presented in this chapter are mostly spectroscopic techniques that provide thousands of data for a single spectrum. Such methods, like multivariate analysis, are required more and more frequently.

Appendix A: Matlab Source Code

Listing 1: Data normalization.

```
dS = std(datac(:,1:12));
dataN = datac;
for i =1: length (dS)
    dataN(:,i) = datac(:,i)./dS(i);
end
datam = zeros (7,12);
for i = 1:7
    [r c] = find(dataN(:,13)==i);
    datam(i ,:) = median(dataN(r,1:12));
end
bar3(datam ,0.5);
box off;
grid on;
xlabel('Variables');
zlabel ('Normalizated value');
ylabel('Maturity level');
view (-48 ,28);
```

Listing 2: Principal component analysis.

```
[coeff,score,latent,tsquared,explained] = pca(dataN(:,1:12));
X=[1:1:12];
bar(X,explained);
set(gca,'yscale','log');
i=0;
value = 0;
while value <99.5 && i<=76
  i=i+1;
  value = sum(explained(1:i));
  text(X(i)+1,explained(i),strcat('CP=',num2str(i),'- accumulated
=', um2str(value)), 'FontSize', 10,'Color','red')
end
box off; grid on;
xlabel('Component');
ylabel('Explained variance(%)');
```

Listing 3: Components to evaluate capacity to discriminate between classes.

```
% Calculating new X using components principals coefficients
coeff=coeff(:,1:i)';
dim=size(dataN);
dataNc=zeros(dim(1),i);
for j=1:dim(1)
    for k=1:i
 dataNc(j,k)=sum(dataN(j,1:12).*coeff(k,:));
 end
end
dataNc=cat(2,dataNc,dataN(:,13));

% Plot of classes in three components
h=figure;
hold on
for t=1:7
 [r c]=find(dataNc(:,5)==t);
 scatter3(dataNc(r,1),dataNc(r,2),dataNc(r,3),'*');
end
hold off
view(25,42)
box off; grid on;
xlabel('Component 1');
ylabel('Component 2');
zlabel('Component 3');
```

Listing 4: KNN example.

```
indices = crossvalind('Kfold',dataNc(:,5),2);
[rt, ct]=find(indices==1);
[rv, cv]=find(indices~=1);
Xt=dataNc(rt,1:i);Xv=dataNc(rv,1:i);
Yt=dataNc(rt,i+1);Yv=dataNc(rv,i+1);

% KNN model training
Mdl = fitcknn(Xt,Yt);

% KNN validation/test
Yp = predict(Mdl,Xv);

% Confusion matrix of models
h=figure;
C = confusionmat(Yv,Yp);
plotconfusion(ind2vec(Yv'),ind2vec(Yp'))
```

References

Ai, Y., Liang, P., Wu, Y., Dong, Q., Li, J.-b., Bai, Y. et al. 2018. Rapid qualitative and quantitative determination of food colorants by both Raman spectra and surface-enhanced Raman scattering (SERS). Food Chemistry 241: 427–433.

Alamprese, C., Amigo, J., Casiraghi, E. and Soren, E. 2016. Identification and quantification of turkey meat adulteration in fresh, frozen-thawed and cooked minced beef by FT-NIR spectroscopy and chemometrics. Meat Science 121: 175–181.

Amodio, M., Ceglie, F., Chaudhry, M.M.A., Piazzolla, F. and Colelli, G. 2017. Potential of NIR spectroscopy for predicting internal quality and discriminating among strawberry fruits from different production systems. Postharvest Biology and Technology 125: 112–121.

Azevedo, J., dos Santos, E. and Domingues, P. 2018. Green analytical chemistry applied in food analysis: alternative techniques. Current Opinion in Food Science 22: 115–121.

Azevedo, M., Seraglio, S., Rocha, G., Balderas, C., Piovezan, M., Gonzaga, L. et al. 2017. Free amino acid determination by GC-MS combined with a chemometric approach for geographical classification of bracatinga honeydew honey (*Mimosa scabrella* Bentham). Food Control 78: 383–392.

Boggia, R., Casolino, M., Hysenaj, V., Oliveri, P. and Zunin, P. 2013. A screening method based on UV–Visible spectroscopy and multivariate analysis to assess addition of filler juices and water to pomegranate juices. Food Chemistry 140(4): 735–741.

Bona, E., Marquetti, I., Link, J., Makimori, G., da Costa, V., Lemes, A. et al. 2017. Support vector machines in tandem with infrared spectroscopy for geographical classification of green arabica coffee. LWT-Food Science and Technology 76: 330–336.

Brito, A., Brito, L., Honorato, F., Pontes, M. and Pontes, L. 2013. Classification of cereal bars using near infrared spectroscopy and linear discriminant analysis. Food Research International 51(2): 924–928.

Cai, W., Guan, G., Pan, R., Zhu, X. and Wang, H. 2018. Network linear discriminant analysis. Computational Statistics & Data Analysis 117: 32–44.

Callao, M. and Ruisánchez, I. 2018. An overview of multivariate qualitative methods for food fraud detection. Food Control 86: 283–293.

Casale, M., Oliveri, P., Casolino, C., Sinelli, N., Zunin, P., Armanino, C. et al. 2012. Characterisation of PDO olive oil chianti classico by non-selective (UV–Visible, NIR and MIR spectroscopy) and selective (fatty acid composition) analytical techniques. Analytica Chimica Acta 712: 56–63.

Castro, W., Oblitas, J., Santa-Cruz, R. and Avila-George, H. 2017. Multilayer perceptron architecture optimization using parallel computing techniques. PloS One 12(12): e0189369.

Chen, H., Tan, C., Lin, Z. and Wu, T. 2017. Detection of melamine adulteration in milk by near-infrared spectroscopy and one-class partial least squares. Spectrochimica Acta Part A: Molecular and Biomolecular Spectroscopy 173: 832–836.

Da Silva, E., Garbelotti, M. and Moita, J. 2006. The application of hierarchical clusters analysis to the study of the composition of foods. Food Chemistry 99(3): 622–629.

Dankowska, A. 2017. Data fusion of fluorescence and UV spectroscopies improves the detection of cocoa butter adulteration. European Journal of Lipid Science and Technology 119(8): 1600268.

De Araújo, C., Dantas, J., De Santana, D., De Oliveira, J., De Macedo, G. and Dos Santos, E. 2017. Baker's yeast invertase purification using aqueous two-phase system–modeling and optimization with PCA/LS-SVM. Food and Bioproducts Processing 101: 157–165.

Delgado, A., Rauh, C., Park, J., Kim, Y., Groß, F. and Diez, L. 2015. Artificial neural networks: Applications in food processing. In Reference Module in Food Science.

D'Archivio, A., Giannitto, A., Maggi, M. and Ruggieri, F. 2016. Geographical classification of Italian saffron (*Crocus sativus* L.) based on chemical constituents determined by high-performance liquid-chromatography and by using linear discriminant analysis. Food Chemistry 212: 110–116.

D'Archivio, A. and Maggi, M. 2017. Geographical identification of saffron (*Crocus sativus* L.) by linear discriminant analysis applied to the UV–visible spectra of aqueous extracts. Food Chemistry 219: 408–413.

D'Archivio, A., Di Pietro, L., Maggi, M. and Rossi, L. 2018. Optimization using chemometrics of HS-SPME/GC–MS profiling of saffron aroma and identification of geographical volatile markers. European Food Research and Technology, pp. 1–9.

Eriksson, L., Trygg, J. and Wold, S. 2014. A chemometrics toolbox based on projections and latent variables. Journal of Chemometrics 28(5): 332–346.

Fan, S., Zhong, Q., Fauhl-Hassek, C., Pfister, M., Horn, B. and Huang, Z. 2017. Classification of Chinese wine varieties using ^{1}H NMR spectroscopy combined with multivariate statistical analysis. Food Control 88: 113–122.

Fayez, Y., Tawakkol, S., Fahmy, N., Lotfy, H. and Shehata, M. 2018. Comparative study of the efficiency of computed univariate and multivariate methods for the estimation of the binary mixture of clotrimazole and dexamethasone using two different spectral regions. Spectrochimica Acta Part A: Molecular and Biomolecular Spectroscopy 194: 126–135.

Georgouli, K., Del Rincon, J. and Koidis, A. 2017. Continuous statistical modelling for rapid detection of adulteration of extra virgin olive oil using mid infrared and Raman spectroscopic data. Food Chemistry 217: 735–742.

Gómez-Narváez, F., Medina-Pineda, Y. and Contreras-Calderón, J. 2017. Evaluation of the heat damage of whey and whey proteins using multivariate analysis. Food Research International 102: 768–775.

Goyeneche, R., Roura, S. and Di Scala, K. 2014. Principal component and hierarchical cluster analysis to select hurdle technologies for minimal processed radishes. LWT-Food Science and Technology 57(2): 522–529.

Granato, D., Santos, J.S., Escher, G., Ferreira, B. and Maggio, R. 2017. Use of principal component analysis (PCA) and hierarchical cluster analysis (HCA) for multivariate association between bioactive compounds and functional properties in foods: A critical perspective. Trends in Food Science & Technology 72: 83–90.

Grassi, S. and Alamprese, C. 2018. Advances in NIR spectroscopy applied to process analytical technology in food industries. Current Opinion in Food Science 22: 17–21.

Hu, J., Ma, X., Liu, L., Wu, Y. and Ouyang, J. 2017. Rapid evaluation of the quality of chestnuts using near-infrared reflectance spectroscopy. Food Chemistry 231: 141–147.

Husson, F., Lê, S. and Pagès, J. 2017. Exploratory Multivariate Analysis by example using R. Chapman and Hall/CRC.

Lawaetz, A., Christensen, S., Clausen, S., Jørnsgaard, B., Rasmussen, S., Andersen, S. et al. 2016. Fast, cross cultivar determination of total carotenoids in intact carrot tissue by Raman spectroscopy and partial least squares calibration. Food Chemistry 204: 7–13.

Lee, I. and Yang, J. 2009. 2.27 - common clustering algorithms. pp. 577–618. *In*: Brown, S.D., Tauler, R. and Walczak, B. (eds.). Comprehensive Chemometrics, Elsevier, Oxford.

Lee, I., We, G., Kim, D., Cho, Y., Yoon, M., Shin, M. et al. 2012. Classification of rice cultivars based on cluster analysis of hydration and pasting properties of their starches. LWT-Food Science and Technology 48(2): 164–168.

López, M., Callao, M. and Ruisánchez, I. 2015. A tutorial on the validation of qualitative methods: From the univariate to the multivariate approach. Analytica Chimica Acta 891: 62–72.

Mabood, F., Jabeen, F., Ahmed, M., Hussain, J., Al Mashaykhi, S., Al Rubaiey, Z. et al. 2017. Development of new NIR-spectroscopy method combined with multivariate analysis for detection of adulteration in camel milk with goat milk. Food Chemistry 221: 746–750.

Manly, B. and Navarro, A. 2016. Multivariate Statistical Methods: A Primer. CRC Press.

Marquetti, I., Varaschim, J., Guimaraés, A., Dos-Santos, M., Valderrama, P. and Bona, E. 2016. Partial least square with discriminant analysis and near infrared spectroscopy for evaluation of geographic and genotypic origin of arabica coffee. Computers and Electronics in Agriculture 121: 313–319.

Mishra, S. and Datta-Gupta, A. 2018. Multivariate data analysis. In Applied Statistical Modeling and Data Analytics, pp. 97–118. Elsevier.

Næs, T., Tomic, O., Greiff, K. and Thyholt, K. 2014. A comparison of methods for analyzing multivariate sensory data in designed experiments—a case study of salt reduction in liver paste. Food Quality and Preference 33: 64–73.

Oliveri, P. and Simonetti, R. 2016. Chemometrics for food authenticity applications. In Advances in Food Authenticity Testing, Chapter 25, pp. 701–728. Elsevier.

Özbalci, B., Boyaci, Í., Topcu, A., Kadılar, C. and Tamer, U. 2013. Rapid analysis of sugars in honey by processing raman spectrum using chemometric methods and artificial neural networks. Food Chemistry 136(3-4): 1444–1452.

Panagou, E., Papadopoulou, O., Carstensen, J. and Nychas, G. 2014. Potential of multispectral imaging technology for rapid and non-destructive determination of the microbiological quality of beef filets during aerobic storage. International Journal of Food Microbiology 174: 1–11.

Probola, G. and Zander, L. 2007. Application of PCA method for characterisation of textural properties of selected ready-to-eat meat products. Journal of Food Engineering 83(1): 93–98.

Puputti, S., Aisala, H., Hoppu, U. and Sandell, M. 2017. Multidimensional measurement of individual differences in taste perception. Food Quality and Preference 65: 10–17.

Reis, N., Botelho, B.G., Franca, A. and Oliveira, L. 2017. Simultaneous detection of multiple adulterants in ground roasted coffee by ATR-FTIR spectroscopy and data fusion. Food Analytical Methods 10(8): 2700–2709.

Sampaio, P., Soares, A., Castanho, A., Almeida, A., Oliveira, J. and Brites, C. 2018. Optimization of rice amylose determination by NIR-spectroscopy using PLS chemometrics algorithms. Food Chemistry 242: 196–204.

Sanahuja, S., Fédou, M. and Briesen, H. 2017. Classification of puffed snacks freshness based on crispiness-related mechanical and acoustical properties. Journal of Food Engineering 226: 53–64.

Sánchez-González, J. and Oblitas-Cruz, J. 2017. Application of weibull analysis and artificial neural networks to predict the useful life of the vacuum-packed soft cheese. Revista Facultad de Ingeniería Universidad de Antioquia 82: 53–59.

Scopus. 2018. Elsevier's Scopus, the largest database of peer-reviewed literature. www.scopus.com (accessed 06 April 2018).

Shafiee, S., Polder, G., Minaei, S., Moghadam-Charkari, N., Van+Ruth, S. and Kus, P.M. 2016. Detection of honey adulteration using hyperspectral imaging. IFAC-PapersOnLine 49(16): 311–314.

Vásquez, N., Magán, C., Oblitas, J., Chuquizuta, T., Avila-George, H. and Castro, W. 2018. Comparison between artificial neural network and partial least squares regression models for hardness modeling during the ripening process of swiss-type cheese using spectral profiles. Journal of Food Engineering 219: 8–15.

Zhang, X., Liu, Y., Li, Y. and Zhao, X. 2017. Identification of the geographical origins of sea cucumber (*Apostichopus japonicus*) in northern china by using stable isotope ratios and fatty acid profiles. Food Chemistry 218: 269–276.

CHAPTER 6

Application of Cluster Analysis in Food Science and Technology

Chapman, J,# Power, A, Chandra, S, Roberts, J and *Cozzolino, D* *

1. Introduction

Food science and any associated technologies are continually evolving due to the large amounts of data either generated by instrumental methods or available in databases. Large amounts of data are produced by rapid analytical techniques and instruments that are able to measure several variables on a large number of samples. These data are deposited in an increasing number of online databases that have grown exponentially (Gross, 2011; Berger et al., 2013). In addition, there are several challenges associated with the growing size of data and these concerns apply to all fields in the sciences. One field which has observed spectacular growth is that of the life sciences, where the advancement of genomics, proteomics and other high-throughput technologies has produced an overwhelming volume of data, more and more often freely available to all researchers. Beside the large number of samples, these data are considered large because they are high-dimensional, i.e., each sample, or instance, of a typical data set contains a large number of degrees of freedom. High-dimensionality makes visualization and exploration of samples and data sets very difficult to manage. To overcome these limitations, a series of techniques that help researchers in visualization, exploration and mining of large datasets have been developed (Van Der Maaten et al., 2009; Hassanien et al., 2013). The focus of this paper will be to discuss the relevance of cluster analysis for food science.

The objective of cluster analysis is to assign observations to groups (clusters) so that observations within each group are similar to one another with respect to variables or attributes of interest where the groups themselves stand apart from one another. In other words, the objective is to divide the observations into homogeneous and distinct groups in order to observe trends or patterns in the dataset. In contrast to other classification issues where each observation is known to belong to one of a number of groups, the objective is to predict the group to which a new observation belongs. Cluster analysis seeks to discover the number and the composition of the groups in the dataset (Van Der Maaten et al., 2009; Hassanien et al., 2013).

The Agri-Chem Group, The School of Health, Medical and Applied Sciences, Central Queensland University, Rockhampton, QLD 4701, Australia.
Current address: School of Sciences, RMIT, Melbourne, Victoria, Australia.
* Corresponding author: daniel.cozzolino@rmit.edu.au

There are a number of clustering methods available. One such method, for example, begins with as many groups as there are observations, and then systematically merges observations in order to reduce the number of groups by one, two, and so on, until a single group containing all observations is formed. Another method begins with a given number of groups and an arbitrary assignment of the observations to the groups. The method then reassigns the observations one by one so that each observation belongs to the nearest group. Cluster analysis is also used to group variables into homogeneous and distinct groups. This grouping of variables approach is used in revising a questionnaire on the basis of responses received to a draft of the questionnaire. The grouping of the questions by means of cluster analysis helps to identify redundant questions and reduce their number.

According to Brereton (2015), chemometric users tend to 'follow the crowd' and use the available software indiscriminately without knowing the principles and fundamentals of each method applied in their research data analysis. In food chemistry studies, principal components analysis (PCA) and hierarchical cluster analysis (HCA) are widely (and, sometimes, improperly) applied as "unsupervised classification" methods (Granato et al., 2017, 2018; Nunez et al., 2015).

2. Supervised and Unsupervised Methods

Methods and techniques used in qualitative analysis or their classification can be categorised into two groups, namely, supervised and unsupervised methods (Otto, 1990; Naes et al., 2002; Gishen et al., 2005; Berrueta et al., 2007; Brereton, 2009, 2015; Granato et al., 2017, 2018; Nunez et al., 2015). Unsupervised approaches attempt to divide the dataset into groups without any predefined categorisation of the samples where patterns or trends among samples are defined during the analysis, without previous knowledge of the patterns or groups (Otto, 1990; Naes et al., 2002; Berrueta et al., 2007; Brereton, 2009, 2015; Granato et al., 2017, 2018; Nunez et al., 2015). Commonly-used unsupervised methods in the field of food analysis are those based in cluster analysis and these are described in detail elsewhere (Otto, 1990; Naes et al., 2002; Berrueta et al., 2007; Brereton, 2009, 2015; Granato et al., 2017, 2018; Nunez et al., 2015).

In contrast, supervised methods are based on *a priori* assignment of objects or samples into class memberships for the creation of a respective mathematical model (training set). Afterwards, unknown objects (a test set or validation) could be predicted based on the models generated (Berrueta et al., 2007). Supervised methods assume that some structure exists in the data and requires the assignment of the samples to pre-specified subgroups using 'artificial' variables to build classification rules which are later used for allocating new and unknown samples to the most probable subgroup (Otto, 1990; Naes et al., 2002; Berrueta et al., 2007; Brereton, 2015; Smyth and Cozzolino, 2012; Granato et al., 2017, 2018; Nunez et al., 2015). Most of the classification or pattern recognition methods reported or used in the food sciences involve the use of supervised learning steps (Otto, 1990; Naes et al., 2002; Berrueta et al., 2007; Brereton, 2009, 2015). Popular supervised methods used in food applications are linear discriminant analysis (LDA), discriminant analysis (DA) and discriminant partial least squares (DPLS or PLS-DA) regression (Berrueta et al., 2007; Brereton, 2009, 2015; Esslinger et al., 2014; Granato et al., 2017, 2018; Nunez et al., 2015).

3. Principal Component Analysis (PCA)

Principal component analysis (PCA) is used as a tool capable of providing an overview of the complexity that exists in multivariate datasets (Bro and Smilde, 2014). This method is generally used for revealing relations between variables and between samples (e.g., patterns), detecting outliers, finding and quantifying patterns, extracting and compressing multivariate datasets (Naes et al., 2002; Brereton, 2009; Cozzolino, 2012; Smyth and Cozzolino, 2012; Bro and Smilde, 2014). Although this technique is extensively used and reported by several authors in many food applications, PCA is not a classification method. It is important to highlight at this point that many references in the literature have previously defined PCA as a classification technique (Naes et al., 2002; Brereton, 2009; Cozzolino, 2012; Smyth and Cozzolino, 2012; Bro and Smilde, 2014; Granato et al., 2017, 2018; Nunez et al., 2015).

PCA employs a mathematical procedure that transforms a set of possibly correlated response variables into a new set of non-correlated variables, called principal components (Bro and Smilde, 2014). PCA can be performed on either a data matrix or a correlation matrix depending on the type of variables being measured (Naes et al., 2002; Brereton, 2009; Smyth and Cozzolino, 2012; Bro and Smilde, 2014). However, in a case where the original variables are nearly non-correlated, nothing can be gained by using a PCA analysis instead of classical statistics. Bro and Smilde (2014) have provided a compressive tutorial on the use of PCA analysis. These authors also discussed some practical aspects for the application of PCA analysis, such as the validation method (e.g., cross validation), pre-processing, definition of the optimal number of components, and issues related with data interpretation and outlier detection (Bro and Smilde, 2014; Brereton, 2009, 2015).

4. Cluster and Hierarchical Cluster Analysis (CA and HCA)

Clustering is used to evaluate where clusters or groups of samples are present or not (Adams, 1995; Beebe et al., 1998; Brereton, 2009, 2015; Granato et al., 2017, 2018; Nunez et al., 2015). In this type of method, two possible outcomes are conceivable, either multiple patterns exist or are present in the data or the samples fall into one pattern (Adams, 1995; Beebe et al., 1998; Brereton, 2009, 2015). HCA is another type of cluster analysis where the inter-point distances between all of the samples and represents the data in a two-dimensional plot called a dendrogram (Adams, 1995; Beebe et al., 1998; Brereton, 2009 and 2015; Granato et al., 2017, 2018; Nunez et al., 2015).

HCA is a clustering method which explores the organization of samples in groups and among these groups depicts a hierarchy (Lee and Yang, 2009). The result of HCA is usually presented in a dendrogram, a plot which shows the organization of samples and its relationships in tree form. There are two main approaches to resolve the grouping problem in HCA, agglomerative or divisive (Lee and Yang, 2009). In agglomerative grouping, each sample is initially considered a cluster, and subsequently pairs of clusters are merged. In divisive approaches, the algorithm starts with one cluster including all samples, recursive splits are then performed. Clustering is achieved by using an appropriate metric of samples distanced (usually, Euclidean, Mahalanobis or Manhattan distance) and then the linkage criterion among groups is applied (Lee and Yang, 2009; Granato et al., 2017, 2018; Nunez et al., 2015). Complete, single and average and Ward's linkage are the more common variants of linkage criterions. Ward's method, based on an optimal value of a target function, is a possible choice (Granato et al., 2015; Granato et al., 2017, 2018; Nunez et al., 2015).

5. Measuring and Reporting Uncertainty in Quality Analysis

The results derived from qualitative analysis must be interpreted using different measurements of uncertainty in a similar way as in quantitative analysis. In a recent publication, Brereton (2015) highlighted several issues associated with the validation and reporting of uncertainty in qualitative analysis. This author highlighted that validation in a qualitative analysis is often misunderstood and consequently, in several applications of these methods, the results reported in the literature can be misleading and overoptimistic (Buco, 1990; Brereton, 2006, 2015; Pulido et al., 2003; Berrueta et al., 2007).

In most of the applications (e.g., food, analytical) where quantitative analysis is used, cross-validation is primarily used to optimise a given model (Pulido et al., 2003; Berrueta et al., 2007; Brereton, 2015; Granato et al., 2017, 2018; Nunez et al., 2015). This analytical approach is valid since the obtained quantitative results are in most of the applications known by the analyst (e.g., researcher, food technologist), such as the concentration of the chemical or property in the food matrix analysed (Pulido et al., 2003; Berrueta et al., 2007; Brereton, 2015). However, in the classification process, the analyst does not always know the answer in advance (e.g., if a wine is from one geographical origin or not) (Pulido et al., 2003; Berrueta et al., 2007; Brereton, 2015).

It is generally accepted that a particular method is suitable to classify samples if it classifies a high percentage of samples into their presupposed group or groups (Pulido et al., 2003; Brereton, 2015).

However, this is not necessarily the most appropriate method to assess the ability of a method to classify unknown samples (Pulido et al., 2003; Brereton, 2015). According to Brereton (2015), this might lead to several issues in relation to the available analytical literature where the aim is to optimise the models. Therefore, a method that apparently predicts a lower percentage of the data into their presupposed groups is felt to be a poorer method than one that predicts a higher percentage, but this is not necessarily the case (Pulido et al., 2003; Berrueta et al., 2007; Brereton, 2009, 2015). One of the key issues in classification is that the suitability of a method to correctly classifying samples depending on the structure of the data (e.g., prior knowledge about the data). In addition, the correct definition of the problem or hypothesis (e.g., if two or more groups are separable and homogeneous is correct), the suitability of the technique used to perform the classification, and issues related with sampling (e.g., sample size, representativeness) are also of great importance (Pulido et al., 2003; Berrueta et al., 2007; Brereton, 2009, 2015).

In quantitative analysis, once the models have been developed, it is imperative to define the overall error obtained as well as to consider that the model error (e.g., calibration) will be the addition of the multiplicative errors of both the sampling and the analytical method. However, in qualitative analysis, the degree of uncertainty cannot be measured in the same way (Pulido et al., 2003; Brereton, 2015; Ellison and Fearn, 2005). Measurements, such as per cent of correct or incorrect classification, are usually reported and used to evaluate the ability of a model (Pulido et al., 2003; Berrueta et al., 2007; Brereton, 2009, 2015). The success of any classification model will be determined by the per cent of correct classification and it is usually agreed that the higher the per cent of correct classification the better the model (Brereton, 2006, 2015). Several authors have reported and described the use of other parameters in order to evaluate the uncertainty of a qualitative method (Pulido et al., 2003; Ellison and Fearn, 2005). Examples of additional measurements of uncertainty reported in the literature are with respect to sensitivity, specificity, false positives, etc. (Pulido et al., 2003; Ellison and Fearn, 2005).

Another important and misunderstood issue in classification is to accept the per cent of classification without any attention to the total number of samples used to develop the models (Pulido et al., 2003; Berrueta et al., 2007; Brereton, 2006, 2015). This issue is of particular importance when qualitative models are interpreted exclusively in the per cent of correct classification where few samples were analysed and used to develop the classification model (Pulido et al., 2003; Berrueta et al., 2007; Brereton, 2006, 2015).

6. Validation

Without a doubt, the validation of the classification models developed is the most important step in either quantitative or qualitative analysis (Badertscher and Pretsch, 2006; Berrueta et al., 2007; Brereton, 2006, 2009, 2015; Westad and Marini, 2015). Model validation implies that an independent set of samples must be used to test the ability of the model to predict a set of unknown samples (Badertscher and Pretsch, 2006; Berrueta et al., 2007; Brereton, 2006, 2015; Westad and Marini, 2015). Unfortunately, in many cases, the ideal of using an external data set cannot be achieved in real life situations. In these cases, cross validation has been the most widely used method. However, several authors have explored and improved this process in using or combining other pre-processing techniques, such as k-fold and repeated k-fold cross validation (leave one out) and jack-knife, among others (Berrueta et al., 2007; Westad and Marini, 2015).

7. Final Considerations and Perspectives

Most of the recent publications and reports in food science targeting issues related with authenticity, contamination, fraud, origin, and traceability of foods are based on qualitative analysis. It seems that this kind of approach can be considered as a purely mathematical or statistical exercise, development of such models can be considered as a complex process that implies an understanding of the food matrix (e.g., biochemical and chemical properties) as well as considering sampling, pre-processing, validation and interpretation of the results in the context of the application.

However, in order to implement and apply these methods in real life situations, care needs to be taken where a proper experimental design, sample selection, training, optimization and validation are key to obtain a successful application. Both PCA and HCA are usually used in parallel using several studies. One could say they are outliers simply by looking at the projection, but this cannot be done as PCA does not "classify" objects.

Neither PCA or HCA creates a "mathematical model" for classification and authentication purposes. Rather, they only project or display the objects under investigation based on selected responses and a grouping of samples may be identified by the user. Moreover, neither PCA nor HCA provides a statistical significance of such similarities (Andrić et al., 2016).

If the aim is to find an association between bioactive compounds and functional properties using HCA, the method may be applied. Although HCA shows the existence of an association between responses, HCA does not provide a measure of the association (qualitative approach). One alternative to overcome this limitation is to calculate the correlation coefficient and provide a quantitative measure of the correlation between responses (Granato et al., 2017). Although PCA and HCA are very useful in studying the data structure and finding similarities among samples, in most cases, linear correlation coefficients would render very similar interpretations of the results. Indeed, it is widely known and recognized that higher levels of phenolic compounds will render a higher antioxidant activity measured by chemical reactions *in vitro* (Guo et al., 2017; Lv et al., 2017). Another main disadvantage of using PCA/HCA in those studies is the real applicability of the observations: It seems that most researchers only use PCA and HCA to increment their data analysis rather than to explain the mechanisms of action and have a strong and in-depth discussion based on a solid hypothesis. In fact, in the field of bioactive compounds, when *in vitro* assays are used, it is somewhat obvious that almost all bioactive compounds, such as carotenoids and phenolic compounds, will exert antioxidant activity. In this case, correlation coefficients should be calculated and the results analysed (Granato et al., 2017, 2018; Nunez et al., 2015).

Acknowledgments

The support of Central Queensland University is acknowledged.

References

Adams, M.J. 1995. Chemometrics in analytical spectroscopy. RSC Spectroscopy Monographs. Edited by N.W. Barnett. The Royal Society of Chemistry, UK. 216 p.

Badertscher, M. and Pretsch, E. 2006. Bad results from good data. Trends in Analytical Chemistry 25: 1131–1138.

Bevilacqua, M., Necatelli, R., Bucci, R., Magri, A.D., Magri, S.L. and Marini, F. 2014. Chemometric classification techniques as tool for solving problems in analytical chemistry. Journal of AOAC International 97: 19–27.

Beebe, K.R., Peel, R.J. and Seasholtz, M.B. 1998. Chemometrics a Practical Guide. John Wiley & Sons Ltd, New York, USA.

Berrueta, L.A., Alonso-Salces, R.M. and Herberger, K. 2007. Supervised pattern recognition in food analysis. Journal of Chromatography A 1158: 196–214.

Brereton, R.G. 2000. Introduction to multivariate calibration in analytical chemistry. The Analyst 125: 2125–2154.

Brereton, R.G. 2006. Consequences of sample size, variable selection, and model validation and optimization, for predicting classification ability from analytical data. Trends in Analytical Chemistry 25: 1103–1111.

Brereton, R.G. 2008. Applied Chemometrics for Scientist. John Wiley & Sons Ltd, Chichester, UK.

Brereton, R.G. 2009. Chemometrics for Pattern Recognition. John Wiley & Sons Ltd, West Sussex, UK.

Brereton, R.G. 2015. Pattern recognition in chemometrics. Chemometrics and Intelligent Laboratory Systems 149(2015): 90–96.

Bro, R. and Smilde, A.K. 2014. Principal component analysis: a tutorial review. Analytical Methods 6: 2812–2831.

Clarke, K.R. and Green, R.H. 1988. Statistical design and analysis for a 'biological effects' study. Mar. Ecol. Prog. Ser. 46: 213–226.

Clarke, K.R. 1993. Non-parametric multivariate analyses of changes in community structure. Aust. J. Ecol. 18: 117–143.

Clarke, K.R., Somerfield, P.J. and Goley, R.N. 2016. Clustering in non-parametric multivariate analysis. J. Expr. Marine Biology and Ecology 483: 147–155.

Cormack, R.M. 1971. A review of classification. J. R. Stat. Soc. Ser. A 134: 321–367.

Cozzolino, D., Cynkar, W.U., Dambergs, R.G., Shah, N. and Smith, P. 2009. Multivariate methods in grape and wine analysis. International Journal of Wine Research 1: 123–130.

Cozzolino, D. 2012. Recent trends on the use of infrared spectroscopy to trace and authenticate natural and agricultural food products. Applied Spectroscopy Reviews 47: 518–530.

Cozzolino, D. 2014. An overview of the use of infrared spectroscopy and chemometrics in authenticity and traceability of cereals. Food Research International 60: 262–265.

Edwards, A.W.F. and Cavalli-Sforza, L.L. 1965. A method for cluster analysis. Biometrics 21: 362–375.

Everitt, B. 1980. Cluster Analysis. second ed. Heinemann, London.

Ellison, S.L.R. and Fearn, T. 2005. Characterising the performance of qualitative analytical methods: Statistics and terminology. Trends in Analytical Chemistry 24: 468–476.

Engel, J., Gerretzen, J., Szymanska, E., Jansen, J.J., Downey, G., Blanchet, L. et al., 2013. Breaking with trends in pre-processing. Trends in Analytical Chemistry 50: 96–106.

Esbensen, K.H. 2002. Multivariate data analysis in practice. CAMO Process AS, Oslo, Norway.

Gishen, M., Dambergs, R.G. and Cozzolino, D. 2005. Grape and wine analysis—enhancing the power of spectroscopy with chemometrics. A review of some applications in the Australian wine industry. Australian Journal of Grape and Wine Research 11: 296–305.

Gonzalez, G.A. 2007. Use and misuse of supervised pattern recognition methods for interpreting compositional data. Journal of Chromatography A 1158: 215–225.

Granato, D., Calado, V.M.A. and Jarvis, B. 2014. Observations on the use of statistical methods in food science and technology. Food Research International 55: 137–159.

Granato, D., Nunes, D.S. and Barba, F.J. 2017. An integrated strategy between food chemistry, biology, nutrition, pharmacology, and statistics in the development of functional foods: A proposal. Trends in Food Science & Technology 62: 13–22.

Granato, D., Santos, J.S., Escher, G.B., Ferreira, B.L. and Maggio, R.M. 2018. Use of principal component analysis (PCA) and hierarchical cluster analysis (HCA) for multivariate association between bioactive compounds and functional properties in foods: A critical perspective. Trends in Food Science and Technology 72: 83–90.

Hawkins, D.M. 2004. The problem of overfitting. Journal of Chemical Informatics Computational Science 44: 1–12.

Khakimov, B., Bak, S. and Engelsen, S.B. 2014. High-throughput cereal metabolomics: Current analytical technologies, challenges and perspective. Journal of Cereal Science 59: 393–418.

Khakimov, B., Gürdeniz, G. and Engelsen, S.B. 2015. Trends in the application of chemometrics to foodomics studies. Acta Alimentaria 44: 4–31.

Kumar, N., Bansal, A., Sarma, G.S. and Rawal, R.K. 2014. Chemometrics tools used in analytical chemistry: An overview. Talanta 123: 186–199.

Lee, I. and Yang, J. 2009. Common clustering algorithms. pp. 577–618. In: Brown, S.D., Tauler, R. and Walczak, B. (eds.). Comprehensive Chemometrics. Oxford, England: Elsevier.

Moller, S.F., von Frese, J. and Bro, R. 2005. Robust methods for multivariate data analysis. Journal of Chemometrics 19: 549–563.

Naes, T., Isaksson, T., Fearn, T. and Davies, T. 2002. A user-friendly guide to multivariate calibration and classification. NIR Publications, Chichester, UK. 420 p.

Nunes, C.A., Alvarenga, V.O., Sant'Ana, A.S., Santos, J.S. and Granato, D. 2015. The use of statistical software in food science and technology: Advantages, limitations and misuses. Food Research International 75: 270–280.

Oliveri, P. and Downey, G. 2012. Multivariate Class Modelling for the Verification of food Authenticity Claims. Trends in Analytical Chemistry 35: 74–86.

Otto, M. 1999. Chemometrics: Statistics and Computer Application in Analytical Chemistry. Wiley-VCH. 314 p.

Palese, L.L. 2018. A random version of principal component analysis in data clustering. Computational Biology and Chemistry 73: 57–64.

Pulido, A., Ruisanchez, I., Boque, R. and Rius, F.X. 2003. Uncertainty of results in routine quality analysis. Trends in Analytical Chemistry 22: 647–654.

Skov, T., Honore, A.H., Jensen, H.M., Naes, T. and Engelsen, S.B. 2014. Chemometrics in foodomics: Handling data structures from multiple analytical platforms. Trends in Analytical Chemistry 60: 71–79.

Szymanska, E., Gerretzen, J., Engel, J., Geurts, B., Blanchet, L. and Buydens, L.M.C. 2015. Chemometrics and qualitative analysis have a vibrant relationship. Trends in Analytical Chemistry 69: 34–51.

Smyth, H. and Cozzolino, D. 2013. Instrumental methods (spectroscopy, electronic nose and tongue) as tools to predict taste and aroma in beverages: Advantages and limitations. Chemical Reviews 113: 1429–1440.

Westad, F. and Marini, F. 2015. Validation of chemometric models: A tutorial. Analytica Chimica Acta 893: 14–23.

CHAPTER **7**

Multiway Statistical Methods for Food Engineering and Technology

Smita S Lele[1] and *Snehasis Chakraborty*[2,*]

1. Introduction

Food systems are a complex mixture of macro, micro and nano molecules, which chemically may be classified as carbohydrates, proteins, lipids, vitamins, minerals and so on. The system may be a solution with suspended particles or an emulsion of clear liquid. During processing, applications of unit operations, like evaporation, pasteurization, sterilization, cooking, extraction, leaching and crystallization, lead to physical, chemical and biological changes to get the desired product and to preserve the food. The data generated in these operations may be related to the performance of the machine and process, product properties, logistics, sensory profile and econometrics, which are often complicated. This leads to the data analytical problems in food systems being more multifaceted because of the biological variations (Bro, 1998). Furthermore, variation in the raw material, if it is a naturally harvested product, makes the system very complex-like an open-ended infinite pool where many ad-hoc data are reported by several researchers and one finds it difficult to arrive at unified solutions (mathematical model) that could be used universally. In this sense, having a diversified methodology to handle the complicated dataset is the real need of the food industry. Handling the data for a response as a function of two variables is easy and conclusive. However, in food systems, the data set generated is often visualized as a function of three or more variables. The examples include process efficiency (temperature × time × compositional parameter), sensory score (treatment condition × attributes × judges), shelf-life (storage temperature × packaging material × spoilage rate) and so on (Bro, 1998). The data set visualizing the influence of three or more factors on the response can be considered as multiway data (Geladi, 1989).

Various tools, such as response surface modelling (RSM), help the engineers to understand effect and interactions (positive and negative synergy) of various independent parameters. In addition, multivariate analysis has emerged as a handy technique to analyse the data series of multiple dimensions at a time. However, these techniques often lead to overfitting of data due to constraints in unfolding techniques. The multiway analysis is the modeling of a data set of the N-way array in which model fitting is conducted

---

[1] Institute of Chemical Technology Mumbai-Marathwada Campus, Jalna, Maharashtra, India – 431203.
  Email: ss.lele@ictmumbai.edu.in
[2] Department of Food Engineering and Technology, Institute of Chemical Technology, Mumbai, India – 400019.
* Corresponding author: sc.chakraborty@ictmumbai.edu.in

to the whole data set rather unfolding to every possible two-way matrix (Giordani et al., 2014). The very first application of multiway analysis was on chromatography where spectra were taken at various combinations of chemicals, solvents and stationary phases acting as three different modes. The retention time served as the actual measurable response and the system acted as a three-way array. Excitation-emission fluorescence spectra, liquid and gas chromatography, process chemometrics and environmental study are other areas where multiway analysis has been applied. Recently, it has been applied in case of food industry problems, like bread baking, classification of lettuce, wine, yogurt, fish oil and sugar, and sensory analysis (Bro and Heimdal, 1996; Bro, 1997, 1998; Munck et al., 1998; Pedersen et al., 2002; Christensen et al., 2005; Preys et al., 2006). The present chapter attempts to give some insights on how multiway analysis can be employed for food processing so that a realistic, more general and robust model with a good degree of confidence can be developed so that the food engineer can predict the system performance (food processing), the quality of end product, chemical, physical, biological and sensory attributes (output – result) when a set of independent parameters (processing conditions) are changed.

1.1 What is multiway data?

Statistically, the two-way data are normally represented by a tabular array that consists of two axes. This signifies the data matrix of the dependent variable is a function of only two variables. For example, while designing a thermal process, the lethality achieved in the thermal treatment is a function of time and temperature. It can be represented as a matrix of lethality, where treatment time and temperature can be placed across the row and column, respectively (Fig. 1).

Now, in the same case, lethality can also be influenced by pH of the sample and the response (lethality) becomes a three-mode or three-way data as a function of pH, time and temperature. This is called multiway data and is very common in food systems or processes where multiway analysis is used as the tool for visualizing the influences of three or more parameters on the dependent variable. For instance, in sensory analysis, the sensory score is the function of the type of samples, number of judges and attributes. In process analysis, the response is a function of batch, variables and time. Apparently, when an experimental design visualizes the influence of three or more factors on the response, the corresponding data set can be considered as multiway data.

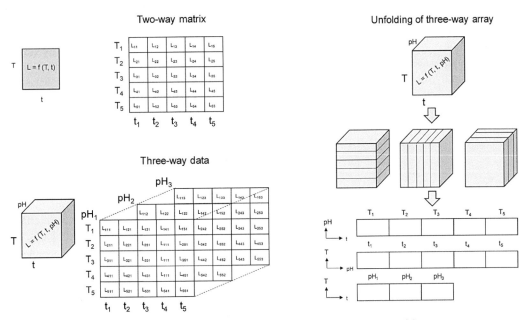

Figure 1: Graphical representation of the two-way matrix and three-way array and unfolding of three-way array to two-way matrix. L, lethality, T, temperature, t, treatment time.

The traditional way of handling the three-way data set is to unfold it to a two-way data set. For the previous example of lethality, in order to visualize the effect of pH through two-way data, the contours of lethality values can be plotted in temperature-time axes, keeping the pH fixed. If there are 3 pH values, the number of matrices or plots will be 3. If the treatment temperature and time are being varied in 5 and 5 levels, respectively, the total number of matrices generated for response will be $3 \times 5 \times 5 = 75$ (Fig. 1). Hence, unfolding the multiway data series through two-way analysis often leads to overfitting of data as the nature of data and model fitted may not be similar in this type of analysis. The way out for this is to analyse or fit the model to the whole dataset in multiway mode itself, without unfolding to every possible two-way matrix. In chemometrics, this type of model fitting or analysis can be performed through principal component analysis (PCA), Tucker 3 model or PARAFAC model. Another way is to apply multilinear partial least square (N-PLS) regression combined with the multiway or N-way analysis.

2. Array Definitions and Properties

Identifying the problem and arranging the data in the proper array are the most important aspects of multiway analysis. Visualizing the properties of the N-way array is critical to understanding the concept of multiway analysis (Kiers, 2000). The following sections focus on a brief discussion about the properties of three-way data set so that the concept can be replicated for N-way dataset (N > 3).

2.1 Modes or slices

Typically, the array of the three-way data is represented by a vector matrix or array (let's say $\underline{X}$) having the indices of x_{ijk}, where, $i = 1, 2, 3, 4, ..., I; j = 1, 2, 3, 4, ..., J; k = 1, 2, 3, 4, ..., K$. The i, j and k are the runs corresponding to three different axes or variables. For a higher array, representing the element of the array may be x_{ijklmn} (for the 6-way array). The predictor variables in the array may be real or complex data. However, in the case of food or chemical systems, the data used are real values, in general. Sometimes this array is also called 'tensor', this is represented by $R^{I \times J \times K}$ for three-way data and $R^{I1 \times I2 \times ... \times IN}$ for N-way data.

In case of a three-way array, the data representation through different modes has been summarized in Fig. 2.

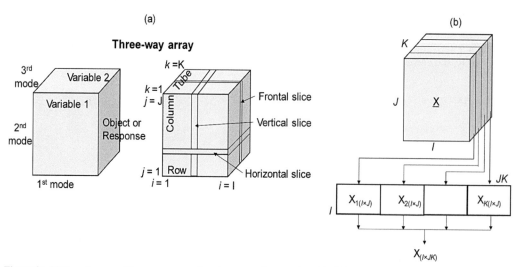

Figure 2: (a) Box diagram of three-way array showing row, column, tube with different modes and slices; (b) unfolding the three-way array into two-way matrix preserving the first mode (Bro, 1998).

Let's say, for the x_{ijk} index, the *i*th index is represented by the vertical axis, and *j*th and *k*th indices correspond to horizontal axis and depth, respectively. These indices or axes are often confused with dimensions. To get rid of this confusion, Tucker (1964) introduced these three different ways of representing array as 'mode'. It may be called as *mode* A, *mode* B and *mode* C. For Nth-way data it is represented as *mode* 1, *mode* 2 to *mode* N. A three-way array is generally considered as three different slices as represented in Fig. 2. A set of horizontal slices can be generated at $i = 1, 2, 3, 4, ..., I$, and denoted as *mode* 1 or *mode* A. The lateral slices are represented by $j = 1, 2, 3, 4, ..., J$ (*mode* 2 or *mode* B), whereas, for frontal slices are placed at $k = 1$ to K (*mode* 3 or *mode* C) (Fig. 2).

The size of the horizontal array is $(J \times K)$, whereas, the same for vertical and frontal slices are $(I \times K)$ and $(I \times J)$, respectively. Instead of slices, the three sets of matrices are often termed as '*fibres*'. For instances, *mode* A subspaces are the vertical fibres of three-way data and *mode* B subspaces are the horizontal fibres; whereas, *mode* C subspaces are the depth fibre ones. The corresponding spaces are termed as R^I, R^J and R^K, respectively.

2.2 Sub-array

Multiway or three-way array is suitably embodied as a group of two-way matrices at different levels of the third variable (Fig. 2b). In an experimental design, the object or response is measured as a function of two different variables (var 1 and 2). When the whole system is represented as a three-way array, it can be denoted as $\underline{X}$ $(I \times J \times K)$. One of the simplified ways of decomposing involves converting it to a set of two-way matrices, X of size $(I \times JK)$. As presented in Fig. 2b, the object can be plotted against a variable (var 1) while keeping the other variable (var 2) constant for each matrix. These are basically the series of linked matrices. Similarly, X may be of $(I \times KJ)$, where, object-var 2 landscape is created at every level of var 1. On a similar note, a matrix can be formed for variable 1 as a function of the other two and the corresponding size will be $(J \times IK)$, $(J \times KI)$. When the variable 2 is plotted against variable 1 and object variable, the matrix size will be $(K \times IJ)$ and $(K \times JI)$, respectively.

2.3 Rank

For the two-way matrix, the minimum number of PCA components (bilinear) required to reproduce the matrix is called as rank. In a three-way array, rank is defined by a minimum number of trilinear or PARAFAC components capable of reproducing the array. For example, if we can reproduce the array by 4-PARAFAC components, the rank of the array will be 4. In linear algebra or two-way matrix, rank pertaining to row, column and matrices are the same. However, in case of PARAFAC model, this is not true. For a 2×2 matrix, the rank is always 2; but for a $2 \times 2 \times 2$ array, the rank is 3 not 2. In case of $3 \times 3 \times 3$ array, the corresponding rank is 5 (Kruskal, 1989).

2.4 Algebra and elementary operations of multiway analysis

Algebra for multiway analysis is performed or solved through tensor analysis using matrix algebra. It is well known that zero-, first-, second-, third- and fourth-order tensor represent scalar, vector, matrix, three- and four-way array, respectively (Fig. 3).

If $\underline{X}$ and $\underline{Y}$ are two same dimensional arrays (I × J × K), then the addition of these two will be according to Eq. 1.

$$z_{ijk} = x_{ijk} + y_{ijk} \quad i = 1,....,I; j = 1,....,J; k = 1,....,K. \tag{1}$$

$x_{ijk}, y_{ijk}, z_{ijk}$, are the elements of array $\underline{X}$, $\underline{Y}$ and $\underline{Z}$, respectively. In this case, the addition and subtraction are commutative and associative in nature.

In the case of outer product, Eq. 2 is applicable, where, x_{ijk}, y_{lmn} and z_{ijklmn} are the elements of $\underline{X}$, $\underline{Y}$ and $\underline{Z}$, respectively.

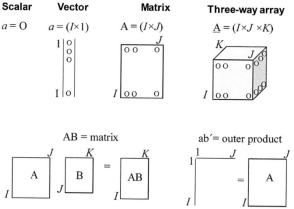

Figure 3: Scalars, vectors, matrices and three-way arrays along with some matrix multiplications and outer product (Smilde et al., 2004).

$$z_{ijklmn} = x_{ijk}y_{lmn} \quad i = 1,....,I; j = 1,....,J; k = 1,....,K;$$
$$l = 1,....,L; m = 1,....,M; n = 1,....,N. \tag{2}$$

Vector operation or vec-operator in the matrix is very crucial. As shown in Fig. 4, if A is $I \times J$ matrix having columns a_j ($j = 1, 2, ..., J$), it should follow the vec-operator rules so that vec A = IJ vector.

If all the column vectors of matrix A are slanted vertically, it will lead to vec A. The simplified rules for vector operation are listed in Fig. 4 (Magnus and Neudecker, 1988). In multiway data analysis, a normal vector multiplication operator may not be enough. In this sense, other matrix operations like Kronecker product, Hadamard product and Khatri–Rao product have been familiarized in the multiway analysis. Tucker models can be written by Kronecker product, whereas Khatri–Rao product is suitable for the PARAFAC model and Hadamard product is applicable for weighted regression (Smilde et al., 2004). For instance, the Kronecker product ($\otimes$) of A ($I \times J$) and B ($K \times M$) leads to the matrix ($IK \times JM$), according to Eq. 3.

$$A \otimes B = \begin{bmatrix} a_{11}B & . & . & . & a_{1J}B \\ . & . & & . \\ . & & . & & . \\ . & & & . & . \\ a_{I1}B & . & . & . & a_{IJ}B \end{bmatrix} \tag{3}$$

Some of the multiplication laws for Kronecker product have been summarized in Fig. 4a. The outer product of two vectors through Kronecker product is also designated as dyads (ab′=a□b′= b□a′). Hadamard product is the matrix operation where element-to-element multiplication is present, as shown in Eq. 4.

$$A * B \begin{bmatrix} a_{11}b_{11} & . & . & . & a_{1J}b_{1J} \\ . & . & & . \\ . & & . & & . \\ . & & & . & . \\ a_{I1}b_{I1} & . & . & . & a_{IJ}b_{IJ} \end{bmatrix} \tag{4}$$

Here, the elements of matrix A and B are given by a_{ij} and b_{ij}, respectively. If we apply Hadamard operator on a square matrix A (having diagonal elements $a_{11}, .., a_{nn}$) with identity matrix I, it will lead to a matrix in which the off-diagonal elements will be zero. Other typical rules for Hadamard product have

(a) Kronecker product

$A \otimes B \otimes C = (A \otimes B) \otimes C = A \otimes (B \otimes C)$

$(A + B) \otimes (C + D) = A \otimes C + A \otimes D + B \otimes C + B \otimes D$

If $A + B$ and $C + D$ exist, $(A \otimes B)(C \otimes D) = AC \otimes BD$

$a \otimes A = aA = Aa = A \otimes a$; when a is a scalar

$(A \otimes B) = A \otimes B$

$a \otimes b = ba = b \otimes a$

$\text{tr}(A \otimes B) = \text{tr}(A)\,\text{tr}(B)$ when A and B are square matrices

$(A \otimes B)^{-1} = A^{-1} \otimes B^{-1}$ if A and B are nonsingular

$r(A \otimes B) = r(A)\,r(B)$

$(A \otimes B)+ = A^+ \otimes B^+$

(b) Hadamard product

$A * B = B * A$

$(A * B) = A * B$

$(A * B) * C = A * (B * C)$

$(A + B) * (C + D) = A * C + A * D + B * C + B * D$

$A * I = \text{diag}(a_{11}, \dots, a_{nn})$, I is identity matrix;

(c) Khatri–Rao product

$(A \odot B) \odot C = A \odot (B \odot C)$

$(T_1 \otimes T_2)(A \odot B) = T_1 A \odot T_2 B$

If A and B are partitioned in K columns,

$(A \odot B)(A \odot B) = (AA) * (BB)$

(d) vec-operator

$\text{vec}(a) = \text{vec } a' = a$

$\text{vec } ab' = b \otimes a$

$\text{vec}(A)'\text{vec}(B) = \text{tr } A'B$; when A & B are of same order

$\text{vec}(ABC) = (C \otimes A)\,\text{vec } B$

Figure 4: Typical rules for vector operations. (a) Kronecker product; (b) Hadamard product; and (c) Khatri–Rao product; and (d) vec-operator; . Superscript '+' and r indicate Moore–Penrose inverse and rank, respectively (Smilde et al., 2004).

been highlighted in Fig. 4b. Another important multiplication used in the multiway analysis is Khatri–Rao product and this is widely utilized in three-way analysis, according to Eq. 5.

$$
\begin{aligned}
A &= [A_1 \quad \cdots \quad A_K] \\
B &= [B_1 \quad \cdots \quad B_K] \\
A \odot B &= [A_1 \otimes B_1 \quad \cdots \quad A_K \otimes B_K]
\end{aligned}
\tag{5}
$$

Here A and B are the smaller composite or block matrices and the blocks or partitions created in both cases are the same. A few important thumb rules for Khatri–Rao product with a partition at k columns are mentioned in Fig. 4c. The concept of dyad can be extended to triad which is the multiplication of three vectors, such as a, b and c, having the dimensions of $(I \times 1)$, $(J \times 1)$ and $(K \times 1)$, respectively. This leads to a three-way array, or triad $(I \times J \times K)$, having the rank of one. This type of multiplication results in a tensor (Δ) like Kronecker product. However, all the tensor product cannot be called a Kronecker product.

2.5 Linearity concepts

The concept of *linearity* is often confused with the term *linear*. Taking an example, a predictor or variable (x_i) has been varied at $i = 1, 2, .., I$ run and the predicted counterpart (y_i) is measured. If the relationship between object and predictor variables is linear in terms of *parameters* and *variables*, the expression can be written as Eq. 6.

$$
y_i = \beta_0 + \beta_1 x_i + e_i \qquad i = 1, 2, \dots, I.
\tag{6}
$$

β_0 and β_1 are the coefficients; e_i stands for an error. Another case may be that the model is linear in terms of *parameters* but not with respect to *variables* (x_i). This type of linearity can be expressed as Eq. 7.

$$
y_i = \beta_0 + \beta_1 x_i + \beta_1 x_i^2 + e_i \qquad i = 1, 2, \dots, I.
\tag{7}
$$

This expression is also linear when the value of the *variable* (x_i) is given or constant.

Moving towards the concept of bilinearity, let's assume a matrix X $(I \times J)$ with x_{ij} as the matrix element and a linear model must be developed for this data set. Therefore, it can be easily represented as a scalar notation, according to Eq. 8.

$$
x_{ij} = \sum_{f=1}^{F} a_{if} b_{jf} + e_{ij}
\tag{8}
$$

Here, x_{ij} is the matrix element of X ($I \times J$); a_{if} and b_{jf} are the similar representatives of a_f and b_f, respectively; a_{if} denotes the ith score of component f; b_{jf} is the jth loading of component f; e_{ij} is the error to matrix component.

Principal component analysis (PCA) is basically a bilinear model as it is linear in both scores and loading. Let's take the two-component PCA model, two different plots can be generated for component 1 and component 2, separately. In each plot, there will be two perpendicular axes of scores and loadings. One extra component will add another plot between scores and loadings. Finally, an error term is added cumulatively to all bilinear plots. We can add easily another dimension to Eq. 8 to make it trilinear and it can be represented by Eq. 9.

$$x_{ijk} = \sum_{f=1}^{F} a_{if} b_{jf} c_{kf} + e_{ijk} \tag{9}$$

Here, x_{ijk} is the distinctive element of array $\underline{X}$ ($I \times J \times K$); c_{kf} is another dimension or element from c_f matrix. In a broader perspective, PARAFAC is nothing but an extension of PCA in case of a three-way array.

3. Data Analysis and Regression

The basic principle of multiway analysis and the method or technique used to arrange the data in suitable arrays/matrices is discussed so far. Now we must understand the regression models developed in multiway analysis. It is typically performed through three-way component approach. Simpler systems can be defined using one-block data and Tuckers models, whereas, for a more complex food system, two-block data might be used.

3.1 Three-way one-block data

The idea for multiway data analysis was first reported by Raymond Cattell (1944). This concept was named the principle of 'parallel proportional profiles' and it is used to find the set of conjoint factors that can be fitted to several data series or matrices simultaneously. While fitting the factors to data series, the dimension weights may be different. This is the same for a three-way array, where a mutual set of variables is prepared for a series of matrices. Tucker (1964) came up with the concept of loading matrix generated from a three-way array and A, B and C are the corresponding three matrices in which a three-way core, G, is there. Further, Tucker (1966) proposed a modified version applying Kronecker product. This is termed as Tucker 3 model. Carroll and Chang (1970) introduced CANDECOMP (Canonical Decomposition) model for the same. The PARAFAC (parallel factor analysis) concept was employed by Harshman (1970). In this chapter, we have focused mainly on PARAFAC and Tucker models, which are the most frequently used multiway models.

3.1.1 Parallel factor analysis (PARAFAC)

PARAFAC models closely resemble CANDECOMP and are basically trilinear decomposition concepts. The more generalized way of representing PARAFAC is summation instead of the matrix product form. Taking a two-way matrix X ($I \times J$) consisting of x_{ij} element based on single value factorization extended to R number of counterparts, the PARAFAC with residual e_{ij} can be expressed as Eq. 10.

$$x_{ij} = \sum_{r=1}^{R} a_{ir} g_{rr} b_{jr} + e_{ij} \tag{10}$$

Here, the orthogonal matrices A ($I \times R$) and B ($J \times R$) are composed of the elements a_{ir} and b_{jr}, respectively; g_{rr} is the component of diagonal or core or singular matrix G ($g_{11}, \ldots, g_{RR}$); R is the largest singular value within X and values are framed in descending order. The matrix notation for this will be as of Eq. 11.

$$X = AGB' + E \tag{11}$$

E is the residual matrix with the elements e_{ij}. Equation 11 can be expressed in a generalized way, where G, being the diagonal matrix, is clubbed within A and B; the corresponding matrix and summation notations are presented in Eqs. 12 and 13.

$$X = AB' + E \tag{12}$$

$$x_{ij} = \sum_{r=1}^{R} a_{ir} b_{jr} + e_{ij} \tag{13}$$

The same concept for the two-way matrix can be extended to the three-way array for $\underline{X}$ ($I \times J \times K$) with R components and element x_{ijk}. The corresponding matrix notation is given by Eq. 14.

$$x_{ijk} = \sum_{r=1}^{R} a_{ir} b_{jr} c_{kr} + e_{ijk} \tag{14}$$

Here, c_{kr} is the component of matrix C ($K \times R$), another added dimension to the two-way matrix as per Eq. 10. e_{ijk} is the unexplained or residual element in the array. With the incorporation of the scalar term g_{rrr}, the expression becomes Eq. 15.

$$x_{ijk} = \sum_{r=1}^{R} g_{rrr} a_{ir} b_{jr} c_{kr} + e_{ijk} \tag{15}$$

While representing in the form of a matrix, three-way PARAFAC model looks like Eq. 16.

$$X_k = AD_k B' + E_k = c_{k1} a_1 b_1' + \dots + c_{kR} a_R b_R' + E_k \tag{16}$$

The kth slice matrix X_k ($I \times J$) from $\underline{X}$ ($I \times J \times K$) has the element x_{ijk} where for each k, $i = 1, 2, .., I$; $j = 1, 2, .., J$. The diagonal matrix D_k consists of elements $c_{k1}, \dots, c_{kR}$ at its kth row; a_r represents the rth column of A ($I \times R$) and b_r represents the rth column of B ($J \times R$) matrix. The diagonal matrix D_k denotes the weight of X_k together with A and B.

The scheme for simultaneous matrix notation for a three-way array with R components through PARAFAC has been summarized in Fig. 5.

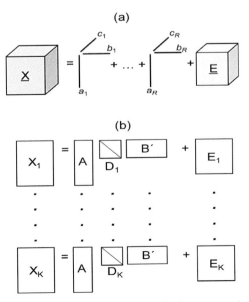

Figure 5: PARAFAC model with (a) R components and (b) simultaneous notation (Smilde et al., 2004).

Both summation and simultaneous matrix notations have been represented. The frontal slices (X_k) possess the same a_r and b_r but with variable weights given by D_k. The same thing can be repeated for two other slices, viz., vertical and horizontal slices. Thus, by solving A, B and C, a unique basis vector orientation, which is absent in PCA, is perceived. While solving the A, B and C using an algorithm, all the factors should be taken simultaneously, not separately, which is common in PCA. This leads to a no rotational freedom to PARAFAC solution. That means, once we get the solution for A, B and C, the values remain unchanged till the residuals are fixed.

3.1.2 Tucker models

Tucker proposed different models for multiway data analysis and now it is named N-mode PCA (Tucker, 1964 and 1966). The detailed description of these models has been discussed by Kroonenberg and de Leeuw (1980). Furthermore, a brief overview with updated Tucker models has been corroborated by Kroonenberg (1983). There are three similar models familiarized in multiway analysis, namely, Tucker3, Tucker2 and Tucker1 models. The concept behind those models is discussed in the following section.

Tucker3 model. Taking the condition of the two-way two-mode matrix, the PARAFAC model is denoted by Eqs. 10 and 11. Tucker3 model provides the flexibility to use a non-diagonal matrix instead of the diagonal matrix used in Eq. 11. The non-diagonal matrix is denoted by $\bar{G}$ having the element $\tilde{g}_{pq}$. Both summation and matrix notation for the Tucker3 model are given by Eqs. 17 and 18.

$$x_{ij} = \sum_{p=1}^{P}\sum_{q=1}^{Q} a_{ip}\tilde{g}_{pq}b_{jq} + e_{ij} \tag{17}$$

$$X = A\tilde{G}B' + E \tag{18}$$

The runs for A and B are p and q, respectively leading to the $\bar{G}$ matrix ($P \times Q$). In the Tucker3 model, it is not necessary for A and B to be of same order. The basic concept of the Tucker3 model has been presented in Fig. 6a.

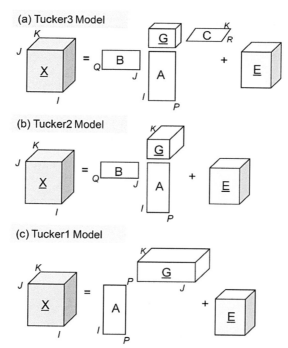

Figure 6: Principle of Tucker models in multiway analysis; (a) Tucker 3 (b) Tucker 2 (c) Tucker 1 model with reduced first mode (Smilde et al., 2004).

Considering the non-diagonal matrix, it is easy to visualize the interaction between factors. In the case of PCA, the score vector interacts only with the corresponding loading vector; whereas, in the Tucker3 model, interaction is not restricted, any loading-scoring vector interaction is possible. The same concept of two-way data is stretched to analyze three-way array also. For instance, if we have three different factors (P, Q and R) with an equal number of mutually exclusive modes, upon applying the Tucker3 model, the element of $\underline{X}$ can be expressed according to Eq. 19.

$$x_{ijk} = \sum_{p=1}^{P}\sum_{q=1}^{Q}\sum_{r=1}^{R} a_{ip}b_{jq}c_{kr}g_{pqr} + e_{ijk} \tag{19}$$

where the component matrices A ($I \times P$), B ($J \times Q$), and C ($K \times R$) consist of a_{ip}, b_{jq}, and c_{kr}, respectively; whereas, g_{pqr} and e_{ijk} are the elements of core array $\underline{G}$ ($P \times Q \times R$) and residual array $\underline{E}$ ($I \times J \times K$), respectively.

On a similar note, in a Tucker3 model, the Kronecker product of a two-way matrix X ($I \times J$) can be represented as Eq. 20.

$$X = AG(C \otimes B)' + E \tag{20}$$

The notations for Eq. 20 are same as of Eq. 16.

Tucker2 and Tucker1 models. In a Tucker3 model, all three modes of a given array (let's take the previous example of $\underline{X}$ ($I \times J \times K$) in ($P \times Q \times R$)) are reduced. In a Tucker2 model, only two modes are reduced (Fig. 6b) and the element of $\underline{X}$ can be expressed as Eq. 21.

$$x_{ijk} = \sum_{p=1}^{P}\sum_{q=1}^{Q} a_{ip}b_{jq}g_{pqk} + e_{ijk} \tag{21}$$

where, a_{ip} signifies the element from A ($I \times P$), b_{jq} denotes the same from B ($J \times Q$), and g_{pqk} is the element from an extended core-array $\underline{G}$ ($P \times Q \times K$). A Tucker2 model possesses rotational freedom and two components can be made orthogonal without compromising the fitting efficiency. On the other hand, a Tucker1 model reduces only one mode. In the array of $\underline{X}$ ($I \times J \times K$), if the second and third modes (A and B) have been set as I, then the third mode is reduced and the corresponding Tucker1 model can be presented as Eq. 22.

$$X_{K \times IJ} = IG_{K \times JP}(I \otimes A)' + E \tag{22}$$

Tucker1 model mimics the two-way PCA on a properly matricized data of $\underline{X}$. Like in Tucker2, the rotational freedom is there for Tucker1 model also (Fig. 6c). The applicability of different Tucker models depends on the number of parameters to be estimated and the risk to be overfitted. In the multiway analysis, the typically Tucker3 model is applied frequently.

3.2 Three-way two-block data

Food processing operations are often considered as a source of three-way data. For instance, an evaporator is operating in a batch mode to concentrate a fruit juice from 10 °Brix to 60 °Brix. In this case, a lot of process variables, like inside pressure, product temperature, the pressure at steam chest, etc., are being monitored during the process at a time interval. The target is to maximize the yield of juice at the end of the batch. If the number of processing factors and time intervals is J and K, respectively, for I batch runs, the data set will look like a three-way array ($\underline{X}$) with the dimension of $I \times J \times K$. The response, total yield, can be collected as vector $y = (I \times 1)$. The regression in this case mainly focuses on developing the connection between $\underline{X}$ and y. Typically, two classes of methods, namely, sequential and simultaneous methods, are used to find out the relation between $\underline{X}$ and y through regression.

3.2.1 Simultaneous method

The same problem for the batch evaporator can be considered as a three-way array ($\underline{X}$) with the corresponding dimension of $I \times J \times K$. The response values y_i within the array are being put in the form of vector y with a dimension of $I \times 1$. The Tucker3 model is used to decompose $\underline{X}$ in ($P \times Q \times R$) as shown in Eq. 20. The matrix A represents the variation in process variables for different batches or objects. Now this variation within objects can be correlated to response vector y through regression according to Eq. 23.

$$y = Aq + e_y \tag{23}$$

e_y is residual and q is the regression coefficients ($P \times 1$) which is calculated using Eq. 24.

$$q = A'y \qquad \text{where } A'A = I \tag{24}$$

This set of procedures mimics the principal component regression analysis. It can also be used in PARAFAC; however, the cross-validation is opted to find the desired components in Tucker3 and PARAFAC (Bro, 1997; Geladi et al., 1998).

Smilde and Kiers (1999) proposed a set of equations to balance X and predicted y. This basically balances the fitting of X with predicting y while selecting A (XW) in a column-space of X ($I \times JK$), as expressed in Eq. 25.

$$\min_W \left[\alpha \parallel X\text{-}AG(C \otimes B)' \parallel^2 + (1-\alpha) \parallel y - Ap \parallel^2 \right] =$$
$$\min_W \left[\alpha \parallel X\text{-}XWG(C \otimes B)' \parallel^2 + (1-\alpha) \parallel y - XWp \parallel^2 \right] \tag{25}$$

W is the free factor optimized at the beginning, followed by other counterparts. The weight α (0 to 1) is optimized like a number of components, as mentioned earlier. The weight value (α) is set highest while fitted for X. Being an extension, this multiway covariates regression is also called principal covariates regression. Multiway arrays apply the similar concept with an arbitrary number of modes. Let's say, $\underline{X}$ and $\underline{Y}$ are two different three-way arrays. Tucker3 model is applicable in this case. Therefore, the multiway modeling of $\underline{X}$ and $\underline{Y}$ coupled with covariate regression can be summarized as Eq. 26.

$$X = AG_X(C_X \otimes B_X)' + E_X$$
$$Y = AG_Y(C_Y \otimes B_Y)' + E_Y$$
$$\min_W \left[\alpha \parallel X\text{-}XWG_X(C_X \otimes B_X)' \parallel^2 + (1-\alpha) \parallel Y\text{-}XWG_Y(C_Y \otimes B_Y)' \parallel^2 \right]$$
$$A = XW \tag{26}$$

3.2.2 Sequential method

The sequential method is initially called the Tucker1-partial least square (PLS) model. Considering the same example of the evaporator mentioned earlier, the regression of y ($I \times 1$) on the three-way array ($\underline{X}$) with the modes $I \times J \times K$ can be matricized first in X ($I \times JK$) through Tucker1 model, followed by PLS. Here, the assumption of Tucker1 is applied, i.e., all the measurements on the same object are represented by first mode itself. If the sequential method predicts y and unfolds X through PARAFAC, the method is called multilinear or N-PLS regression (Bro, 1996; Faber and Bro, 2002). Take a three-way array ($\underline{X} = I \times J \times K$) for a sequential method, where w^J ($J \times 1$) and w^K ($K \times 1$) stands for J- and K-mode weighing vectors, respectively. After introducing an arbitrary vector s ($I \times 1$), the solution can be provided according to Eq. 27 (Smilde, 1997).

$$t = X(w^J \otimes w^K) \qquad \text{where } \parallel w^J \parallel = \parallel w^K \parallel = 1$$
$$\min_s \left[\parallel X - s(w^J \otimes w^K)' \parallel^2 \right] \tag{27}$$

Further, N-PLS use Eq. 27 and provide the solution as represented in Eq. 28.

$$\max_{w^J, w^K} \left[\text{cov}(t, y) \middle| t = X(w^J \otimes w^K); \| w^J \| = \| w^K \| = 1 \right] \tag{28}$$

If b is the estimated coefficient for y on t, the generalized one component trilinear PARAFAC model can be solved as Eq. 29.

$$X = t(w^J \otimes w^K)' + E_X \tag{29}$$
$$y = tb + e_y$$

Now, V is the weighing matrix which is written according to Eq. 30 in which the total PLS solution has been summed up.

$$T = XV$$
$$V = [w_1(I - w_1 w_1')w_2 \dots (I - w_1 w_1')(I - w_2 w_2') \dots (I - w_{R-1} w_{R-1}')w_R]$$
$$w_1 = (w_1^K \otimes w_1^J) \text{ and } w_R = (w_R^K \otimes w_R^J)$$
$$X = TW' + E_X; W = [w_1^K \otimes w_1^J | \dots | w_R^K \otimes w_R^J] \tag{30}$$
$$y = Tb_R + e_y$$
$$\max_{w_R^J, w_R^K} \text{cov}(t_r, y^{r-1}); r = 1 \dots R$$

The W matrix is not orthogonal, therefore, X is not the fit from TW' using the least square method. This entire series of steps is called deflating through the sequential method of regression. There are quite a few literary works which used both simultaneous and sequential method of regression in case of three-way array (De Jong, 1998; Gurden et al., 2001); moreover, it is believed that the ability to predict for a simultaneous method is better, even though it depends on the nature of data being analyzed.

4. Algorithm and Tools

The model fitting techniques which are used frequently in the multiway analysis are solutions based on the alternating least square (ALS). The eigenvalue also serves as the backbone behind multiway analysis. Alternating least square (ALS) technique is basically iterative in nature which means a closed loop continues till the solution and predictability converge together. This ensures that there will be a convergence. On the other hand, the optimization based on eigenvalues leads to an approximated fitting. As discussed earlier, in multiway analysis, PARAFAC model is regularly used; moreover, Tucker3 models are also equally important for the same. Model fitting basically refers to estimating the coefficients of the suggested expression. In this sense, both PARAFAC and Tucker3 must undergo a series of steps in order to come up with the solution, generally abbreviated as algorithm (Andersson and Bro, 1998). In the following sections, the algorithm required for PARAFAC and Tucker3 models have been elaborated in detail.

4.1 PARAFAC algorithm

The alternating least square (ALS) is commonly used in PARAFAC models. Recalling Fig. 4, the PARAFAC model is represented as Eq. 31.

$$X_{(I \times JK)} = A(C \odot B)' + E$$
$$\min_{A,B,C} \| X_{(I \times JK)} - A(C \odot B)' \| \tag{31}$$

A can be estimated through optimization on B and C using the condition given in Eq. 32.

$$\min_A \| (X - AZ') \|^2$$

$$Z = C \odot B = [D_1 B' D_2 B' ... D_k B'] \tag{32}$$

where, D_k is the diagonal matrix having a kth row of C. Now, for a given value of B and C, this becomes a two-way array and A can be found out using least square model (Eq. 33).

$$X = AZ' + E$$

$$A = X(Z')^+ = XZ(Z'Z)^{-1} \tag{33}$$

Here, Z has a full column rank. The solution for B and C can be found in the following similar way. The overall algorithm for PARAFAC-ALS method has been presented in Fig. 7.

It initiates the B and C, then A can be computed accordingly. In a similar way, the convergence is tested for calculating B and C. However, it is clear from Fig. 7 that calculating Z for a large array seems to be tedious because the data array must be continuously rearranged. Frontal slices can be taken for the formulation of the algorithm which leads to computing only XZ and Z′Z from B, C and X (Carroll and Chang, 1970; Harshman, 1970). This can be rewritten as Eq. 34.

$$XZ = [X_1 X_2 ... X_k][D_1 B' D_2 B' ... D_k B'] = X_1 BD_1 + X_2 BD_2 + ... + X_k BD_k$$

$$Z'Z = (C \odot B)'(C \odot B) = (C'C) * (B'B) \tag{34}$$

$$A = XZ(Z'Z)^{-1} = \left(\sum_k^K X_k BD_k \right)[(C'C) * (B'B)]^{-1}$$

Once the update on A is there from Eq. 34, the iteration stops when the relative deviation between previous and current iterations becomes less than 0.0001%. Initialization of the PARAFAC-ALS

Given $\underline{X} = (I \times J \times K)$; Sought dimension R

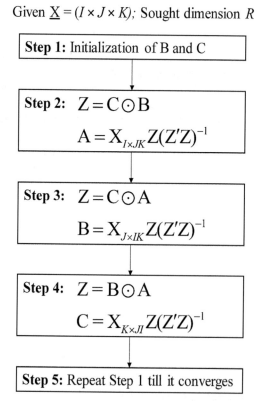

Figure 7: A typical algorithm used for PARAFAC-ALS in multiway analysis (Smilde et al., 2004).

algorithm is one of the crucial factors. To reach a faster optimization, a good parameter initialization is required. Otherwise, it may lead to difficult situations like slow converged interim solution (called swamp) or intrinsically correlated loadings. If the problem is not properly understood, it may also mislead the initialization. Then, the process needs to be started again with good approximation and the model should be refitted.

4.2 Tucker3 algorithm

The algorithm used for Tucker3 model emphasizes developing the core arising from the loading matrix and $\underline{X}$. It is crucial to keep in mind that in Tucker3 model, there exist interactions between every mode and the vector component. This is absent in case of the PARAFAC model. Therefore, for orthogonal loading matrix, the expression for Tucker3 core will be different from the PARAFAC (Fig. 8).

In case of the Tucker3 model, a least square method is commonly used and a typical algorithm for orthogonal loading vectors has been summarized in Fig. 9.

Introduction of $\underline{G}$ in the case of data array $\underline{X}$ (loading matrix) is very important while using the Tucker3 model. Finding out A from fixed B and C follows the minimization step, as discussed in Eq. 35.

$$AG(C' \otimes B') = AA'X(C \otimes B)(C' \otimes B')$$

to minimize

$$\| \{X-AA'X(C \otimes B)(C' \otimes B')\}(C \otimes B) \|^2$$

Being $(C \otimes B)$ orthogonal,

$$\min \| X(C \otimes B) - AA'X(C \otimes B)(C' \otimes B')(C \otimes B) \|^2$$

$$= \min \| X(C \otimes B) - AA'X(C \otimes B) \|^2$$

$$= \min \| (I-AA')X(C \otimes B) \|^2$$

(35)

The matrix $(I-AA')$ represents the orthogonal residual projection of column space A. Now, minimizing the residual square, A can be found for the first P left singular vectors of X(C⊗B) (Fig. 9). This same method can be extended to compute B and C while keeping Q left singular vectors of X(C⊗A) and taking R orthogonal left singular vectors of X(B⊗A), respectively. Putting the values of A, B and C automatically calculates G, which is the sum of square interim. In the algorithm, the loop closes when the relative difference between two consecutive iterations is very small, like 0.0001%, as of PARAFAC-ALS. A restricted Tucker3-ALS is applied when there are non-orthogonal loading vectors in the data

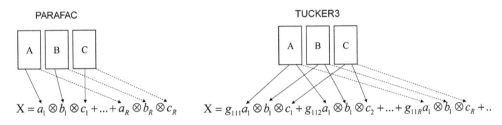

PARAFAC

| A | B | C |

$$X = a_1 \otimes b_1 \otimes c_1 + ... + a_R \otimes b_R \otimes c_R$$

TUCKER3

| A | B | C |

$$X = g_{111}a_1 \otimes b_1 \otimes c_1 + g_{112}a_1 \otimes b_1 \otimes c_2 + ... + g_{11R}a_1 \otimes b_1 \otimes c_R + ..$$

For Tucker3 model

$$X_{I \times JK} = AW' + E_{I \times JK}$$

$$W' = (A'A)^{-1}A'X_{(I \times JK)}$$

For noiseless case

$$X_{I \times JK} = A(A'A)^{-1}A'X_{(I \times JK)}$$

$$G_{P \times QR} = (A'A)^{-1}A'X_{(I \times JK)}[(C'C)^{-1}C' \otimes (B'B)^{-1}B']$$

For orthogonal loadings

$$G_{P \times Q \times R} = A'X(C \otimes B)$$

Figure 8: Difference between PARAFAC and Tucker3 models for calculating core expression in multiway analysis; G is the core array (P × Q × R) matricized to (P × QR) form, whereas, the data array is X and it is represented in (I × JK) (Smilde et al., 2004).

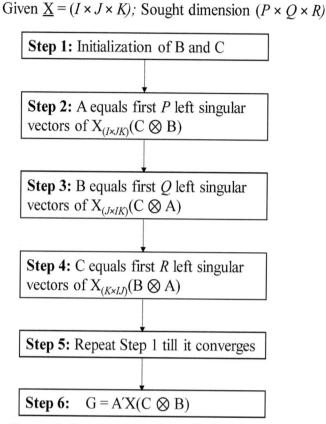

Given $\underline{X} = (I \times J \times K)$; Sought dimension $(P \times Q \times R)$

Step 1: Initialization of B and C

Step 2: A equals first P left singular vectors of $X_{(I \times JK)}(C \otimes B)$

Step 3: B equals first Q left singular vectors of $X_{(J \times IK)}(C \otimes A)$

Step 4: C equals first R left singular vectors of $X_{(K \times IJ)}(B \otimes A)$

Step 5: Repeat Step 1 till it converges

Step 6: $G = A'X(C \otimes B)$

Figure 9: A typical algorithm used for Tucker3-ALS for orthogonal loading vectors in multiway analysis (Smilde et al., 2004).

array. Like all other algorithms, initialization of B and C plays a crucial role in executing the loop for Tucker3-ALS also. Initially, for calculating A, it is reserved as row-mode, keeping other two as combined column-mode. In addition, the loading and core parameters remain fixed.

4.3 Tools for multiway analysis

There are several statistical tools available for multiway analysis. One of these platforms is MATLAB (Matlab 2018a, The Mathworks, Inc., Natick, MA, USA). A freeway toolbox with detailed algorithm has been developed by Andersson and Bro (2000). It is called 'N-way toolbox for MATLAB'. In addition, the partial least squares (PLS) toolbox (Eigen Vectors Research Inc., WA, USA) can help to plot the results of the N-way toolbox. Once these two tool boxes are patched up with MATLAB, it seems very easy to perform PARAFAC model. The function and MATLAB command for this is 'parafac(array name, number of components);'. Rasmus Bro has also provided a series of video lectures on how to perform multiway analysis and these can be found out at www.models.life.ku.dk. Coppi (1994) has also briefed about the introduction to multiway analysis. In other programming language 'R', multiway data analysis can be performed using package 'multiway' developed by Nathaniel E. Helwig (2017). In case of 'R', the command - 'parafac' is used for fitting 3- and 4-way array through the PARAFAC1 model, whereas, the command 'tucker' is able to fit multiway (3- or 4-way) Tucker Factor Analysis model. In addition, the command - 'sca' can perform different Simultaneous Component Analysis models.

5. Applications in Food Processing

The earlier sections provide the information about the theoretical concept of multiway analysis. However, it is crucial to discuss the application of multiway analysis for real time data from the food industry. In the last decades, multiway analysis has been applied in various food systems, like wine, sugar, bread, fish oil, lettuce and vinegar. A summary of these applications has been presented in Table 1. Most of the studies applied this technique to fluorescence data and came up with robust and flexible predictive model. One of the applications of multiway analysis in food-related problems have been discussed further in detail.

5.1 Case study

Effect of high-pressure processing conditions on phenolic content in pineapple juice.

5.1.1 Background

High-pressure processing is one of the emerging nonthermal technologies being applied on fruit products. Being a nonthermal treatment, it has the unique ability to retain the thermosensitive phytochemicals within the processed samples. The typical process variables for a batch high-pressure process are applied pressure, temperature and treatment time. In this study, the pressure level was varied at 200, 300, 400, 500 and 600 MPa, respectively. The temperatures during the treatment were set at 30, 40, 50, 60 and 70°C, the dwell time varied between 1 second, 5, 10, 15 and 20 min. The total phenolic content in the juice for each of the combinations was quantified. Every experimental run was performed in duplicate and analyzed in triplicate. The data set from the experiment is nothing but an array of $5 \times 5 \times 5 \times 6$ data. The multiway analysis specifically PARAFAC model was applied to this data set.

5.1.2 Pre-processing of data

The data set of $5 \times 5 \times 5 \times 6$ array was processed in such a way that it should form a three-way array. For this, the total phenolic content (TPC) was arranged in matrix form for a fixed temperature of 30°C. The matrix was of pressure (5) × dwell time (5) for each of 6 replications containing phenolic content as the matrix element (Fig. 10).

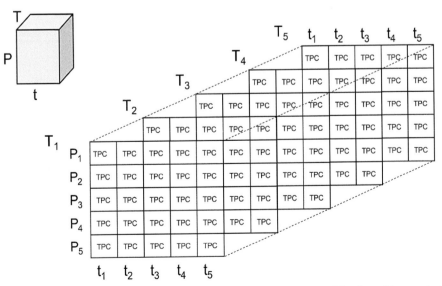

Figure 10: Graphical representation of the three-way array formed from the experimental data for multiway analysis of total phenolic content (TPC) for each replication. T, temperature; P, pressure and t, dwell time.

Table 1: Summary of the recent studies on multiway analysis applied in food systems.

Food matrix	Objective	Response	Variables	Tools used	Major inferences	Reference
Iceberg lettuce	To quantify the enzymatic browning	Polyphenol oxidase (PPO) activity	different substrates, O_2, CO_2 concentration, pH and temperature	PARAFAC and N-PLS	• PARAFAC model for PPO activity was a good fit. • N-PLS model relates the UV/VIS spectra data efficiently than HPLC data.	Bro and Heimdal (1996)
Bread	Correlating the sensory data	Sensory profile	Types, attributes, judges	PARAFAC, PLS	• Multiway analysis proved to be better than two-way unfolding. • PARAFAC modelling was able to classify the variation in data from judges.	Bro (1997)
Sugar	Modelling of the quality of refined sugar	Fluorescence spectra for amino-N concentration	Excitation, emission, intensity, sample types	PARAFAC, PLS, Tucker models	• PARAFAC model was able to quantify the response as function of variables adequately. • PLS models provided better solution than PARAFAC with fewer components.	Bro (1997)
Beetroot sugar	To model the interaction between the latent functional factors for sugar quality	Fluorescence spectra for ash content and colour	Excitation, emission, intensity, sample types, temperature, flow, and pH	PARAFAC, EEM	• 4-component PARAFAC model was developed. • A correlation was established between the raw material, processing condition and product quality.	Munck et al. (1998)
Fish oil	Screening of dioxin contamination in fish oil	Fluorescence spectra	Excitation, emission, intensity, sample types	PARAFAC and N-PLSR	• Local regression model was developed comparable with reference standard. • Excitation and emission optima were modelled through four mode PARAFAC.	Pedersen et al. (2002)
Yogurt	Modelling the oxidative stability of yogurt during storage	Fluorescence spectra for riboflavin and tryptophan	Excitation, emission, intensity, sample types	PARAFAC, N-PLS	• Meaningful PARAFAC model was developed with substituted missing values. • Prediction from the PARAFAC-PLS model was very precise.	Christensen et al. (2005)
Wine	Correlate polyphenolic composition with the sensory profile	Sensory profile	Polyphenolic compounds and sample variety	CCSWA, PLS	• CCSWA able to correlate the three sensory properties with the phenolic profile. • Astringency was adequately modelled with phenolic composition.	Preys et al. (2006)
Sherry vinegars	Classification of Sherry vinegars according to ageing	Fluorescence spectra	Excitation, emission, intensity, sample types	PARAFAC, SVM, EEM	• EEM and SVM prediction were robust and reproducible. • PARAFAC coupled with SVM classified the Sherry vinegar samples quite effectively.	Callejón et al. (2012)

Honey	Characterization and classification of honey	Fluorescence spectra	Excitation, emission, intensity, sample types	PARAFAC, PLS-DA, EEM	• A six component PARAFAC model was developed with chemically meaningful data. • The developed PLS-DA model obtained from PARAFAC was able to accurately classify and detect fake honey samples.	Lenhardt et al. (2015)
Cereal flours	Characterization of cereal flours	Fluorescence spectra	Excitation, emission, intensity, sample types	PARAFAC, PLS-DA, EEM	• 4-component PARAFAC successfully modelled the fluorescence data and decomposed the EEM. • Variations in indicator concentration were also analysed by PARAFAC as influenced by botanical origin of flour samples.	Lenhardt et al. (2017)
Wine vinegars	Characterization and authentication of Spanish PDO wine vinegars	Fluorescence spectra	Excitation, emission, intensity, sample types	PARAFAC, PLS-DA, SVM	• PARAFAC and SVM modelling successfully classified the wine vinegar samples. • SVM methodology had better precision than PLS-DR method.	Rios-Reina et al. (2017)

After that, it was transformed into a 3-way array once all the 5-temperature data have been merged together. The array was named as $\underline{X}$. The phenolic content represents the mixture of different phenolic compounds.

5.1.3 Analysis and inferences

Once the data had been arranged properly in order to get a 3-way array, the PARAFAC algorithm was run through MATLAB 2018a (The Mathworks, Inc., Natick, MA, USA) using N-way and partial least squares (PLS) tool boxes (Eigen Vectors Research Inc., WA, USA). The screen appeared immediately after the simulation or run has been presented in Fig. 11.

The separate window showed an image of 10 graphs merged across four columns and rows, respectively. The figure at the top left corner of Fig. 11 is the line plot of loadings for three different modes (M1, M2 and M3). Unlike PCA, this loading plot can be changed with respect to the individual mode by clicking on it. Loading plot says that one component is almost fixed for second mode (M2) but another one varies abruptly along five temperature conditions (in the horizontal axis).

It is important to mention here, that the models from PARAFAC for varying components are not nested, unlike the same in the principal component analysis (PCA). If PCA is applied to the same set of data, it will show a one component model in addition to two other components nested together. If 10 components are being chosen initially for PCA, only the first 3 components can do the job describing the data set. However, in the multiway analysis it is not the situation. In the case of PARAFAC, a model taking a single component is not a subset of a 3-component model. Therefore, the selection of a number of components plays a crucial role in the multiway analysis. In this study, the number of components has been set to 3, which means the total phenolic content in the sample is basically the mixture of three major phenolic compounds and we want to model the influence of three parameters (pressure, temperature and time) on each of the three major compounds. As the PARAFAC components are non-orthogonal, the shape of the loading curves changes with the varying number of components selected.

In Fig. 11, along the parameter column below the loading plot, there is a scatter plot which describes the same thing as of loading but between two factors for a single mode. For instance, it is showing the score of factors 1 and 2 in M2 mode (Fig. 11). By clicking on the white part of each plot, mode and factors can be changed. Just below the scatter plot is the influence plot, which shows the sum of squares against the outliers in mode M1. At the bottom of the parameter column, it is showing the variation in components. As the components are not orthogonal, variances due to components are not additive. This is reflected in the plot (Fig. 11). For each component, two bars are there in the 'variation per component' plot. The first bar represents the total variance of component 1 whereas the second one signifies the unique variation in that component which has not been correlated by any other components in the analysis. It is obvious that the unique variation for any component should be less than that of the actual variation within it. The difference between the two variations is very high, which represents that the 3-component PARAFAC model is not well fit for these components. For that reason, the PARAFAC model has been rebuilt, taking the number of components 2 and the output figures have been shown in Fig. 12.

In the second column of Fig. 12, the residuals have been plotted together. It starts with the residuals of the sum of squares in mode 1. This is nothing but the three-way array of the original dataset shown in a specific mode. Besides, the residuals, quantified as the difference between actual and predicted ones, have been presented in different modes. Residual normal probability plot is another representation of the fitting. It shows that the data set is slightly biased in both sides of the predicted line and the residuals are not randomly distributed. The average residuals and the residual landscape of S3 in the three-dimensional domain have been presented thereafter. It seems that, by reducing the number of components, the structured variations in the residuals have been increased. In the third column of Fig. 11, the modeled and raw data for the sample 2 have been presented, which also signifies the selection of components influencing the trend. In the last column of Fig. 12, core consistency has been plotted. As this is a 2-component model, the core array is $2 \times 2 \times 2$. For the PARAFAC model, it should be a super-diagonal of ones for the first two core components and the rest should ideally be zero. This is reflected in the graph. The dotted line is the ideal one and the dots are the estimated elements, so the core consistency is 100% signifying a nice model. The corresponding percentage variation in the data set is 99%.

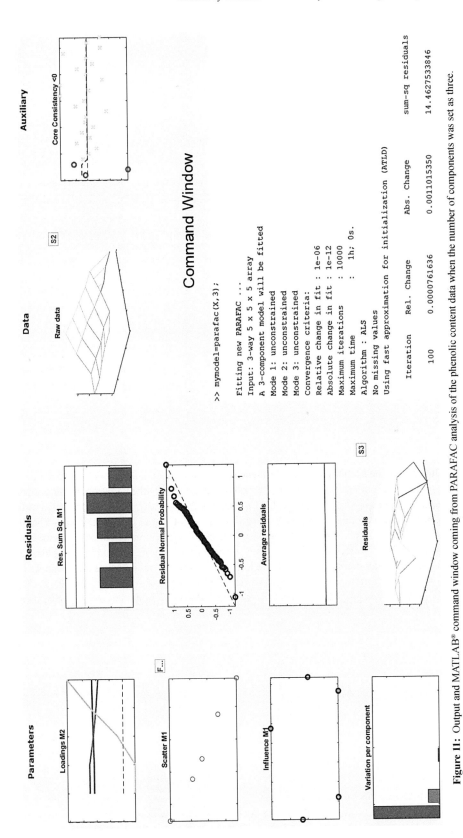

Figure 11: Output and MATLAB® command window coming from PARAFAC analysis of the phenolic content data when the number of components was set as three.

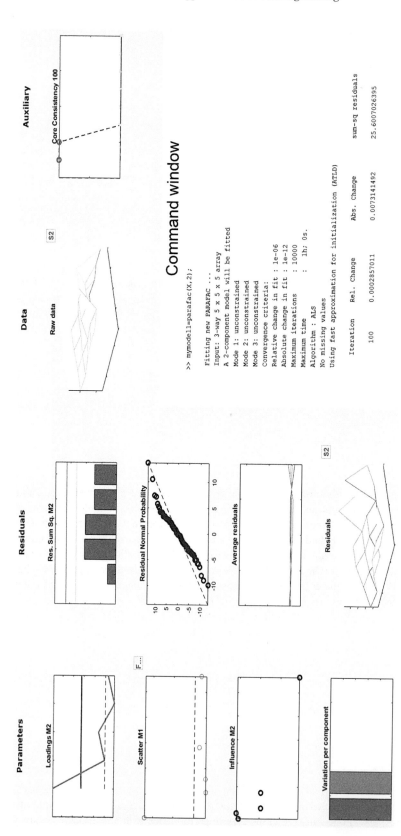

Figure 12: Output and MATLAB® command window coming from PARAFAC analysis of the phenolic content data when the number of components was set as two.

5.2 Other possible applications

5.2.1 Sensory analysis

Sensory analysis is one of the important methods used in the food processing sectors. For food product development, this is one of the crucial steps to come up with a decision. For example, a mixed fruit-vegetable juice has to be developed by mixing pomegranate, papaya and carrot juices. To find out the optimum combination, a series of trials must be conducted. Let's say, applying mixture design for the three components of the beverage, 15 runs must be conducted. The number of sensory panelists for judging the beverage is 16. The number of attributes to be assessed is 7; appearance, aroma, consistency, mouthfeel, taste, aftertaste, cloud stability and the 9-point Hedonic scale will be used for the scoring. Now the dataset will become a three-way array of sensory scores, i.e., $15 \times 16 \times 7$ arrays. It can be easily decomposed to the two-way matrix of 15×112 data set. However, applying the PARAFAC model to the three-way array will lead to a better conclusion in this type of noisy data (sensory scores).

5.2.2 Fluorescence data

In chemometrics, multiway analysis has been frequently applied on fluorescence spectra (Bro, 1999). One of the possible extensions may be applied in the sugar industry. For instance, the discoloration in sugar from raw beet or sugar cane during its production is one of the major concerns in the sugar industry. The reason behind the discoloration may be enzymatic or non-enzymatic browning and formation of color precursors. PARAFAC is suitable to be applied in this case. The required data set will be the different fluorescence spectra taken at varying excitation and emission wavelengths. This leads to a three-way array of spectra data (excitation × emission × number of samples). PARAFAC can decompose the data correlating the scores and quality or process parameters responsible for producing different sugar varieties. The analysis will emphasize about the chemical components responsible for discoloration arising from the PARAFAC components. The chemical components can act as an indicator and the production process can be monitored accordingly.

5.2.3 Second order calibration

In food analysis, we often need the calibration curve to quantify a compound, like phytochemicals, amino acids, etc. Typically, different concentrations of the standard or pure compounds in model solutions or buffer are dissolved and corresponding spectra or absorbance values are recorded. Then, the regression curve between the concentration of standard compound and absorbance generates the calibration curve. This is called first order calibration. The situation can be visualized for the calibration curve of gallic acid for phenolic compounds taking the absorbance at 750 nm. Now, if the aqueous solution of gallic acid has some caffeine in it, it will be difficult to get the absorbance or spectrum at 750 nm due to the interference of the caffeine. Using the PARAFAC model, the spectra can be decomposed and the concentration of both caffeine and gallic acid can be quantified. This requires exploring the spectra at different concentrations of caffeine and gallic acid dissolved separately and combinations in the solvent. This type of calibration is called second-order calibration.

6. Concluding Remarks

The multiway analysis in chemometrics and analytical chemistry has been practiced for the last two decades. However, in the food processing sectors, the applications are very limited so far. Needless to say, that if one does rounding off without any assumptions and addresses each and every parameter across the entire range of operation (processing), almost a near perfect prediction of the output is possible. Nevertheless, the data handling could be too complex, the convergence and computational times could be too long. Here, is a fine difference in approach of a scientist-mathematician or statistician and an engineer or technologist. If my system can be represented by the simplest mathematical model (if not straight line may be a polynomial series) and if one can predict the system performance with a reasonable accuracy and unified approach—then seldom will a food processing industry opt for the more complex model. One reason for this is that, at end of the day, the sensory attributes and the biological aspect of the food system will always have some uncertainty and subjective evaluation (sensory). This could be one of the reasons why not many researchers have reported the application of multiway analysis for food processing. We have made a humble attempt to narrate how this tool could be used in the food industry, particularly for processing where making microbially safe food with minimum loss of natural wholesomeness, taste, flavor and nutrition is the biggest challenge.

Undoubtedly, multiway analysis through PARAFAC modeling has the potential to produce a better solution for a wide range of food industry problems. One advantage of PARAFAC modeling is the robust and efficient insight it generates from the complex data set. Analysing the real-time data for different food operations through multiway method can lead to a significant conclusion and cost-effective technique. Being a mixture of several components (carbohydrate, protein, lipids, etc.), in food systems, multiway analysis can serve as a better tool to provide the insights on the individual contribution of each component present in the mixture. In all these situations, visualizing the problem and identifying the real data are the major concerns to arrive at a conclusion. On the other hand, multiway analysis has proved its credential to get rid of this type of problems. The efficient handling of the multitude problems keeps the door open for multiway analysis to be applied in food sectors.

References

Anderson, T.W. 1971. An Introduction to Multivariate Statistical Analysis. John Wiley & Sons, Inc. New York.

Andersson, C.A. and Bro, R. 1998. Improving the speed of multi-way algorithms: Part I. Tucker3. Chemometr. Intell. Lab. Syst. 42: 93–103.

Andersson, C.A. and Bro, R. 2000. The N-way toolbox for MATLAB. Chemometr. Intell. Lab. Syst. 52: 1–4.

Bro, R. 1996. Multiway calibration. Multi-linear PLS. J. Chemometr. 10: 47–61.

Bro, R. and Heimdal, H. 1996. Enzymatic browning of vegetables. Calibration and analysis of variance by multiway methods. Chemometr. Intell. Lab. Syst. 34: 85–102.

Bro, R. 1997. PARAFAC. Tutorial and applications. Chemometr. Intell. Lab. Syst. 38: 149–171.

Bro, R. 1998. Multi-way analysis in the food industry. Models, algorithms, and applications. Ph.D. thesis, University of Amsterdam, Amsterdam, The Netherland. http://www.mli.kvl.dk/staff/foodtech/brothesis.pdf.

Bro, R. 1999. Exploratory study of sugar production using fluorescence spectroscopy and multi-way analysis. Chemometr. Intell. Lab. Syst. 46: 133–147.

Callejón, R.M., Amigo, J.M., Pairo, E., Garmón, S., Ocaña, J.A. and Morales, M.L. 2012. Classification of Sherry vinegars by combining multidimensional fluorescence, PARAFAC and different classification approaches. Talanta 88: 456–462.

Carroll, J.D. and Chang, J. 1970. Analysis of individual differences in multidimensional scaling via an N-way generalization of 'Eckart–Young' decomposition. Psychometrika 35: 283–319.

Cattell, R.B. 1944. Parallel proportional profiles and other principles for determining the choice of factors by rotation. Psychometrika 9: 267–283.

Christensen, J. and Becker, E.M. and Frederiksen, C.S. 2005. Fluorescence spectroscopy and PARAFAC in the analysis of yogurt. Chemometr. Intell. Lab. Syst. 75: 201–208.

Coppi, R. 1994. An introduction to multiway data and their analysis. Computational Statistics Data Analysis 18: 3–13.

De Jong, S. 1998. Regression coefficients in multilinear PLS. J. Chemometr. 12: 77–81.

Faber, N.K.M. and Bro, R. 2002. Standard error of prediction for multiway PLS: 1. Background and a simulation study. Chemometr. Intell. Lab. Syst. 61(1-2): 133–149.

Geladi, P. 1989. Analysis of multi-way (multi-mode) data. Chemometr. Intell. Lab. Syst. 7(1-2): 11–30.

Geladi, P. and Xie, Y.L., Polissar, A. and Hopke, P.K. 1998. Regression on parameters from three-way decomposition. J. Chemometr. 12: 337–354.

Giordani, P., Kiers, H.A. and Del Ferraro, M.A. 2014. Three-way component analysis using the R package ThreeWay. J. Statistical Software 57(7): 1–23.

Gurden, S.P., Westerhuis, J.A., Bro, R. and Smilde, A.K. 2001. A comparison of multiway regression and scaling methods. Chemometr. Intell. Lab. Syst. 59: 121–136.

Harshman, R.A. 1970. Foundations of the PARAFAC procedure: Models and conditions for an 'explanatory' multi-modal factor analysis. UCLA Working Papers in Phonetics 16: 1–84.

Helwig, N.E. 2017. [Package 'multiway'] Component Models for Multi-Way Data, version 1.0-4, Repository CRAN. https://cran.r-project.org/web/packages/multiway/multiway.pdf.

Kiers, H.A. 2000. Towards a standardized notation and terminology in multiway analysis. J. Chemometr. 14(3): 105–122.

Kroonenberg, P.M. 1983. Three-mode Principal Component Analysis. Theory and Applications. DSWO Press. Leiden.

Kroonenberg, P.M. and de Leeuw, J. 1980. Principal component analysis of three-mode data by means of alternating least squares algorithms. Psychometrika 45: 69–97.

Kruskal, J.B. 1989. Rank, decomposition, and uniqueness for 3-way and N-way arrays. pp. 8–18. *In:* Coppi, R. and Bolasco, S. (eds.). Multiway Data Analysis. Elsevier, Amsterdam.

Lenhardt, L., Bro, R., Zeković, I., Dramićanin, T. and Dramićanin, M.D. 2015. Fluorescence spectroscopy coupled with PARAFAC and PLS DA for characterization and classification of honey. Food Chem. 175: 284–291.

Lenhardt, L., Zeković, I., Dramićanin, T., Milićević, B., Burojević, J. and Dramićanin, M.D. 2017. Characterization of cereal flours by fluorescence spectroscopy coupled with PARAFAC. Food Chem. 229: 165–171.

Magnus, J.R., Neudecker, H. 1988. Matrix Differential Calculus with Applications in Statistics and Econometrics. John Wiley & Sons Ltd., Chichester. UK.

Munck, L., Nørgaard, L., Engelsen, S.B., Bro, R. and Andersson, C.A. 1998. Chemometrics in food science—a demonstration of the feasibility of a highly exploratory, inductive evaluation strategy of fundamental scientific significance. Chemometr. Intell. Lab. Syst. 44(1-2): 31–60.

Pedersen, D.K., Munck, L. and Engelsen, S.B. 2002. Screening for dioxin contamination in fish oil by PARAFAC and N-PLSR analysis of fluorescence landscapes. J. Chemometr. 16(8–10): 451–460.

Preys, S., Mazerolles, G., Courcoux, P., Samson, A., Fischer, U., Hanafi, M. et al. 2006. Relationship between polyphenolic composition and some sensory properties in red wines using multiway analyses. Analytica Chimica Acta 563(1-2): 126–136.

Ríos-Reina, R., Elcoroaristizabal, S., Ocaña-González, J.A., García-González, D.L., Amigo, J.M. and Callejón, R.M. 2017. Characterization and authentication of Spanish PDO wine vinegars using multidimensional fluorescence and chemometrics. Food Chem. 230: 108–116.

Smilde, A., Bro, R. and Geladi, P. 2004. Multi-way Analysis with Applications in the Chemical Sciences. John Wiley & Sons Ltd., Chichester. UK.

Smilde, A.K. 1997. Comments on multilinear PLS. J. Chemometr. 11: 367–377.

Smilde, A.K. and Kiers, H.A.L. 1999. Multiway covariates regression models. J. Chemometr. 13: 31–48.

Tucker, L. 1964. The extension of factor analysis to three-dimensional matrices. pp. 110–182. *In:* Frederiksen, N. and Gulliksen, H. (eds.). Contributions to Mathematical Psychology. Holt, Rinehart Winston, New York, USA.

Tucker, L. 1966. Some mathematical notes on three-mode factor analysis. Psychometrika 31: 279–311.

CHAPTER 8

Application of Multivariate Statistical Analysis for Quality Control of Food Products

Soumen Ghosh and *Jayeeta Mitra**

1. Introduction

Multivariate Data Analysis is a special statistical technique used to analyze more than one data or response variable. It has the ability to process information in a meaningful way. The available information is stored in the databases in rows and columns. Generally, in the case of a multivariate model, if the model is studied for each individual variable, the statistical power of the process is lost. In univariate studies, the inter-relationship between the variables cannot be done. The simultaneous study accounts covariance and the relationship between the variables also in the multivariate case. The covariance is the mean value of the product of the deviations of two varieties from their respective means. Multivariate analysis is generally used in consumer and market research, quality control analysis of food, beverages, pharmaceutical, agricultural and chemical products, etc. It has a further contribution in research and development work related to any process engineering. In food processing, it can be used to study spectroscopic analysis and sensory evaluation of final products. Nowadays, PAT (Process Analysis technology) is used in the automated food and beverage pilot plant in order to get faster information and more precise data. Over the past few decades, PAT has been recognised as the most efficient tool to continuously monitor and control the overall processing line and quality parameters of ingredients and finished products for in/on line analysis (Beebe et al., 1993; Huang et al., 2008; Liu et al., 2001; Blanco and Villarroya, 2002). It helps to avoid sample preparation and processing errors, because we take the process from the lab to an in/on line process, since it is the most robust and time saving method (Kueppers and Haider, 2003).

MVA (Multivariate Analysis) tools are used in design, data acquisition and analysing online/at line/in line PAT to reproduce the process under realistic conditions. Apart from simulation, it helps in continuously monitoring the process and properly assessing the conditions at any time, even detecting early signs of failure in the process (Roggo et al., 2007).

2. Different Methods of Multivariate Analysis

In general, multivariate studies are done based on three aspects: Classification of individuals, dimension reduction and causal relationship. Some of the important multivariate techniques are given in Fig. 2.

Agricultural and Food Engineering Department, Indian Institute of Technology Kharagpur.
  Email: gsoumen2@gmail.com
* Corresponding author: jayeeta.mitra@agfe.iitkgp.ac.in

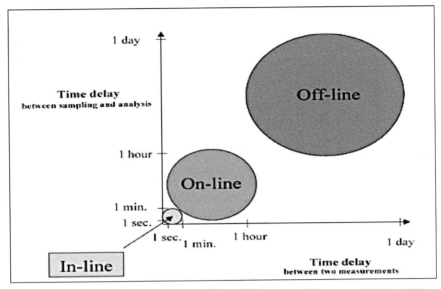

Figure 1: Time Saving during switching to off to on/in line process (Kueppers and Haider, 2003).

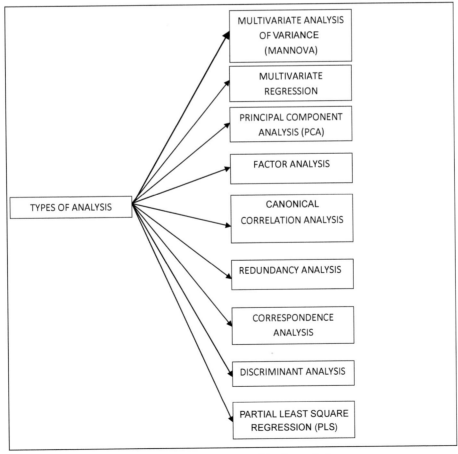

Figure 2: Some important multivariate analysis.

Classification of groups is done based on similarity between the sample data through Cluster Analysis (CA), and the difference between each cluster is found by discriminant analysis. Reduction of variables is achieved through principal component analysis (PCA), factor analysis (FA) or canonical correlation analysis. In the case of PCA, the linear combinations of variables are taken in order to form a new variable. From each new variable, we have to extract principal components, which will carry more than 90 to 95% of the information of all the variables taken together.

However, in case of factor analysis, variables are grouped according to a common factor. The influence of the factors is taken into account for dimension reduction. Path analysis or redundancy analysis are used to check the dependency among response variables. If a subset of variables affects the subset of other variables, we will use one way MANNOVA instead of ANNOVA, because in the multivariate case the number of response variables is more than one (Roggo et al., 2007). To establish the linear relationship between the variables, multivariate linear regression model, which is known as ordinary least square regression (OLS), is used. Linear square regression (LSR) model cannot be used for noisy and correlated data (Johnson and Wichern, 2007; Mark and Workman, 2003) because a fairly high degree of correlation among variables may lead to an unstable inverted matrix. Other common MVA algorithms, like partial least square method (PLS), are used to avoid the problems related to noise and correlations (Romía and Bernàrdez, 2009). Some common algorithms for developing calibration models are given in Table 1 (Mark and Workman, 2003).

After developing any calibration model, the most important steps are to check the accuracy of the model and obtain validation. Without validation, a calibration model cannot be justified. Underfitting and overfitting are also a problem for calibration of a model. In order to avoid overfitting of the model, validation is applied. In most of the multivariate cases identifying the number of variables is so important to restrict too much redundancy in independent variables; otherwise it may lead to overfitting. In case of underfitting, the fewer components are taken into account, which are unable to explain the overall variability in the model (Haaland and Thomas, 1988).

Table 1: Common algorithm for developing calibration model (Mark and Workman, 2003).

Method	Features
Least squares (LS)	Easy to solve, used for simple sample sets, large prediction errors.
Classical least squares (CLS)	More suitable for large number of variables, cannot be used for mixture of samples.
Inverse least squares (ILS)	Used for any number of variables, flexible method and applied for indirect calibration only, basis for multivariate regression.
Principal component regression (PCR)	Used for full spectral method, uses inverse regression, requires knowledge of PCA.
Partial least square regression (PLSR)	Similar to PCR, but compatible with inverse and indirect calibration, combines ILS and CLS, effect of collinearity.
Artificial neural network (ANN)	Limited number of input variables can be handled, compatible with inverse and indirect calibration, flexible method.

3. Importance of Multivariate Analysis in Quality Control

Quality assurance and safety is a major concern in the Industries. However, sometimes it is very difficult to collect samples for analysis, as safety may be compromised and breaks in a process may induce unnecessary changes. In that case, MVA tools are combined with some process analysis technologies (PAT) which will analyse the quality changes in raw material and the final products under a steady condition; whether it is batch or continuous process. This reduces the analysis time for the off-line laboratory method. For any quality testing, it should be kept in mind that the exact purpose of analysis should not be diluted by sampling error. If proper sampling is not done, it will not represent the bulk sample property, and this may lead to sampling error. So, the overall error is the combined effect of sampling error and analytical error.

The combination of instrumental methods with MVA has a key role in manufacturing and production plants. Sensors equipped with a multivariate data analysis tool can monitor and predict the chemical composition of food and agricultural commodities with high accuracy. The calibration model produces a relationship between the instrument and the analyte. The infrared spectrum with MVA techniques (PLS, PCA) can easily analyse the chemical composition, which cannot be easily detected by the targeted chemicals. This method has become more popular, overtaking traditional chemical and chromatographic methods, as it requires minimal/no sample preparation. Though much more attention has been given for the interpretation of different calibration models, there are still difficulties in understanding the fundamentals and the applicability of the model due to the knowledge gaps regarding spectroscopy, chemistry, biochemistry and physics involved in it. However, these instrumental methods have disadvantages due to lack of formal education in both instrumental as well as multivariate techniques in the food industry (Cozzolino and Ares, 2013).

Among most MVA techniques, here we will discuss: CA, PCA and PLS. These techniques are highly important in relation to the quality analysis of the product.

4. Cluster Analysis (CA)

Cluster analysis is a statistical technique that tries to make an individual cluster or group from heterogeneous multidimensional objects to homogeneous group. It is basically an exploratory data analysis (EDA) technique. There are two types of cluster analysis technique which are known as the hierarchical and non-hierarchical methods. In case of hierarchical method, the cluster follows a tree shape structure which is known as dendogram. In case of non-hierarchical method, K-means clustering approach can be applied for a large set of data. Here we are trying to give an example with *RStudio* software.

Example 1

There are 3 types of wine taken. Each type of wine has 30 samples and 6 attributes (Alcohol content, Malic acid, Ash, Magnesium, Total phenol, Flavonoids) (Næs et al., 2002). How these attributes affect these 3 types of wine to form individual cluster; we will analyse this with K-means clustering method. The data is given in the Table 2.

In above problem, there are 3 types of wine having 6 attributes. In case of K means clustering we have to standardized the attribute data with respect to mean = 0 and with a standard deviation = ±1. Here, we have 3 types of attributes. So we are taking K= 3 (number of clusters). In our case, after importing our data file in *RStudio* software we run the programme. Here the attributes are assigned by numeric values, which are float in general. The characteristics of attributes are assigned by 'Double' in programming language. The code is given as follows:

```
> view(wine)
> wine.f=wine
> wine.f$`Type of wine` <- NULL
> view(wine.f)
> wine.stand <- scale(wine[-1])
> view (wine.stand)
> results <- kmeans(wine.stand,3)
> attributes(results)
> results$centers
> table(wine$`Type of wine`, results$cluster)
> plot(wine[c(""Alcohol",Malic acid","Ash","Magnesium","Total phenols","Flavonoids")], col = results$cluster)
```

Table 2: Collected sample data.

Type of wine	Alcohol	Malic acid	Ash	Magnesium	Total phenols	Flavonoids
A	13.86	1.35	2.27	98	2.98	3.15
A	14.1	2.16	2.3	105	2.95	3.32
A	14.12	1.48	2.32	95	2.2	2.43
A	13.75	1.73	2.41	89	2.6	2.76
A	14.75	1.73	2.39	91	3.1	3.69
A	14.38	1.87	2.38	102	3.3	3.64
A	13.63	1.81	2.7	112	2.85	2.91
A	14.3	1.92	2.72	120	2.8	3.14
A	13.83	1.57	2.62	115	2.95	3.4
A	14.19	1.59	2.48	108	3.3	3.93
A	13.64	3.1	2.56	116	2.7	3.03
A	14.06	1.63	2.28	126	3	3.17
A	12.93	3.8	2.65	102	2.41	2.41
A	13.71	1.86	2.36	101	2.61	2.88
A	12.85	1.6	2.52	95	2.48	2.37
A	13.5	1.81	2.61	96	2.53	2.61
A	13.05	2.05	3.22	124	2.63	2.68
A	13.39	1.77	2.62	93	2.85	2.94
A	13.3	1.72	2.14	94	2.4	2.19
A	13.87	1.9	2.8	107	2.95	2.97
A	14.02	1.68	2.21	96	2.65	2.33
A	13.73	1.5	2.7	101	3	3.25
A	13.58	1.66	2.36	106	2.86	3.19
A	13.68	1.83	2.36	104	2.42	2.69
A	13.76	1.53	2.7	132	2.95	2.74
A	13.51	1.8	2.65	110	2.35	2.53
A	13.48	1.81	2.41	100	2.7	2.98
A	13.28	1.64	2.84	110	2.6	2.68
A	13.05	1.65	2.55	98	2.45	2.43
A	13.07	1.5	2.1	98	2.4	2.64
B	12.33	0.99	1.95	136	1.9	1.85
B	12.7	3.87	2.4	101	2.83	2.55
B	12	0.92	2	86	2.42	2.26
B	12.72	1.81	2.2	86	2.2	2.53
B	12.08	1.13	2.51	78	2	1.58
B	13.05	3.86	2.32	85	1.65	1.59
B	11.84	0.89	2.58	94	2.2	2.21
B	12.67	0.98	2.24	99	2.2	1.94
B	12.16	1.61	2.31	90	1.78	1.69
B	11.65	1.67	2.62	88	1.92	1.61

Table 2 contd. ...

... Table 2 contd.

Type of wine	Alcohol	Malic acid	Ash	Magnesium	Total phenols	Flavonoids
B	11.64	2.06	2.46	84	1.95	1.69
B	12.08	1.33	2.3	70	2.2	1.59
B	12.08	1.83	2.32	81	1.6	1.5
B	12	1.51	2.42	86	1.45	1.25
B	12.69	1.53	2.26	80	1.38	1.46
B	12.29	2.83	2.22	88	2.45	2.25
B	11.62	1.99	2.28	98	3.02	2.26
B	12.47	1.52	2.2	162	2.5	2.27
B	11.81	2.12	2.74	134	1.6	0.99
B	12.29	1.41	1.98	85	2.55	2.5
B	12.37	1.07	2.1	88	3.52	3.75
B	12.29	3.17	2.21	88	2.85	2.99
B	12.08	2.08	1.7	97	2.23	2.17
B	12.6	1.34	1.9	88	1.45	1.36
B	12.34	2.45	2.46	98	2.56	2.11
B	11.82	1.72	1.88	86	2.5	1.64
B	12.51	1.73	1.98	85	2.2	1.92
B	12.42	2.55	2.27	90	1.68	1.84
B	12.25	1.73	2.12	80	1.65	2.03
B	12.72	1.75	2.28	84	1.38	1.76
C	13.49	3.59	2.19	88	1.62	0.48
C	12.84	2.96	2.61	101	2.32	0.6
C	12.93	2.81	2.7	96	1.54	0.5
C	13.36	2.56	2.35	89	1.4	0.5
C	13.52	3.17	2.72	97	1.55	0.52
C	13.62	4.95	2.35	92	2	0.8
C	12.25	3.88	2.2	112	1.38	0.78
C	13.16	3.57	2.15	102	1.5	0.55
C	13.88	5.04	2.23	80	0.98	0.34
C	12.87	4.61	2.48	86	1.7	0.65
C	13.32	3.24	2.38	92	1.93	0.76
C	13.08	3.9	2.36	113	1.41	1.39
C	13.5	3.12	2.62	123	1.4	1.57
C	12.79	2.67	2.48	112	1.48	1.36
C	13.11	1.9	2.75	116	2.2	1.28
C	13.23	3.3	2.28	98	1.8	0.83
C	12.58	1.29	2.1	103	1.48	0.58
C	13.17	5.19	2.32	93	1.74	0.63

Table 2 contd. ...

... Table 2 contd.

Type of wine	Alcohol	Malic acid	Ash	Magnesium	Total phenols	Flavonoids
C	13.84	4.12	2.38	89	1.8	0.83
C	12.45	3.03	2.64	97	1.9	0.58
C	14.34	1.68	2.7	98	2.8	1.31
C	13.48	1.67	2.64	89	2.6	1.1
C	12.36	3.83	2.38	88	2.3	0.92
C	13.69	3.26	2.54	107	1.83	0.56
C	12.85	3.27	2.58	106	1.65	0.6
C	12.96	3.45	2.35	106	1.39	0.7
C	13.78	2.76	2.3	90	1.35	0.68
C	13.73	4.36	2.26	88	1.28	0.47
C	13.45	3.7	2.6	111	1.7	0.92
C	12.82	3.37	2.3	88	1.48	0.66

From the above example, it has been seen that type A has 2 misleading data, Type B has 4 misleading data and the Type C has also 4 misleading plots. Under cluster 1, 2 and 3 rest of the 80 sample data are well defined.

5. Principal Component Analysis (PCA)

Principal component analysis is a dimension reduction technique, where p number of X variables (manifest variable) is reduced to q number of independent z variables. For this case, $q \leq p$. Principal component analysis is basically an orthogonal transformation, where the transformed variable is Z. We are not giving the details of principal component here. We will just discuss the application with an example in "RStudio" software. After transformation, the variability of Z will be as follows:

$Var (Z_1) \geq Var (Z_2) \geq \text{--------------} \geq Var (Z_q)$

And, $Z = A^T X; A^T A = I$

The variability of all PC's can be found out by scree plot, where the variability of the contributing principal components can be considered up to elbow point. Rest of the principal components have very less variability effect. So, they can be neglected. In principal component analysis, there is no role of dependent variable (Y). Only principal components are formed with a linear combination of independent variables.

Example 2

In continuation, we will find out the principal components from the above examples

The code is given as follows:

```
> wine.f=wine
> wine.f$`Type of wine` <- NULL
> View(wine.f)
> pc <- princomp(wine.f, cor = TRUE, scores = TRUE)
> Summary(pc)
> plot(pc)
> plot(pc , type="lines")
> biplot(pc)
> attributes(pc)
> pc$loadings
```

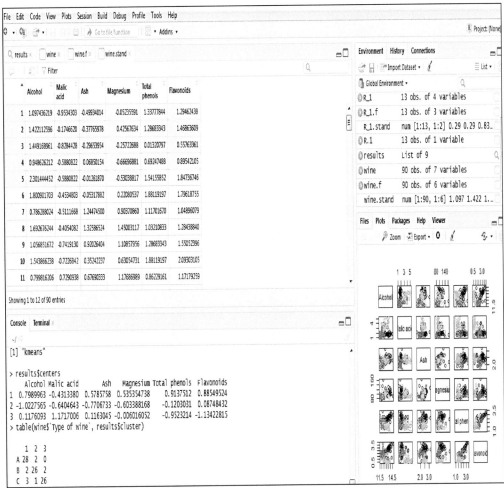

Figure 3: Distribution sample under algorithm (k = 3) for each type of wine [RStudio].

Figure 5 represents the histogram and scree plot of each principal component. From Fig. 6, it can be seen that the cumulative variance is more than 90% for the first four principal components. So, we can consider the first 4 principal components. Figure 7 represents the biplot of PC1 and PC2 for 90 sample data. Here, the six variables are plotted as vectors. From this, we have seen attributes like Ash and Alcohol vary in the same direction. This means that they are highly correlated. The value of correlation between any two variables can be found out by the cosine of the angle between them. If the angle is small, the correlation will be higher. Whereas, the variation of phenol content and flavonoids in same direction. From this figure, we also get the idea about the input potential of malic acid variation. Figure 8 represents the loading of each attribute in principal components. Loading represents the Eigen vectors. It has been observed, in the case of PC2, that there is no effect of Total phenols, whereas in case of PC4, the effect of total phenol and flavonoids are negligible.

6. Partial Least Square Regression (PLSR)

Partial least squares regression (PLSR) is a statistical technique which is related to principal components regression. It develops a linear regression model by projecting the observed variables and the projected variables to a new space. It is used to find the fundamental relations between two matrices (X and Y).

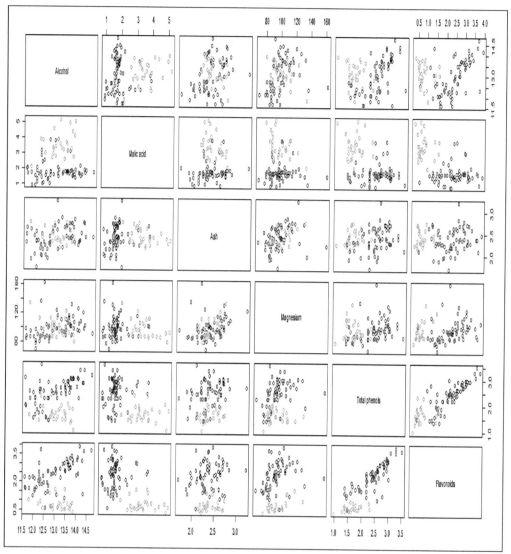

Figure 4: Cluster plot for each combination of any two attributes [RStudio].

Color version at the end of the book

A PLS model will try to find the multidimensional direction in the X space that explains the maximum multidimensional variance direction in the Y space. Instead of finding the major variation in X and Y, it looks for a direction in both which is good for correlating the X score with the Y score. PLS regression is particularly suited when the variables in the predictors' matrix are higher than the matrix of observations, and when there is multicollinearity among X values.

The general multivariate PLS model is given as follows:

$$X = TP^T + E$$

$$Y = UQ^T + F$$

where, X is an **n × m** matrix of predictors, Y is an **n × p** matrix of responses, T and U are n × l matrices that are, respectively, projections of X (the X score, component or factor matrix) and Y (the Y scores), and P and Q are, respectively, m × l and p × l orthogonal loading matrices.

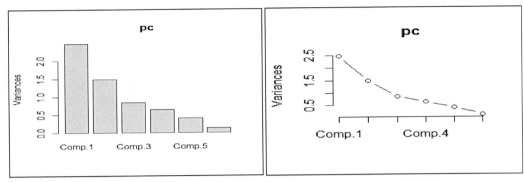

Figure 5: Histogram and scree plot of variance of each of the components [RStudio].

```
Importance of components:
                        Comp.1    Comp.2    Comp.3    Comp.4    Comp.5
Standard deviation     1.5730479 1.2218177 0.9229018 0.8015623 0.63816961
Proportion of Variance 0.4124133 0.2488064 0.1419580 0.1070837 0.06787674
Cumulative Proportion  0.4124133 0.6612197 0.8031777 0.9102613 0.97813808
                        Comp.6
Standard deviation     0.36217608
Proportion of Variance 0.02186192
Cumulative Proportion  1.00000000
```

Figure 6: Cumulative variance of principal components [RStudio].

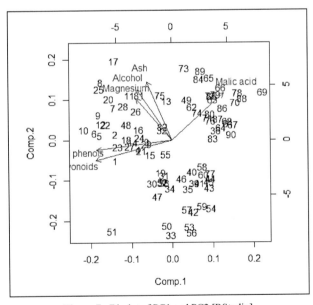

Figure 7: Bi-plot of PC1 and PC2 [RStudio].

```
Loadings:
                Comp.1 Comp.2 Comp.3 Comp.4 Comp.5 Comp.6
Alcohol         -0.264  0.495  0.625  0.198  0.502
Malic acid       0.392  0.459  0.322  0.162 -0.695  0.150
Ash             -0.189  0.573 -0.282 -0.740
Magnesium       -0.274  0.409 -0.612  0.617
Total phenols   -0.574         0.177        -0.461 -0.642
Flavonoids      -0.580 -0.197  0.139        -0.224  0.744
```

Figure 8: Loading of each attribute in principal components [RStudio].

Example 3

In a blueberry jam production process, the blueberries are collected from 3 harvest places (P) and at 3 harvesting times (T) for each. 6 instrumental parameters are taken: L value, a value, b value, solubility, absorbance and acidity. 5 sensory parameters are taken: Colour, redness, sweetness, thickness and springiness (https://www.youtube.com/watch?v=Qt3Vv5KsnpA). With a PLS regression model one can establish relations between instrumental and sensory parameters. The data is given in Table 3.

We have used XLSTAT to solve the problem. The step by step discussions have been given below. Here, we are seeing the correlation of scores of PC-1 and PC-2 on t axes between explanatory and dependable variables. We will interpret result from score and loading plot.

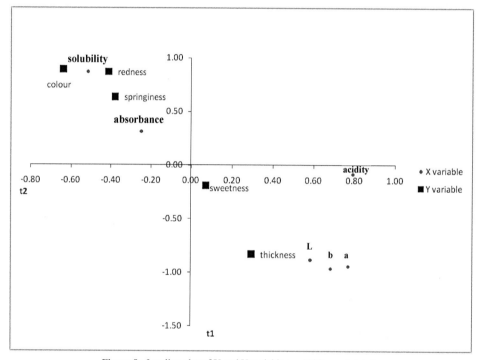

Figure 9: Loading plot of X and Y variables on t1/t2 axes [xlstat].

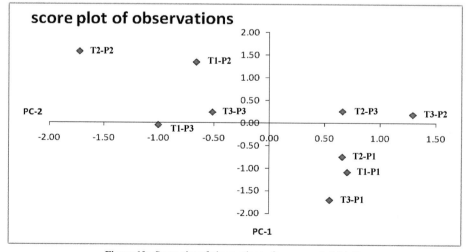

Figure 10: Score plot of observations of components [xlstat].

Table 3: Collected sample data.

	L	a	b	Solubility	Absorbance	Acidity	Colour	Redness	Sweetness	Thickness	Springiness
T1-P1	38.250	17.860	9.000	1.836	8.300	26.360	3.970	4.440	5.130	4.560	2.710
T2-P1	39.360	16.040	9.150	1.820	7.800	25.610	3.630	4.330	4.860	4.360	2.690
T3-P1	38.452	18.900	9.400	2.172	4.650	25.400	3.190	4.860	3.970	4.930	2.320
T1-P2	34.930	12.890	3.580	5.237	8.100	25.930	8.420	8.027	4.080	3.180	3.210
T2-P2	34.680	9.360	4.100	4.605	6.920	21.510	7.790	7.630	4.630	3.720	2.950
T3-P2	37.120	15.410	7.040	3.413	7.520	33.770	6.680	7.083	3.760	3.280	2.610
T1-P3	36.820	14.390	6.860	2.205	10.100	19.040	6.140	6.050	3.610	3.630	2.290
T2-P3	35.380	14.300	6.740	2.673	8.010	32.710	4.160	6.920	3.560	3.770	2.360
T3-P3	37.295	13.270	6.980	2.749	8.410	22.630	6.020	5.260	3.270	4.100	2.740

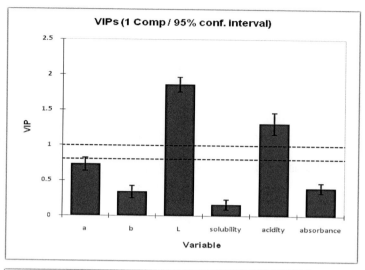

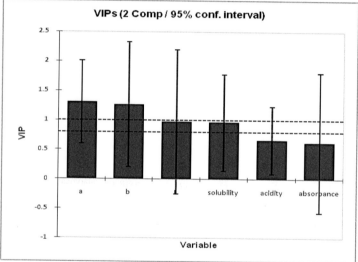

Figure 11: VIPs of variables for component 1 and component 2 [xlstat].

From the above loading plot, we can see that L, a and b are located close to each other. There is a high correlation between them. We have also seen than sweetness is comparatively an irrelevant variable, as its loading is plotted near (0, 0). We can also see the correlation between X and Y variables. L, a and b are correlated in terms of thickness and oppositely correlated with colour and the redness of different jams.

The coordinates of the blueberry jams in the space of the t coordinates are displayed in the above figure. It can be inferred that the products are satisfactorily identified. We have seen that a particular harvesting place (P1), with respect to different harvesting time, has no significant effect on scores. This means that the blueberry jam produced from a particular harvesting place P1 and P2 have similar kinds of sensory and instrumental attributes. We can also see that the variation of first component is lower than the second component.

The following figure shows the Variable Importance for the Projection (VIPs) for each explanatory variable, for an increasing number of components. This allows us to identify which are the explanatory variables that significantly contribute to the models. In the models having one component, it can be seen that a, b, solubility and absorbance have a low influence on the models. However, with two components, acidity and absorbance have less influence.

From Fig. 12 and Fig. 13 we have also seen the model equation and the goodness of fit.

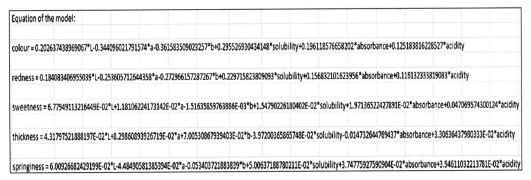

Figure 12: Model equation [xlstat].

Observations	9.000
Sum of weights	9.000
DF	7.000
R^2	0.751
Std. deviation	1.017
MSE	0.805
RMSE	0.897

Figure 13: Goodness of fit of the model [xlstat].

References

Beebe, Kenneth R., Wayne W. Blaser, Robert A. Bredeweg, Jean Paul. Chauvel, Richard S. Harner, Mark. LaPack et al. 1993. Anne. Process analytical chemistry. Analytical Chemistry 65(12): 199–216.

Blanco, M. and Villarroya, I.N.I.R. 2002. NIR spectroscopy: A rapid-response analytical tool. TrAC Trends in Analytical Chemistry 21(4): 240–250.

Cozzolino, Daniel and Gastón Ares. 2013. The use of correlation, association and regression to analyse processes and products. Mathematical and Statistical Methods in Food Science and Technology 19–30.

Haaland, David, M. and Edward V. Thomas. 1988. Partial least-squares methods for spectral analyses. 1. Relation to other quantitative calibration methods and the extraction of qualitative information. Analytical Chemistry 60(11): 1193–1202.

https://www.youtube.com/watch?v=Qt3Vv5KsnpA

https://gist.github.com/tijptjik/9408623#file-wine-csv

Huang, H., Yu Haiyan, Xu Huirong and Ying Yibin. 2008. Near infrared spectroscopy for on/in-line monitoring of quality in foods and beverages: A review. Journal of Food Engineering 87(3): 303–313.

Johnson, Richard, A. and Dean W. Wichern. 2007. Applied Multivariate Statistical Analysis.

Kueppers, Stephan and Markus Haider. 2003. Process analytical chemistry—future trends in industry. Analytical and Bioanalytical Chemistry 376(3): 313–315.

Liu, Yen-Chun, Feng-Sheng Wang and Wen-Chien Lee. 2001. On-line monitoring and controlling system for fermentation processes. Biochemical Engineering Journal 7(1): 17–25.

Mark, Howard and Jerry Workman Jr. 2003. Statistics in Spectroscopy. Elsevier.

Naes, T., Isakson, T., Fearn, T. and Davies, T. 2002. A user friendly guide to multivariate calibration and classification. NIR publications (ISBN 0-9528666-2-5).

Romía, M. Blanco and Alcalà Bernàrdez, M. 2009. Multivariate calibration for quantitative analysis. Infrared spectroscopy for food quality analysis and control, 1st edn. Academic Press, New York.

Roggo, Yves, Chalus, P., Maurer, L., Lema-Martinez, C., Edmond, A. and Jent, N. 2007. A review of near infrared spectroscopy and chemometrics in pharmaceutical technologies. Journal of Pharmaceutical and Biomedical Analysis 44(3): 683–700.

CHAPTER **9**

Importance of Normality Testing, Parametric and Non-Parametric Approach, Association, Correlation and Linear Regression (Multiple & Multivariate) of Data in Food & Bio-Process Engineering

Soumen Ghosh and *Jayeeta Mitra**

1. Introduction

Over the past few decades, quantitative approaches have become popular for analysing overall quality in food, chemical, and biotechnological industries (Samaniego-Esguerra et al., 1991). In process engineering, statistical methods like regression and correlation are used in the measurement of chemical and physical properties (rheology), sensory analysis, drying kinetics related modeling, any type of conjugate heat and mass transfer related problems, microbial inactivation and measurement and any kind mechanical and bio-chemical processing related to grain, dairy, meat and aqua products. From different literature studies, it has been concluded that the use of statistical techniques is important in food and bio-process engineering. It has been seen that most of the industrial personnel and workers often face difficulties in choosing the right and ideal tests, or use the relevant tests in unsuitable conditions. However, many researchers are often unable to use important data interpretation techniques and statistical concepts before comparing mean values (Granato et al., 2014). This may happen due to the following reasons: To avoid tedious calculations, misinterpretation and misuse of statistical procedure. It is difficult to solve all analytical issues related to food processing industries with a single technique. Modern instrumental techniques, such as chromatography, texture analysis, spectroscopy, electronic nose, electronic tongues and other sensors (pressure, temperature, thermal conductivity, moisture content), are combined with multivariate analysis in order to have more advantages over other physical and chemical instrumental methods (Cozzolino and Ares, 2013). The objectives of this chapter are: (1) to introduce some concepts related to parametric

Agricultural and Food Engineering Department, Indian Institute of Technology Kharagpur.
* Corresponding author: jayeeta.mitra@agfe.iitkgp.ac.in

and non-parametric statistics used in process engineering; (2) to give an idea of some statistical analysis based on association, testing of normality and homoscedasticity, linear regression (MLR, MVLR) and correlation and (3) to discuss some reviewed examples of regression modeling.

2. Some Review of Literature, Where Correlation and Regression Play an Important Role

Here are some research works (given below), where the primary statistical analysis plays an important role in analysing the experimental data and, based on that, conclusions have been drawn.

Moisture sorption isotherm plays an important role for the long-term storage of dried and packaged foods. As the permeability of the packaging film is main concern for shelf life enhancement of the product, the use of GAB model is more appropriate than all other models to determine the sorption isotherm. It has been observed that the nonlinear regression analysis based on standard GAB equation helps in minimizing the experimental error and influenced the resulting constants, rather than transformed GAB model (Samaniego-Esguerra et al., 1991). The unfavorable condition has been avoided by weighting the square of residuals with the variance of data at each a_w (water activity) level.

A dimension-less correlation between Nusselt, Prandtl and Reynolds number is developed in order to predict the heat transfer co-efficient in a Newtonian fluid during E-O-E (end over end) rotation of the can. It has been seen that L/D (length/diameter) does not significantly affect HTC (heat transfer co-efficient). An iteration procedure has been used to simulate time-temperature profile using the correlation developed for heat transfer coefficient (Anantheswaran and Rao, 1985).

The kinetics of the colour change of double concentrated tomato paste during heating has been studied to measure Hunter 'L', 'a' and 'b' values characterize the colour; and colour difference (ΔE), saturation index (SI) and 'a/b' ratio. The effect of colour change has been studied in both the first and second phase of thermal treatment, which is significant, statistically (Barreiro et al., 1997).

3. Background of Basic Statistics Applied in Food Technology & Engineering

Scientists need to find out much more information using correct statistical tools from the result. Generally, the analysed results are often limited to descriptive statistics (mean, median, minimum, maximum values, standard deviation or coefficient of variation). Descriptive statistics along with other statistical tests, such as correlation, regression, and comparison of mean values, are often based on the random use of statistical software that may or may not be useful for industrial and research purposes. Sometimes, the application of a basic inferential statistical test is necessary prior to any further application or analysis. It is important to understand how to plan the experimental work for relevant data analysis, understand the consequences based on the result and then draw conclusions. It is obvious, that the statistical quality of data must be checked regardless of experimental design. Sometimes poor quality data produces a misleading conclusion. The inferential statistics gives two basic tests for checking the statistical quality of data:

1) Checking for normality whether the data is normally distributed (ND) or not.
2) Checking for homoscedasticity whether there is uniformity in variance or not.

The concept of normal distribution can be well understood by the bell-shaped curve given in (Fig. 1). The rule of thumb is that 67% values lie within mean (μ) $\pm$ 1 standard deviation(σ), 95% values lie within mean (μ) $\pm 2\sigma$ and 99.7% values lie within mean (μ) $\pm 3\sigma$. If a large number of data sets are randomly drawn from a population, we cannot say all the data are identical as there is uncertainty in measurements. If only a few samples were analysed, the results may be randomly distributed within higher and lower limits. For larger data sets for n (sample size) ≥ 20, there is a higher chance to obtain normally distributed data.

Uniformity of variances is highly important for data sets of two or more. Due to high variability of two data sets, sometimes many standard parametric statistical tests cannot be used to compare the

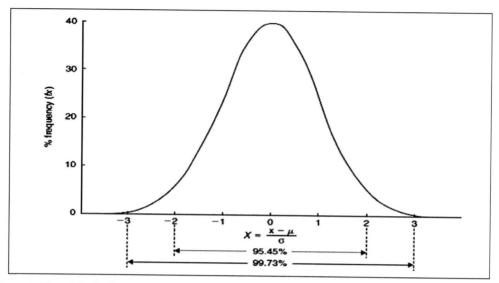

Figure 1: Normal distribution curve (population) showing the percentage of values lies within first, second and third standard distribution, with respect to mean (Jarvis, 2016).

means. For example, two different data sets drawn from a population with respect to the same mean value having a standard deviation of 0.20 and 0.40. It implies the variation is 4 times higher in the second case. Such difference in variation ultimately concludes that there is a significant difference between the two distributions.

The distribution of measured variable data may be either continuous or discrete. In some situations, experimental variables may be 'nominal', 'categorical' or 'ordinal'. It is important to understand data set designations and special nonparametric procedures are required to analyze these different types of data.

Over the past 40–50 years, process analytical technologies (PAT) has been considered as the most efficient and advanced tool for controlling and continuously monitoring the overall process and quality parameters of raw ingredients and finished products for in/on/off line analysis (Beebe et al., 1993; Huang et al., 2008; Liu et al., 2001; Blanco and Villarroya, 2002). To avoid sample preparation and unnecessary errors, we move from a laboratory to an in/on line process, as it is more robust and time-saving (Kueppers and Haider, 2003).

In most cases, MVA (multivariate analysis) is used for line processes to simulate the process under realistic environment, early detection of failure, continuously monitor the process and properly asses the conditions at any time (Roggo et al., 2007).

Nowadays, various kinds of statistical packages and software tools are available to perform the right test. An analyst can choose from various statistical packages based on their current need.

Statistical packages, such as SAS, Microsoft Excel, SPSS, Minitab, Design-Expert and Prisma, are available for analysing the result.

4. Normality and Homoscedasticity

4.1 Checking for normality—Why is it important?

Normality study of experimental data is important for further analysis of variance (ANOVA), correlation, simple and multiple regression and t-tests. If the data is not found to be normal after the normality test, any conclusion from a statistical test will not be valid (Shapiro and Wilk, 1965). Generally, for a given data set (X_1, X_2, X_3, ----------, X_n), checking of normality is based on hypothesis test. Generally, null hypothesis (H_0) accounts for the observed data that is normally distributed with respect to mean and

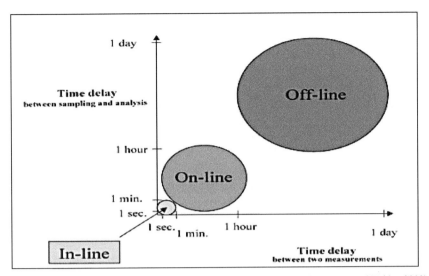

Figure 2: Saving of time during switching from off to on/in line process (Kueppers and Haider, 2003).

variance, alternate hypothesis (H_a), on the other hand, states that the distribution is not normal. It has been observed that the most of the chemical constituents and contaminants are normally well-distributed, whereas microbiological data is not. The microbial colony generally follows lognormal distribution. The number of cells in a dilute suspension follows Poisson distribution. Over dispersion in other dried and infant foods can be specified well by negative binomial and beta-poisson distribution (Jongenburger, 2012). Hence, the microbiological data requires mathematical transformation before statistical analysis (Jarvis, 2016). In practice, there are two ways of checking normality: Graphically (Q-Q plot, histogram plot) or by numerical methods (Anderson-Darling test, Shapiro-Wilk test, etc.). Of these two tests, the graphical method should not be used for the small data sets due to the lack of quantitative information. Among different numerical approaches, the SW test (Shapiro-Wilk test) shows higher efficiency in a Monte Carlo simulation (Razali and Wah, 2011).

4.2 Checking for homoscedasticity—Why is it important?

Sometimes the researcher forgets to apply proper methods to validate the statistical data. Variance of homogeneity, which is known to be like homoscedasticity, should be checked graphically or by numerical method (Montgomery, 2009). The equity of variance needs the following hypothesis: H_0: $\sigma_1^2 = \sigma_2^2 = \cdots = \sigma_j^2$ and H_a: $\sigma_1^2 \neq \sigma_j^2$ at least for one pair (1, j). Incorrect assumption of normality of variances during the goodness of fit measured by pearson coefficient of correlation analysis technique may lead to misuse of parametric tests. Several tests are there to check the equity of variances, such as Cochran, Bartlet, Levene and F-test. When the variances are not equal, one may use either a non-parametric approach or parametric statistics, after transforming the dependable variables by Box-Cox, logarithmic, square root or inverse transformation in order to obtain a constant variance of the residues.

5. Use of Parametric and Non-Parametric Statistics in Food Science

Depending on sample size, distribution and homogeneity, samples and treatments can be compared using parametric and non-parametric statistics. If the data are normally distributed, as shown by any normality test or equity of variances by Levene or F-test, parametric statistics, such as student's t-test, can be used to compare mean values between two samples. For more than 2 samples, ANNOVA can be used (Fig. 3).

In non-parametric procedures, ranking of data is more important than the actual values. The data are arranged in an ascending order with integer values from 1 to n (where n = total sample size) (Hollander et

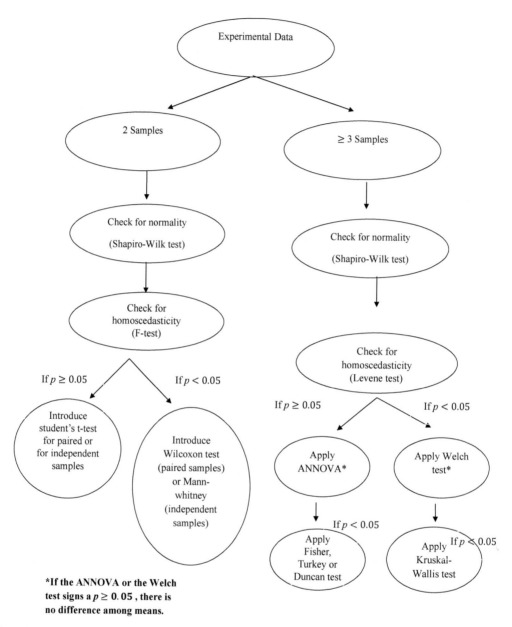

Figure 3: Statistical steps for comparing two or more variables in a relation to a quantitative response variable (Granato et al., 2014).

al., 2013). The non-parametric method is useful when there is no reliable underlying scale for the original data (Siegal, 1956). The main disadvantage of the non-parametric procedure is the need to discard the information regarding actual values. As information is discarded, non-parametric procedures can never be as strong as their parametric counterparts (Hollander et al., 2013).

6. Association

Generally, analysis of association means how the nominal variables are associated with each other. Nominal variable can be transformed by dividing a higher level of measurements into two or more categories. Though the nominal variables can be more than two categories, the 'two-by-two' case is easier to explain. Generally, Pearson's chi-square test is used with two sets of nominal data to check the significance of association between the two variables. The test is performed to check the independence between two variables. Two nominal variables are displayed in a special table, known as a contingency table; this is also known as cross tabulation. Sample size must be sufficient to allow a cell count of 5 under each category in order to perform chi square test for checking the significance of association (Bower, 2009). In two-by-two nominal variables, four cells are found in a contingency table. The chi square test generally tries to find whether the counts differ between the individual category or not. If there is sufficient deviation from the expected frequency found, it may indicate significant dependence. To measure the strength of association, calculation of coefficient is important. Cramer's V (available in SPSS and MegaStat) is most useful to check the association. The range of Cramer's V goes from zero (no association) to positive unity (maximum association). The value ranges from –1 to +1: 0.00–0.20 (no association), 0.20–0.60 (weak association), 0.60–0.80 (moderate association) and 0.80–1.00 (strong association). Cramer's V is calculated based on chi-square statistic (Upton and Cook, 2002). In Excel, the simplified formula is given as follows:

$$V = \sqrt{\frac{X^2}{n \times (k-1)}}$$

where, n = grand total and k = lower value among the variable counts for rows and columns.

Example 1

A consumer survey is done to check the association between the gender and the low-calorie ice-cream. A sample of consumer survey is taken, regarding whether they like low calorie ice cream over conventional or not.

	A	B	C	D	E	F	G	H	I
1	OBSERVED VALUE								
2			MALE	FEMALE	TOTAL	PERCENTAGE OF CHOICE			
3		LOW FAT (YES)	31	42	73	0.73			
4		LOW FAT (NO)	19	8	27	0.27			
5		TOTAL	50	50	100				
6									
7	EXPECTED VALUE		MALE	FEMALE					
8		LOW FAT (YES)	36.5	36.5					
9		LOW FAT (NO)	13.5	13.5					
10								Cramer's V	0.24777 (ASSOCIATION IS PRESENT)
11									
12			P VALUE	0.013223091					
13			Df	1					
14			chi-squared	6.139015728					
15									
16				p<0.05, so null hypothesis is rejected.					

Figure 4: Association analysis of two nominal variables (excel).

Generally, null hypothesis says that there is no association present between the gender and the low-calorie ice cream. Using excel, we are going to justify the above question with a suitable example.

From the above example, it has been found that the p value = .01322 < 0.05. This means the variables are not independent. Using the output from the analysis and the formula above, Cramer's V = 0.247. So, the study found that gender has a significant association with other variables.

7. Correlation

The possible association between two continuous variables can be measured by a method known as correlation. The correlation coefficient (r) is the measurement of degree of association between two variables (Oliveira et al., 2015). For ordinal or metric data, Spearman's correlation coefficient is appropriate, whereas, for ratio or interval data, pearson's (product moment) correlation coefficient is (r) is applicable (Bower, 2009). The value of r (x, y) is given by the following formula for a bivariate data:

$$R\ (x, y) = \frac{COV(X,Y)}{\sqrt{\sigma x^2 \sigma y^2}}$$

where, σ^2 = variance

The association between two variables ranges from –1 to +1. If the absolute value of r ($|r|$) = 0, it indicates that there is no correlation between the variables. Generally, zero correlation is shown by a cloud of points with no obvious trend. From a scatter plot, if the two variables show increment in positive direction, this is known as positive correlation. If one increases and other one decreases, then it is known as negative correlation. Based on some arbitrary scale, the strength of correlation is measured by the following criteria: Perfect ($|r|$ = 1.0)) strong ($0.80 \le |r| \le 1$)) moderate ($0.50 \le |r| \le 0.80$)) weak ($0.10 \le |r| \le 0.50$), very weak ($0.10 \le |r|$) (Granato et al., 2010).

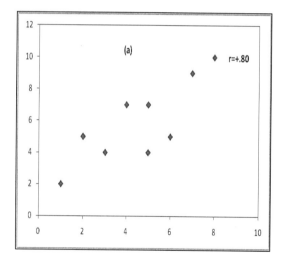
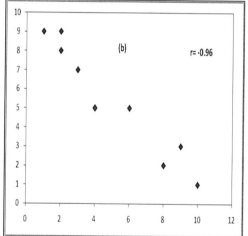

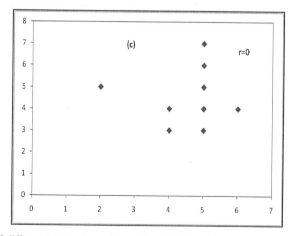

Figure 5: Scatter diagram of different levels of correlation (Excel) [(a) positive correlation, (b) negative correlation, (c) negative correlation].

When the sample size (n) is ≥ 30 and the data does not conform to a normal distribution, then a nonparametric approach is used, known as spearman's correlation coefficient. Spearman's correlation coefficient does not require the assumption that the relationship between two variables is linear (Granato et al., 2014). This may be marked by curvilinear appearance on the scatter diagram when there is no linear relationship between the variables. The data can be transformed to a linear pattern in order to avoid misleading in correlation. Some important features must be present in correlation studies (Bower, 2009).

- Adequate range of values
- Independent readings (no multi co-linearity)
- Linear relationship

The magnitude of correlation coefficient does not imply any statistical significance other than degree of association between variables. Sometimes, if the sample size is large, a scatter graph is useful for finding the outlier which could compromise the mean value. Once identified, corrective measures can be taken to minimize the error.

Generally, the associated measure is obtained between the variables by squaring the correlation coefficient, which is known as coefficient of determination (R^2). The variation in one variable by can be explained by the variation in the other variable. For multivariate case, if the number of variables is p and the observation under each variable is p, the correlation coefficient is expressed in the form of a Correlation matrix (p × p), given as follows:

$$R_{(p\times p)} = \begin{bmatrix} 1 & r_{12} & \cdots & r_{1p} \\ r_{21} & 1 & \cdots & r_{2p} \\ \cdots & \cdots & \ddots & \vdots \\ r_{p1} & r_{p2} & \cdots & 1 \end{bmatrix}$$

Only from correlation values it is difficult to judge statistical significance of correlation. Only p-value can justify the statistical significance of it.

Correlation analysis is performed by following steps:

- Collect data
- Draw scatter plot
- Calculate the coefficient
- Calculate statistical significant

Example 2

In a solar drying process, drying rate and moisture content data is collected for an elapsed time period. We need to justify whether there is a statistically significant correlation between these two variables or not.

From the above example, we can see that the correlation between drying rate and moisture is 0.9845, which is found from the correlation matrix (Pearson's correlation coefficient). From Regression, it was found that the p-value = $1.5E{-}06 < 0.05$, which is statistically significant within 95% confidence interval.

Correlation is widely-used in food science and technology for testing the relations between instrumental, sensory and consumer acceptance and also in nutrition in order to compare food intake with health-related issues. The main application of correlation is that it is an initial point for predictive method of regression.

8. Regression

Regression analysis is the extended part of correlation analysis. Generally, regression is the establishment of a linear relationship between predictor (independent) variable and response (dependent) variable. There are several types of regression based on the number of predictor and response variables. If there is one predictor and one response variable is known, it is simple linear regression. If more than one

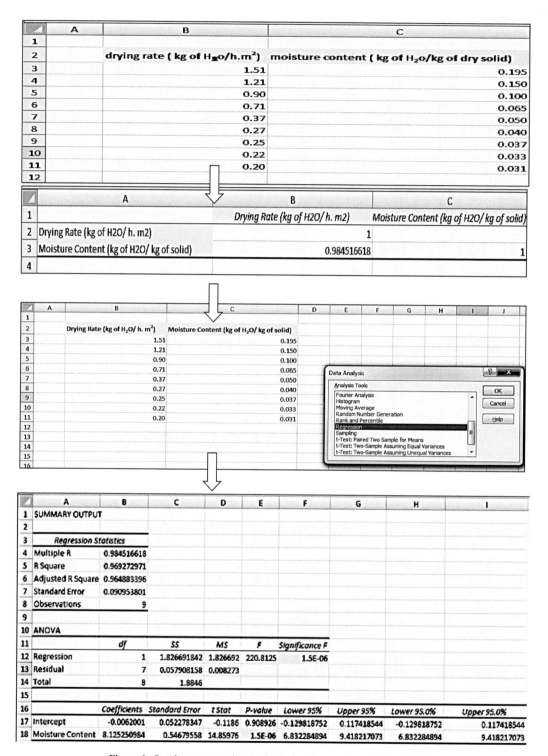

Figure 6: Step by step procedure to check significance of correlation (Excel).

predictor variable present, it is known as a multiple linear regression model (MLR). When more than one independent and dependable variable is present in a model, it is known as multivariate linear regression (MVLR).

In food science, regression is used in several places. Regression is used in microbial death kinetics in order to see the effect of treatment times (or temperatures) on the heat resistivity of microorganisms in order to calculate thermal effects (D-values) under certain conditions. Sometimes, regression model can be used also to establish calibration curves for continuous data sets in chemical and bio-chemical analysis (Granato et al., 2014). There are certain assumptions in the regression model. The distribution of errors should conform to a normal distribution with a mean = 0 and the variance of errors should be homoscedastic. The predictor variable should be linearly independent. The significant and the lack of fit of a regression model can be tested through F-test and one-way ANNOVA. Mathematically, regression is expressed as one variable being a function of another variable. For a simple linear regression, the regression equation is given as follows:

$$Y = \beta_0 + \beta_1 X + \varepsilon$$

where, β_0 = intercept, β_1 = slope of X, ε = error.

Here, our main goal is to minimize the error with the plotting of a best-fit line by using the method of least square. Residuals are the variation between the observed values and predicted values. For a multiple linear equation, the regression equation is given as follows (Johnson and Wichern, 2002).

$$Y = \beta_0 + \beta_1 X_1 + \beta_2 X_2 + \cdots + \beta_r X_r + \varepsilon$$

where, $X_1, X_2, ---, X_r$ = predictor variables

For a multivariate linear regression, the regression equation for n independent observation of Y with associated values of X_i (i = 1 to p) is given as follows (Johnson and Wichern, 2002):

$$Y_1 = \beta_0 + \beta_1 X_{11} + \beta a_2 X_{12} + \cdots + \beta_r X_{1p} + a$$
$$Y_2 = \beta_0 + \beta_1 X_{21} + \beta_2 X_{22} + \cdots + \beta_r X_{2p} + \varepsilon_2$$
$$\vdots \qquad \qquad \vdots$$
$$Y_n = \beta_0 + \beta_1 X_{n1} + \beta_2 X_{n2} + \cdots + \beta_r X_{np} + \varepsilon_n$$

if there are **q** number of Y variables and under each variable there are **n** number of observations. And if there are **p** no of independent X variables, then we can write as follows:

Data matrix = Design matrix × Coefficient matrix + Error matrix

---

where,

$$\text{Data matrix} = \mathbf{Y}_{n \times q} = \begin{bmatrix} y_{11} & y_{12} & \cdots & y_{1q} \\ y_{21} & y_{22} & \cdots & y_{2q} \\ \vdots & \vdots & \vdots & \vdots \\ y_{i1} & y_{i2} & \cdots & y_{iq} \\ \vdots & \vdots & \vdots & \vdots \\ y_{n1} & y_{n2} & \cdots & y_{nq} \end{bmatrix} \quad \text{Design matrix} = \mathbf{X}_{n \times (p+1)} = \begin{bmatrix} 1 & x_{11} & \cdots & x_{1p} \\ 1 & x_{21} & \cdots & x_{2p} \\ \vdots & \vdots & \vdots & \vdots \\ 1 & x_{i1} & \cdots & x_{ip} \\ \vdots & \vdots & \vdots & \vdots \\ 1 & x_{n1} & \cdots & x_{np} \end{bmatrix}$$

$$\text{Coefficient matrix} = \boldsymbol{\beta}_{(p+1) \times q} = \begin{bmatrix} \beta_{10} & \beta_{20} & \cdots & \beta_{q0} \\ \beta_{11} & \beta_{21} & \cdots & \beta_{q1} \\ \vdots & \vdots & \vdots & \vdots \\ \beta_{1k} & \beta_{2k} & \cdots & \beta_{qk} \\ \vdots & \vdots & \vdots & \vdots \\ \beta_{1p} & \beta_{2p} & \cdots & \beta_{qp} \end{bmatrix} \quad \text{Error matrix} = \boldsymbol{\varepsilon}_{n \times q} = \begin{bmatrix} \varepsilon_{11} & \varepsilon_{12} & \cdots & \varepsilon_{1q} \\ \varepsilon_{21} & \varepsilon_{22} & \cdots & \varepsilon_{2q} \\ \vdots & \vdots & \vdots & \vdots \\ \varepsilon_{i1} & \varepsilon_{i2} & \cdots & \varepsilon_{iq} \\ \vdots & \vdots & \vdots & \vdots \\ \varepsilon_{n1} & \varepsilon_{n2} & \cdots & \varepsilon_{nq} \end{bmatrix}$$

Applying OLS (Ordinary least square) method to the model equation in order to minimize the sum of square errors, we can get the value of $\hat{\beta}$ (estimated parameter). This is given as follows:

$$\hat{\beta} = (X^T X)^{-1} X^T Y$$

And estimated error can be given as:

$$\hat{\varepsilon} = Y - \hat{Y} \text{ [Where, } \hat{Y} = X\hat{\beta} \text{]}$$

According to test assumptions, the distribution of point estimator $(\hat{\beta})$ and estimated error $(\hat{\varepsilon})$ will be multivariate normal. To check the effect of each individual independent variable, we have to use Wilk's lambda test. Here, we only discuss the testing of each point estimator $(\hat{\beta})$. The effect of parameters can be tested from the following hypothesis:

$H_0: \beta_{kj} = 0$ (β has no certain contribution on Y)

$H_A: \beta_{kj} \neq 0$ (β has no certain contribution on Y)

Using t-distribution under hypothesis testing, we can say:

$$t_{calculated} = \frac{\hat{\beta}_{kj} - E(\hat{\beta}_{kj})}{SE(\hat{\beta}_{kj})!}$$

Here, $SE(\hat{\beta}_{kj})$ = standard error.

To satisfy the null hypothesis $t_{calculated} < t_{tabulated} (\alpha/2)$ at (n-p-1) degree of freedom (Where, $\alpha = 0.05$). For overall fit of MVLR, the following relationship is important:

> $SSCP_T$ *(total sum of squares and cross products)* $= SSCP_R$ *(regression sum of squares and cross products)* $+ SSCP_E$ *(regression sum of squares and cross products)*

It can also be represented by the following equation:

$$Y^T Y = \hat{Y}^T \hat{Y} + \hat{\varepsilon}^T \hat{\varepsilon}$$

Generally, a statistical significance study of regression analysis is required in order to build an appropriate model. The accuracy and the validation of the model are highly important. The standard deviation of the estimated model should be as small as possible. P-value obtained from F-test of ANNOVA should be < 0.05 to have statistical significance. Generally, a model is considered to be excellent if $R^2 > 0.90$ (Montgomery et al., 2011). This is one criterion to evaluate goodness of fit. However, sometimes it also depends upon the type of analysis. For sensory evaluation, the coefficient of determination is good if $R^2 > 0.60$ (Granato et al., 2014).

In general, we often make the common mistake of using R^2 to compare the models. The value of coefficient of determination will be always higher if the order of the model is increased (R^2 value will be higher in case of polynomial model rather than linear model). It is obvious due to the presence of more terms in higher order, but it does not mean that the higher value coefficient will ensure the better model than the lower order model. An analysis of the degrees of freedom comes into the picture. Another criterion, known as adjusted R^2 (R^2_{adj}), is used for this reason. This term adjusts the explanatory terms relative to the number of data points, which is less than or equal to R^2. Higher values of adjusted R^2 show the best model (Granato et al., 2014). This is significant in the case of multiple linear regression studies.

Sometimes testing of assumptions for normality comes into the picture. The residual error should be normal, independent and homoscedastic. The total sum of residual errors should be 0. The plot of residual error against predictor variable should not show curvature or funneling effect. In that case, transformation of dependable variable is needed. Generally, constant variability on both sides of the line, with respect to mean, is preferred. For testing of model adequacy, these are very important steps to perform.

Example 3

We want to check, for a particular drying process, whether any significant effect of moisture content, sample temperature and relative humidity on drying rate is present or not (using MLR).

From the above example, we have taken three independent variables: Moisture content, sample temperature and relative humidity. Among these, we can clearly see that the p-value of intercept and moisture content is significant as it is < 0.05. However, the effect of sample temperature and relative humidity is not significant since the p-value < 0.05, though the sample temperature has a high correlation with drying rate. If we omit the significance of sample temperature and relative humidity, the regression equation is given as:

$$drying\ rate = 1.14482767 + 7.051 \times moisture\ content$$

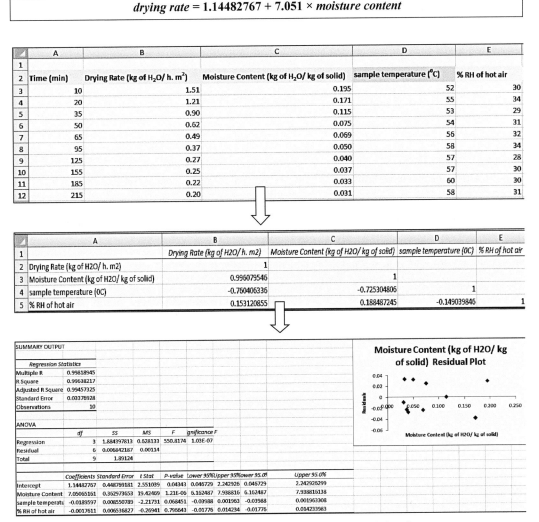

Figure 7: Step by step procedure to carry out MLR example (excel).

From the residual plot, it has been seen that there is a random pattern in scatter plot. The residual plot with respect to first predictor variable (moisture content) shows equity in variance either side of the zero line, which also perfectly matches the assumption criteria for residuals.

In food science, regression model analysis is applicable to sensory assessment. A researcher found a linear relation between raw and cooked fat (Rimal et al., 2001). A simple linear regression is used to predict the concentration of sucrose "ideal" and most "preferred sweetness" (Bower and Boyd, 2003).

In multivariate analysis, most effective procedures for defining an accurate model are PCA (principle component analysis), Clustering and FA (factor analysis). Here, however, we will only discuss multivariate multiple linear regression models.

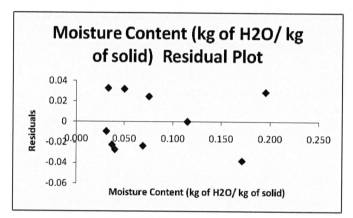

Figure 8: Residual plot for regression analysis (excel).

Example 4

From agricultural by-product waste, pulp fiber is collected and the paper is made from it. There are 10 observations of pulp fiber characteristics: X_1 = arithmetic fiber length, X_2 = long fiber fraction,

A	B	C	D	E	F	G	H	I	J	K	L	M	N	O	P	Q
	Y_1	Y_2	Y_3	X_0	X_1	X_2	X_3	Y_1_cap	Y_2_cap	Y_3_cap	Error ($\varepsilon 1$)	Error ($\varepsilon 2$)	Error ($\varepsilon 3$)			
	21.3120	7.0390	5.3260	1	35.2390	36.9910	1.0570	20.76653	6.888116	5.05964	0.5455	0.1509	0.2664			
	21.2060	6.9790	5.2370	1	35.7130	36.8510	1.0640	21.24889	6.987229	5.283993	-0.0429	-0.0082	-0.0470			
	20.7090	6.7790	5.0600	1	39.2200	30.5860	1.0530	20.58036	6.782127	4.973873	0.1286	-0.0031	0.0861			
	19.5420	6.6010	4.4790	1	39.7560	21.0720	1.0500	20.23525	6.74694	4.81666	-0.6933	-0.1459	-0.3377			
	20.4490	6.7950	4.9120	1	32.9910	36.5700	1.0490	20.12134	6.800631	4.760414	0.3277	-0.0056	0.1516			
	16.4410	6.3150	2.9970	1	2.8450	84.5540	1.0080	16.81894	6.527625	3.217103	-0.3779	-0.2126	-0.2201			
	16.2940	6.5720	3.0170	1	1.5150	81.9880	0.9980	16.04755	6.400682	2.859605	0.2464	0.1713	0.1574			
	20.2890	7.7190	4.8660	1	2.0540	8.7860	1.0810	20.21507	7.783832	4.826607	0.0739	-0.0648	0.0394			
	17.1630	7.0860	3.3960	1	3.0180	5.8550	1.0330	17.04234	7.050519	3.350803	0.1207	0.0355	0.0452			
	20.2890	7.4370	4.8590	1	17.6390	28.9340	1.0700	20.61774	7.354299	5.000301	-0.3287	0.0827	-0.1413			

Where, X_0=1 is the independent varriable correspondence to β_0

		10	209.99	372.187	10.463			523.1048	0.184428	-0.3479	-491.1873			193.694	69.322	44.149
	(x^Tx)=	209.99	7059.21	6774.239	221.4877		$(x^Tx)^{-1}$=	0.184428	0.000476	-6.2E-05	-0.183627		x^Ty=	4294.904	1446.108	1032.333
		372.187	6774.239	20263.29	384.5633			-0.3479	-6.2E-05	0.000388	0.3199322			6954.684	2509.768	1522.774
		10.463	221.4877	384.5633	10.95367			-491.187	-0.18363	0.319932	461.75647			203.0716	72.60921	46.38485

		-51.3553	-8.49456	-28.4769
β_CAP= $(x^Tx)'x'y$=		0.048837	-0.01447	0.022338
		0.018224	-0.00188	0.008103
		65.9666	15.10142	30.69979

Y_cap = Xβ_CAP

Y_1 = -51.3553+0.048837 X_1+0.018224 X_2+65.9666 X_3
Y_2 = -8.49456-0.01447 X_1-0.00188 X_2+15.10142 X_3
Y_3 = - 28.4769+0.022338 X_1+0.008103X_2+ 30.69979X_3

Figure 9: Multivariate regression model equation (excel).

Distribution of β

Cov (β) = (X'X)⁻¹.Σ						1.235549171	0.276462	0.618929653			
			Σ = cov(ε) =			0.276462281	0.131035	0.149859587			
						0.618929653	0.14986	0.314345614			
(X'X)' =	523.1048	0.184428	-0.3479	-491.1872758							
	0.184428	0.000476	-6.2E-05	-0.183626988							
	-0.3479	-6.2E-05	0.000388	0.319932182							
	-491.187	-0.18363	0.319932	461.7564743							

Cov (β) =

646.3217	144.6188	323.7651	0.227869801	0.050987372	0.114147927	-0.42985	-0.096181216	-0.215325601	-606.886	-135.7948	-304.01037
144.6188	68.545	78.39227	0.050987372	0.024166501	0.027638296	-0.09618	-0.045587042	-0.052136144	-135.7948	-64.36268	-73.609122
323.7651	78.39227	164.4357	0.114147927	0.027638296	0.057974117	-0.21533	-0.052136144	-0.109360826	-304.0104	-73.60912	-154.40257
0.22787	0.050987	0.114148	0.00058833	0.000131643	0.000294715	-7.6E-05	-1.70704E-05	-3.82164E-05	-0.22688	-0.050766	-0.1136522
0.050987	0.024167	0.027638	0.000131643	6.23947E-05	7.13585E-05	-1.7E-05	-8.09087E-06	-9.25321E-06	-0.050766	-0.024062	-0.0275183
0.114148	0.027638	0.057974	0.000294715	7.13585E-05	0.000149682	-3.8E-05	-9.25321E-06	-1.94095E-05	-0.113652	-0.027518	-0.0577223
-0.42985	-0.09618	-0.21533	-7.62901E-05	-1.70704E-05	-3.82164E-05	0.00048	0.000107347	0.000240322	0.3952919	0.0884492	0.1980155
-0.09618	-0.04559	-0.05214	-1.70704E-05	-8.09087E-06	-9.25321E-06	0.000107	5.08791E-05	5.81885E-05	0.0884492	0.0419223	0.0479449
-0.21533	-0.05214	-0.10936	-3.82164E-05	-9.25321E-06	-1.94095E-05	0.00024	5.81885E-05	0.000122056	0.1980155	0.0479449	0.1005693
-606.886	-135.795	-304.01	-0.226880173	-0.050765936	-0.113652188	0.395292	0.088449181	0.198015514	570.52283	127.65825	285.79477
-135.795	-64.3627	-73.6091	-0.050765936	-0.024061547	-0.027518265	0.088449	0.041922287	0.047944905	127.65825	60.506221	69.198634
-304.01	-73.6091	-154.403	-0.113652188	-0.027518265	-0.057722338	0.198016	0.047944905	0.100569278	285.79477	69.198634	145.15112

Figure 10: Distribution of parameter (excel).

X_3 = fine fiber fraction and the paper properties; Y_1 = elastic modulus, Y_2 = stress at failure, Y_3 = Burst strength. With these response variables, develop a multivariate regression model (MVLR) (Johnson and Wichern, 2002).

In Fig. 8, the above problem is solved in order to find out the regression equations. From Fig. 11, the distribution of parameter (β) and covariance structure have been found out. From hypothesis testing, it has been found out that only β = 65.9666 and 30.69979 has significant contribution on Y as they reject null hypothesis. It has been assumed for this case that, the normality, linearity and independence test for error are satisfied also.

Parameter testing

t_tab₆(0.025) = 2.4469

B_jk =	-51.3553	-8.49456	-28.4769		SE(B_jk) =	25.42	8.279	12.82
	0.048837	-0.01447	0.022338			0.024	0.00789	0.22
	0.018224	-0.00188	0.008103			0.022	0.00713	0.011
	65.9666	15.10142	30.69979			23.88	7.78	12.05

t_calculated (For, E{B_jk} = 0) =	-2.02	-1.03	-2.22
	2.03	-1.83	0.10
	0.83	-0.26	0.74
	2.76	1.94	2.55

Figure 11: Parameter testing (excel).

References

Anantheswaran, R.C. and Rao, M.A. 1985. Heat transfer to model Newtonian liquid foods in cans during end-over-end rotation. Journal of Food Engineering 4(1): 1–19.

Barreiro, J.A., Milano, M.F. and Sandoval, A.J. 1997. Kinetics of colour change of double concentrated tomato paste during thermal treatment. Journal of Food Engineering 33(3-4): 359–371.

Beebe, K.R., Blaser, W.W., Bredeweg, R.A., Chauvel, J.P., Harner, R.S., LaPack, M. et al. 1993. Process analytical chemistry. Analytical Chemistry 65(12): 199–216.

Blanco, M. and Villarroya, I.N.I.R. 2002. NIR spectroscopy: A rapid-response analytical tool. TrAC Trends in Analytical Chemistry 21(4): 240–250.

Bower, J.A. and Boyd, R. 2003. Effect of health concern and consumption patterns on measures of sweetness by hedonic and just-about-right scales. Journal of Sensory Studies 18(3): 235–248.

Bower, J.A. 2009. Association, correlation and regression. pp. 137–149. *In*: Statistical Methods for Food Science: Introductory Procedures for the Food Practitioner, 2nd edition (ISBN: 978-1-118-54162-3).

Cozzolino, Daniel and Gastón Ares. 2013. The use of correlation, association and regression to analyse processes and products. Mathematical and Statistical Methods in Food Science and Technology 19–30.

de Oliveira, C.C., Calado, V.M., Ares, G. and Granato, D. 2015. Statistical approaches to assess the association between phenolic compounds and the *in vitro* antioxidant activity of *Camellia sinensis* and *Ilex paraguariensis teas*. Critical Reviews in Food Science and Nutrition 55(10): 1456–1473.

Granato, Daniel et al. 2010. Physical stability assessment and sensory optimization of a dairy-free emulsion using response surface methodology. Journal of Food Science 75(3).

Granato, Daniel, Verônica Maria de Araújo Calado and Basil Jarvis. 2014. Observations on the use of statistical methods in food science and technology. Food Research International 55: 137–149.

Hollander, Myles, Douglas A. Wolfe and Eric Chicken. 2013. Nonparametric Statistical Methods. Vol. 751. John Wiley & Sons.

Huang, H., Yu, H., Xu, H. and Ying, Y. 2008. Near infrared spectroscopy for on/in-line monitoring of quality in foods and beverages: A review. Journal of Food Engineering 87(3): 303–313.

Jarvis, Basil. 2016. Statistical aspects of the microbiological examination of foods. Academic Press.

Johnson, Richard A. and Dean Wichern. 2002. Multivariate Analysis. John Wiley & Sons, Ltd.

Jongenburger, Ida. 2012. Distributions of microorganisms in foods and their impact on food safety.

Kueppers, Stephan and Markus Haider. 2003. Process analytical chemistry—future trends in industry. Analytical and Bioanalytical Chemistry 376(3): 313–315.

Liu, Yen-Chun, Feng-Sheng Wang and Wen-Chien Lee. 2001. On-line monitoring and controlling system for fermentation processes. Biochemical Engineering Journal 7(1): 17–25.

Montgomery, Douglas C. 2009. Introduction to Statistical Quality Control. John Wiley & Sons (New York).

Montgomery, Douglas C., George C. Runger and Norma Faris Hubele. 2011. Engineering statistics. Wiley.

Razali, Nornadiah Mohd and Yap Bee Wah. 2011. Power comparisons of Shapiro-Wilk, Kolmogorov-Smirnov, Lilliefors and Anderson-Darling tests. Journal of Statistical Modeling and Analytics 2(1): 21–33.

Rimal, A., Fletcher, S.M., McWatters, K.H., Mishra, S.K. and Deodhar, S. 2001. Perception of food safety and changes in food consumption habits: a consumer analysis. International Journal of Consumer Studies 25(1): 43–52.

Roggo, Y., Chalus, P., Maurer, L., Lema-Maryinez, C., Edmond, A. and Jent, N. 2007. A review of near infrared spectroscopy and chemometrics in pharmaceutical technologies. Journal of Pharmaceutical and Biomedical Analysis 44(3): 683–700.

Samaniego-Esguerra, Christine M., Ian F. Boag and Gordon L. Robertson. 1991. Comparison of regression methods for fitting the GAB model to the moisture isotherms of some dried fruit and vegetables. Journal of Food Engineering 13(2): 115–133.

Shapiro, Samuel Sanford and Martin B. Wilk. 1965. An analysis of variance test for normality (complete samples). Biometrika 52(3/4): 591–611.

Siegal, Sidney. 1956. Nonparametric Statistics for the Behavioral Sciences. McGraw-hill.

Upton, G. and Cook, I. 2002. Reliability. A Dictionary of Statistics.

CHAPTER 10

Regression Analysis Methods for Agri-Food Quality and Safety Evaluations Using Near-Infrared (NIR) Hyperspectral Imaging

Chandra B Singh[1] and *Digvir S Jayas*[2]*

1. Introduction

The agri-food processing industry continues to address the increasing consumer demand for high quality, safe, nutritious, minimally-processed food products through the development of innovative food processing technologies and improved processing methods. However, it is quite complex to understand the interactions among the developed food products, processes and associated technologies which are critical to optimizing the food production and processing. To optimize the processes in food systems, statistical approaches and mathematical modelling are used as powerful analytical tools. Researchers in food science and engineering often use regression analyses to establish a relationship between measured/ observed variables (also called predictors/factors), which are relatively easy to collect, and predicted variables called responses, which are labour intensive and difficult to measure. A simple linear regression (SLR) model is built by linearly regressing the measured/observed variables against a response variable in order to predict the properties, which are difficult to measure. Simple linear regression (SLR) is rarely used in solving practical food engineering research problems as there is more than one predictor variable involved in the experimental design. However, it is important to understand the concept of SLR, as it is still fundamentally relevant in multiple linear regression (MLR). In MLR, more than one predictor variable (multiple) is used in regression analysis to predict the behaviour of response. Simple MLR models are based on the assumption of non-collinearity. However, in many applications, such as NIR spectroscopy, measured variables (e.g., absorbance, reflectance) at hundreds of wavelengths are likely to be highly collinear. In such scenarios, the simple MLR would fail to construct an accurate predictive model. Linear regression methods based on least square criteria give highly unstable and unreliable

[1] School of Engineering, University of South Australia, Adelaide, SA, Australia.
  Email: Chandra.Singh@unisa.edu.au
[2] Biosystems Engineering, University of Manitoba, Winnipeg, MB, Canada.
* Corresponding author: Digvir.Jayas@umanitoba.ca

regression coefficients and predictions if collinear variables are used (Naes et al., 2002). Principal component analysis (PCA) and partial least squares (PLS) are used as effective tools to construct accurate predictive models when there are many predictors with high collinearity. Both PCA and PLS linearly transform original collinear variables into few non-collinear independent components which are used in regression models. The PCA is a well-known multivariate data analysis technique that can identify and eliminate variables with insignificant effect on the response (Singh, 2009). The PCA can transform large data sets (variables) into few independent orthogonal principal components that capture the maximum variance of the original data (predictor variables). These independent components are used as predictors to develop principal component regression (PCR) models. In PCR, the components are selected based on the percentage of variability captured from original data by respective components which may not have significant information to develop relationships with response variables. In PLS, this issue is tackled by selecting components based on maximum co-variance between predictors and response variables. The PCA and PLS have been successfully used as multivariate methods for calibration of multivariate models, batch processes control, quality control variables that are difficult to measure online, detection of faults and process irregularities, treatment of missing data, and in industrial predictive modelling (Godoy et al., 2014).

Hyperspectral imaging has been extensively researched in agriculture and food for quality evaluation of grains (Singh, 2009; McGoverin et al., 2011; Xing et al., 2011), fruits (Chen et al., 2015; Li et al., 2018), vegetables (Ariana et al., 2006) and meat products (Kamruzzaman et al., 2011). Hyperspectral imaging extends the capabilities of NIR spectroscopy in spatial domain by allowing spectral analysis of each pixel of the scanned sample. However, this is achieved at the cost of a huge increase in data size (spectral). For example, a 640 (H) by 512 (V) size area scan hyperspectral sensor used to scan a food sample in NIR region will collect 327680 (640 × 512 = 327680) spectra instead of a single spectrum collected from NIR spectrophotometer in the same NIR region. Such large increase in hyperspectral data adds redundancy and makes it difficult to handle and analyse the data and develop regression models. The statistical approaches discussed in this chapter are used to eliminate redundancy in hyperspectral data and develop accurate regression models.

The principles and limitations of SLR, MLR, PCR, and PLS are discussed in the following sections. Readers are expected to have basic knowledge of linear algebra and matrices which are critical for understanding of this topic. Near-infrared spectroscopy and hyperspectral imaging were used as applications of regression methods for agri-food quality and safety evaluations in this chapter.

2. Regression Methods

2.1 Simple linear regression (SLR)

The simplest form of regression is linear regression analysis, which develops a linear relationship between two variables. One variable (x) is called the predictor, explanatory, or independent variable and the other variable (y) is referred to as the response or dependent variable. The relationship between two variables is developed in the form of the best fitting regression line that fits the data best with minimum prediction error for the observed data set. This approach ensures that the prediction error for each piece of data is as small as possible, which is determined by using the criteria of least square error (LSE). The prediction errors of data points are squared to make the value "absolute", otherwise the total sum of the negative and positive errors could make the sum zero.

In simple linear regression (SLR), there is only one independent variable (x) whose values are collected at m observations which form a column vector of size ($m \times 1$). The values of dependent variable or response (y) are measured corresponding to each observation, these also form a column vector of length ($m \times 1$).

The simple regression model is built as shown in regression Eq. 1:

$$y_i = b_0 + b_1 x_i \tag{1}$$

where, y_i is the predicted response for each observation i (i = 1, 2,...m), b_0 and b_1 are intercept and slope of the fitted line, respectively.

Thus, the sample residual error (observed (O) – predicted response (y)) of each response can be calculated by Eq. 2 as:

$$e_i = O_i - y_i \tag{2}$$

The *LSE* for m observations can be calculated by Eq. 3 as:

$$LSE = \sum_{i=1}^{i=m}(O_i - y_i)^2 \tag{3}$$

LSE is minimized in order to determine the coefficients. Other criteria, such as error variance, coefficient of determination, and correlation coefficient are also used to find the best fitted regression lines.

In Eq. 1, the regression coefficients, b_0 and b_1, are estimated from a sample and not from a true population. However, researchers are keen to establish such a relationship for a true population. The SLR model that estimates the population regression coefficients from the predictors and mean of response is given by Eq. 4 as:

$$E(y_i) = \beta_0 + \beta_1 x_i + \varepsilon_i \tag{4}$$

where, $E(y_i)$ is the predicted response of population mean and β_0 and β_1 are population regression coefficients. Since, it is nearly impossible to collect all the predictor values from a true population, population regression coefficients β_0 and β_1 are estimated from sample regression coefficients b_0 and b_1, respectively. Also, it is important to note that each response will be different than the predicted population mean response, so an error term is introduced into population SLR model in Eq. 4. In SLR model it is assumed that errors ε_i are independent, normally distributed and have equal variances. An example in Fig. 1 shows the relationship between counted (insect fragments) and predicted insect fragments in semolina by regression analysis with 95% confidence intervals (Bhuvaneswari et al., 2011).

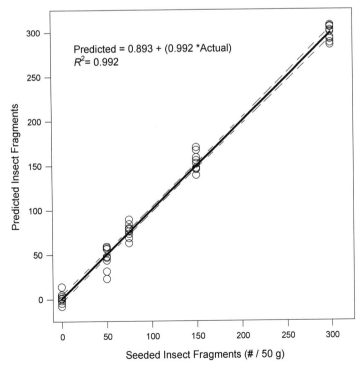

Figure 1:   Correlation between actual (seeded insect fragments) and predicted levels of insect fragments in semolina samples by partial least square (PLS), regression line with 95% confidence intervals (Bhuvaneswari et al., 2011).

2.2 Multiple linear regression (MLR)

Despite being simple, using one independent variable in SLR analysis runs the risks of high inaccuracy due to the potential instrumentation error in measuring predictor variable, poor signal, or non-collinearity. Therefore, in practical food analysis applications, more than one independent observation is recorded corresponding to each response variable. In this case, the dependent variable or response (y) is still a vector of length ($m \times 1$) but independent variable or predictor x is a full matrix of size ($m \times k$) with each element denoted as x_{ik}, where k is the number of predictor variables for each of m observations ($i = 1, 2.. $ m). In this case, the regression analysis can be performed using multiple linear regression (MLR).

The MLR model can be represented by Eq. 5 as:

$$y_i = b_0 + b_1 x_{i1} + + b_2 x_{i2} + b_3 x_{i3} + \cdots b_k x_{ik} \tag{5}$$

where, y_i is the predicted response for each observation i ($i = 1, 2.. $ m) and $b_0, b_1, b_2, .. b_k$ are intercept and slopes of the predictors, respectively.

The MLR model for estimating response from a population can be expressed as:

$$E(y_i) = \beta_0 + \beta_1 x_{i1} + \beta_2 x_{i2} + \beta_3 x_{i3} + \cdots \beta_k x_{ik} + \varepsilon_i \tag{6}$$

where, $E(y_i)$ is the predicted response of population mean, β_0 is intercept and $\beta_1, \beta_2, \beta_3, .. \beta_k$ are estimated population regression coefficients.

Equation 6 can be simplified into a matrix for easy understanding of analyses and to incorporate large data sets with k predictor variables for $i = 1, 2, ...m$ observations:

$$y_1 = \beta_0 + \beta_1 x_{11} + \beta_2 x_{12} + \cdots \beta_k x_{1k} + \varepsilon_1 \tag{7}$$

$$y_2 = \beta_0 + \beta_1 x_{21} + \beta_2 x_{22} + \cdots \beta_k x_{2k} + \varepsilon_2 \tag{8}$$

$$y_m = \beta_0 + \beta_1 x_{m1} + \beta_2 x_{m2} + \cdots \beta_k x_{mk} + \varepsilon_m \tag{9}$$

$$\begin{bmatrix} y_1 \\ y_2 \\ \vdots \\ y_m \end{bmatrix} = \begin{bmatrix} 1 & x_{11} & x_{12} & \cdots & x_{1k} \\ 1 & x_{21} & x_{22} & \cdots & x_{2k} \\ \vdots & \vdots & \vdots & \vdots & \vdots \\ 1 & x_{n1} & x_{n2} & \cdots & x_{nk} \end{bmatrix} \begin{bmatrix} \beta_0 \\ \beta_1 \\ \beta_2 \\ \vdots \\ \beta_k \end{bmatrix} + \begin{bmatrix} \varepsilon_1 \\ \varepsilon_2 \\ \vdots \\ \varepsilon_m \end{bmatrix} \tag{10}$$

Equation 10 can be generalized as:

$$Y = X\beta + \varepsilon \tag{11}$$

The regression coefficient parameters (β) can be estimated from Eq. 12 as:

$$\beta = (X^T X)^{-1} X^T Y \tag{12}$$

where X^T is transpose of X and superscript -1 shows the inverse of a matrix.

If all the predictor variables can be controlled to select discrete values of predictors, orthogonality can be easily enforced so that matrix $X^T X$ becomes a diagonal matrix and regression coefficients β can be easily calculated from Eq. 12 (Rajalahti and Kvalheim, 2011). However, if the number of predictors exceed the measured responses, it may cause a co-linearity problem and the inverse of $X^T X$ cannot be calculated. To resolve the co-linearity issue, and if the number of predictor variables is greater than the experimental responses, PCR and PLS regression are preferred as alternative methods.

2.3 Principal Component Regression (PCR)

The co-linearity problem in large size predictor variable data can be resolved by PCR. Instead of using the original predictor values in regression analysis, orthogonally transformed principal component (PC) scores from PCA are used.

2.3.1 Principal component analysis (PCA)

In PCA, the data matrix (composed of predictors) is orthogonally transformed into fewer independent PCs that explain maximum variance in original data. In mathematical terms, the data matrix is decomposed into few simpler matrices using the score vector S_a and the loading vector p_a. In latent variable space, S_a are the vectors of size $m \times 1$ that carry the information about samples in variable space and p_a are the vectors of size $k \times 1$ that carry information about variable in object space (Rajalahti and Kvalheim, 2011). The decomposition of matrix X into scores and loadings for A number of PCs is represented by:

$$X = SP^T + \varepsilon = \sum_{a=1}^{A} s_a p_a^T = s_1 p_1^T + s_2 p_2^T + \cdots \ldots s_A p_A^T + \varepsilon \tag{13}$$

where, S is a score matrix of size $m \times A$ which can be thought of as transformation of original $X(m \times k)$ variables into new coordination systems called loadings P, which have dimensions of k by A. Score matrix S captures maximum variation in X. Transpose of loadings (P^T) is a matrix of dimension $A \times k$.

Scores and loading vectors of matrix X can be calculated by using the concept of eigenvectors and eigenvalues explained in the following section.

2.3.2 Eigenvectors and eigenvalues

Eigenvectors and eigenvalues are fundamental concepts used in PCA and other statistical data analyses methods. Eigenvectors are considered a special case of multiplication of two matrices of compatible sizes as per the rules of linear algebra. For example, if you multiply a square matrix M of size $b \times b$ by a column vector z of size $b \times 1$, the multiplication results in a scaler (λ) times the original vector z. The column vector in this transformation (multiplication) is called eigenvector and has been transformed from its original position (co-ordinates) by a scaler value called the eigenvalue.

The eigenvector and eigenvalue relationship is represented by:

$$Mz = \lambda z \tag{14}$$

Eigenvectors and eigenvalues have specific characteristics: Eigenvectors can only be determined for square matrices and not every square matrix has eigenvectors. The maximum number of eigenvectors that can be found for a square matrix is equal to its dimension, i.e., matrix M ($b \times b$) can have maximum b eigenvectors, all the eigenvectors are orthogonal (perpendicular) to each other and have unit length, and eigenvalue (λ) is always a scalar.

In PCA, the loading vectors of matrix X are same as eigenvectors of $X^T X$ and λ is an eigenvalue of $X^T X$. The first PC is the loading vector or eigenvector with the highest eigenvalue and the second PC is the eigenvector with second highest value, and so on. Therefore, principal components are ranked and selected based on the highest to the lowest eigenvalues (absolute) or based on the percentage of the cumulative sum of squares or the total sum of square of all eigenvalues. The PC loadings and scores are also used as data visualization tools to look for similarities and patterns in multidimensional data.

Scores can be calculated from loadings as:

$$s_a = Xp_a \tag{15}$$

similarly,

$$S = XP \tag{16}$$

As can be observed, Eq. 16 is the inverse of generalized Eq. 13.

The score matrix S for A selected components is used for developing linear PCR model similar to MLR described in Section 2.2. It is important to observe that, if we use scores of all the principal components in matrix S instead of the selected most significant component, it will be equivalent to the full original matrix X and the regression model will be simply an MLR.

In PCA, covariation and correlation matrix of X^TX are often used to calculate eigenvectors and eigenvalues. Readers are expected to have knowledge of calculating covariance and correlation from a matrix. Converting X^TX to a covariance matrix removes absolute average levels and it may be advantageous where average values of predictors are of little or no concern (Geladi and Grahn, 1996). Correlation matrix of X^TX is calculated by variable-wise mean-centering, followed by scaling with standard deviation. Sometimes scaling may help to reduce the noise in variables. Usually, researchers try these methods in regression modelling to see if they are helpful or not in improving the model performance as data characteristics can vary across the experiments.

2.4 Partial least square (PLS) regression

Partial least square (PLS) regression is suited for situations where there are many variables (predictors) but a limited number of samples or observations (Hoskuldsson, 1988). PLS is more robust than MLR and PCR as model parameters do not change significantly when new calibration samples are used (Geladi and Kowalski, 1986). The PLS model selects a relatively small number of variables in order to develop a stable prediction model. Similar to PCA, PLS also selects principal components but uses the maximum covariance between response (y) and predictor (x) variable instead of variance as the selection criteria. First, weight vectors are calculated, then scores are calculated from weight vectors.

Normalized weight vector (w) is calculated as covariance:

$$W^T = \frac{Y^TX}{\|Y^TX\|} \tag{17}$$

The scores can be calculated from weight vectors by projecting X on weight vector as:

$$s_a = Xw_a \tag{18}$$

similarly.

The PLS loadings are obtained by project variable matrix on weight vector as:

$$p_a^T = \frac{s_a^TX_a}{s_a^Ts_a} \tag{19}$$

In PLS regression, each successive component is evaluated for its prediction ability and part of variable matrix X explained by scores and loadings by respective components are subtracted before next set of scores and loading is calculated. The iterations continue until a predefined stopping criterion is met or convergence occurs. This method of convergence is called nonlinear iterative partial least squares (NIPALS) (Wold et al., 1987; Geladi and Kowalski, 1986). As per the above description, the matrix X is substituted with X_{new} (Eq. 20), next, a set of loadings and scores are calculated using X_{new}, and iteration continues until convergence occurs.

$$X_{new} = X_a - s_ap_a^T \tag{20}$$

The NIPALS decomposition algorithm (Eqs. 17–20) can also be used for convergence of PCA, except that weight vectors are same as loadings in PCA (Rajalahti and Kvalheim, 2011).

The PLS equations above describe the regression model for one response variable, however, PLS regression is capable of handling several response variables simultaneously. Regression with several response variables is similar, but instead of maximizing the covariance between response variables and linear function of predictors, covariance between two linear functions, both in response and predictor, is optimized. PLS regression using multiple response variables is described in detail by Höskuldsson (1988).

3. Applications

The applications section is reproduced from the unpublished Ph.D. dissertation of the lead author. Quotes around the text are direct reproductions from Singh (2009).

3.1 Near-infrared (NIR) spectroscopy

"Near-infrared (NIR) spectroscopic instruments record the absorption of NIR radiation by a material at a specified wavelength range with a narrow resolution". Figure 2 shows an NIR mean reflectance spectra of healthy and midge-damaged wheat kernels for different locations in western Canada. In NIR region, molecules absorb energy (radiation) which causes vibration at specific frequencies, these are related to distinct chemical compositions of materials (Murray and Williams, 1990). "Reflected or transmitted light is collected by a spectrometer (detector) and related to the energy absorbed by the sample by transforming the reflected or transmitted radiation into absorbance". The NIR spectra consist of several overtones and combination bands of fundamental vibrations that occur in the NIR region. Several peaks in NIR spectra are related to the concentration of a specific constituent of the analysed material. However, broad and highly overlapping peaks in NIR spectra make spectral analysis and extraction of chemical and physical information very challenging. Chemometric methods, such as PCA, PCR and PLS are widely used as qualitative and quantitative analytical tools in NIR spectroscopy. Various applications of NIR spectroscopy for the grain quality evaluation given in Table 1 are discussed below.

"Delwiche and Masie (1996) classified kernels of five wheat classes (hard white, hard red spring (HRS), hard red winter (HRW), soft red winter (SRW), and soft white) using visible and NIR reflectance

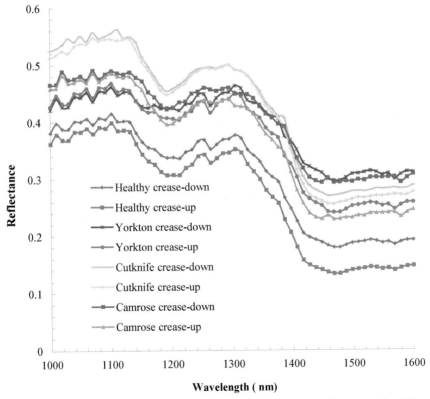

Figure 2: Near-infrared (NIR) mean reflectance spectra of healthy and midge-damaged wheat kernels for different locations in western Canada (Singh et al., 2009).

Table 1: Examples of applications of PCA, PCR, and PLS in near-infrared (NIR) spectroscopy in grain quality analysis.

Product	Objective	Wavelength range (nm)	Data analysis*	Reference
Wheat	Cultivar classification	551–750, 1120–2476	PLS, MLR	Delwiche and Masie (1996)
Wheat	Color classification	500–1900	PLS	Wang et al. (1999); Ram et al. (2002)
Wheat	Hardness	1100–2498 400–1700	MLR, PCA, PLS	Delwiche and Norris (1993); Maghirang and Dowell (2003)
Wheat	Vitreousness	400–1700	PLS	Dowell (2000); Wang et al. (2002)
Wheat	Insect	400–2500	PLS	Maghirang et al. (2003)
Soybean	Fungal damage	490–1690	PLS	Wang et al. (2003)
Corn	Mycotoxins	500–1700	PLS	Pearson et al. (2001)
Corn	Carotenoid	400–2498	PLS	Berardo et al. (2004)
Wheat	Protein composition	400–2498	PLS	Wesley et al. (2001)
Cereal foods	Fat content	1100–1700	PLS	Kays et al. (2005)

* PLS: Partial least square regression; MLR: Multiple linear regression; PCA: Principal component analysis.

characteristics obtained from single kernels. Their PLS-based binary calibration models using both VIS and NIR regions gave an accuracy of 78–99%. Wang et al. (1999) classified wheat kernels based on kernel colours. The PLS calibration model using absorbance spectra in the region 500–1700 nm correctly classified 99.8% and 98.4% kernels in a calibration set and a test set, respectively". Ram et al. (2002) classified wheat kernels using PLS for 490–750 nm wavelength region. "Delwiche and Norris (1993) developed a calibration model to classify the hard red wheat into HRS and HRW wheat, based on growing season. They applied Mahalanobis discriminant analysis to the scores of principal components and achieved a maximum accuracy of 95%. Maghirang and Dowell (2003) measured hardness of bulk wheat by calibrating the spectral data of single kernels in the range of 400–1700 nm. They analyzed the spectra by applying PLS on the averaged spectra and more than 97% wheat kernels were accurately classified as soft or hard".

"Vitreousness of hard wheat is the glossy or shiny appearance of the wheat kernel. Vitreousness affects the milling performance of durum wheat (semolina yield) and quality of some products from semolina (e.g., pasta). Dowell (2000) used NIR spectroscopy (400–1700 nm) to classify single wheat kernels into vitreous and non-vitreous classes. Classification algorithms were developed by PLS using mean-centred diffuse reflectance spectra. He also investigated the effect of protein, starch and hardness on vitreousness and on classification by excluding wavelength regions 750–1250 nm and 1400–1700 nm (significant for protein, starch and hardness) from calibration and found that information outside these wavelength regions is useful for vitreousness classification. Wang et al. (2002) further sub-classified HRS wheat in dark hard vitreous (DHV) and non-dark hard vitreous (NDHV) kernels by analysing the spectra from single DHV and NDHV kernels obtained using a diode-array spectrometer." Their PLS model gave good prediction results.

"Maghirang et al. (2003) detected insect infestation by different growth stages (pupae, large larvae, medium larvae and small larvae) of both dead and live rice weevils using a spectrometer in the wavelength range of 400–1700 nm". The PLS calibration model was evaluated by coefficient of determination (r^2), standard error of cross validation (SECV), and beta coefficient. They used SECV to determine the optimum number of independent variables in developing a calibration model. Beta coefficients with higher values in the plot indicated the significant wavelengths.

Wang et al. (2003) used NIR spectroscopy to segregate healthy and fungal-damaged soybean seeds and identify various types of fungal damage. A two-class PLS model accurately detected over 99% of fungus-damaged soybean seeds. Pearson et al. (2001) used NIR spectroscopy to detect aflatoxin contamination in corn kernels. They used the standard AflaTest affinity chromatography procedure to measure the aflatoxin content in corn for reference data. They developed classification models using PLS and discriminant analyses. Their models accurately identified nearly 95% of corn kernels contaminated with either high (> 100 ppb) or low (< 10 ppb) levels of aflatoxin.

Compositional analysis of cereal grains and grain products is essential quality evaluation criteria. "Berardo et al. (2004) measured the carotenoid concentrations in maize by using NIR spectroscopy and collecting reference values by high performance liquid chromatography (HPLC)". They developed a PLS calibration model to predict carotenoid concentrations with NIR spectroscopy. Predicted results were very satisfactory in comparison to the HPLC method (r^2 value ranged from 0.82 for lutein to 0.94 for zeaxanthin). "In another study, Wesley et al. (2001) developed a model to predict the protein composition (gliadin and glutenin) in wheat flour using NIR spectroscopy. Reference gliadin and glutenin content in wheat was determined by size-exclusive, high-performance liquid chromatography (SE-HPLC) technique and then a PLS calibration model was developed and results were compared. However, both of their methods did not give very good results. The model performance was affected by the high co-linearity of total protein content to gliadin and glutenin content. Kays et al. (2005) predicted the fat content of several intact cereal food samples by NIR spectroscopy. They collected the reference data by solvent extraction method and developed PLS calibration models".

3.2 Hyperspectral imaging

Hyperspectral imaging is an imaging technique that collects the full spectra of each pixel of the scanned sample. This spatially distributed spectral information of scanned material, termed as a hypercube, is described by a three-dimensional matrix of size m × n × λ, where m and n are the spatial dimensions (pixels) in x and y coordinates and λ is the spectral dimension (wavelength) (Fig. 3). Dimensionality of hypercube is reduced by applying image processing (in spatial dimension) combined with analytical tools in chemometrics (in spectral dimension) and distinct featural information is extracted in order to develop classification and calibration models (Grahn and Geladi, 2007; Geladi and Grahn, 1996). Multivariate image analysis (MVI), which uses PCA as the data reduction and feature extraction tool, is popularly used in hyperspectral imaging. The hyperspectral imaging data (hypercube) is reshaped into a two-dimensional array by sequentially rearranging all the spatial information (intensities) at each of the λ wavelengths into a column, as shown in Fig. 4 for scanned grain kernels. This results in a (m × n) × λ sized two-dimensional array, where m × n is the total number of pixels in a sample. The PCA, as explained in the previous section, is then applied to the reshaped two-dimensional data and each pixel

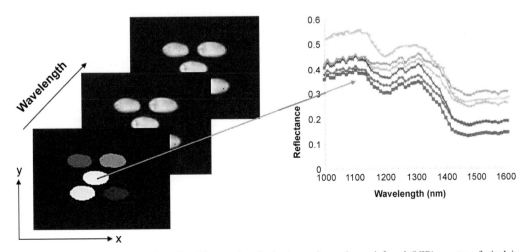

Figure 3: Schematic of a three-dimensional hypercube of wheat samples and near-infrared (NIR) spectra of pixel in hypercube.

Color version at the end of the book

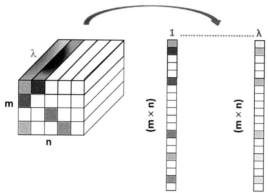

Figure 4: Reshaping three-dimensional hypercube into two-dimensional matrix for principal component analysis (PCA).

Color version at the end of the book

(row) becomes a sample and wavelength (column) becomes a variable. The PCA transforms this matrix into eigenvalues, eigenvectors and PC scores which are selectively used for calibration and classification.

Table 2 summarizes applications of PCA, PCR and PLS in hyperspectral imaging for quality evaluation of agriculture and food products. Hyperspectral imaging requires image processing and spectral pre-treatment techniques before applying PCA and PLS, however, the scope of this chapter is limited to PCA, PCR, and PLS only and further reading of the listed references is suggested for spectral and spatial pre-processing details. Singh (2009) investigated the application of hyperspectral imaging for detecting insect, mold, midge and sprout damage in wheat. In hyperspectral imaging, the important task is to identify the most significant wavelengths. The PCA applied to reshaped data of single wheat kernels (hypercube) was used to identify the most significant wavelengths. The most significant wavelengths were selected based on the highest factor loadings of the first PC, which captured over 90% variance from original hypercube, and used for feature extraction (maximum, minimum, mean, median, standard deviation, and variance). Figure 5 shows the plot of first PC factor loadings of sound and fungus-damaged wheat kernels. The extracted featured were used in discriminant analysis in order to detect damaged wheat kernels. The PCA is also used as data visualization technique such as score mapping or cluster analysis. Figure 6 shows the first PC score image of healthy and sprouted and midge-damaged wheat kernels. The sound kernels have solid contours and a germ that shows relatively higher and concentrated intensity is

Table 2: Examples of applications of PCA, PCR, and PLS in near-infrared (NIR) hyperspectral imaging for quality and safety evaluation of agri-food products.

Product	Analysis	Wavelength range (nm)	Classification method*	Reference
Wheat	Insects, molds, midge and sprout damage	400–900, 1000–1600	PCA	Singh (2009)
Semolina	Insect fragment	900–1700	PLS	Bhuvaneswari et al. (2011)
Strawberry	Quality attributes	400–1000	PLS, MLR	ElMasry et al. (2007)
Apple	Surface contamination	450–851	PCA	Kim et al. (2002)
Apple	Bruise detection	900–1700	PCA	Lu (2003)
Cucumber	Chilling injury	450–950	PCA	Liu et al. (2006)
Poultry	Contamination	400–2498, 430–900	PCA	Lawrence et al. (2003)
Wheat	Vitreousness	650–1100	PLS	Gorretta et al. (2006)
Maize	Moisture & protein	750–1090	PCR, PLS	Cogdill et al. (2004)
Mushroom	Freeze damage	400–1000	PCA	Gowen et al. (2009)

* PCA: Principal component analysis; PLS: Partial least square; MLR: Multiple linear regression; PCR: Principal component regression.

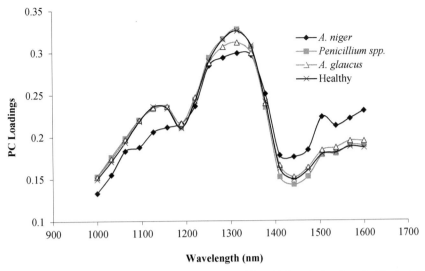

Figure 5: First principal component (PC) loadings of healthy and fungal infected wheat kernels (Singh et al., 2007).

Color version at the end of the book

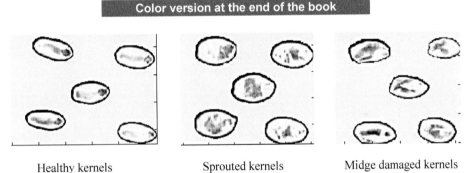

| Healthy kernels | Sprouted kernels | Midge damaged kernels |

Figure 6: First principal component (PC) scores images of healthy, sprouted, and midge-damaged wheat kernels (Singh et al., 2009).

Color version at the end of the book

clearly visible; however, in sprouted and midge-damaged wheat kernels, the germ does not show a similar pattern. Score plots were also used to separate mold-damaged wheat kernels from healthy wheat tissues using infrared hyperspectral data (Singh et al., 2011) (infrared is not listed in Table 2). Figure 7 shows score plot of first PC1 (82% variance) and second PC2 (8% variance) with clear separation of healthy wheat from damaged wheat.

Bhuvaneswari et al. (2011) used hyperspectral imaging to detect insect fragments in semolina and developed a PLS calibration model. The 15 most important wavelengths were selected based on the highest beta coefficient values that significantly contributed to the PLS regression. A plot of beta coefficients is shown in Fig. 8. These 15 PLS components captured more than 99% variation from the original hyperspectral spectral data and were used in the prediction model development. The predicted insect fragments values were close to the actual seeded insect fragment counts and yielded a high degree of linear fit ($R_2 = 0.99$). "ElMasry et al. (2007) used hyperspectral imaging to develop PLS-based calibration models to quantitatively predict moisture content, total soluble solids, and pH in strawberry. They also developed MLR prediction models by using β coefficients from PLS analysis to select the optimum number of wavelengths. Prediction accuracy of PLS and MLR models was determined by correlation coefficient, standard error of calibration (SEC), and standard error of prediction (SEP). Both models gave good prediction results." Kim et al. (2002) used VIS and NIR reflectance imaging to detect faeces-contaminated apples. They applied PCA to background removed images and successfully discriminated

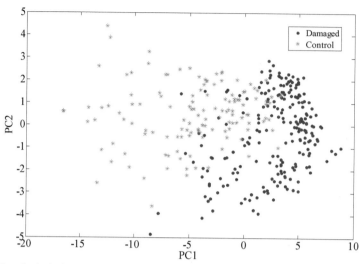

Figure 7: Score plot of principal component 1 (PC1) (82% variance) and principal component 2 (PC2) (8% variance) for fungal damaged and healthy (control) wheat samples (Singh et al., 2011).

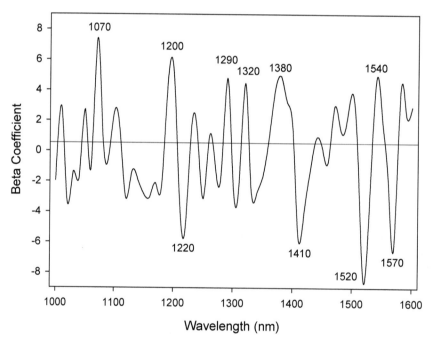

Figure 8: Beta coefficients showing wavelengths that contributed most to the partial least-square model predicting insect fragments in flour (Bhuvaneswari et al., 2011).

uncontaminated apple surface area from contaminated surface area using PC score images. "Lu (2003) detected the bruises on the apple surface using NIR hyperspectral imaging. A new transformed image was developed by multiplication of first and third PC images. Mean pixel values from this transformed image corresponding to the dark and light areas in minimum noise fraction (MNF) image were calculated and compared. Liu et al. (2006) detected chilling injury in cucumbers by hyperspectral imaging technique, using both spectral and image analysis. Reflectance difference, reflectance ratio, two-class PCA models (with pre-treatments) on region of interest (ROI) spectra were used for classification. Lawrence et al.

(2003) used hyperspectral imaging to detect faecal and ingesta contamination on poultry carcasses and compared the results with NIR spectroscopic analysis. They applied PCA to spectral data and plotted PCA scores and loadings to get significant wavelengths and then extracted the images corresponding to these wavelengths from the hypercube for analysis." Qualitative and quantitative analyses were done using histogram stretching and faecal thresholding, respectively. Gorretta et al. (2006) used NIR hyperspectral imaging to determine vitreousness of durum wheat. The wheat kernels were classified by applying PLS-factorial discriminant analysis (FDA) algorithm to pre-processed mean spectral data with maximum classification accuracy of 94%. Cogdill et al. (2004) developed a calibration model to predict corn kernel moisture and oil content using NIR hyperspectral imaging and PCR and PLS. Gowen et al. (2009) used hyperspectral imaging to detect freeze damage in mushroom samples. The dimensionality of the data was reduced by PCA and discriminant analysis classification algorithms were developed.

4. Summary

A simple linear regression (SLR) model is useful with one predictor variable. If there is more than one variable (with known linear relationship with response) and the number of predictor variables is relatively small compared to the samples (response), multiple linear regression (MLR) can be used to develop calibration models. However, if the number of variables is larger than the sample with collinearity issue, principal component regression (PCR) using principal component (PC) scores obtained using principal component analysis (PCA) or partial least square (PLS) should be used. The PCA is best suited for selecting the most significant variables (e.g., wavelength in spectroscopy and hyperspectral imaging) whereas PLS is used for developing robust calibration models for prediction of constituents in agri-food products. Various applications of PCA, PCR and PLS in NIR spectroscopy and hyperspectral imaging show their effectiveness as multivariate data analysis and calibration tools.

Acknowledgements

Some sections of this chapter are based on our previously published manuscripts and the Ph.D. dissertation of the lead author, which are cited in the text and listed in the References.

References

Ariana, D.P., Lu, R. and Guyer, D.E. 2006. Near-infrared hyperspectral reflectance imaging for detection of bruises on pickling cucumbers. Comput. Electron. Agric. 53: 60–70.

Berardo, N., Brenna, O.V., Amatoa, A., Valotia, P., Pisacanea, V. and Motto, M. 2004. Carotenoids concentration among maize genotypes measured by near infrared reflectance spectroscopy (NIRS). Innov. Food Sci. Emerg. Technol. 5: 393–398.

Bhuvaneswari, K., Varadharaju, N., Fields, P.G., White, N.D.G., Sarkar, A.K., Singh, C.B. et al. 2011. Image analysis for detecting insect fragments in semolina. J. Stored Prod. Res. 47: 20–24.

Chen, S., Zhang, F., Ning, J., Liu, X., Zhang, Z. and Yang, S. 2015. Predicting the anthocyanin content of wine grapes by NIR hyperspectral imaging. Food Chem. 172: 788–793.

Cogdill, R.P., Hurburgh, C.R., Rippke, G.R., Bajic, S.J., Jones, R.W., McClelland, J.F. et al. 2004. Single kernel maize analysis by near-infrared hyperspectral imaging. Trans. ASAE 47: 311–320.

Delwiche, S.R. and Norris, K.H. 1993. Classification of hard red wheat by near-infrared diffuse reflectance spectroscopy. Cereal Chem.70: 29–35.

Delwiche, S.R. and Massie, D.R. 1996. Classification of wheat by visible and near-infrared reflectance from single kernels. Cereal Chem. 73: 399–405.

Dowell, F.E. 2000. Differentiating vitreous and nonvitreous of durum wheat kernels by using near-infrared spectroscopy. Cereal Chem. 77: 155–158.

ElMasry, G., Wang, N., ElSayed, A. and Ngadi, M. 2007. Hyperspectral imaging for nondestructive determination of some quality attributes for strawberry. J. Food Eng. 81: 98–107.

Geladi, P. and Kowalski, B.R. 1986. Partial least-squares regression: A tutorial. Anal. Chim. Acta 185: 1–17.

Geladi, P. and Grahn, H. 1996. Multivariate image analysis. Chichester, UK: John Wiley and Sons.

Godoy, J.L., Vega, J.R. and Marchetti, J.L. 2014. Relationships between PCA and PLS-regression. Chemometr. Intell. Lab. Syst. 130: 182–191.

Gorretta, N., Roger, J.M., Aubert, M., Bellon-maurel, V., Campan, F. and Roumet, P. 2006. Determining vitreousness of durum wheat kernels using near infrared hyperspectral imaging. J. Near Infrared Spectrosc. 14: 231–239.

Gowen, A.A., Taghizadeh, M. and O'Donnell, C.P. 2009. Identification of mushrooms subjected to freeze damage using hyperspectral imaging. J. Food Eng. 93: 7–12.

Grahn, H. and Geladi, P. 2007. Techniques and Applications of Hyperspectral Image Analysis. New York: John Wiley and Sons.

Höskuldsson, A. 1988. PLS regression methods. J. Chemom. 2: 211–228.

Kamruzzaman, M., Elmasry, G., Sun, D.W. and Allen, P. 2011. Application of NIR hyperspectral imaging for discrimination of lamb muscles. J. Food Eng. 104: 332–340.

Kays, S.E., Archibald, D.D. and Sohn, M. 2005. Prediction of fat in intact cereal food products using near-infrared reflectance spectroscopy. J. Sci. Food Agric. 85: 1596–1602.

Kim, M.S., Lefcourt, A.M., Chao, K., Chen, Y.R., Kim, I. and Chan. D.E. 2002. Multispectral detection of faecal contamination on apples based on hyperspectral imagery: Part I. Application of visible and near-infrared reflectance imaging. Trans. ASAE 45: 2027–2037.

Lawrence, K.C., Windham, W.R., Park, B. and Buhr, R.J. 2003. A hyperspectral imaging system for identification of faecal and ingesta contamination on poultry carcasses. J. Near Infrared Spectros. 11: 269–281.

Li, B., Cobo-Medina, M., Lecourt, J., Harrison, N.B., Harrison, R.J. and Cross, J.V. 2018. Application of hyperspectral imaging for nondestructive measurement of plum quality attributes. Postharvest Biol. Technol. 141: 8–15.

Liu, Y., Chen, Y.R., Wang, C.Y., Chan, D.C. and Kim, M.S. 2006. Development of hyperspectral imaging technique for the detection of chilling injury in cucumbers; spectral and image analysis. Appl. Eng. Agric. 22: 101–111.

Lu, R. 2003. Detection of bruise on apples using near-infrared hyperspectral imaging. Trans. ASAE 46: 523–530.

Maghirang, E.B. and Dowell, F.E. 2003. Hardness measurement of bulk wheat by single kernel visible and near-infrared reflectance spectroscopy. Cereal Chem. 80: 316–322.

Maghirang, E.B., Dowell, F.E., Baker, J.E. and Throne, J.E. 2003. Automated detection of single wheat kernels containing live or dead insect using near-infrared reflectance spectroscopy. Trans. ASAE 46: 1277–1282.

McGoverin, C.M., Engelbrecht, P., Geladi, P. and Manley, M. 2011. Characterisation of non-viable whole barley, wheat and sorghum grains using near-infrared hyperspectral data and chemometrics. Anal. Bioanal. Chem. 401: 2283–2289.

Murray, I. and Williams, P.C. 1990. Chemical principles of near-infrared technology. pp. 17–34. *In*: P.C. Williams and K.H. Norris (eds.). Near-infrared Technology in the Agricultural and Food Industries. St. Paul, MN: American Association of Cereal Chemists.

Naes, T., Isaksson, T., Fearn, T. and Davies, T. 2002. A user-friendly guide to multivariate calibration and classification. Chichester, UK: NIR Publications.

Pearson, T.C., Wicklow, D.T., Maghirang, E.B., Xie, F. and Dowell, F.E. 2001. Detecting aflatoxin in single corn kernels by using transmittance and reflectance spectroscopy. Tran. ASAE 44: 1247–1254.

Rajalahti, T. and Kvalheim, O.M. 2011. Multivariate data analysis in pharmaceutics: A tutorial review. Int. J. Pharm. 417: 280–290.

Ram, M.S., Dowell, F.E., Seitz, L. and Lookhart, G. 2002. Development of standard procedures for a simple, rapid test to determine wheat colour class. Cereal Chem. 79: 230–237.

Singh, C.B. 2009. Detection of insect and fungal damage and incidence of sprouting in stored wheat using near-infrared hyperspectral and digital colour imaging. PhD dissertation, University of Manitoba, Winnipeg, Canada.

Singh, C.B., Jayas, D.S., Borondics, F. and White, N.D.G. 2011. Synchrotron-based infrared imaging study of compositional changes in stored wheat due to infection with *Aspergillus glaucus*. J. Stored Prod. Res. 47: 372–377.

Singh, C.B., Jayas, D.S., Paliwal, J. and White, N.D.G. 2007. Fungal detection in wheat using near-infrared hyperspectral imaging. Trans. ASABE 50: 2171–2176.

Singh, C.B., Jayas, D.S., Paliwal, J. and White, N.D.G. 2009. Detection of sprouted and midge-damaged wheat kernels using near-infrared hyperspectral imaging. Cereal Chem. 86: 256–260.

Wang, D., Dowell, F.E. and Lacey, R.E. 1999. Single kernel color classification by using near-infrared reflectance spectra. Cereal Chem. 76: 30–33.

Wang, D., Dowell, F.E. and Dempster, R. 2002. Determining vitreous subclasses of hard red spring wheat using visible/near-infrared spectroscopy. Cereal Chem. 79: 418–422.

Wang, D., Dowell, F.E., Ram, M.S. and Schapaugh, W.T. 2003. Classification of fungal-damaged soybean seeds using near-infrared spectroscopy. Int. J. Food Prop. 7: 75–82.

Wesley, I.J., Larroque, O., Osborne, B.G., Azudin, N., Allen, H. and Skerritt, J.H. 2001. Measurement of gliadin and glutenin content of flour by NIR spectroscopy. J. Cereal Sci. 34: 125–133.

Wold, S., Esbensen, K. and Geladi, P. 1987. Principal component analysis. Chemometr. Intell. Lab. Syst. 2: 37–52.

Xing, J., Symons, S., Hatcher, D. and Shahin, M. 2011. Comparison of short-wavelength infrared (SWIR) hyperspectral imaging system with an FT-NIR spectrophotometer for predicting alpha-amylase activities in individual Canadian Western Red Spring (CWRS) wheat kernels. Biosyst. Eng. 108: 303–310.

CHAPTER 11

Partial Least Square Regression for Food Analysis

Basis and Example

Wilson Castro,[1] Jimy Oblitas,[2] Edward E Rojas[3] and Himer Avila-George[4],*

1. Introduction

1.1 Food analysis: From univariate to multivariate analysis

Nowadays, the analysis of different compounds of interest in foods (e.g., anthocyanins, phenols, etc.) has utilized univariate inferential methods, such as spectroscopy, refractometry, and others, reading one variable or characteristic at a time and, based on this, predicting properties of interest. For example, in spectroscopy, one of the most commonly used techniques is read absorbance, transmittance or reflectance value of a sample at one specific wavelength which corresponds to one pick in a wavelength range, see Fig. 1a. This value is compared with a set of values from known concentrations solutions or pattern solutions using a simple regression, see Fig. 1b.

The Association of Official Analytical Chemists (AOAC) collects a significant number of this method, such as AOAC 2012.10, used for Vitamin A - Acetate determination at 325 nm, AOAC 2005-02, for monomeric anthocyanin determination, expressed as cyanidin-3-glucoside, at 520 and 720 nm, and AOAC 965.17, for determination of phosphorus at 400 nm. However, these methods are not capable of managing high complexity systems and interactions between compounds, as in food systems. In this sense, purification and comparison with standard solutions of known concentrations is required as a previous step.

In the last decades, other methods capable of handling highly complex systems have been employed in order to extract as much information as possible. So, as explained by Miller and Miller (2010), multi-channel array detectors and small sensor arrays have made multi-parameter data acquisition and multianalyte measurement possible, extending their applications to several areas of the food industry, especially in those applications where analysis of several variables at a time can reduce timeouts.

---

[1] Facultad de Ingeniería, Universidad Privada del Norte. Cajamarca, Cajamarca 06002, Peru.
   Email: wilson.castro@upn.edu.pe
[2] Departamento de Tecnología de Alimentos, Universidad de Lleida. Lleida 25198, Spain.
[3] Vicepresidencia de Investigación, Universidad Nacional Intercultural de Quillabamba, Quillabamba, Cusco 08741, Peru.
[4] Centro Universitario de los Valles, Universidad de Guadalajara. Ameca, Jalisco 46600, Mexico.
* Corresponding author: himer.avila@academicos.udg.mx

However, a large number of measured variables increases the complexity of the system, as overlapping of constituent profiles produces problems of high dimensionality and collinearity (Hyötyniemi, 2001; Vega-Vilca and Guzmán, 2011).

One example of the multi-parameter data acquisition technologies is the hyperspectral image (HSI), which retrieves a spectrogram of samples over broad wavelength ranges; then, these profiles and chemometric analysis are used to build models and facilitate visualization of chemical or classified images. In this analysis, each wavelength becomes a variable, to be measured and evaluated according to its selected relevance. However, as shown in Fig. 2, this kind of image typically presents a high level of overlapping due to collinearity, which introduces difficulties in the analysis that cannot be handled by ordinary univariate regression.

So, modeling the relationships between variables of highly complex systems is commonly performed through linear regression models, such as Multi-linear Regression (MLR) or Partial Least Square Regression (PLSR). Although MLR is a well-known model that presents a high level of maturity and is applied in most fields of research, it is unable to differentiate between variables according to their predictive capacity in the model. Such a disadvantage makes MLR generate oversized models or variables that often introduce undesired noise to the model.

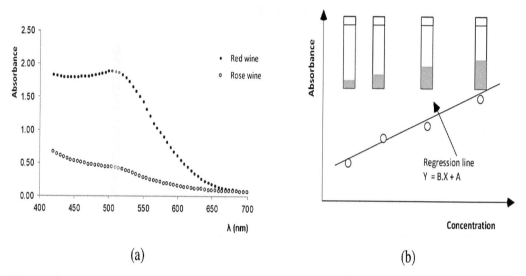

(a)   (b)

Figure 1: Absorbance analysis example. (a) Absorbance spectra profiles for red and rose wines. (b) Absorbance vs concentration.

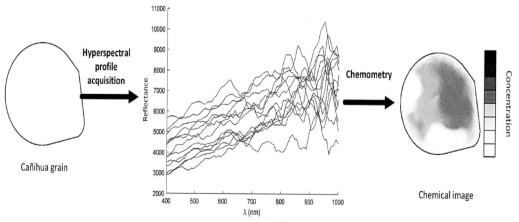

Figure 2: Generalized sequence for HSI analysis.

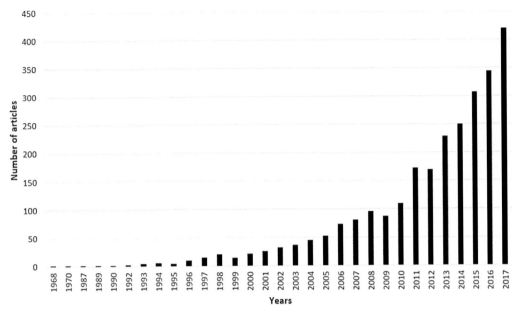

Figure 3: Articles published using PLSR method found in the *Scopus* database.

The more recent PLSR constitutes a comparatively new method of constructing linear regression equations and was developed by H. Woldd in 1975. It was developed for analysis in economic sciences, but currently shows great potential in chemometric analysis (Hervé, 2010; Vega-Vilca and Guzmán, 2011). This potential is mainly related to its capacity to reduce the number of variables in complex systems. This advantage triggered an exponential growth in the number of research products that reported employing the PLSR model in one way or another, and, at this date, more than 2,700 works can be found in the *Scopus* database using PLSR as a search term, see Fig. 3.

2. PLSR Applications in Food Engineering

PLSR is considered useful for constructing prediction equations when there are many explanatory variables and comparatively few training samples (Garthwaite, 1994); mainly because of this method is it possible to combine the advantages from Principal Component Analysis (PCA) and the multiple regression analysis.

Most of the initial applications of PLSR have been in analytical chemistry, particularly food chemistry (Cramer, 1993). This field's relatively new multivariate analytical techniques, such as Near Infrared Spectroscopy (NIRS), Hyperspectral Imaging (HSI), and Dielectric Spectroscopy (DS), among others, made it possible to collect massive amounts of information from parameters measured along a specific range of values. So, it is normal to find applications in food engineering using different signals, such as those presented in Table 1, and in which spectroscopy techniques, such as NIR, HSI, FT-IR, and others, represent an essential part of them.

As can be seen, according to the increase on the number and nature of the sensors employed for food analysis, the applications and researches on PLSR models also present a significant increase, mainly arguing the robustness of the method.

Therefore, a variety of works were generated, mainly for the prediction of properties, such as those listed in Table 1. Furthermore, some works focus on the comparison of variants of the general PLSR method, e.g., Colombani et al. (2012) compared PLSR against PLSR dispersed in the genomic selection of French dairy cattle, Vásquez et al. (2018) compared the viability of use of PLSR and Artificial Neuronal Network (ANN) to model hardness in cheese during ripening, and Zhai et al. (2013) compared the ability of minimum square regression methods and vector machine regression methods (SVMR) to estimate

Table 1: PLSR application in food analysis using different techniques.

Material	Measuring	Techniques	R^2	Source
Rabbit meat	Lipid oxidation	RS [360–740] nm	0.87	Cifuni et al., 2016
Melon	Texture	GC-MS [40 450] m/z	-	Dos-Santos et al., 2013
Standard	Aspartame, cyclamate, saccharin, and acesulfame-K	FTRMS [2850–200000] nm	0.96–0.99	Duarte et al., 2017
Olive oil	Degradation of carotenoids	VIS-RM-S [3225–14285] nm	0.99	Rasha et al., 2010
Quinoa grains	Moisture, protein, fat and ashes contents	NIR-T-S [850–1048] nm	0.39–0.92	Encina-Zelada et al., 2017
Cheese	Moisture and inorganic salt contents	DS [0.3–3.0] GHz	0.92–0.94	Everard et al., 2006
Cheese	Texture profile analysis/ meltability	MIRS [2500–15625] nm	0.68–0.94	Fagan et al., 2007
Apples	Soluble solids and firmness	VIS-NIR-T-S [650 920] nm	0.81–0.95	Fan et al., 2009
Apples	Carbaryl pesticide	SERS [5000–50000] nm	0.98	Fan et al., 2015
Fruit samples	Diphenylamine residue	FS [300–500] nm	-	Farokhcheh and Alizadeh, 2013
Brazilian soybean	Composition	FT-NIR-S [1000–2500] nm	0.50–0.81	Ferreira et al., 2013
Salmonella enterica	Serotypes discrimination	FT-IR-S [2500–16666] nm	0.73–0.99	Preisner et al., 2012

Signal

DS = Dielectric spectroscopy
FS = Fluorescence spectroscopy
FT-IR-S = Fourier transform infrared spectroscopy
FT-NIR-S = Fourier transform near infrared spectroscopy
FTRMS = Fourier transform Raman spectroscopy
GC-MS = Gas chromatography coupled to mass spectrometry
RS = Reflectance spectroscopy
NIR-T-S = Near infrared transmission spectroscopy
MIRS = Mid infrared spectroscopy
SERS = Surface-enhanced Raman spectroscopy
VIS-NIR-T-S = Visible and near-infrared transmittance spectroscopy
VIS-RM-S = Visible Raman spectroscopy

nitrogen (C_N), phosphorus (C_P) and potassium (C_K) content present in the leaves of various plants, using visible and near-infrared (Vis-NIR) spectroscopy. Therefore, based on the increasing importance of this technique, this chapter presents the information required to understand and apply PLSR.

In this chapter, we discuss the principal characteristics of PLSR as one of the most extensively employed statistical tools in chemometrics. The chapter broaches the topic of actual food analysis systems and their capacity to simultaneously measure a significant number of variables, often using technology based on multi-channel array detectors or small sensors. This strategy increases the complexity of the system under study and possible existing relationships, and requires the use of statistical tools to determine the existing relationships. Then, the similarity between PLSR and Multiple Linear Regressions models is exposed, mainly when all variables are used or how these techniques are employed to reduce the high dimensionality, by selecting the most relevant variables. This reduction action is commonly based on variance differences or creation of new variables, such as in Principal Component Analysis. Details of both processes are explained in the document in the same way as its implementation through the Nonlinear Iterative Partial Least Squares (NIPALS) algorithm. Finally, an example of PLSR application for water activity prediction of quinoa grains is detailed, and the code used for its implementation in MatLab 2015 is provided.

3. PLSR Basis

The objective of the PLSR technique is to capture most of the information in predictor variables X or block X that is useful for predicting the Y or dependent variables (Cramer, 1993; Garthwaite, 1994), see Fig. 4. This configuration, where usually $J > I$ have been recently called the "Small I large J problem," has become common in the areas of chemometrics, bioinformatics, image analysis, among others (Hervé, 2010).

High dimensionality problems can be solved by selecting relevant variables, through step-wise methods commonly based on variance differences or creating new variables, such as in *PCA* method. The latter is performed to find new variables named *principal components*, which can explain the variance of the variables X.

Technically, pretreated X is decomposed in a number of *latent variables (LV)*, using singular value decomposition in loading (P) and score (T) vectors. This decomposition is summarized in Fig. 5 and the follow equations.

$$\overline{X} = \hat{X} + E_X$$

$$\overline{X} = TP^T + E_X$$

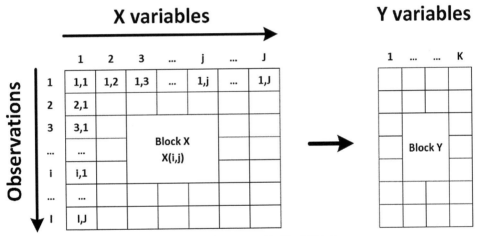

Figure 4: PLSR data arranged in X and Y blocks.

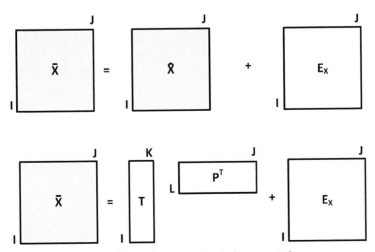

Figure 5: Principal component analysis representation.

where:

- $\bar{X}$ = pretreated X, dimension $[I \times J]$
- $\hat{X}$ = predicted X, $[I \times J]$
- T = score vector, $[I \times K]$
- P^T = transposed loading vector, $[L \times J]$
- E_X = X - error, $[I \times J]$

On the other hand, relationship $Y = f(X)$ is decomposed in scores and loadings sets, see Fig. 6 and the following equations.

$$\hat{X} = TP^T + E_X = \sum_{i=1}^{I} t_i p_i^T + E_{iX}$$

$$\hat{Y} = UQ^T + E_Y = \sum_{i=1}^{I} u_i q_i^T + E_{iY}$$

In addition to the variables used for X-decomposition:

$\hat{Y}$ = predicted Y, $[I \times N]$ (with $N = 1$ for one response and $N > 1$ for multiple responses).
U = Y-scores vector, $[I \times K]$
Q^T = Y-transposed loading factors, $[L \times N]$
E_Y = Y-Error, $[I \times N]$

Likewise, a first step in determining the relevant variables is to relate both score vectors, U and T, in a linear function using the next equation.

$$U = BT + E_U$$

where B is the regression coefficient and E_U error for U; then, replacing this in the previous equation, the new equation for Y predictions is

$$\hat{Y} = BTQ^T + E_Y^*.$$

The new error E_Y^* gathers the errors of the decomposition of $\hat{Y}$ and the linear model.

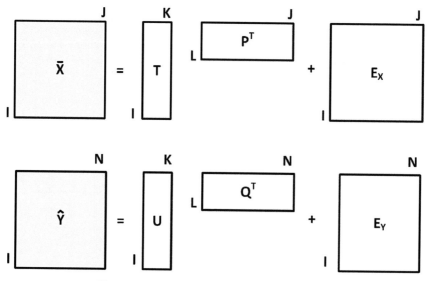

Figure 6: Partial least square regression representation.

4. PLSR Algorithm

According to Esposito and Russolillo (2013), the more common algorithm for PLSR is named Nonlinear Iterative Partial Least Squares (NIPALS). This is an iterative loop used for each component when Y contains only one response variable. In every iteration, Y-scores, X-weights, X-scores, and Y-weights are calculated as a sequence of bivariate regressions.

In this way, according to Tang et al. (2014) and Vega-Vilca and Guzmán (2011), the NIPALS algorithm could be explained in the following steps.

1. Pre-treatment, X and Y must be normalized, using standard deviation, and centered. $E_{Y(0)} = Y$; $E_{X(0)} = X$; $v = 1$ (counter of latent variables).

2. $w_v^T = E_{Y(v-1)}^T E_{X(v-1)} / E_{Y(v-1)}^T . E_{Y(v-1)}$

3. Normalization of w; $w_v^T = w_v^T / \| w_v^T \|$

4. $t_v = E_{X(v-1)} w_v / w_v^T w_v$

5. $p_v^T = t_v^T E_{X(v-1)} / t_v^T t_v$

6. $t_v = t_h \| p_v^T \|$

7. $w_v^T = w_v^T \| w_v^T \|$

8. Normalization of p; $p_v^T = p_v^T / \| p_v^T \|$

9. $b_v = E_{Y(v)}^T t_v / t_v^T t_v$

10. Calculation of X-residuals, $E_{X(v)} = E_{X(v-1)} - t_v p_v^T$

11. Calculation of Y-residuals, $E_{Y(v)} = E_{Y(v-1)} - b_v t_v$

12. $v = v + 1$; if $v < J$ go to Step 1

5. Performance Evaluation of PLSR

Researchers, such as Hervé (2010) and Sun et al. (2017), among others, explain that the quality of the prediction in PLSR models must be evaluated through similarity between response values (Y) and predictions (Y_{pred}). However, these values, specifically (Y_{pred}), could be affected by the representability of the dataset used for calibration/modeling and validation/testing (Esposito and Russolillo, 2013).

At this point, the more important action is to build a dataset with representative capacity and reduce the effect of subjectivity, which can be performed using the *cross-validation* process (*CVP*). *CVP* splits datasets into training (or modeling/calibration) set and test (or validation) dataset. The schematic procedure for *CVP* is shown in Fig. 7.

CVP divides the block of predictors (X) and response (Y) in K folds, then $\forall folds <> fold_i$ a PLSR model is created and calibrated. Next, the model is used to calculate Y_{pred} for $fold_i$. The process is repeated for all folds and the results make the matrix of Y_{pred}. Finally, matrices of Y and Y_{pred} are used for model performance evaluation.

Then, to measure robustness and predictability of multivariate regressions, two characteristic parameters are calculated Coefficient of Determination (R^2), see equation

$$R^2 = \frac{\sum_{i=1}^{n}\left(\widehat{y}_i - y_i\right)^2}{\sum_{i=1}^{n}\left(\widehat{y}_i - y\right)^2},$$

and Root Means Squared Error (*RMSE*), see equation

$$RMSE = \sqrt{\frac{1}{n}\sum_{i=1}^{n}\left(\widehat{y}_i - y\right)^2}.$$

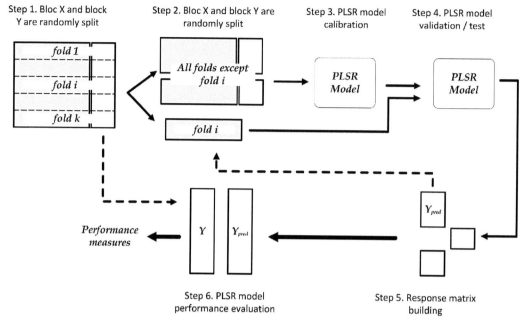

Figure 7: Schematic procedure for cross-validation process.

where $\hat{y}_i$ and y_i are the values of the ith register for Y_{pred} and Y, respectively. y is the average values Y, and n is the numbers of elements of Y.

6. Variable Reduction Techniques

As explained by Wold and Sjöström (2001), high dimensionality is a problem when building models, mainly due to some variables that could add noisy or irrelevant information. Likewise, the removal of highly correlated variables produces better prediction and a more straightforward process. For this reason, it is necessary for our empirical models to determine which of our variables X must be removed, based on their predictable power. This is performed through a strict test of the predictive significance of each PLS component.

Additionally, the use of a shortened number of key-variables reduces the needs of computational resources in the model's calibration, and, hence, the time for analysis, which allows for the construction of online systems for quality evaluation (Sun et al., 2017). Then, according to Liu et al. (2014), high dimensionality reduction can be scheduled by the Fig. 8.

This scheme explains that:

1. The problem starts with a block of predictor variables X, dimension $[I \times J]$, where I = number of observations and J = number of variables. Commonly, X has intrinsic dimensionality j (where $j < J$).
2. Establishing of the conditions which will be used to decide when the loop stops (maximum R^2, minimum RMSE, maximum variance between variables, etc.).
3. Generation of variables subset; each subset can contain more than one variable and there is not a specific rule to prepare a subset.
4. Subset evaluation, testing the subset in $X - Y$ relationship modeling. If the condition is met, go to step 5, if not, return to step 3.
5. Finally, the new block x is created using variables of subset.

According to Mehmood et al. (2012), the general sequence, previously mentioned, can be implemented in three main variable methods, which are detailed in the Table 2.

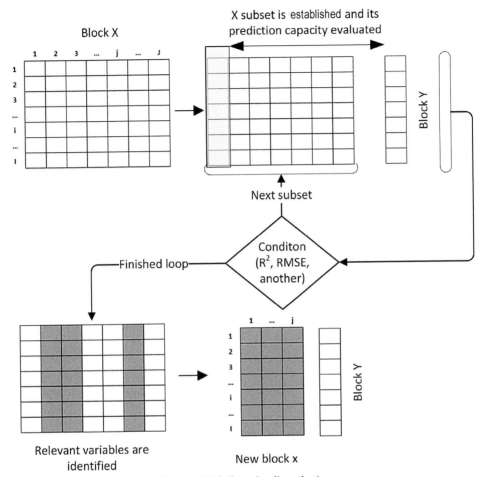

Figure 8: High dimensionality reduction.

Table 2: Methods for wavelength selection in PLSR modeling.

Main method	Basis	Specific method
Filter	Based in subset's quality and independently of classification function.	Loading weights (w). Beta coefficient (β). Variable importance (VIP).
Wraper	Both subset and function are evaluated in order to determine prediction capacity.	Genetic algorithm (GA). Uninformative variable elimination (UVE). Backward variable elimination PLS (BVE). Sub-window permutation analysis (SwPA). Iterative predictor weighting (IPW). Regularized elimination procedure COVPROC. Interval partial least-squares (IPLS).
Embedded	Learn algorithm is used for optimization of prediction capacity determination.	Interactive Variable selection (IVS). Soft-threshold (ST). Sparce Powered (P).

Each method uses a different technique to get the best subset; for instance, *filter* methods use differences or similitude between specific characteristics, such as a correlation between subsets and into subset; however, redundancy in variable selection is not detected due to a relationship between subsets. In *wrapper* methods, the use of a learning algorithms to evaluate subset combinations and their interactions is common. However, this could demand high computational cost. Finally, for the recent proposal, *embedded* methods, both selection and evaluation of subset use learning algorithms. In the last two cases, techniques of learning machines, such as a genetic algorithm or decision tree, among others, can be used.

More information about the different methods selection of variables, advantages and disadvantages are found in the works of Sun et al. (2017), Saeys et al. (2007), and Ladha and Deepa (2011).

At this point, it is observed that a variety of methods have been developed for the selection of relevant variables, but, as comment Vásquez et al. (2018), sometimes different methods produce different results for the same condition, which can be seen in Table 3.

For this reason, in order to detect the best method for a specific application, different methods should be evaluated.

Table 3: PLSR application in food using different signals.

Source	Sample	Parameter	Signal	Method	Latent variables	R^2
Jia et al. (2017)	Chicken	pH	VNIR [400–900] nm	CARS	20	
Yang et al. (2018)	Chicken	pH	VNIR [400–900] nm	PC_1 Score	12	
Su et al. (2014)	Beef	Fat	NIR [1000–1800] nm	MPRESS	7–13	
		Moisture			7–14	
ElMasry et al. (2013)	Beef	Fat	NIR [900–1700] nm	MPRESS	7	0.84
		Moisture			8	0.89

7. PLSR Application Example

In the following part, we present an application example for PLSR. In this example, the water activity of the Andean grain variety of quinoa, with high humidity concentrations, is modeled using functions and scripts implemented in MatLab 2015.

MatLab mathematical software (Matrix Laboratory), currently available for MacOS, Windows and Linux, is widely used for the analysis of multidimensional systems, such as those analyzed by HSI, NIR and DP, among others. Likewise, scripting is done in a friendly environment and, being a high-level language, enables quick prototyping of programs for the user.

The reason previously commented was the reason for using this software and its *PLS_Toolbox* to exemplify the use of the PLSR models. In this way, the reader can get a free trial version of this software through its web page.[1] Similarly, scripts and functions developed for this example are listed at the end of the chapter.

8. Methodology

The methodology for this application example is organized in six steps, as summarized in Fig. 9 and detailed in the following paragraphs.

9. Quinoa Grains

Quinoa, Fig. 10, is an ancient grain of Andean origin and an important food crop for the Inca civilizations. It is still used by people living in rural regions, particularly in Peru and Bolivia (Escribano et al., 2017; Navruz-Varli and Sanlier, 2016).

10. Data Extraction

The principal steps in this stage are explained below:

- *Water activity (a_w) establishing.* The system for establishing isopiestic conditions consisted of desiccators thermally stabilized at 25°C with different saturated salt solutions covering the range

---

[1] https://www.mathworks.com/campaigns/products/trials.html

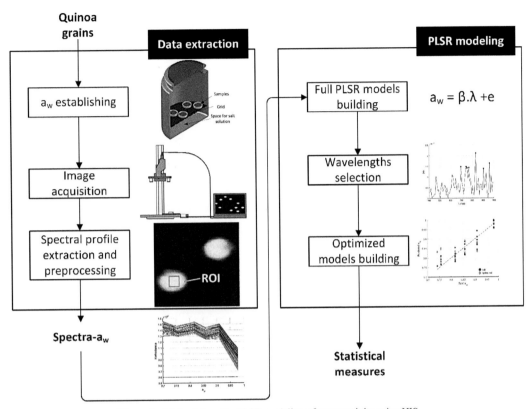

Figure 9: Methodology for PLSR modeling of water activity using HIS.

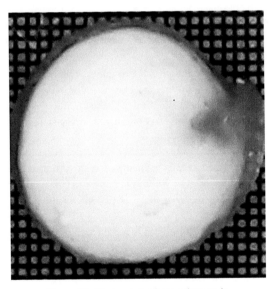

Figure 10: Photograph of one quinoa grain.

of 0.75 to 0.98, and crystalline thymol to prevent microbial growth (Cruz et al., 2010; Greenspan, 1977; Wani and Kumar, 2016). Five grams of quinoa were placed in each desiccator, similar to those reported by Cruz et al. (2010), and moisture was measured gravimetrically every week until constant weight around forty days.

- *Image acquisition.* To acquire images one HSIs an HSI acquisition system (Pika XC; Resonon Inc. Montana, USA) controlled using Resonon Inc.'s SpectrononPro 2.62 software. This system operates in reflectance mode with line-by-line scanning in the range of 400–1000 nm, obtaining intensity images at 8 nm intervals; then, using the system, hyperspectral images of ten grains for a_w were acquired, see Fig. 11.

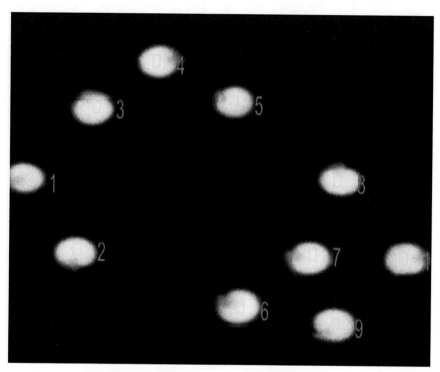

Figure 11: Grayscale image of Quinoa grains' HSI at 712 nm.

The next step consisted in using the equation below and images of white (*W*) and dark (*D*) references which were acquired from a Teflon pattern (reflectance value ~ 99.9%) and by blocking the lens (reflectance ~ 0.0%).

$$R_c = C\left(\frac{R-D}{W-D}\right)$$

where R is the reflectance of the raw image, R_c is the image with corrected reflectance, D is reflectance in the dark reference image, W is reflectance of white reference and C is a correction factor to show images in the full dynamic range with values of 4,096 or 16,384 levels for images of 12 or 14 bits, respectively (Zhang et al., 2016).

- *Spectral profile extraction and preprocessing.* In the center of each grain square ROIs of 20 × 20 pixels were manually selected with the SpectrononPro software, and the spectral profiles from each pixel were acquired. Next, spectra were smoothed, using a second-order Savitzky-Golay filter with twenty frames, and finally divided randomly in training and validation datasets, fifty percent for each set.

Then, all spectra were saved in an excel sheet, named DataPLSRaw.xlsx, which is provided with this chapter; its structure is shown in Fig. 12a and median spectral profiles are calculated and plotted using Listing 1, see Fig. 12b.

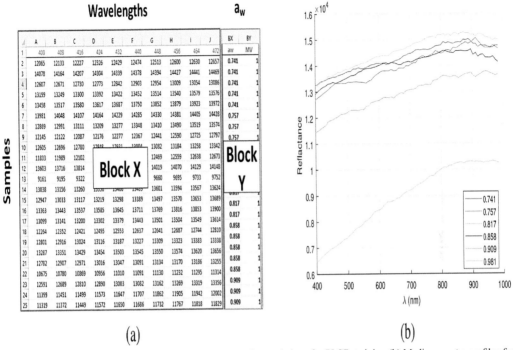

(a)   (b)

Figure 12: Data extraction example. (a) Data arrangement in excel sheet for PLSR training (b) Median spectra profiles for each a_w value.

11. PLSR Modeling

11.1 Full PLSR models building

At this stage, both block X and block Y are used for PLSR modeling, using the previously mentioned *PLS_Toolbox* utilizing Listing 2; using this and the validation dataset, the obtained results are shown in Fig. 13.

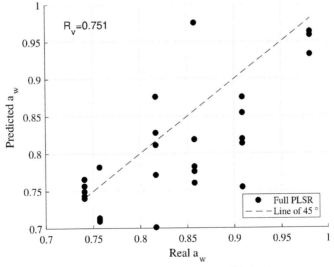

Figure 13: Full PLSR model results using validation dataset.

At this point, however, all wavelengths which could introduce noise into the model were used, for this reason, in the following step, the more relevant wavelengths were selected. For this purpose, the code shown in Listing 3 is used.

11.2 Wavelength selection

The selection of a smaller subset of specific bands that describe the variation of a hypercube factor is necessary to eliminate redundant information provided by contiguous wavelengths (Burger and Gowen, 2011; Liu et al., 2014).

This method requires two steps to be implemented using the code presented in Listing 4, and detailed in the following point: (1) *Establishment of latent variables number*. This requires statistical measures maximization or minimization, see Fig. 14a, which is performed through cross-validation. (2) *Relevance*

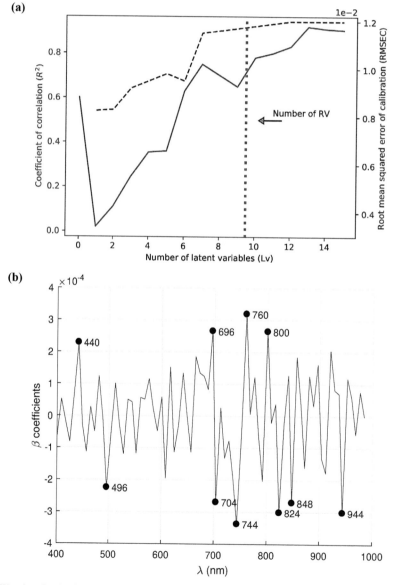

Figure 14: Wavelength selection. (a) Latent variables versus R_c^2 and $RMSE_c$; (b) β coefficients and RW (blue points).

determination. Using the β coefficient method, the number of relevant wavelengths (RW) was set up according to relevance as seven for each variety, which is shown in Fig. 14b.

In this case, RW is mainly placed in two specific ranges near to 460 nm and over 700 nm; then, these wavelengths are used for optimized model building in the next step.

11.3 Optimized models building

Finally, selecting only the relevant wavelengths in the modeling and validation dataset, new PLSR models are built using, in this case, Listing 5. The results, identically that for full PLSR model using a validation dataset, for an optimized model is summarized in Fig. 15.

In this case, the PLSR model can predict water activity in quinoa grain with high a_w values (over 0.75) with an accuracy above 75%; likewise, the use of relevant variables can slightly increase the determination coefficient and strongly reduce the *RMSE* when validation dataset is used, see Table 4.

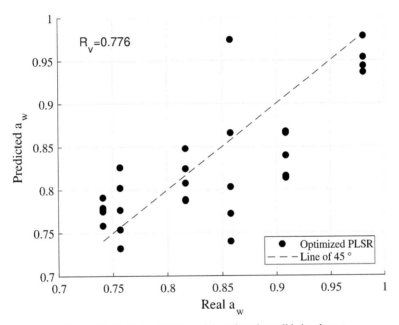

Figure 15: Optimized PLSR model results using validation dataset.

Table 4: Statistical measures for PLSR models.

Model	R	RMSE
Full	0.751	0.020
Optimized	0.776	0.009

Appendix: Matlab Source Code

Listing 1: Loading data and median spectra calculation.

```
%clear data in memory
clc;clear
%reading data from excel worksheet
data = xlsread('DataPLSRaw.xlsx','data','B1:BY61');
dim=size(data);
%first reading spectra for modeling
[f c]=find(data(:,dim(2))==1);dataM= data(f,:);
%calculating median spectra profiles
Valaw=unique(dataM(:,dim(2)-1));
hold on
for i=1:length(Valaw)
   [r c]=find(dataM(:,dim(2)-1)==Valaw(i));
   Valaw_m=median(dataM(r,1:dim(2)-2));
   waves=(1:1:dim(2)-2)*8+392;
   plot(waves,Valaw_m);
end
hold off
legend({'0.741','0.757','0.817','0.858','0.909','0.981'},'Location','best' )
xlabel('\lambda (nm)');ylabel('Reflactance');box off; grid on
```

Listing 2: PLSR model using all waves in spectra profiles.

```
%clear data in memory
clc;clear
%reading data from excel worksheet
data = xlsread('DataPLSRaw.xlsx','data','B1:BY61');
dim=size(data);
%first reading spectra for modeling
[f c]=find(data(:,dim(2))==1);dataM= data(f,:);
%calculating median spectra profiles
Valaw=unique(dataM(:,dim(2)-1));
hold on
for i=1:length(Valaw)
   [r c]=find(dataM(:,dim(2)-1)==Valaw(i));
   Valaw_m=median(dataM(r,1:dim(2)-2));
   waves=(1:1:dim(2)-2)*8+392;
   plot(waves,Valaw_m);
end
hold off
legend({'0.741','0.757','0.817','0.858','0.909','0.981'},'Location','best' )
xlabel('\lambda (nm)');ylabel('Reflactance');box off; grid on
```

Listing 3: PLSR model using all waves in spectra profiles.

```
%Dividing data in set of modeling and validation
X=data(2:dim(1),1:74);
Y=data(2:dim(1),75);
indices=data(2:dim(1),76);

%data for modeling and validation
[rm, cm]=find(indices==1);
[rv, cv]=find(indices==2);
Xm=X(rm,:);Xv=X(rv,:);
Ym=Y(rm,:);Yv=Y(rv,:);

[n,p] = size(Xm); %where n=samples, p=variables
NVlat = 20; % least than number of variables

%PLSR analysis
[XL,YL,Xsc,Ysc,beta,PLSVar,PLSmse,stats] = plsregress(Xm,Ym);

%calculating predicted a_w for modeling and validation
Ypv=[ones(n,1),Xv]*beta;
mdl = fitlm(Ym,Ypv);
r_cv=(mdl.Rsquared.Ordinary)^0.5;

%Plotting full PLSR with validation dataset
g=figure;
hold on
   plot(Yv, Ypc,'ok','MarkerFaceColor','k')
   plot([min(Yv) max(Yv)],[min(Yv) max(Yv)],'--k')
hold off
ylim([0.7 1]);xlim([0.7 1])
xlabel('Real a_w','FontSize',8);
ylabel('Predicted a_w','FontSize',8);
legend('Full PLSR','Line of 45','Location','southeast')
text(0.72, 0.97, strcat('R_{v}=',num2str(r_cv,3)),'fontsize',12)
a = get(gca,'XTickLabel');
set(gca,'XTickLabel',a,'FontName','Times','fontsize',12)
grid on; box off
```

Listing 4: Beta coefficient method for relevant variables selection.

```
%Dividing data in set of modeling and validation
X=data(2:dim(1),1:74);
Y=data(2:dim(1),75);
indices=data(2:dim(1),76);

%data for modeling and validation
[rm, cm]=find(indices==1);
[rv, cv]=find(indices==2);
Xm=X(rm,:);Xv=X(rv,:);
Ym=Y(rm,:);Yv=Y(rv,:);
```

Listing 4 contd. ...

... Listing 4 contd.

```
[n,p] = size(Xm); %where n= samples, p= variables
NVlat = 20; % least than number of variables

%PLSR analysis
[XL,YL,Xsc,Ysc,beta,PLSVar,PLSmse,stats] = plsregress(Xm,Ym);

%calculating predicted a_w for modeling and validation
Ypv=[ones(n,1),Xv]*beta;
mdl = fitlm(Ym,Ypv);
r_cv=(mdl.Rsquared.Ordinary)^0.5;

%Plotting full PLSR with validation dataset
g=figure;
hold on
    plot(Yv, Ypc,'ok','MarkerFaceColor','k')
    plot([min(Yv) max(Yv)],[min(Yv) max(Yv)],'--k')
hold off
ylim([0.7 1]);xlim([0.7 1])
xlabel('Real a_w','FontSize',8);
ylabel('Predicted a_w','FontSize',8);
legend('Full PLSR','Line of 45','Location','southeast')
text(0.72, 0.97, strcat('R_{v}=',num2str(r_cv,3)),'fontsize',12)
a = get(gca,'XTickLabel');
set(gca,'XTickLabel',a,'FontName','Times','fontsize',12)
grid on; box off
```

Listing 5: Beta coefficient method for relevant variables selection.

```
%Optimized PLSR model
Xmr=Xm(:,I(1:10));
[XL,YL,Xsc,Ysc,beta_r,PLSVar, PLSRmseo, PLSstat] = plsregress(Xmr,Ym,10);

%Model evaluation
Xvr=Xv(:,I(1:10));
Ypo=[ones(n,1),Xvr]*beta_r;
mdl = fitlm(Yv,Ypo);
r_ov=(mdl.Rsquared.Ordinary)^0.5;

% Plotting optimized model with validation dataset
figure
hold on
    plot(Yv, Ypo,'ok','MarkerFaceColor','k')
    plot([min(Yv) max(Yv)],[min(Yv) max(Yv)],'--k')
hold off
ylim([0.7 1]); xlim([0.7 1])
xlabel('Real a_w','FontSize',8);
ylabel('Predicted a_w','FontSize',8);
legend('Optimized PLSR','Line of 45','Location','southeast')
text(0.72, 0.97, strcat('R_{v}=',num2str(r_ov,3)),'fontsize',12)
a = get(gca,'XTickLabel');
set(gca,'XTickLabel',a,'FontName','Times','fontsize',12)
grid on; box off
```

References

Burger, J. and Gowen, A. 2011. Data handling in hyperspectral image analysis. Chemom. Intell. Lab. Syst. 108(1): 13–22.

Cifuni, G., Contè, M. and Failla, S. 2016. Potential use of visible reflectance spectra to predict lipid oxidation of rabbit meat. Journal of Food Engineering 169: 85–90.

Colombani, C., Croiseau, P., Fritz, S., Guillaume, F., Legarra, A., Ducrocq, V. et al. 2012. A comparison of partial least squares (PLS) and sparse PLS regressions in genomic selection in French dairy cattle. Journal of Dairy Science 95(4): 2120–2131.

Cramer, R. 1993. Partial least squares (PLS): Its strengths and limitations. Perspectives in Drug Discovery and Design 1(2): 269–278.

Cruz, R., Pelayo, C., Ferreira, L. and Cladera-Olivera, F. 2010. Adsorption isotherms of pinhão (*Araucaria angustifolia* seeds) starch and thermodynamic analysis. J. Food Eng. 100(3): 468–473.

Dos-Santos, N., Bueso, K. and Fernández-Trujillo, J. 2013. Aroma volatiles as biomarkers of textural differences at harvest in non-climacteric near-isogenic lines of melon. Food Research International 54(2): 1801–1812.

Duarte, L., Paschoal, D., Izumi, C., Dolzan, M., Alves, V., Micke, G. et al. 2017. Simultaneous determination of aspartame, cyclamate, saccharin and acesulfame-K in powder tabletop sweeteners by FT-Raman spectroscopy associated with the multivariate calibration: PLS, iPLS and siPLS models were compared. Food Research International 99: 106–114.

ElMasry, G., Sun, D. and Allen, P. 2013. Chemical-free assessment and mapping of major constituents in beef using hyperspectral imaging. Journal of Food Engineering 117(2): 235–246.

Encina-Zelada, C., Cadavez, V., Pereda, J., Gómez-Pando, L., Salvá-Ruíz, B., Teixeira, J. et al. 2017. Estimation of composition of quinoa (*Chenopodium quinoa Willd.*) grains by near-infrared transmission spectroscopy. LWT-Food Science and Technology 79: 126–134.

Escribano, J., Cabanes, J., Jiménez-Atiénzar, M., Ibañez-Tremolada, M., Gómez-Pando, L.R., García-Carmona, F. et al. 2017. Characterization of betalains, saponins and antioxidant power in differently colored quinoa (*Chenopodium quinoa*) varieties. Food Chem. 234: 285–294.

Esposito, V. and Russolillo, G. 2013. Partial least squares algorithms and methods. Wiley Interdisciplinary Reviews: Computational Statistics 5(1): 1–19.

Everard, C., Fagan, C., O'Donnell, C., O'Callaghan, D. and Lyng, J. 2006. Dielectric properties of process cheese from 0.3 to 3 ghz. Journal of Food Engineering 75(3): 415–422.

Fagan, C., Everard, C., O'Donnell, C., Downey, G., Sheehan, E., Delahunty, C. et al. 2007. Prediction of processed cheese instrumental texture and meltability by mid-infrared spectroscopy coupled with chemometric tools. Journal of Food Engineering 80(4): 1068–1077.

Fan, G., Zha, J., Du, R. and Gao, L. 2009. Determination of soluble solids and firmness of apples by VIS/NIR transmittance. Journal of Food Engineering 93(4): 416–420.

Fan, Y., Lai, K., Rasco, B. and Huang, Y. 2015. Determination of carbaryl pesticide in Fuji apples using surface-enhanced Raman spectroscopy coupled with multivariate analysis. LWT-Food Science and Technology 60(1): 352–357.

Farokhcheh, A. and Alizadeh, N. 2013. Determination of diphenylamine residue in fruit samples using spectrofluorimetry and multivariate analysis. LWT-Food Science and Technology 54(1): 6–12.

Ferreira, D., Lima, J. and Poppi, R. 2013. Fourier transform near-infrared spectroscopy (FT-NIRS) application to estimate Brazilian soybean [*Glycine max* (L.) Merrill] composition. Food Research International 51(1): 53–58.

Garthwaite, P. 1994. An interpretation of partial least squares. Journal of the American Statistical Association 89(425): 122–127.

Greenspan, L. 1977. Humidity fixed points of binary saturated aqueous solutions. Journal of Research of the National Bureau of Standards 81(1): 89–96.

Hervé, A. 2010. Partial least squares regression and projection on latent structure regression (PLS regression). Wiley Interdisciplinary Reviews: Computational Statistics 2(1): 97–106.

Hyötyniemi, H. 2001. Multivariate regression: Techniques and tools. Helsinki University of Technology.

Jia, B., Yoon, S., Zhuang, H., Wang, W. and Li, C. 2017. Prediction of pH of fresh chicken breast fillets by VNIR hyperspectral imaging. Journal of Food Engineering 208: 57–65.

Ladha, L. and Deepa, T. 2011. Feature selection methods and algorithms. International Journal on Computer Science and Engineering 3(5): 1787–1797.

Liu, D., Sun, D. and Zeng, X. 2014. Recent advances in wavelength selection techniques for hyperspectral image processing in the food industry. Food and Bioprocess Technology 7(2): 307–323.

Mehmood, T., Hovde, K., Snipen, L. and Sæbo, S. 2012. A review of variable selection methods in partial least squares regression. Chemometrics and Intelligent Laboratory Systems 118: 62–69.

Miller, J. and Miller, C. 2010. Statistics and chemometrics for analytical chemistry. Pearson Education.

Navruz-Varli, S. and Sanlier, N. 2016. Nutritional and health benefits of quinoa (*Chenopodium quinoa Willd.*). J. Cereal. Sci. 69: 371–376.

Preisner, O., Cardoso, J., Guiomar, R., Machado, J. and Almeida, J. 2012. Discrimination of salmonella enterica serotypes by Fourier transform infrared spectroscopy. Food Research International 45(2): 1058–1064.

Rasha, E., Donfack, P. and Materny, A. 2010. Assessment of conventional and microwave heating induced degradation of carotenoids in olive oil by vis Raman spectroscopy and classical methods. Food Research International 43(3): 694–700.

Saeys, Y., Inza, I. and Larrañaga, P. 2007. A review of feature selection techniques in bioinformatics. Bioinformatics 23(19): 2507–2517.

Su, H., Sha, K., Zhang, L., Zhang, Q., Xu, Y., Zhang, R. et al. 2014. Development of near infrared reflectance spectroscopy to predict chemical composition with a wide range of variability in beef. Meat Science 98(2): 110–114.

Sun, Y., Liu, Y., Yu, H., Xie, A., Li, X., Yin, Y. et al. 2017. Non-destructive prediction of moisture content and freezable water content of purple-fleshed sweet potato slices during drying process using hyperspectral imaging technique. Food Analytical Methods 10(5): 1535–1546.

Tang, L., Peng, S., Bi, Y., Shan, P. and Hu, X. 2014. A new method combining LDA and PLS for dimension reduction. PloS One 9(5): e96944.

Vásquez, N., Magán, C., Oblitas, J., Chuquizuta, T., Avila-George, H. and Castro, W. 2018. Comparison between artificial neural network and partial least squares regression models for hardness modeling during the ripening process of swiss-type cheese using spectral profiles. Journal of Food Engineering 219: 8–15.

Vega-Vilca, J. and Guzmán, J. 2011. Regresión PLS y PCA como solución al problema de multicolinealidad en regresión múltiple. Revista de Matemática Teoría y Aplicaciones 18(1): 9–20.

Wani, S.A. and Kumar, P. 2016. Moisture sorption isotherms and evaluation of quality changes in extruded snacks during storage. LWT-Food Sci. Technol. 74: 448–455.

Wold, S., Sjöström, M. and Eriksson, L. 2001. PLS-regression: A basic tool of chemometrics. Chemometrics and Intelligent Laboratory Systems 58(2): 109–130.

Yang, Y., Zhuang, H., Yoon, S., Wang, W., Jiang, H. and Jia, B. 2018. Rapid classification of intact chicken breast fillets by predicting principal component score of quality traits with visible/near-infrared spectroscopy. Food Chemistry 244: 184–189.

Zhai, Y., Cui, L., Zhou, X., Gao, Y., Fei, T. and Gao, W. 2013. Estimation of nitrogen, phosphorus, and potassium contents in the leaves of different plants using laboratory-based visible and near-infrared reflectance spectroscopy: Comparison of partial least-square regression and support vector machine regression methods. International Journal of Remote Sensing 34(7): 2502–2518.

Zhang, R., Li, C., Zhang, M. and Rodgers, J. 2016. Shortwave infrared hyper-spectral reflectance imaging for cotton foreign matter classification. Computers and Electronics in Agriculture 127: 260–270.

CHAPTER 12

Mathematical Modelling of High Pressure Processing in Food Engineering

Deepak Kadam,[1] Surajbhan Sevda,[2] Namrata Tyagi[3] and Chetan Joshi[1,]*

1. Introduction

Unlike temperature, pressure is a thermodynamic state variable, hence, it significantly influences the physical, mechanical and chemical kinetics as well as often inducing a new chemical potential phenomenon. The application of pressure processing in food for preservation was first studied by Hite (1899) and Bridgman (1912). Since then, many studies were made in order to understand the biological effect of high pressure on food spoilage microorganisms with a great deal of attention in food, pharmaceutical, and biotechnology industries. High pressure (up to several hundreds of MPa) force applied perpendicular to the surface of an object per unit area is widely utilised in bioengineering, chemical engineering and food engineering applications. The main aim of these processes is to increase the shelf life and improve the retention of organoleptic and nutritional content as compared to conventional treatment.

Currently, consumers are looking for safe and healthy diet food, minimally processed and without any additives. Therefore, in order to meet the needs of the consumers, the use of several new technologies using lower temperatures has increased significantly in the past year. High-pressure processing (HPP) is a procedure in which solid or liquid foods, with or without packaging, are subjected to elevated pressure, between 1000 to 8000 bars (14500 to 116000 PSI) (IFT and USFDA, 2014). The process temperature during high-pressure processing can be anywhere in the range of 0 to 100, or above, leading to inactivation of enzymes and various foodborne microorganisms. HPP reduces thermal degradation and exhibits minimal changes in the sensory characteristics of food, retaining its taste, appearance, texture and nutritional aspects. HPP at ambient and refrigerated temperatures causes a minimal effect on heat-sensitive food products and eliminates the undesireable thermally-induced flavours.

The organism of concern, product characteristics, and nature of the final product significantly affects HPP of the food product. Microorganisms vary in their sensitivity to high pressures. Inactivation of pathogenic and spoilage microorganisms involves structural degradation as well as an increase in permeability of the cell wall, causing fluids to be forced outside the cells. Two principles govern the

[1] Food Engineering & Technology Department, Institute of Chemical Technology, Matunga, Mumbai 400 019, India.
[2] Department of Biosciences and Bioengineering, IIT Guwahati, Assam – 781039, India.
[3] Catalysts Biotechnologies Pvt. Ltd., Sahibabad, UP-201010.
* Corresponding author: chetanudct@gmail.com

effect of HPP technology. The first governing principle is that of Le Chatelier's, according to whom, any phenomenon (phase transition, chemical reaction, chemical reactivity or change in molecular configuration) accompanied by a decrease in volume will be enhanced by pressure. Secondly, the pressure is instantaneously and uniformly transmitted independently of the size and the shape of the food product, this is also known as isostatic pressure (Ramos and Smith, 2009). Mathematical modeling is a unique method which helps to correlate the effect of different process parameters within a system. Furthermore, it minimizes the number of trials that need to be carried out in order to evaluate the effect of various process parameters on the quality and safety of the process. The objective of this chapter is to focus on modelling and simulation of the effect of the combination of high pressure and thermal treatments on different food processes, as well as on certain enzymes.

2. Mathematical Modelling of Microbial and Enzymatic Inactivation

Mathematical modelling is a tool for studying the effect of different process parameters of a system on the outcome of a process. In general, it is the fastest and least expensive way to minimize the number of experiments required to determine the influence of several parameters on the quality and safety of the process (Smith et al., 2014; Joshi and Singhal, 2016; Joshi and Singhal, 2017).

Various mathematical models have been proposed, depending on the type of food (liquid or solid) and the process conditions (initial temperature, maximum pressure, etc.). For example, a heat transfer model with conduction effect is sufficient for a solid food product (Infante et al., 2009; Otero et al., 2007), whereas convection effect coupled with the heat transfer equation is used for a liquid food product (Infante et al., 2009). However, in the case of a high-pressure freezing process, the model with solidification has to be considered. Moreover, in some cases, the interpretation of process outcome based on the features of the processed food is also required (i.e., enzyme inactivation, microbial load and product of the chemical reaction).

For microbial and enzyme inactivation in high pressure processed food product, uniformity in temperature is required, which affects uniform pressure distribution. In addition, the thermal history of the product under high pressure is critical in optimizing and homogenising high-pressure processes. Therefore, a considerable body of research is focused on developing heat transfer models that simulate the combination of high pressure and thermal treatments on food products.

Numerous research works carried out in the field of HPP have already proven that it causes the inactivation of foodborne microorganisms and enzymes. However, several parameters, like the type of microorganism involved and the enzymes present, affect the inactivation efficiency of high pressure processing (Cheftel and Dumay, 1996; Smelt, 1998).

The native state of the enzyme is stabilized by covalent bonds, hydrogen bonds, Van der Waals interactions, hydrophobic interactions and electrostatic interactions. High-pressure processing of enzyme inactivation is an immensely complex phenomenon. In the high-pressure environment, the mechanism of enzyme inactivation can be hypothesized similar to protein denaturation (Considine et al., 2008; Hendrickx et al., 1998).

HPP induces many changes in the microbial cell, including alteration in the cell membrane and its membrane protein, enzyme and ribosome, as well as the genetic mechanisms responsible for cellular function (Considine et al., 2008). Prokaryotic gram-positive bacteria having a thick layer of peptidoglycan offer a greater resistance towards pressure in comparison to gram-negative bacteria. On the other hand, yeast and molds are much more pressure sensitive (except *Byssochlamys* and *Talaraomyces* sp.).

Bacterial spores, such as *C. botulinum*, are one of the most pressure-resistant microorganisms. Dealing with them requires the addition of heat, and this presents several challenges regarding the design of an efficient high-pressure processing process.

Changes in microbial count as a function of time at a fixed temperature (T) and pressure (P) can be explained by using the first-order kinetic model (Doğan and Erkmen, 2004; Erkmen and Doğan, 2004; Smith et al., 2014).

$$\begin{cases} \dfrac{dN(t)}{dt} = -kN(t), \qquad t \geq 0 \\ N(0) = N_0 \end{cases}$$

2.1

where $N(t)$ is the microbial count at time t, N_0 (cfu g^{-1}) is the initial microbial count and κ is the *inactivation rate constant* [min^{-1}]. The solution to (2.1) is given by

$$ln\left(\dfrac{N(t)}{N_0}\right) = -\kappa t \qquad \text{or, equivalently, } N(t) = N_0 e^{-\kappa t}$$

2.2

Change in the microbial count can also be calculated using the concept of decimal reduction time [D, min], which is the time required for a 90% reduction in the microbial count.

From Eq. (2.2)

$$log\left(\dfrac{N(t)}{N_0}\right) = log(exp(-\kappa t)) = -\kappa t log(e) = -\kappa t \dfrac{\ln(e)}{\ln(10)} = \dfrac{-\kappa t}{\ln(10)}$$

2.3

As D is the time required for $N(t) = N_0/10$

$$D = \dfrac{\ln(10)}{\kappa} log\left(\dfrac{N_0}{N_0/10}\right) = \dfrac{\ln(10)}{\kappa}$$

2.4

From (2.3) and (2.4), the microbial count reduction as a function of time can be expressed as

$$t = D log\left(\dfrac{N_0}{N(t)}\right), \text{or equivalently, } log\left(\dfrac{N(t)}{N_0}\right) = \dfrac{-t}{D} \quad t \geq 0$$

2.5

with these parameters, we can estimate the microbial count reduction after t [min] of processing at constant pressure P and temperature T.

Modelling of the population N (t; P) for a fixed temperature

If we now want to study a pressure (P) and temperature (T) dependent process (arbitrary pressure and temperature, in a suitable range), it is extremely important to identify the kinetic parameters that define the influence of pressure (P) and temperature (T) on the inactivation rate.

If we follow (2.1), we will be able to find the equation for temperature T and pressure P dependent inactivation rate, i.e., $\kappa(T)$ and $\kappa(P)$.

The temperature dependency is given by Arrhenius' equation:

$$\kappa(T) = \kappa_{Tref} exp\left(\left(\dfrac{-E_a}{R}\right)\left(\dfrac{1}{T} - \dfrac{1}{T_{ref}}\right)\right),$$

2.6

where $\kappa(T)$ [min^{-1}] is the inactivation rate for an arbitrary temperature T[K], T_{ref} [K] is a reference temperature, κ_{Tref} [min^{-1}] is the inactivation rate at reference temperature, E_a is an activation energy[3] [J/mol] and R = 8314 [J/(mol K)] is a universal gas constant

$$K(P) = \kappa p$$

2.7

From these equations, we can quantify how pressure affects the inactivation rate constant.

Furthermore, the concept of *volume of activation*, ΔV^* (cm^3/mol) (Difference between the partial molar volumes of the transition state and the sum of the partial volumes of the reactants at the same temperature and pressure) must be introduced.

The volume of activation, ΔV^*, can further be obtained by the pressure dependence of the inactivation rate constant of a reaction, defined by the equation:

$$\Delta V^* = -RT\left(\frac{d\ln\kappa}{dP}\right)_T$$
2.8

where P is the pressure (MPa), R(8.314 cm² MPa mol⁻¹ K⁻¹) is the universal gas constant, T is the temperature (K) and κ [min⁻¹] is the *inactivation rate constant*.

From Eq. (2.8), a formula for the inactivation rate constant κ can be derived as follows

$$\left(\frac{d\ln\kappa}{dP}\right)_T = -\frac{\Delta V^*}{RT}$$
2.9

while working on processes involving bacteria, it is difficult to give a physical meaning to ΔV^*. However, the identification of apparent ΔV^* is essential as it helps to characterize the dependence of the inactivation rate constant.

The Eq. (2.9) can further be written using mathematical integration from reference pressure P_{ref} to arbitrary pressure P (based on Eyeing's equation)

$$\kappa(P) = \kappa p_{ref} exp\left(\frac{-\Delta V^*(P - P_{ref})}{RT}\right),$$
2.10

where $\kappa(P)$ [min⁻¹] is the inactivation rate for arbitrary pressure [MPa], κp_{ref} [min⁻¹] is the inactivation rate constant at the reference pressure (p_{ref}), and ΔV^* is the activation volume. From Eqs. (2.6) and (2.10), we can now construct a model for $N(t; T)$ and $N(t; P)$, respectively. The Equation for Inactivation rate $\kappa(T, P)$ can now be written as:

$$\kappa(T,P) = \kappa_{Tref,Pref} \exp\left(-B\left(\frac{1}{T} - \frac{1}{T_{ref}}\right)\right)\exp(-C(p - P_{ref})),$$
2.11

where $\kappa(T, P)$ [min⁻¹] is the inactivation rate for temperature T [K] and pressure P [MPa], and $\kappa_{Tref, Pref}$ [min⁻¹], B [K] and C [MPa] are kinetic constants that represent the influence of pressure and temperature on $\kappa(T, P)$. Represent the resulting mathematical model obtained by coupling Eqs. (2.1) and (2.11) as follows:

$$\begin{cases} \kappa(T,P) = \kappa_{Tref,Pref} \exp\left(-B\left(\frac{1}{T} - \frac{1}{T_{ref}}\right)\right)\exp(-C(p - P_{ref})), T,P \text{ in adequate range} \\ \dfrac{dN(t;T,P)}{dt} = -\kappa(T,P)N(t;T,P), \qquad\qquad t \geq 0 \\ \qquad\qquad N(0;T,P) = N_0 \\ \text{[Solution: } N(t;T,P) = N_0 \exp(-\kappa(T,P)t)] \end{cases}$$
2.12

Model for enzyme inactivation can similarly be obtained by substituting $N(t; T, P)$ by $A(t; T, P)$ and N_0 by A_0.

- 'N' as the dependent function of temperature (T) and pressure (P) (For enzymatic inactivation this model are equivalent).

$$log\left(\frac{D(T)}{D_{T_{ref}}}\right) = -\frac{T - T_{ref}}{z_T}$$
2.12

$$log\left(\frac{D(P)}{D_{P_{ref}}}\right) = -\frac{P - P_{ref}}{z_p} \qquad \text{2.13}$$

where $D(T)$ and $D(P)$ is a decimal reduction of time at temperature T [K] and pressure P [MPa]; T_{ref} [K] and P_{ref} [MPa] are the reference temperature and pressure, respectively; z_T [K] and z_p [MPa] are the thermal and pressure resistant coefficients needed to accomplish a 1-log-cycle reduction or decimal reduction time value D [min]. On incorporating Eqs. (2.12) and (2.13) in model Eq. (2.5), we can now construct the model to investigate $N(t; T)$ and $N(t; P)$, respectively. The mathematical model for $N(t; P)$

$$\begin{cases} D(P) = D_{P_{ref}} 10^{-\frac{P - P_{ref}}{z_p}}, & P \text{ in the adequate range} \\[2mm] \dfrac{dN(t;P)}{dt} = -\dfrac{\ln(10)}{D(P)} N(t;P), & t \geq 0 \\[2mm] N(0;P) = N_0 \\[2mm] [solution: N(t;P) = N_0 10^{-\frac{t}{D_{P_{ref}}} 10^{\frac{P - P_{ref}}{z_p}}}] \end{cases} \qquad \text{2.14}$$

For this model, the kinetic parameters we need to determine are D_{Pref}, P_{ref} and z_p. When coupling model (2.1) with Eq. (2.6) and model (2.2) with Eq. (2.12) or (2.13) and applying them to microbial population reduction rate data over the same temperature range, the correlation between the coefficients E_a and z_T can be obtained.

The following relationships are satisfied:

$$E_a = \frac{\ln(10)RT^2}{z_T} \quad and \quad \Delta V^* = -\frac{\ln(10)RT}{z_p} \qquad \text{2.15}$$

All the models and equations derived herewith mainly focus on the microbial and enzymatic inactivation under isothermal and isobaric conditions, in which the same pressure and temperature are applied during the process. From the industrial perspective, these models are not as realistic as in industrial processes with the increase in pressure; there is a rise in temperature due to the generation of adiabatic heat. Therefore, from the previous studies, we obtain the following mathematical model that allows us to predict microbial and enzymatic inactivation after the dynamic process (temperature and pressure change with time).

In a dynamic process, where temperature T and pressure P change with time, there will not be any constant. For example, $k(T)$, $k(P)$, $k(T, P)$, $D(P)$ or $D(T)$.

If $P(\bullet)$, $T(\bullet)$: $[0, t_f] \rightarrow \mathbb{P}$ are known functions, we can construct the complete model. For example, if we are considering model Eq. (2.11), we just need to change the constants T and P. For the functions $T(t)$ and $P(t)$

$$N(t; T(\bullet), P(\bullet)) = N_0 \exp\left[-\int_0^{t_f} \kappa(T(T), P(T)) \, dT\right] \qquad \text{2.16}$$

This model systematically demonstrates the influence of known temperature and pressure treatment on the microbial and enzymatic inactivation. However, in the food processing industry, prior profiles are often not known, i.e., only the pressure profile is known before the process, the temperature profile is not known. Hence, we have to construct a model that can predict the temporal and spatial distribution of temperature T of the food sample under processing conditions. The model can be constructed using the heat transfer model.

3. Mass and Heat Transfer Model for High Pressure

HP processing is a "Non-thermal" technique in which food products are sealed in high barriers, flexible pouches or plastic containers and are placed in a steel cylindrical vessel containing pasteurised fluid as a pressure transmitting media. A typical HP treatment consists of a compression step, a holding phase at target pressure, and a decompression step.

Initially, the food packages, pressure medium and vessel all are in equilibrium. During compression, samples are pasteurised by using a hydro pneumatic pump, which slowly increases (at a fixed rate) the pressure value until the target pressure is achieved. After holding the product at the target pressure for a fixed time, the vessel is decompressed by releasing the pressure-transmitting fluid. The process is usually performed under laminar conditions in order to avoid any turbulence appearing in the pasteurised media and in the liquid food. These two conditions (cylindrical geometry and laminar flow) propose the use of an axis-symmetry model (Fig. 1) (Infante et al., 2009).

Let us consider the following two-dimensional sub-domains (see Fig. 1)

- Ω_F: Domain that contains the food sample.
- Ω_C: Cap of the sample holder (typically a rubber cap).
- Ω_p: Domain occupied by pasteurizing medium.
- Ω_s: Domain of the steel that surrounds the domain mentioned above.

The domain in the (r, z)-coordinate is a rectangle $\Omega = [0, L] \times [0, H]$,

Axis symmetry is defined by $\{0\} \times (0, H)$

The whole domain is defined as $\overline{\Omega} = \overline{\Omega_F \cup \Omega_C \cup \Omega_p \cup \Omega_s}$ 3.1

- The boundary of Ω, is denoted by Γ, where we can differentiate.
- $\Gamma_r \subset \{L\} \times (0, H)$, where the temperature is known.
- $\Gamma_{up} = [0, L] \times \{H\}$, where heat transfer within the room in which the equipment is located.
- $\Gamma \setminus \{\Gamma_r \cup \Gamma_{up}\}$, with zero heat flux, either by axial symmetry or by isolation of the equipment.

Finally, super index ([]*) is used to denote Ω^*, Ω_F^*, Ω_C^*, Ω_p^*, Ω_s^*, Γ^*, Γ_r^* and, Γ_{up}^*, as a three-dimensional domain generated by rotating Ω, Ω_F, Ω_C, Ω_p, Ω_s, $\Gamma(\{0\} \times (0, H))$, Γ_r and Γ_{up} along the axis symmetry (in the 3D version of the problem), respectively.

In the following section, we describe the equation for the mathematical model, distinguishing two significant cases: Solid- and liquid-type foods.

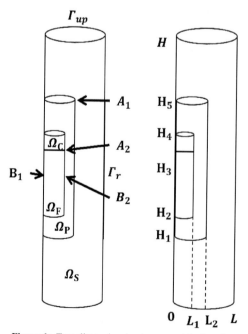

Figure 1: Two-dimensional axis-symmetric domain.

3.1 Solid food products

(a) Heat transfer by conduction

In the case of solid food products, the starting point is the heat conduction equation for temperature T (K). For a stationary, homogeneous and isotropic solid body with heat generation inside, the heat equation can be written as:

$$\rho C_p \frac{\partial T}{\partial t} - \nabla \cdot (\kappa \nabla T) = \beta \frac{dP}{dt} T, \ in \ \Omega^* \times (0, t_f) \qquad 3.2$$

where, ρ is the density [kg/m^3], C_p is the specific heat capacity [J/kg K], κ the thermal conductivity [W/mK], β is the thermal expansion coefficient [K^{-1}], $P(t)$ is the pressure [MPa] as a function of time and Ω is the special domain which contains the food product, and t_f is the final time(s).

The right-hand side of Eq. (3.2) refers to the heat generated due to the change in pressure. Here, $P = P(t)$ is the pressure (Pa) applied by the equipment and $\beta = \beta(T, P)$ is given by

$$\beta = \begin{cases} \text{thermal expansion coefficient (K}^{-1}\text{) of the food in } \Omega^*_F, \\ \text{thermal expansion coefficient (K}^{-1}\text{) of the pressurising fluid in } \Omega^*_p, \\ 0, \text{ elsewhere} \end{cases} \qquad 3.3$$

This term results from the following equation (valid for isotropic processes)

$$\frac{\Delta T}{\Delta P} = \frac{\beta TV}{MC_P} = \frac{\beta T}{\rho C_p} \qquad 3.4$$

where ΔT (K) denotes the temperature change due to the pressure change ΔP (Pa), V (m^3) is the volume and M (kg) is the mass. In order to solve the conductive heat transfer Eq. (3.2), we must specify initial and boundary conditions as per the HP machinery used for the solution of the problem.

We have used the same conditions as reported by Otero et al. (2007) for a pilot plant unit (ACB GEC Alsthom, Nantes, France).

$$\begin{cases} \kappa \dfrac{\partial T}{\partial n} = 0 & on \ \Gamma_0^* \times (0, t_f), \\ \kappa \dfrac{\partial T}{\partial n} = h(T_{amb} - T) & on \ \Gamma_{up}^* \times (0, t_f), \\ T = T_r & on \ \Gamma_r^* \times (0, t_f), \\ T(0) = T_0 & in \ \Omega^*, \end{cases} \qquad 3.5$$

where n is the outward unit normal vector on the boundary of the domain, T_0 is the initial temperature T_r (K) is the fixed temperature on Γ_r^*, T_{amb} (K) is the ambient temperature and h (W m^{-2} K^{-1}) is the heat transfer coefficient. By using cylindrical coordinates and axial symmetry, system (3.2), (3.5) may be rewritten as the following two-dimensional case:

$$\begin{cases} \rho C_p \dfrac{\partial T}{\partial t} - \dfrac{1}{r}\dfrac{\partial}{\partial r}\left(rk\dfrac{\partial T}{\partial r}\right) - \dfrac{\partial}{\partial z}\left(rk\dfrac{\partial T}{\partial z}\right) = \beta\dfrac{dT}{dt}T & in \ \Omega^* \times (0, t_f) \\ k\dfrac{\partial T}{\partial n} = 0 & on \ \Gamma_0 \times (0, t_f), \\ \kappa\dfrac{\partial T}{\partial n} = h(T_{amb} - T) & on \ \Gamma_{up} \times (0, t_f), \\ T = T_r & on \ \Gamma_r \times (0, t_f), \\ T(0) = T_0 & in \ \Omega \end{cases} \qquad 3.6$$

This model is suitable when we apply HP process on a "large sample", i.e., when the filling ratio of the food sample inside the vessel is much higher than that of the pressurizing medium. Otero et al., (2007), has described the model which has been validated with several comparisons between numerical and experimental results. Becker and Fricke (1999) had earlier reported that for "small sample", i.e., when the filling ratio of the food inside the vessel is not significantly higher than that of the pressure medium, the output of the above model differs significantly from the actual experimental results. Therefore, they improved this model by incorporating convection effects of the pressurizing medium. Equation (3.6) may be used, while dealing with small samples, i.e., highly viscous fluids at low temperatures, where convection effects may be negligible. This would reduce the time of computation without any significant loss in prediction.

(b) Heat transfer by conduction and convection

Non-homogeneous temperature distribution induces a non-homogeneous density distribution in the pressurizing medium and, consequently, a buoyancy fluid motion, i.e., free convection in which the fluid motion may strongly influence the temperature distribution. In order to address this fact, a non-isothermal flow model is considered and we assume that the fluid velocity field **u** (m/s) satisfies the Navier–Stokes equations for compressible Newtonian fluid under Stokes' assumption.

The resulting system of equations is

$$
\begin{cases}
\rho C_p \dfrac{\partial T}{\partial t} - \nabla \cdot (\kappa \nabla T) + \rho C_p u \cdot \nabla T = \beta \dfrac{dP}{dt} T & \text{in } \Omega^* \times (0, t_f), \\[2mm]
\rho \dfrac{\partial u}{\partial t} - \nabla \cdot \eta(\nabla u + \nabla u') + \rho(u \cdot \nabla)u = -\nabla p - \dfrac{2}{3}\nabla(\eta\nabla \cdot \mathbf{u}) + \rho g & \text{in } \Omega_p^* \times (0, t_f), \\[2mm]
\dfrac{\partial \rho}{\partial t} + \nabla \cdot (\rho u) = 0 & \text{in } \Omega_p^* \times (0, t_f), \\[2mm]
\kappa \dfrac{\partial T}{\partial n} = 0 & \text{on } \Gamma_0^* \times (0, t_f), \\[2mm]
\kappa \dfrac{\partial T}{\partial n} = h(T_{amb} - T) & \text{on } \Gamma_{up}^* \times (0, t_f), \\[2mm]
T = T_r & \text{on } \Gamma_r^* \times (0, t_f) \\[1mm]
u = 0 & \text{on } \Gamma_p^* \times (0, t_f) \\[1mm]
T(0) = T_0 & \text{in } \Omega^* \\[1mm]
u(0) = 0 & \text{in } \Omega^* \\[1mm]
p = 10^5 & \text{at } A_1 \times (0, t_f)
\end{cases}
\qquad 3.7
$$

where **g** is the gravity vector (m/s²), $\eta = \eta(T; P)$ is the dynamic viscosity (Pa s), $p = p(x; t)$ is the pressure generated by the mass transfer inside the fluid, and $P + p$ is the total pressure (Pa) in the pressurizing medium Ω_p^*. A_1 is a corner point of Γ_p^*, which is the boundary of Ω_p^*. Infante et al. (2009) had earlier pointed out that the right hand side of the first equation of (3.7) could have been written as $\beta \dfrac{d(P + p)}{dt} T$, but, it was assumed that internal heat generation due to mass transfer is negligible.

On the right hand side of the second equation of (3.7), we have written ∇p, since $P = P(t)$ depends only on time and, thus, $\nabla(P + p) = \nabla p$. As in the previous section, the density $\rho = \rho(T, P)$ is a known state function uniquely determined except for additive constants. A_1 such condition means that the total pressure $(P + p)$ at this point is equal to the equipment pressure P plus the atmospherical pressure.

As in Section 3.1a for the conductive heat transfer model, system (3.7) can also be rewritten as an equivalent 2D problem by using cylindrical coordinates.

(c) Liquid-type food

For liquid-type food, we proposed a model which includes heat transfer by conduction and convection. The convection occurs in both the region of food sample (Ω_F) and pressurising medium regions (Ω_p), in which two different velocity fields, U_F and U_p, are considered, respectively (Infante et al., 2009).

The pressurizing medium and the food are separated by the sample holder and do not mix. It is assumed that both the pressurizing fluid and the liquid food product are compressible and Newtonian fluids.

Hence, the governing equations are:

$$
\begin{cases}
\rho C_p \dfrac{\partial T}{\partial t} - \nabla \cdot (\kappa \nabla T) + \rho C_p u \cdot \nabla T = \beta \dfrac{dP}{dt} T & in\ \Omega^* \times (0,\,t_f), \\[2ex]
\rho \dfrac{\partial u_F}{\partial t} - \nabla \cdot \eta (\nabla u_F + \nabla u_F^t) + \rho (u_F \cdot \nabla) u_F = -\nabla p - \dfrac{2}{3} \nabla (\eta \nabla \cdot u_F) + \rho g & in\ \Omega_F^* \times (0, t_f), \\[2ex]
\rho \dfrac{\partial u_p}{\partial t} - \nabla \cdot \eta (\nabla u_p + \nabla u_p^t) + \rho (u_p \cdot \nabla) u_p = -\nabla p - \dfrac{2}{3} \nabla (\eta \nabla \cdot u_p) + \rho g & in\ \Omega_p^* \times (0, t_f), \\[2ex]
\dfrac{\partial \rho}{\partial t} + \nabla \cdot (\rho u_F) = 0 & in\ \Omega_F^* \times (0,\,t_f) \\[2ex]
\dfrac{\partial \rho}{\partial t} + \nabla \cdot (\rho u_p) = 0 & in\ \Omega_p^* \times (0,\,t_f)
\end{cases}
\qquad 3.8
$$

With point boundary and initial conditions

$$
\begin{cases}
\kappa \dfrac{\partial T}{\partial n} = 0 & on\ \Gamma_0^* \times (0,\,t_f), \\[2ex]
\kappa \dfrac{\partial T}{\partial n} = h(T_{amb} - T) & on\ \Gamma_{up}^* \times (0,\,t_f), \\[1.5ex]
T = T_r & on\ \Gamma_r^* \times (0,\,t_f), \\[1.5ex]
u_F = 0 & on\ \Gamma_F^* \times (0,\,t_f), \\[1.5ex]
u_p = 0 & on\ \Gamma_p^* \times (0,\,t_f), \\[1.5ex]
T(0) = T_0 & in\ \ \Omega^* \\[1.5ex]
\boldsymbol{u}_F(0) = 0 & on\ \ \Omega_F^* \\[1.5ex]
\boldsymbol{u}_p(0) = 0 & on\ \ \Omega_p^* \\[1.5ex]
p = 10^5 & at\ A_1 \times (0, t_f) \\[1.5ex]
p = 10^5 & at\ A_2 \times (0, t_f)
\end{cases}
\qquad 3.9
$$

where Γ_F^* and Γ_p^* denote the boundary and A_1 and A_2 denote the corner point of Ω_F^* and Ω_p^*, respectively.

In this process, convection plays an important role for both liquid food product and pressurising fluid. Thus, neglecting its effect (as in the big solid case) could lead to entirely different results from the actual experimental thermal behaviour for both large and small samples.

(d) Simplified model

Infante et al. (2009) have proposed several simplified models in order to reduce the computational complexity and checked the validity of such models. Moreover, simplified models yield similar results and are proven to be faster and easier to implement. These models can, thus, be used in the implementation of freezing models and three-dimensional models.

(e) For large and small solid food product

For a large solid-type food product, we consider a simplified version of model (3.6) with constant coefficients by setting the thermo-physical parameters *Cp, k, β* and *ρ* to their mean value in the range of temperature and pressure considered in the process (Infante et al., 2009).

For a small solid-type food, we included convection effects in the pressurising medium (3.7). Furthermore, considering constant thermo-physical properties, a simplified model based on the Boussinesq approximation is made (Infante et al., 2009). More precisely, C_p, k, β and η are considered to be coefficient constant and are set to their mean values ($\overline{C}_p, \overline{k}, \overline{\beta}, \overline{\rho}$ and $\overline{\eta}$, respectively) in the range of temperature and pressure considered in the process.

With the simplifications, system (3.7) can be written as

$$
\begin{cases}
\overline{\rho C_p}\dfrac{\partial T}{\partial t} - \nabla\cdot(\overline{\kappa}\nabla T) + \overline{\rho C_p}u\cdot\nabla T = \overline{\beta}\dfrac{dP}{dt}T & \text{in } \Omega^* \times (0, t_f), \\[2mm]
\overline{\rho}\dfrac{\partial u}{\partial t} - \nabla\cdot\overline{\eta}\nabla^2 u + \overline{\rho}(u\cdot\nabla)u = -\nabla p - \rho g & \text{in } \Omega_P^* \times (0, t_f), \\[2mm]
\nabla.(u) = 0 & \text{in } \Omega_P^* \times (0, t_f), \\[2mm]
\overline{\kappa}\dfrac{\partial T}{\partial n} = 0 & \text{on } \Gamma_0^* \times (0, t_f), \\[2mm]
\overline{\kappa}\dfrac{\partial T}{\partial n} = h(T_{amb} - T) & \text{on } \Gamma_{up}^* \times (0, t_f), \\[2mm]
T = T_r & \text{on } \Gamma_r^* \times (0, t_f), \\[2mm]
u = 0 & \text{on } \Gamma_p^* \times (0, t_f), \\[2mm]
\quad\quad T(0) = T_0 & \text{in } \Omega^* \\[2mm]
\quad\quad u(0) = 0 & \text{in } \Omega^* \\[2mm]
p = 10^5 & \text{at } A_1 \times (0, t_f)
\end{cases}
$$

3.10

(f) For a liquid food product

For a liquid-type food, we describe a simplified model based on the Boussinesq approximation, this time for both the liquid food and pressurising medium (this approximation is denoted by LB in (Infante et al., 2009)). Thus, system (3.8) is simplified to:

$$
\begin{cases}
\overline{\rho C_p}\dfrac{\partial T}{\partial t} - (\overline{\kappa}\nabla^2 T) + \overline{\rho C_p}u\cdot\nabla T = \overline{\beta}\dfrac{dP}{dt}T & \text{in } \Omega^* \times (0, t_f), \\[2mm]
\overline{\rho}\dfrac{\partial u_F}{\partial t} - \overline{\eta}\nabla^2 u_F + \overline{\rho}(u_F\cdot\nabla)u_F = -\nabla p + \rho g & \text{in } \Omega_F^* \times (0, t_f), \\[2mm]
\overline{\rho}\dfrac{\partial u_P}{\partial t} - \overline{\eta}\nabla^2 u_P + \overline{\rho}(u_P\cdot\nabla)u_P = -\nabla p + \rho g & \text{in } \Omega_P^* \times (0, t_f), \\[2mm]
\nabla.(\rho u_F) = 0 & \text{in } \Omega_F^* \times (0, t_f), \\[2mm]
\nabla.(\rho u_P) = 0 & \text{in } \Omega_P^* \times (0, t_f),
\end{cases}
$$

3.11

with boundary and initial conditions given by (3.9).

4. High-Pressure Shift Freezing Processes

Freezing is a popular food preservation technique with an extended implementation area that ensures high quality of stored product during prolonged storage. However, the larger ice crystals that form during freezing can damage the food product, this along with other undesirable textural and organoleptic properties. Therefore, it is important to develop a process that can create a homogeneous mixture of small ice crystals. The recent industry research goals are set to improve known freezing methods and develop new techniques to overcome the hurdles.

With recent advances in high-pressure technology in food, processing efforts were made in order to study the effect of this process on freezing. On experimentation, researchers were able to effectively decrease the freezing and melting point of water to a minimum of –22°C at 207.5 MPa (Smith et al., 2012). This led to the development of several high pressure freezing-thawing processes.

According to the phase diagram of water, three different high pressure freezing processes that can be distinguished by the way of phase transitions are possible, viz., high pressure assisted freezing (HPAF), high pressure induced freezing (HPIF) and high pressure shift freezing (HPSF). In high-pressure assisted freezing process, the phase transition occurs under a constant pressure, higher than the atmospheric pressure, while the temperature is lowered to the freezing point. In HPIF, the phase transition is initiated by an initial pressure increase and then continues at constant pressure. However, in high-pressure shift freezing, phase transition occurs by a pressure change that causes metastable conditions and instantaneous ice production. This is attained via instantaneous pressure release throughout the product during expansion and, thus, the subsequent temperature reduction. In this process, large-scale super-cooling and high ice nucleation velocities are observed. Furthermore, studies have shown that HPSF produces small granular ice crystals, which in turn indicates homogeneous nucleation throughout the product. Compared to the classical freezing process, the freezing times are significantly reduced in HPSF and among the three high pressure freezing processes HPSF is found to be advantageous (Smith et al., 2012).

In HPSF process, the sample is cooled under pressure to a temperature of less than 0°C, while still being in a non-frozen state. After achieving the desired temperature, the pressure is released, causing either rapid expansion or slow expansion. Next is the phase transition, which occurs due to the pressure release that induces uniform super-cooling throughout the whole sample due to the isostatic nature of pressure. Uniform nuclei formation takes place and the latent heat is released, raising the temperature of the sample to its freezing point. Further freezing is then carried out at constant pressure. Previous reports have mainly explored the advantages of HPSF over other methods with respect to texture and structure of food products. Denys et al. (1997) found out that the nucleation is faster and uniform in HPSF owing to better textural quality of the food products when compared to conventional freezing. HPSF has been successfully applied in the processing of fruits, pork, lobster and tofu, among other products (Smith et al., 2012).

Furthermore, A-B shows compression, B-C shows precooling and C-D pressure release. At point D, the sample is super-cooled and no latent heat has been released yet. For D-E rapid expansion, latent heat is released and the product temperature rises up to the corresponding atmospheric freezing point. 1-2 slow expansion as well the latent heat is released. However, the temperature increases only up to the corresponding freezing point under pressure. At 2-E, the remaining pressure is released and the sample temperature remains on the melting line.

a. Modelling for HPSF process

Two-dimensional axially symmetric mathematical model has been proposed for solid-type food with large and small filling sample versus pressurising media ratio. The model is derived from an enthalpy formulation based on volume fractions dependent on pressure and temperature as well as convection effects in the pressurising fluid. In the following HP process, we first present a general heat transfer model in order to stimulate a HPSF process. Then, we modify the model based on the solidification process, by taking into account the enthalpy formulation at non-constant pressure. Additionally, we also present the equations which can be used in the calculation of the amount of ice instantaneously produced just after expansion and during the rest of the HPSF process, as well as the liquid volume fraction. So here, we represent the complete models used to simulate a HPSF process.

i. Heat transfer in a general HP process

The geometry of the HP device considered for the model of HPSF process was derived in section three (see Fig. 1). In this section, we are going to study a large and small solid-type of food, therefore, a model takes into account of conduction effect presented in Section a(i) and also included convection effect in the pressurising medium presented in Section a(ii), respectively.

ii. Heat transfer for large solid-type food

$$
\begin{cases}
\overline{\rho C p}\dfrac{\partial T}{\partial t} - \dfrac{1}{r}\dfrac{\partial}{\partial r}\left(r\overline{k}\dfrac{\partial T}{\partial r}\right) - \dfrac{\partial}{\partial z}\left(\overline{k}\dfrac{\partial T}{\partial z}\right) = \overline{\beta}\dfrac{dP}{dt}T & \quad in \ \ \Omega^{*}\times(0,t_{f}) \\[2mm]
\overline{\kappa}\dfrac{\partial T}{\partial n} = 0 & \quad on \ \ \Gamma_{0}\times(0,t_{f}), \\[2mm]
\overline{\kappa}\dfrac{\partial T}{\partial n} = h(T_{amb}-T) & \quad on \ \ \Gamma_{up}\times(0,t_{f}), \\[2mm]
T = T_{r} & \quad on \ \ \Gamma_{r}\times(0,t_{f}), \\[2mm]
T(0) = T_{0} & \quad in \ \ \Omega
\end{cases}
\qquad 4.1
$$

iii. Heat transfer for a small solid food product

The heat transfer for small solid food product can be modelled by the following system

$$
\begin{cases}
\overline{\rho C_{p}}\dfrac{\partial T}{\partial t} - \nabla\cdot(\overline{\kappa}\nabla T) + \overline{\rho C_{p}}u\cdot\nabla T = \overline{\beta}\dfrac{dP}{dt}T & \quad in \ \ \Omega^{*}\times(0,t_{f}), \\[2mm]
\overline{\rho}\dfrac{\partial u}{\partial t} - \nabla\cdot\overline{\eta}\nabla^{2}u + \overline{\rho}(u\cdot\nabla)u = -\nabla p + \rho g & \quad in \ \ \Omega_{p}^{*}\times(0,t_{f}), \\[2mm]
\nabla\cdot(pu) = 0 & \quad in \ \ \Omega_{p}^{*}\times(0,t_{f}), \\[2mm]
\overline{\kappa}\dfrac{\partial T}{\partial n} = 0 & \quad on \ \ \Gamma_{0}^{*}\times(0,t_{f}), \\[2mm]
\overline{\kappa}\dfrac{\partial T}{\partial n} = h(T_{amb}-T) & \quad on \ \ \Gamma_{up}^{*}\times(0,t_{f}), \\[2mm]
T = T_{r} & \quad on \ \ \Gamma_{r}^{*}\times(0,t_{f}), \\[2mm]
u = 0 & \quad on \ \ \Gamma_{p}^{*}\times(0,t_{f}), \\[2mm]
\quad\quad T(0) = T_{0} & \quad in \ \ \Omega^{*} \\[2mm]
\quad\quad u(0) = 0 & \quad in \ \ \Omega^{*} \\[2mm]
p = 10^{5} & \quad at \ \ A_{1}\times(0,t_{f})
\end{cases}
\qquad 4.2
$$

Equation (4.2) can also be rewritten as an equivalent 2D problem by using cylindrical coordinates (as described in Fig. 2). The numerical experiments considered in this Eqs. (4.1) and (4.2) were carried out using the 2D version of the corresponding equations.

b. Modelling a solidification process using the enthalpy formulation

The phenomenon of solidification is associated with many practical applications in the food industry. In general, the factors responsible for the solidification process are to control the position of the solid/liquid interface and to deal with the release of latent heat, which evolves over a very small temperature range. Crank (1987) explained the *front-tracking* and *fixed-grid* methods for solving the 'moving-boundary' problem. Additionally, the former methods may be used when there is a distinct phase change and a smooth continuous front (Voller et al., 1990). If the solid/liquid interface do not move smoothly with time

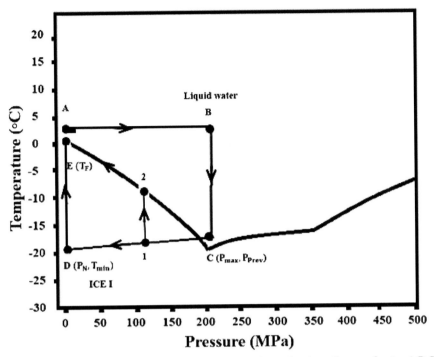

Figure 2: The process of high pressure shifting process is represented over the phase diagram of water. A-B-C-D-E and A-B-C–1-2-E represents the rapid and slow expansions, respectively.

(i.e., interface becomes less distinct), it is often difficult or even impossible to track the moving boundary directly. It may generate sharp peaks or even disappear. Hence, it is necessary to reformulate the problem in such a way that the new equations are applicable over the whole of the fixed domain so that it becomes unnecessary to track the solid liquid interface. These are called the fixed grid methods.

In the fixed-grid method, latent heat evolution is taken into account when determining the energy equation by defining either an apparent specific heat, a total enthalpy or a heat source term (Voller et al., 1990). Followed by the numerical solution can also be carried out on a spatial grid that remains fixed during the calculations. Moreover, the numerical solution of the phase change can be attained through simple modifications of existing heat transfer numerical methods and/or software. To model HPSF processes, a combination of such methods, adapted to the case of non-constant pressure can be used (Smith et al., 2012).

A pure, non-convecting material obeying Fourier's law of heat conduction, conservation of enthalpy *can be written as*

$$\frac{\partial e}{dt} = \nabla \cdot (k \nabla T) + \frac{\partial P}{\partial t} \qquad 4.3$$

where e is the enthalpy per unit volume (J m^{-3}).

Voller et al. (1990) derived the enthalpy equation from Eq. (4.3) under constant pressure. Whereas, the pressure term on the right-hand side of Eq. (4.3) was neglected in the enthalpy model of the HPSF process (Norton et al., 2009).

Enthalpy is a function of both temperature and pressure at non-constant pressure conditions and, from thermodynamics, we may derive the following relation

$$de = \rho C_p dT + (1 - \beta T) dP \qquad 4.4$$

By integrating the equation above within boundary conditions, the following is obtained using an expression for the enthalpy

$$e(T, P) = e(T_{ref}, P_{ref}) + \int_{Pref}^{T} \rho C_p d\theta + \int_{Pref}^{P} (1 - \beta T) d\Pi \qquad 4.5$$

where T_{ref} is the reference temperature, P_{ref} is the reference pressure, T is the current temperature and P is the current pressure.

Based on the phase fractions, we have derived an enthalpy model defining 'mixture enthalpy'. It explains the mixed-phase region composed of the solid part (i.e., ice crystals) and the solid and liquid part of the food that has not yet frozen (referred to as liquid).

$$e(T, P, g_l) = e(T_{ref}, P_{ref}, g_{l,ref}) + (1 - g_l)\int_{Tref}^{T} \rho_s C_{\rho s} d\theta + g_l \int_{Tref}^{T} \rho_l C_{\rho l} d\theta + (g_l - g_a)\rho_l \lambda + (1 - g_l)\int_{Pref}^{P} (1 - \beta_s T)$$
$$d\Pi + \int_{Pref}^{P} (1 - \beta_l T)d\Pi \qquad 4.6$$

where g_l and g_a are the volume fraction of liquid and solid parts in the food (i.e., volume fraction of water in the food is $(g_l - g_a)$) and λ is the latent heat of freezing of water (J/kg). T_{ref} is the temperature (K) at which all the latent heat has been released. We consider, $g_{l,ref} = g_a$ (typically $Tref = -40°C = 233.15$ K) and that $P_{ref} = P_{atm}$ (Pa), the atmospheric pressure.

Similarly, a 'mixture conductivity' is defined as:

$$k = (1 - g_l) k_s + g_l k_l \qquad 4.7$$

In Eq. (4.6), solid and liquid phases are denoted as subscript [ ]$_s$ and [ ]$_l$, respectively. Taking the total derivative of (4.6) gives:

$$de = C_{vol} dT + (1 - \beta_{Vol} T)dP + \delta_e \, dg_l \qquad 4.8$$

where $C_{vol} = (1 - g_l)\rho_s C_{\rho s} + g_l \rho_s C_{\rho l}$, $\delta_e = \int_{Tref}^{T} (\rho_l C_{\rho l} - \rho_s C_{\rho s}) \, d\theta + \int_{Pref}^{P} T(\beta_s - \beta_l)d\Pi + \rho_l \lambda$ and $\beta_{vol} = (1-g_l)$ $\beta_s + g_l \beta_l$.

Inserting (4.8) in to the enthalpy Eq. (4.3) lead to:

$$C_{vol} \frac{\partial T}{\partial t} - \nabla \cdot (k\nabla T) = \beta_{vol} T \frac{\partial P}{\partial t} - \delta e \frac{\partial g_l}{\partial t} \qquad 4.9$$

For a HPSF process with rapid expansion (see ABCDE in Fig. 2), typically, the pressure profile is as follows:

$$P(t) = \begin{cases} \dfrac{P_{max} - P_{atm}}{t_{p_1}}t + P_{atm} & t < t_{p_1} \\[2ex] P_{max} & t_{p_2} < t < t_{p_3} \\[2ex] -\dfrac{P_{max} - P_{atm}}{t_{p_1} - t_{p_2}}(t - t_{p_2}) + P_{max} & t_{p_2} < t < t_{p_3} \\[2ex] P_{max} & \text{elsewhere} \end{cases} \qquad 4.10$$

where t_{p_1} and t_{p_2} are the time taken to reach maximum pressure (P_{max}) and the time at which the pressure is released, respectively. t_{p_3}, on the other hand, is the time at which all pressure has been released ($P_{atm} \approx 0.1$ MPa).

Experiments conducted by (Otero and Sanz, 2006) suggest that, when the pressure is rapidly released ($t \in (t_{p_2}, t_{p_3})$), nucleation is first delayed until the pressure is close to atmospheric, when ice suddenly nucleates and forms in a small time interval. Therefore, we assume that there is no ice formation while the pressure is changing (until $= t_{p_3}$).

During this stage, Eq. (4.9) becomes:

$$\rho_l C_{\rho_l} \frac{\partial T}{\partial t} - \nabla \cdot (k \nabla T) = \beta_l T \frac{\partial P}{\partial t}, \qquad t \leq t_{p_3} \qquad\qquad 4.11$$

Once the pressure has been released, ice suddenly nucleates and forms in some small time interval (ε), where we assume that the change in ice fraction owing to the drop in pressure is some known function of time, hence, (4.9) becomes:

$$C_{vol} \frac{\partial T}{\partial t} - \nabla \cdot (k \nabla T) = \delta e \frac{\partial g_l}{\partial t} \qquad t \in (t_{p_3}, t_{p_3} + \varepsilon) \qquad\qquad 4.12$$

where $\delta e = \int_{Tref}^{T} (\rho_l C_{\rho_l} - \rho_s C_{\rho_s}) d\theta + \rho_l \lambda$, since $P = P_{atm}$.

After this point and until the end of the process, we assume that the rest of the ice is computed as a function of temperature and is expressed as:

$$\left[C_{vol} + \delta e \left(\frac{\partial g_l}{\partial T} \right) \right] \frac{\partial T}{\partial t} - \nabla \cdot (k \nabla T) = 0, \qquad t \geq t_{p_3} + \varepsilon \qquad\qquad 4.13$$

The model given by Eq. (4.9) gives an acceptable relationship between theoretical and experimental observations. All the parameters mentioned in the model except g_l can easily be found in literature. Using this model, one can also explain various characteristics of HPSF and, thus, it is possible to derive a formula for g_l.

c. Deriving an expression for the volume fraction g_l for a gel food simile

i. Supercooling reached after expansion

Otero and Sanz (2006) have determined the extent of super cooling, which is an important factor in the dynamics of a freezing process ΔT_{se} (°C), as the difference between the lowest temperature at the sample centre just before nucleation (T_{min}) and the sample freezing pressure P_N, at temperature T_F, where the nucleation takes place. At P_{atm}, in conventional freezing, supercooling and nucleation occurs only at the surface of the sample. In the HPSF experimental process, at the time of pressure release, a metastable state is attained throughout the sample before nucleation (Otero and Sanz, 2006).

Extensive supercooling (Fig. 2) shows that pressure/temperature coordinates of the food sample move to (P_N, T_{min}) after expansion from (P_{max}, T_{prev}). In addition, minimum temperature (T_{min}) is reached at nucleation pressure P_N after rapid pressure released can be find out according to the Eq. (3.4) and is accounted for in our heat transfer model. However, the prediction of (T_{min}) using (3.4) implies anticipation of the nucleation pressure P_N and this is, in general, very difficult, owing to the stochastic nature of the nucleation phenomenon.

ii. Modelling the amount of ice formed instantaneously after expansion

As shown in Fig. 2, the pressure release in HPSF processes can be divided into two different phases (Otero and Sanz, 2006).

In the first phase, a percentage of water is instantaneously frozen under the metastable phase.

The latent heat raises the sample temperature to the corresponding freezing point. The amount of ice formed at P_N can be calculated from the following equation

$$m_i \lambda(P_N) = (m_i \overline{C_{P_i}} + (1 - m_i) \overline{C_{P_W}}) \Delta T_{sc} \qquad\qquad 4.14$$

where m_i is the mass fraction of ice formed after expansion, $\lambda(P_N)$ (J/kg) is the latent heat at nucleation pressure P_N, $\overline{C_{P_i}}$ (J/kg °C) is the specific heat capacity of ice at the nucleation pressure, $\overline{C_{P_W}}$ (J/kg °C) is the specific heat capacity of water at the nucleation pressure and ΔT_{sc} (°C) is the extent of supercooling.

Latent heat of freezing of water (J/kg) as a function of pressure P (MPa) can be estimated by the following equation (Otero and Sanz, 2006):

$$\lambda(P) = 3.114 \times 10^{-3}P^3 - 1.292P^2 - 3.379 \times 10^2 P + 3.335 \times 10^5 \qquad 4.15$$

From Eqs. (4.14) and (4.15), we can calculate the percentage of instantaneously frozen water m_i, taking into account the extent of supercooling attained and the latent heat released at nucleation pressure (P_N). However, this is just a theoretical evaluation and is not sufficient in modelling purposes because during expansion, experimental data of temperature and nucleation pressure during expansion (T_{min}, P_N) is required.

In order to overcome this problem (Otero and Sanz, 2006), proposed a simplified method for HPSF processes with rapid expansions. Assume that nucleation occurs at atmospheric pressure (P_{atm}). On this basis, the amount of ice formed after expansion can be determined by using experimental temperature and pressure values immediately prior to expansion (P_{max}, T_{prev}), instead of (T_{min}, P_N). They do as follows: The minimum temperature (T_{min}) after expansion is calculated using (3.3), Eq. (4.15) is used to calculate the latent heat released at atmospheric pressure, the corresponding supercooling is attained as $\Delta T_{sc} = T_F - T_{min}$ and, finally, we introduced these values in (4.14) to determine the percentage of ice (m_i) instantaneously formed in HPSF experiments with rapid expansions. We use this simplification in our simulations.

iii. Modelling the mass fraction of ice of the rest of the process

After calculating the amount of ice formed immediately after expansion, we need an equation to calculate how the rest of the ice is formed as a function of temperature at atmospheric pressure (P_{atm}). The temperature is raised to the corresponding freezing point at atmospheric pressure (P_{atm}) due to the release of latent heat.

The mass fraction of ice f_s as a function of temperature can be calculated with the following expression:

$$f_s(T) = 1 - x^0/x(T) \qquad 4.16$$

where x_0 and $x(T)$ are the initial mass fraction of solids in the food and the mass fraction of solids in the food (generated after the initial reference instant) at temperature T, respectively. Take $'m'$ as the initial mass of solids, m_{w_0} as the initial mass of water and $m_w(T)$ as the mass of water in the food at temperature T, then

$$x_0 = \frac{m}{m + m_{w_0}}, \quad x(T) = \frac{m}{m + m_w(T)}$$

And, therefore

$$f_s(T) = \frac{mass\ pf\ ice}{total\ mass} = 1 - x_0/x(T) = \frac{m + m_w(T)}{m + m_{w_0}} = \frac{m_{w_0} - m_w(T)}{m + m_{w_0}}$$

In Rahman et al. (2010), the following extended Clausius–Clapeyron equation is presented in order to calculate the freezing point depression $(T_w - T)$ of gel, as a function of mass fraction of solids (x):

$$Tw - T = -\frac{\alpha_w}{\gamma_w}\ln\left(\frac{1 - x - Bx}{1 - x - Bx + Ex}\right) \qquad 4.17$$

where T_w and T are the freezing point of water and food at atmospheric pressure (i.e., 0°C), respectively. α_w is the molar freezing point constant of water (1860 kg K/mol), γ_w is the molecular weight of water (18 kg/mol), E is the molecular weight ratio of water and solids (γ_w/γ_s) and B is the ratio of unfrozen water to total solids. E and B are model parameters that were estimated by Rahman et al. (2010) using Statistical Analysis Software non-linear regression. From Eq. (4.17), we can calculate the freezing point depression, depending on the mass fraction of solids as a function of T

$$x(T) = \frac{e^{\left(\frac{\gamma w}{aw}\right)(T-T_w)} - 1}{e^{\left(\frac{\gamma w}{aw}\right)(T-T_w)}(1+B-E)-1-B}$$

4.18

Substituting (4.18) into (4.16), we finally obtain the equation for the mass fraction of ice as a function of temperature

$$f_s(T) = \frac{e^{\left(\frac{\gamma w}{aw}\right)(T-T_w)} - 1 - x_0(e^{\left(\frac{\gamma w}{aw}\right)(T-T_w)} - (1+B-E)-1-B}{e^{\left(\frac{\gamma w}{aw}\right)(T-T_w)} - 1}$$

4.19

iv. Expression for the volume fractions

In this case, we work with volume fractions instead of mass fractions. The relation between them is given in the following expression. In a liquid and solid phases mixture, the volume fractions (g_l, g_s) as functions of the mass fractions (f_l, f_s) are:

$$g_l = \frac{f_l}{f_l\rho_s + f_s\rho_l} \quad and \quad g_s = \frac{f_s}{f_s\rho_l + f_l\rho_s}$$

4.20

The density of the mixture can be written as $\rho = \rho_l g_l + \rho_s g_s$ and the relationships between the mass and volume fractions in liquid and solid phases are:

$$f_l = \frac{\rho_l g_l}{\rho} \quad and \quad f_s = \frac{\rho_s g_s}{\rho}$$

As $\rho = \rho_l g_l / f_l$ and also $\rho = \rho_l g_l + \rho_s g_s = g_l(\rho_l - \rho_s) + \rho_s$,

$$g_l = \frac{f_l\rho_s}{f_s\rho_l + f_l\rho_s} \quad and \quad g_s = 1 - g_l = 1 - \frac{f_l\rho_s}{f_s\rho_l + f_l\rho_s} = \frac{f_s\rho_l}{f_s\rho_l + f_l\rho_s}$$

The solid and liquid volume fractions for our model depend on temperature and time. In a typical freezing process, a solid and liquid volume fractions model is only dependent on temperature. Therefore, in a HPSF process, we have information *a priori* about when the sample starts to freeze. We know that, until the pressure has been completely released, the sample remains unfrozen (i.e., at $= t_{p_3}$), and, at that point, there is a percentage of ice immediately formed (m_i), the mass of which we can calculate using (4.13) and (4.14). The formation of the rest of the ice, $f_s(T)$, is determined as a function of temperature (4.19). We take $x_0 = f_b + (f_l - f_b)m_i$, where f_b and f_l are the mass fractions of solids and liquids in the food, respectively.

Finally, we have that

$$g_l(t,T) = \begin{cases} 1 & t \leq t_{p_3} \\ 1 - \left(\frac{m_i\rho_l}{m_i\rho_l + (1-m_i)\rho_s}\right)\dfrac{t-t_{p_3}}{\varepsilon} & t_{p_3} < t < t_{p_3} + \varepsilon \\ 1 - \dfrac{f_l(T)\rho_l}{f_s(T)\rho_l + (1-f_l(T))\rho_s} & t \geq t_{p_3} + \varepsilon \end{cases}$$

4.21

where ε is the minor interval in which the mass fraction m_i of ice is immediately formed after the pressure release.

d. Resulting model for HPSF process

We focus on two different situations: HPSF of a large solid-type food and a small solid-type food

i. For a large solid-type food

$$
\begin{cases}
C_{vol}\dfrac{\partial T}{\partial t} - \dfrac{1}{r}\dfrac{\partial}{\partial r}\left(\mathrm{rk}\dfrac{\partial T}{\partial r}\right) - \dfrac{\partial}{\partial z}\left(\mathrm{rk}\dfrac{\partial T}{\partial z}\right) = \beta_{vol}\dfrac{dP}{dt}T - \delta e\dfrac{\delta g_1}{\delta t} & \text{in } \Omega^* \times (0, t_f), \\[2ex]
\mathrm{k}\dfrac{\partial T}{\partial n} = 0 & \text{on } \Gamma_0 \times (0, t_f), \\[2ex]
\mathrm{k}\dfrac{\partial T}{\partial n} = h(T_{amb} - T) & \text{on } \Gamma_{up} \times (0, t_f), \\[2ex]
T = T_r & \text{on } \Gamma_r \times (0, t_f), \\[2ex]
T(0) = T_0 & \text{in } \Omega
\end{cases}
\tag{4.22}
$$

ii. For small solid-type food

$$
\begin{cases}
C_{vol}\dfrac{\partial T}{\partial t} - \nabla\cdot(\kappa\nabla T) + \rho C_p u\cdot\nabla T = \beta\dfrac{dP}{dt}T - \delta_e\dfrac{\delta g_1}{\delta t} & \text{in } \Omega^* \times (0, t_f), \\[2ex]
\rho\dfrac{\partial u}{\partial t} - \nabla\cdot\eta\nabla^2 u + \rho(u\cdot\nabla)u = -\nabla p + \rho g & \text{in } \Omega_p^* \times (0, t_f), \\[2ex]
\nabla\cdot(\rho u) = 0 & \text{in } \Omega_p^* \times (0, t_f), \\[2ex]
\kappa\dfrac{\partial T}{\partial n} = 0 & \text{on } \Gamma_0^* \times (0, t_f), \\[2ex]
\kappa\dfrac{\partial T}{\partial n} = h(T_{amb} - T) & \text{on } \Gamma_{up}^* \times (0, t_f), \\[2ex]
T = T_r & \text{on } \Gamma_r^* \times (0, t_f), \\[2ex]
u = 0 & \text{on } \Gamma_p^* \times (0, t_f), \\[2ex]
\quad T(0) = T_0 & \text{in } \Omega^* \\[1ex]
\quad u(0) = 0 & \text{in } \Omega^* \\[1ex]
p = 10^5 & \text{at } A_1 \times (0, t_f)
\end{cases}
\tag{4.23}
$$

5. Conclusion

Mathematical modelling is one of the most useful tools in studying the effect of various systems and parameters on the outcome of a process with minimum time and expense. HP is a promising food processing technique and it depends on both temperature and pressure. The temperature evolution ensures a uniform distribution of pressure and temperature effect and can also avoid altering the food properties. In this chapter, we have described, modelled and analysed the problems related to High-Pressure processes in Food Engineering. Furthermore, a wide array of problems can be studied, depending on the type of food, desired effects of the process and size and geometry of the high-pressure vessel. For large solid-type food, a heat transfer model with only conduction effects was developed, whereas for liquid-type food, convective effects were also considered.

One of the major goals for any food-processing unit is to ensure the quality and safety of the product via inactivation of enzymes and microorganisms. This chapter explained a bacterial and enzyme inactivation model along with an example. This will make it easy to calculate the final bacterial count in a food sample at a given temperature and pressure.

Various simplified one and two-dimensional models for predicting the temperature profile inside a large solid food were discussed in this chapter. These models make it easier to do the approximation of the full model with less computational time. Further, models for a high-pressure shift freezing process for solid foods were developed based on the enthalpy formulation at non-constant pressure and volume phase fractions. The temperature profile throughout the process, as well as the plateau times, can be predicted with this model. The conventional and High Pressure Shift Freezing process were also modeled, assuming the homogeneous distribution of ice crystals of the same size and shape. The model was solved and validated for salty water and ice cream.

References

Becker, B.R. and Fricke, B.A. 1999. Food thermophysical property models. International Communications in Heat and Mass Transfer 26(5): 627–636. http://doi.org/10.1016/S0735–1933(99)00049-4.

Bridgman, P.W. 1914. The coagulation of albumen by pressure. J. Biol. Chem. 19(1): 511–512.

Cheftel, J.C. and Dumay, E. 1996. Effects of high pressure on dairy proteins: A review. High Pressure Bioscience and Biotechnology Proceedings of the International Conference on High Pressure Bioscience and Biotechnology 13(3): 299–308. http://doi.org/http://dx.doi.org/10.1016/S0921-0423(06)80050-X.

Considine, K.M., Kelly, A.L., Fitzgerald, G.F., Hill, C. and Sleator, R.D. 2008. High-pressure processing—Effects on microbial food safety and food quality. FEMS Microbiology Letters 281(1): 1–9. http://doi.org/10.1111/j.1574-6968.2008.01084.x.

Crank, J. 1987. Free and Moving Boundary Problems. New York, NY: Oxford Science Publications.

Denys, S., Van Loey, A.M., Hendrickx, M.E. and Tobback, P.P. 1997. Modeling heat transfer during high-pressure freezing and thawing. Biotechnology Progress 13(4): 416–423. http://doi.org/10.1021/bp970022y.

Doğan, C. and Erkmen, O. 2004. High pressure inactivation kinetics of Listeria monocytogenes inactivation in broth, milk, and peach and orange juices. Journal of Food Engineering 62(1): 47–52. http://doi.org/10.1016/S0260-8774(03)00170-5.

Erkmen, O. and Doğan, C. 2004. Kinetic analysis of Escherichia coli inactivation by high hydrostatic pressure in broth and foods. Food Microbiology 21(2): 181–185. http://doi.org/10.1016/S0740-0020(03)00055-8.

Hendrickx, M., Ludikhuyze, L., Van Den Broeck, I. and Weemaes, C. 1998. Effects of high pressure on enzymes related to food quality. Trends in Food Science and Technology 9(5): 197–203. http://doi.org/10.1016/S0924-2244(98)00039-9.

Hite, B.H. 1899. The effects of pressure in the preservation of milk. West Verginia Agricultural Experiment Station, Morgantown 58: 15–35.

Infante, J.A., Ivorra, B., Ramos, A.M. and Rey, J.M. 2009. On the modelling and simulation of high pressure processes and inactivation of enzymes in food engineering. Mathematical Models and Methods in Applied Sciences 19(12): 2203–2229. http://doi.org/10.1142/S0218202509004091.

Norton, T., Delgado, A., Hogan, E., Grace, P. and Sun, D.W. 2009. Simulation of high pressure freezing processes by enthalpy method. Journal of Food Engineering 91(2): 260–268. http://doi.org/10.1016/j.jfoodeng.2008.08.031.

Otero, L. and Sanz, P.D. 2006. High-pressure-shift freezing: Main factors implied in the phase transition time. Journal of Food Engineering 72(4): 354–363. http://doi.org/10.1016/j.jfoodeng.2004.12.015.

Otero, L., Ramos, A.M., de Elvira, C. and Sanz, P.D. 2007. A model to design high-pressure processes towards an uniform temperature distribution. Journal of Food Engineering 78(4): 1463–1470. http://doi.org/10.1016/j.jfoodeng.2006.01.020.

Rahman, M.S., Al-Saidi, G., Guizani, N. and Abdullah, A. 2010. Development of state diagram of bovine gelatin by measuring thermal characteristics using differential scanning calorimetry (DSC) and cooling curve method. Thermochimica Acta 509(1-2): 111–119. http://doi.org/10.1016/j.tca.2010.06.011.

Rajha, H.N., Boussetta, N., Louka, N., Maroun, R.G. and Vorobiev, E. 2014. A comparative study of physical pretreatments for the extraction of polyphenols and proteins from vine shoots. Food Research International 65(PC): 462–468. http://doi.org/10.1016/j.foodres.2014.04.024.

Ramos, M.A. and Smith, N. 2009. Mathematical models in food engineering. XXI Congreso de Ecuaciones Diferenciales Y Aplicadas 2009: 1–8.

Smelt, J.P.P. 1998. Recent advances in the microbiology of high pressure processing. Trends in Food Science & Technology 9(4): 152–158. http://doi.org/10.1016/S0924-2244(98)00030-2.

Smith, N.A.S., Peppin, S.S.L. and Ramos, A.M. 2012. Generalized enthalpy model of a high-pressure shift freezing process. Proceedings of the Royal Society A: Mathematical, Physical and Engineering Sciences 468(2145): 2744–2766. http://doi.org/10.1098/rspa.2011.0622.

Smith, N.A.S., Mitchell, S.L. and Ramos, A.M. 2014. Analysis and simplification of a mathematical model for high-pressure food processes. Applied Mathematics and Computation 226: 20–37. http://doi.org/10.1016/j.amc.2013.10.030.

Voller, V.R., Swaminathan, C.R. and Thomas, B.G. 1990. Fixed grid techniques for phase change problems: A review. International Journal for Numerical Methods in Engineering 30(4): 875–898. http://doi.org/10.1002/nme.1620300419.

CHAPTER 13

Food Process Modeling and Optimization by Response Surface Methodology (RSM)

Narjes Malekjani[1] and *Seid Mahdi Jafari*[2,*]

1. Introduction

In the current world of energy and market competition, the main purpose of any industrial practice is achieving the optimum production conditions. Optimization means finding the best quality criteria (product and process efficiency) while saving time and cost (Baş and Boyacı, 2007; Sevda et al., 2012; Witek-Krowiak et al., 2014; Nugent et al., 2017). Exploring the optimum point for food production is not an easy practice because several variables affect the quality of such complex biological products. So, identification of the effects of individual factors and also the combination and interaction of more than one factor are crucial in locating the optimum condition which can only be achieved by performing experiments (Nwabueze, 2010). In the past, optimization experiments in food processes were performed by investigating the effect of one-variable changes on a response while all others were held at a constant level (one-variable analysis or one factor at a time). This method couldn't take the interaction between the variables and their effects on the response into account. Also, it required a large number of experiments to be conducted, leading to high costs and long experimentation time (Bezerra et al., 2008; Witek-Krowiak et al., 2014). Nowadays, multivariate analysis is used to resolve these problems. In this method, the number of experiments is decreased while the effect of independent variables and their relationships on process response is well understood (Witek-Krowiak et al., 2014). The RSM is a collection of statistical and mathematical methods established on the fit of a polynomial model to the data that must depict the behavior of a data set with the purpose of making statistical predictions. The approach is useful for data modeling, optimizing, designing, developing, and improving processes where a response or responses are affected by several variables (Kaushik et al., 2006; Ghorbannezhad et al., 2016). This method was first used for a real (physical) system optimization by Box and Wilson (1951) and now it is used in many industrial and research applications in the fields of food, biological and clinical sciences, chemical engineering, physics and even social sciences (Baş and Boyacı, 2007; Nicolai and Dekker, 2009). Before using the RSM approach, the correct experimental design must be selected that will designate which treatments should be done in the experimental region being studied. For this purpose, experimental designs for

[1] Department of Food Science and Technology, Faculty of Agricultural Sciences, University of Guilan, Rasht, Iran.
[2] Department of Food Materials and Process Design Engineering, Gorgan University of Agricultural Sciences and Natural Resources, Gorgan, Iran.

quadratic response surfaces, such as three-level factorial, central composite and Box-Behnken, should be applied (Bruns et al., 2006). Generally speaking, RSM is an appropriate approach which can be widely used to optimize food industry processes. However, some researchers are not familiar with this approach. This misusing can be attributed to performing inappropriate experimental design, inappropriate screening of independent variables, and improper selection of levels for independent variables. Although there are a lot of valuable books and review articles regarding the fundamentals and application of RSM (Gunsts et al., 1996; Khuri, 2006; Baş and Boyacı, 2007; Sarabia and Ortiz, 2009; Barton, 2013; Myers et al., 2017; Yolmeh and Jafari, 2017), in this chapter, we focus on the practical applications of RSM in the food industry, along with a case study, after explaining the basic definitions and procedure for applying this method.

2. Basic Principles of RSM

There are some important terms that are frequently used in RSM analysis; before explaining the basic principles of RSM we represent these terms in Fig. 1. In conventional factorial data analysis, the presence of the effects of independent variables (factors) on dependent variables (response) is determined, while in RSM analysis the value of the response is predicted based on factors. As we mentioned earlier, RSM gives us a mathematical model which determines the optimum combinations of the factors in order to optimize the response. The model describes the relationship between factors and interaction between the combination of factors and response within the entire experimental domain (Nwabueze, 2010).

The general equation describing the relationship between response (y) and input variables (X1, X2,…, Xk) is given in Eq. (1):

$$y = f(X_1, X_2, …, X_k) + \varepsilon \tag{1}$$

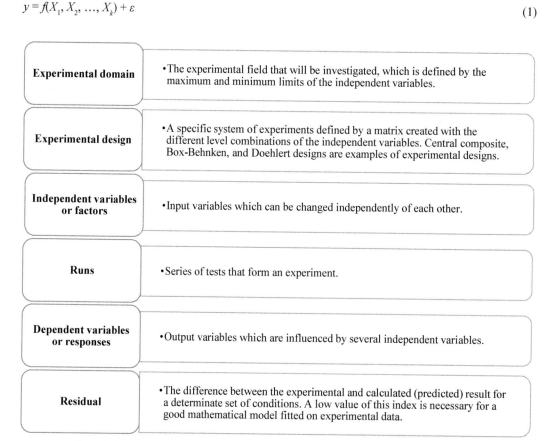

Experimental domain	• The experimental field that will be investigated, which is defined by the maximum and minimum limits of the independent variables.
Experimental design	• A specific system of experiments defined by a matrix created with the different level combinations of the independent variables. Central composite, Box-Behnken, and Doehlert designs are examples of experimental designs.
Independent variables or factors	• Input variables which can be changed independently of each other.
Runs	• Series of tests that form an experiment.
Dependent variables or responses	• Output variables which are influenced by several independent variables.
Residual	• The difference between the experimental and calculated (predicted) result for a determinate set of conditions. A low value of this index is necessary for a good mathematical model fitted on experimental data.

Figure 1: Frequently used terms in RSM analysis (Bezerra et al., 2008; Yolmeh and Jafari, 2017).

where k is the number of input variables and ε is a statistical term which accounts for other sources of variability not accounted for, such as measurement errors, noise, etc. X_1, X_2,..., X_k are called natural variables. These variables have different measurement units or ranges, so they must be normalized before doing regression analysis. Usually, in RSM analysis, the natural variables are transformed into coded variables between the ranges of –1 to 1. The equation used for producing coded variables is Eq. (2).

$$x = \frac{X - [X_{max} + X_{min}]/2}{[X_{max} - X_{min}]/2} \qquad (2)$$

where x represents the coded variable and the subscript indices of *min* and *max* show the minimum and maximum values of natural variables. The actual response function isn't usually defined, so selecting the appropriate regression function is a critical step which will significantly influence the responses.

3. RSM Implementation Procedure

The procedure for implementing RSM analysis is presented briefly in Fig. 2. Details of each analysis step are described in the following subsections.

3.1 Definition of the case and selection of independent variables and their levels

Numerous input variables usually affect the output response of a system or process in food and biological materials, but taking the individual effects of each of these parameters on process response is impossible due to the high costs involved and the number of runs required. Therefore, the first and most crucial step in RSM analysis is the determination of the factors which have the largest influence on the response and their ranges. This process is called "screening". This step can be done using full or fractional factorial designs (for 2–4 factors) and fractional factorial or Plackett-Burman design (for 5 or more factors). In such designs, only main effects are estimated; interactions between independent variables are usually considered insignificant and are neglected. The Plackett-Burman design type is a two-level fractional factorial screening design for studying N-1 variables using N runs, where N is a multiple of 4 (Lundstedt et al., 1998; Myers et al., 2016; Yolmeh and Jafari, 2017). Minimum run equireplicated resolution V design is another method of design of experiments which has the advantage of high resolution and reduced number of experiments and considers 2 and 3-level interactions (Witek-Krowiak et al., 2014).

Afterward, the levels of the selected variables should be identified carefully. Inappropriate selection of independent variable levels would directly lead to an incorrect optimization of the process (Nwabueze, 2010; Witek-Krowiak et al., 2014).

3.2 Design of experiments (DOE)

Selection of experimental design is another important step in RSM analysis. The basic goal of RSM is to guide the experimenter in finding the optimum points (Nwabueze, 2010). DOE determines the points where the response should be evaluated. Some of the most used designs in RSM analysis in the food industry are full (or fractional) factorial designs (FFD), Box–Behnken design (BBD), central composite design (CCD), Plackett–Burman design (PBD) and Doehlert Matrix (D) (Sablani et al., 2006; Erdogdu 2008; Witek-Krowiak et al., 2014; Yolmeh and Jafari, 2017), which are discussed in the following subsections.

3.2.1 Full factorial design (FFD)

In this design, all of the possible combinations of input variables are considered. FFD does not exhibit good efficiency in the modeling of second or higher order polynomial functions, especially when the number of factors is higher than 2. FFD is appropriate to obtain a lot of information about the main effects

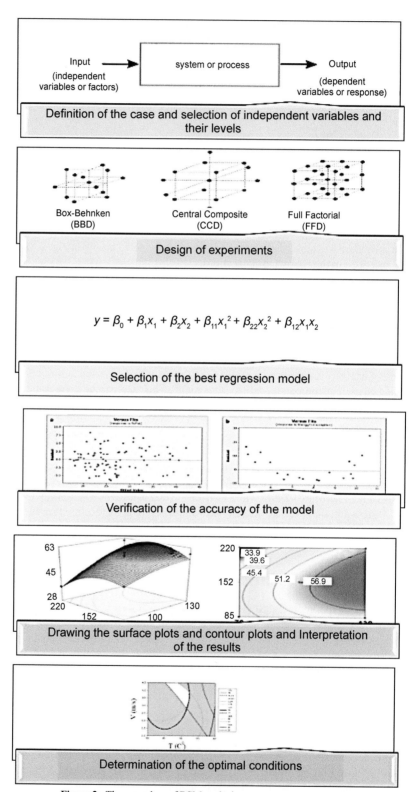

Figure 2: The procedure of RSM analysis.

in a proportionately small numbers of runs. However, FFD is not appropriate to evaluate the interaction between factors due to deficiency of this design to provide information about interactions. It is the most labor-intensive experiment design (Sablani et al., 2006; Yolmeh and Jafari, 2017). If the number of design variables becomes large, a fraction of a full factorial design can be used at the cost of estimating only a few combinations among variables. This is called fractional factorial design and is usually used for screening important design variables. For a 3^n factorial design, a $(1/3)^p$ fraction can be constructed, resulting in 3^{n-p} points. For instance, if $p = 1$ in a 3^3 design, the result is a one-third fraction which is called 3^{3-1} design (Montgomery, 2017).

3.2.2 Central composite design (CCD)

The central composite design consists of factorial points, a central point, and axial points which are at a distance α from the central point. CCD can predict linear and quadratic models with high quality. It provides a reasonable amount of information for testing lack-of-fit, while not involving an unusually large number of experimental runs. Also, CCD is appropriate to study factors with three and/or five levels. However, a CCD considers extreme points, which is not advisable for special processes like the extraction of a compound sensitive to high temperatures and pressures. The experimental points can be imagined like points on the surface of a ball that has a cube inside it (Myers et al., 2016). Experimental runs in CCD can be estimated by Eq. (3):

$$N = k^2 + 2k + c_0 \tag{3}$$

where k is the number of factors and C_0 is the number of replications in central point.

α values can be calculated by $\alpha = 2^{(k-p)/4}$; quintuplet levels are applied for all factors ($-\alpha$, -1, 0, $+1$, $+\alpha$). CCD is categorized into three types depending on the location of axial points. Circumscribed central composite (CCC), inscribed central composite (CCI) and face-centered composite (CCF). α value is equal to 1 in face-centered composite, which means that this design has only three levels and the design space is then a cube, rather than a sphere. Comparison between the domain of interest with the domain of operability helps in finding the appropriate CCD design (Witek-Krowiak et al., 2014; Myers et al., 2016).

3.2.3 Box-behnken design (BBD)

BBD is a three-level design as an alternative to FFD which is labor intensive. It has been developed for fitting second order polynomial response surfaces. BBD has fewer experiment numbers than CCD and FFD (Table 1, 13 in the case of BBD compared to 27 and 15 for FFD and CCD, respectively). So, applying this design is popular in food processes for financial reasons. In a BBD, the experimental points are situated on a hypersphere, equidistant from the central point. It can be imagined that the experimental points lay on the center and surface of a ball that is inside a cubic box (in contrast to central composite design, where the cube was located inside the ball). The design points never exceed the cube's surface. The input variables are also not set simultaneously at their extreme levels (vertices of the cube, where high levels of factors involved in the process are difficult to implement, such as high temperatures and pressures next to each other). This approach avoids performing optimization under extreme conditions. In addition, BBD is not appropriate for studying factors with more than three levels. In other words, only triplex levels (-1, 0, $+1$) are applied. This design may contain some regions with poor quality for response prediction (Witek-Krowiak et al., 2014; Myers et al., 2016). The number of experimental points in this method is calculated using Eq. (4):

$$N = 2k(k-1) + C_0 \tag{4}$$

where k is the number of factors and C_0 is the number of replications in the central point (Said and Amin, 2016).

Table 1. Some experimental matrices for RSM designs based on three level variables.

Experiment no.	Full factorial design (FFD)			Box-Behnken design (BBD)			Central composite design (CCD)			
	x_1	x_2	x_3	x_1	x_2	x_3	x_1	x_2	x_3	Factorial points
1	−1	0	0	−1	−1	0	−1	−1	−1	
2	−1	0	−1	1	−1	0	1	−1	−1	
3	−1	−1	−1	−1	1	0	−1	1	−1	
4	−1	−1	0	1	1	0	1	1	−1	
5	−1	0	1	−1	0	−1	−1	−1	1	
6	−1	1	0	1	0	−1	1	−1	1	
7	−1	1	1	−1	0	1	−1	1	1	
8	−1	−1	1	1	0	1	1	1	1	
9	−1	1	−1	0	−1	−1	$-\alpha$	0	0	Axial points
10	0	0	0	0	1	−1	α	0	0	
11	0	0	−1	0	−1	1	0	$-\alpha$	0	
12	0	−1	−1	0	1	1	0	$-\alpha$	0	
13	0	−1	0	0	0	0	0	0	$-\alpha$	
14	0	0	1				0	0	α	
15	0	1	0				0	0	0	Central points
16	0	1	1							
17	0	−1	1	0	0	0				
18	0	1	−1	0	0	−1				
19	1	0	0	0	−1	−1				
20	1	0	−1	0	−1	0				
21	1	−1	−1	0	0	1				
22	1	−1	0	0	1	0				
23	1	0	1	0	1	1				
24	1	1	0	0	−1	1				
25	1	1	1	0	1	−1				
26	1	−1	1							
27	1	1	−1							

3.2.4 Plackett–Burman design (PBD)

Plackett–Burman design (Table 2) is a helpful and economical design for the identification of the most important input variables in the pre-experimental phase, when the behavior of the system is not completely understood. In other words, it is often considered as a useful screening method. Only N = k+1 experiments are needed for this design. The number of experiments is the same as the number of parameters and the degree of freedom in this design is equal to zero. One of the limitations of this design is its inability to predict interactions between factors in second-order polynomial models. On the other hand, the number of experiments in this type of design should be a multiple of 4. To avoid this shortcoming, dummy variables are usually used (Witek-Krowiak et al., 2014; Myers et al., 2016).

3.2.5 Doehlert design (DD)

This design, which is sometimes called Uniform Shell Design, is based on a simple design. In this technique, the simplex points are subtracted from each other, forming a hexagon called a Doehlert design.

Table 2: Plackett-Burman Design with 12 experiments for 11 factors.

	X_1	X_2	X_3	X_4	X_5	X_6	X_7	X_8	X_9	X_{10}	X_{11}
1	+1	+1	+1	+1	+1	+1	+1	+1	+1	+1	+1
2	−1	+1	−1	+1	+1	+1	−1	−1	−1	+1	−1
3	−1	−1	+1	−1	+1	+1	+1	−1	−1	−1	+1
4	+1	−1	−1	+1	−1	+1	+1	+1	−1	−1	−1
5	−1	+1	−1	−1	+1	−1	+1	+1	+1	−1	−1
6	−1	−1	+1	−1	−1	+1	−1	+1	+1	+1	−1
7	−1	−1	−1	+1	−1	−1	+1	−1	+1	+1	+1
8	+1	−1	−1	−1	+1	−1	−1	+1	−1	+1	+1
9	+1	+1	−1	−1	−1	+1	−1	−1	+1	−1	+1
10	+1	+1	+1	−1	−1	−1	+1	−1	−1	+1	−1
11	−1	+1	+1	+1	−1	−1	−1	+1	−1	−1	+1
12	+1	−1	+1	+1	+1	−1	−1	−1	+1	−1	−1

The response surface is calculated by the minimum number of experiments and the addition of a few experiments can easily help in exploring the adjacent domain (Lundstedt et al., 1998). This design is very flexible and fully sequential and can be upgraded to a (k+1) factor design with the addition of a few experiments (Witek-Krowiak et al., 2014).

3.3 Selection of the best regression model in RSM

The mathematical form of response surfaces is represented by regression equations called models (Nwabueze, 2010). These models could be first or second-order depending to the relationship between response (y) and the input variables (x_1, x_2, ...x_n). If this relationship is linear, the first order model suits; otherwise, second order model can predict non-linearity of the system. First order model in general form is represented by Eq. (5). For k independent variables the equation is as Eq. (6).

$$y = \beta_0 + \sum_{i=1}^{k} \beta_i x_i + \varepsilon \tag{5}$$

$$y = \beta_0 + \beta_1 x_1 + \beta_2 x_2 + \beta_3 x_3 + \cdots + \beta_k x_k + \varepsilon \tag{6}$$

where $\beta_0, \beta_1, \beta_2 \ldots \beta_k$ are the intercept and regression coefficients corresponding to the factors $x_1, x_2, \ldots, x_k$ and ε is the random error which has been introduced before. This model yields straight linear surface plots (Khuri and Mukhopadhyay, 2010). In the case of a complex relationship between response and factors where the curvature is observed in the system, second order or polynomial model (Eqs. (7) and (8)) are applied. Depending on the type of interactions between variables, this model yields dumbbell, inclined or twisted-shaped plots. As this model can predict nonlinear terms, it is usually preferred over first order model (Nwabueze, 2010).

$$y = \beta_0 + \sum_{i=1}^{k} \beta_i x_i + \sum_{i=1}^{k} \beta_{ii} x_i^2 + \sum_{1 \leq i \leq j}^{k} \beta_{ij} x_i x_j + \varepsilon \tag{7}$$

$$y = \beta_0 + \beta_1 x_1 + \beta_2 x_2 + \cdots + \beta_k x_k + \beta_{12} x_1 x_2 + \beta_{13} x_1 x_3 + \cdots + \beta_{1k} x_1 x_k + \beta_{23} x_2 x_3 + \beta_{24} x_2 x_4 + \cdots + \beta_{2k} x_2 x_k + \cdots + \beta_{k-1k} x_{k-1} x_k + \beta_{11} x_1^2 + \beta_{22} x_2^2 + \cdots + \beta_{kk} x_k^2 + \varepsilon \tag{8}$$

where $\beta_0, \beta_i, \beta_{ii}$ and β_{ij} are corresponding coefficients to intercept, linear, quadratic and interaction terms. The second order model is preferred and widely used in several RSM analyses because of its flexibility, easy estimation of model coefficients (usually by using least square method) and the proved capability in solving RSM problems in many cases. Sometimes, models with higher orders might be used, however, using models with lower order, yielding a suitable fit over experimental data, is always preferred (Nwabueze, 2010). After selecting the order of the model, an analysis should be performed to determine the model coefficients and their statistical significance, the mean response, and the optimum conditions which yield the best response (Khuri, 2006).

3.4 Verification of the accuracy of the model

After designing the experiments and fitting the measured responses into the selected model, the accuracy of the response model should be checked. Determination of statistical significance of the model is done using analysis of variance (ANOVA). The total sum of squares of y (SST), regression sum of squares (SSR), and error sum of squares of the model (SSE), respectively, are defined by Eqs. (9), (10) and (11).

$$SST = \sum_{i=1}^{n} (y_i - \bar{y})^2 \tag{9}$$

$$SSE = \sum_{i=1}^{n} (y_i - \hat{y}_i)^2 \tag{10}$$

$$SSR = \sum_{i=1}^{n} (\hat{y}_i - \bar{y})^2 \tag{11}$$

where n is the number of observations, y_i is the ith observation, $\bar{y}$ is the mean value of all observations, and $\hat{y}_i$ is the predicted response. The total sum of squares is comprised of the regression sum of squares plus error sum of squares. So:

SSR = SST–SSE $\tag{12}$

Dividing the sum of squares with the respective degrees of freedom yields respective mean squares (MSs) or variances. The effects of individual model coefficients are determined through their respective sum of squares (Erdogdu, 2008; Mäkelä, 2017). The ANOVA analysis is presented in Table 3, where p is the number of coefficients of the model and m is the numbers of levels used in the investigation (Bezerra et al. 2008). Determination of F value (Table 3) is the most important part of ANOVA. If $\frac{MS_{regression}}{MS_{residuals}} > F_{\gamma,k,n-p}$ the model is suitable. The value of $F_{\gamma,k,n-p}$ may be looked up from the Tables of the F distribution or calculated by computer software.

Table 3. The general form of analysis of variance (ANOVA) in RSM.

Parameter	Sum of the square	Degree of freedom	MS	F value
Regression	SSR	p-1	$\dfrac{SSR}{p-1}$	$\dfrac{MS_{regression}}{MS_{residuals}}$
Residuals	SSE	n-p	$\dfrac{SSE}{n-p}$	
Lack of fit	SSE_L	m-p	$\dfrac{SSE_L}{m-p}$	$\dfrac{MS_{Lack\ of\ fit}}{MS_{Pure\ error}}$
Pure error	SSE_P	n-m	$\dfrac{SSE_P}{n-m}$	
Total	SST	n-1		

3.4.1 Lack-of-fit test

Another F-test which can determine the adequacy of the model is the lack-of-fit test. This test can determine the integrity of the fitted regression model and the accuracy of the model order. Lack-of-fit involves rejection or acceptance of a null hypothesis (H_0) which states that the regression model is correct. The error sum of squares is comprised of two parts SSE = SSE_P + SSE_L where SSE_P and SSE_L are related to the sum of squares due to pure error and lack-of-fit, respectively. A lack-of-fit test requires at least one set of repeated observations. If there are q sets of repeated experiments containing r_1, r_2, …, r_q observations, then Eq. (13) can calculate the sum of squares due to the pure error:

$$SSE_P = \sum_{i=1}^{q}\sum_{j=1}^{r_i} (y_{ij} - \bar{y}_i)^2 \tag{13}$$

where, y_{ij} is the jth observation in the ith set of repeated experiments. And;

$$SSE_L = SSE - SSE_P \tag{14}$$

As it is shown in Table 3, respective mean squares (MSs) or variances are calculated by dividing the sum of squares with the respective degrees of freedom. Also, the influence of each model's coefficients can be calculated using their respective sum of squares. Same as the previous section, if $\dfrac{MS_{Lack\ of\ fit}}{MS_{Pure\ error}} >$ $F_{\gamma,k,n-p}$, the regression model cannot adequately fit the experimental data and a more appropriate model should be selected. Otherwise, the model is suitable.

3.4.2 Coefficient of determination (R^2)

Coefficient of determination is another parameter which is commonly used to confirm the general predictive capability of the fitted model. However, R^2 index alone cannot demonstrate the accuracy of the model, because the index is a measure for the amount of the decreasing changeability of response achieved by using the regressor variables in the model. In addition, increasing a variable to the model will increase R^2 without considering the statistical significance of the additional variable. Thus, a high value of R^2 is not necessarily indicative of the accuracy of the model (Yolmeh and Jafari, 2017). R^2 is calculated using Eq. (15):

$$R^2 = \frac{SSR}{SST} = 1 - \frac{SSE}{SST} \tag{15}$$

The value of R^2 is between 0 and 1. The adjusted R^2 which won't increase with the addition of not statistically significant variables is a better indicator of model accuracy (Eq. (16)):

$$R_{adj}^2 = 1 - \frac{SSE/(n-p)}{SST/(n-1)} = 1 - \left(\frac{n-1}{n-p}\right)(1-R^2) \tag{16}$$

For the models which are utilized for prediction, the predictive quality can be determined using the coefficient of determination for prediction (R_{pred}^2) as Eq. (17) (Erdogdu, 2008).

$$R_{pred}^2 = 1 - \frac{PREES}{SST} \tag{17}$$

where

$$PRESS = \sum_{i=1}^{n} (y_i - \hat{y}_{(i)})^2 \tag{18}$$

PRESS represents the prediction error sum of squares. $\hat{y}_{(i)}$ is the value for y_i which is predicted by the regression model when y_i has been omitted and the remaining n-1 observations were applied.

3.4.3 Absolute average deviation (AAD)

The absolute average deviation (AAD) is another measure of accuracy. AAD is the average of the absolute deviations from a central point and is a summary index of statistical dispersion or variability, which is calculated by Eq. (19).

$$AAD = \{[\sum_{i=1}^{p} (|yi_{exp} - yi_{cal}|/yi_{exp})]/p\} \times 100 \tag{19}$$

where p, yi_{exp}, and yi_{cal} represent the number of experiments, experimental, and calculated responses, respectively. Both R^2 and AAD should be evaluated to check the accuracy of the model. R^2 must be near 1 and the AAD between the estimated and observed data must be as low as possible. The suitable values of R^2 and AAD imply that the fitted model depicts the correct behavior of the process and it can be successfully used for the optimization of food processes (Ghorbannezhad et al., 2016).

3.4.4 Residual plots

Briefly, it can be concluded that, if the regression is significant and the lack of fit is non-significant, the model is well fitted to the experimental data. Moreover, residual analyses (plots of observed-predicted values versus the factors and responses) would be a valuable criterion for this assessment. Using residual plots, we can assess whether the observed error (residuals) is consistent with the stochastic error. The residuals should not be either systematically high or low. In addition, the residuals should be centered on zero throughout the range of fitted values. In other words, the model is correct on average for all fitted values. Further, in ordinary least squares context, random errors are assumed to produce residuals that are normally distributed. Therefore, the residuals should fall in a symmetrical pattern and have a constant spread throughout the range (Myers et al., 2016). The non-random pattern in the residuals indicates that the deterministic portion (predictor variables) of the model is not capturing some explanatory information, which is "leaking" into the residuals. The graph could represent several ways in which the model is not explaining all that is possible. Possibilities include missing variables and/or higher order term of a variable in the model to explain the curvature, and/or interaction between terms already in the model. Identifying and fixing the problem so that the predictors now explain the information that they missed before should produce a good-looking set of residuals (Myers et al., 2016). Various criteria (R^2, AAD, residual analysis, and lack-of-fit) are used to evaluate the suitability of response surface models for fitting the data, which are reviewed here. In almost all RSM studies, R^2 has been used to evaluate the general predictive capability of the fitted model. However, R^2 is not a sufficient index for this evaluation, so other criteria should be investigated. Residual analysis and lack-of-fit evaluations have not been applied in some papers; however, lack-of fit value was inappropriate in some of them (the P value of lack-of-fit should be insignificant ($P > 0.05$) for a suitable fitting of response surface models). Unfortunately, AAD criterion has not been considered in any of the papers reviewed here even though it is a useful value to assay the suitability of fitting the response surface model, as mentioned earlier (Ghorbannezhad et al., 2016).

3.5 The surface plots and contour plots

The visualization of the predicted model equation can be accomplished by the response contour and surface plots, as shown in Fig. 3. These plots are theoretical three-dimensional outputs of the RSM approach that indicate the relationship between the dependent and independent variables. They have been plotted after removing the insignificant terms in the response model using related software.

The contour and surface plots are used to describe and visualize the mental picture of changes in response by changes in the independent variables. The contour plot is a two-dimensional screen of the surface plot, in which ranges of constant dependent variables are drawn in the plane of the independent variables. It must be noted that these plots display the visualization of estimated responses, not the real responses (Baş and Boyacı, 2007; Myers et al., 2016). There are different types of patterns which may occur in contour and surface plots. One of the important issues in optimization of surface plots is the location of the stationary point. The stationary point is the minimum or maximum point of a second-order regression model, where the first derivative of the equation is equal to zero. If the aim of the process is an only the determination of the effects of factors without optimization, the stationary point isn't determined (Baş and Boyacı, 2007).

3.5.1 Simple maximum pattern "Peak"

This plot of this case shows a symmetrical hill shape representing a maximum point in the center of the contour. The minimum response has the same shape with a depression in the center of the contour plot. In such plots, if the contours are circular shaped, the effects of the factors are independent (no interaction),

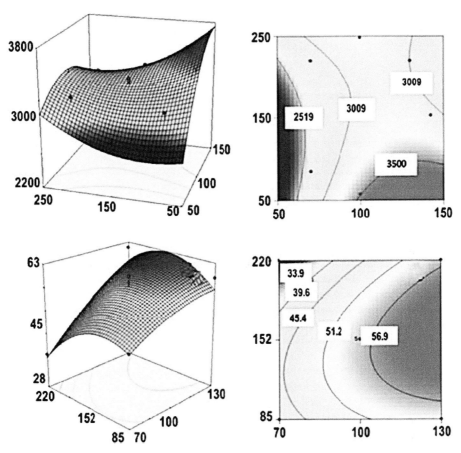

Figure 3: Response surface (left) and contour (right) plots of the predicted model equation.

Color version at the end of the book

otherwise, the elliptically shaped contours mean the interaction between factors (Gardiner and Gettinby, 1998).

3.5.2 Minimax pattern "Saddle"

The saddle point or stationary point is located near the central region of both plots. In this case, the optimum point cannot be considered as a maximum or minimum because it depends on the direction of travel from the central point, meaning that the center is maximum in one direction and minimum in another one (Gardiner and Gettinby, 1998).

3.5.3 Stationary ridge pattern "Hillside"

In this case a plot with stationary ridge pattern is observed. This plot is shaped like an arch. In such plots, there are many possible factor settings that maximize the response, meaning that there is flexibility in selecting the operating conditions (Myers et al., 2016).

3.5.4 Rising ridge pattern

The rising ridge plot shows that the maximum point is located outside of the experimental domain (Gardiner and Gettinby, 1998). The plot in the opposite direction, in which the minimum point is located outside the experimental data, is called the falling ridge. In contrast to a stationary ridge pattern, this plot shows that conducting additional experiments outside the experimental region might be helpful. Falling or rising ridge patterns are often a sign of the incorrect selection of the experimental design region (Myers et al., 2016).

3.6 Determination of the optimal conditions

The optimum point (maximum or minimum) could be calculated through the first derivate of the mathematical equation (Eq. (20)) which depicts the response surface:

$$y = \beta_0 + \beta_1 x_1 + \beta_2 x_2 + \beta_{11} x_1^2 + \beta_{22} x_2^2 + \beta_{12} x_1 x_2 \tag{20}$$

By calculating $\Delta y / \Delta x_1$ and $\Delta y / \Delta x_2$ and setting them to zero, the optimum point could be found:

$$\Delta y / \Delta x_1 = \beta_1 + 2\beta_{11} x_1 + \beta_{12} x_2 = 0 \tag{21}$$

$$\Delta y / \Delta x_2 = \beta_2 + 2\beta_{22} x_2 + \beta_{12} x_1 = 0 \tag{22}$$

Equations (21) and (22) are solved in order to determine the values of X_1 and X_2. These values reveal the coded value of the factors, which give the maximum or minimum response. For optimization of the process, the target point has to be found in the ranges of studied factors. Generally, in the food industry, researchers are interested in the optimization of several responses concurrently, which is more complicated than the optimization of processes with only one response. For this purpose, a desirability function, which is a multi-criteria methodology, is used. This technique makes a desirability function for each individual response. The desirability function approach is one of the most widely-used methods in the industry for the optimization of multiple-response processes. It is based on the idea that the "quality" of a product or process which has multiple quality characteristics, with one of them outside of some "desired" limits, is completely unacceptable. The method finds operating conditions of X that provide the "most desirable" response values. For each response $Y_i(x)$, a desirability function $d_i(Y_i)$ assigns numbers between 0 and 1 to the possible values of Y_i, with $d_i(Y_i) = 0$ representing a completely undesirable value of Y_i and $d_i(Y_i) = 1$ representing a completely desirable or ideal response value. The individual desirability values are then combined using the geometric mean, which gives the overall desirability (D).

$$D = (d_1(Y_1) d_2(Y_2) \dots d_k(Y_k))^{1/k} \tag{23}$$

where k represents the number of responses. It should be noted that, if any response Yi is completely undesirable, then the overall desirability is zero. Applying desirability functions in food processes has several advantages, such as cost-effectiveness, efficiency, and objectivity in optimizing the multiple-response processes (Bezerra et al., 2008). Measurement of responses under the predicted optimum conditions and comparison of the observed responses with the responses predicted through the model is another important point. The difference between observed and predicted responses must be as low as possible; this finding reveals that the predicted optimal conditions are valid (Myers et al., 2016).

4. Common Mistakes in Performing RSM

It has been shown that there is great potential for applying RSM in food industry processes. However, this technique is not always successful and there are several common mistakes made when using it:

Correct choice of the range of independent variables: A certain amount of preliminary work or experience is needed in order to select the appropriate range of each factor, which directly affects the success of RSM optimization. As it was stated before, the stationary point in RSM is a very important parameter. In most RSM optimizations in literature, unsuitable determination of independent variables has led to no optimum point and no maximum, or minimum point has been considered as the optimum point. If the stationary point is not determined, the study cannot be called an optimization study (Baş and Boyacı, 2007; Yolmeh and Jafari, 2017).

Correct selection of the polynomial model: A second order equation is commonly used in RSM methodology. This mistake is often made because most of the RSM packages use second-order model and it is easier to use a second-order equation since it only has one stationary point (Baş and Boyacı, 2007).

5. Advantages and Limitations of RSM

There are several advantages, compared to other classical methods, which can be considered when dealing with RSM analysis. First of all, other techniques can only analyze one variable at a time, while several variables can be analyzed simultaneously in RSM method. Secondly, the number of experiments is reduced significantly in the RSM analysis, and a high amount of information about the behavior of the system is achieved. This reduction in the number of experiments leads to time-saving. Another merit of RSM is the possibility of determining the effect of interactions between factors on responses. One of the limitations of RSM is that it is not suitable for all types of food processes. According to Bas and Boyaci (2007), RSM is only helpful for changes which can be fitted to a second order equation. Although there are some methods to overcome this limitation, like logarithmic transformations or other linearization methods, they might not be helpful in all cases and are also time-consuming and difficult to be determined. RSM is also not suitable for problems considering reaction kinetics (Baş and Boyacı, 2007). Also, another problem of RSM is the difficulty in the estimation of the accuracy of the prediction. Finally, it is a local analysis method and the developed response is only reliable in the studied range of factors (Anandharamakrishnan, 2017).

6. Application of RSM in Different Food Technologies

In this section, we have reviewed some RSM studies published in different fields of food industry processes during the last few years. Generally, these processes involve CCD and BBD designs.

6.1 Optimization of the extraction processes

In recent years, numerous studies have been published regarding the extraction of oil, protein, phenolic compounds, pigments, polysaccharides, hydrocolloids, etc., by applying RSM in order to optimize the process conditions. There are several techniques for extraction of various bioactive compounds such as maceration, ultrasonic-assisted extraction (UAE), microwave-assisted extraction (MAE), supercritical fluid extraction (SFE), and enzymatic extraction (Wang et al., 2012; Khazaei et al., 2015; Sarfarazi et al., 2015). In these techniques, several factors, including time, pH, temperature, reagents concentration, irradiation time, flow rate, and solvent to solid ratio affect the extraction yield and process efficiency, which can be considered as independent variables in RSM. We have classified the recent extraction studies into 12 groups, based on the target compound, and each group involves different independent variables and specific objectives; studies of phenolic compounds and polysaccharides comprise the major works in this field. Maximizing the extraction yield (MEY) has been the main target for these studies. Quality of analyte (target compound) is also important in addition to the yield, which RSM can introduce in the optimal conditions for multiple-response processes. Among independent variables studied in the optimization of extraction processes, temperature, time, pH, and solid to solvent ratio were most widely used. Also, the CCD was the major design in the extraction studies by RSM. Randomization of experiments and replication on central point are important criteria that should be used in RSM designs to validate RSM modeling.

6.2 Optimization of the drying processes

Drying is one of the most important techniques used to improve the shelf life of foods, such as fruits and vegetables, which have a high moisture content (Kumar et al., 2014). There are various techniques for drying and dehydration of foods, such as hot air drying, osmotic dehydration, microwave-vacuum drying, fluidized bed drying, IR-assisted drying and microwave assisted hot air drying (Madamba, 2002). Producing a good quality dried product is dependent on several factors, such as temperature, time, thickness, airspeed, vacuum rate, microwave power, the concentration of an osmotic solution, etc. (Burande et al., 2008). Thus, the optimization of these process conditions is important in terms of improving the food quality and minimizing the process costs. Statistical approaches, such as RSM, can introduce the optimal conditions for the drying process based on several process responses, such as water loss, a solid gain, final moisture, color, and rehydration ratio. Temperature, time, osmotic solution concentration and thickness of the product have been widely used as independent variables in these optimizing studies on drying processes. In most of these studies, CCD has been applied in order to investigate and optimize the drying conditions. Also, compared with Section 6.1 (extraction processes), a higher number of papers considering drying optimization have implemented validation of predicted optimum conditions. Madamba (2002) has employed RSM appropriately because of the randomization, RSM modeling graphical presentation, modeling validity criteria and adequate validation of predicted optimum conditions.

6.3 Application of RSM in the food formulation processes

Since several factors are considered in the formulation of food products, researchers are seeking the optimization of each response (factor) which could be performed by RSM, as this approach is useful for multi-response processes. RSM has been used for optimizing formulations of different foods, such as functional short dough biscuits (Gallagher et al., 2003), soy-coffee beverage (Felberg et al., 2010), low-calorie mixed fruit jelly (Acosta et al., 2008), extruded snack food (Thakur and Saxena, 2000), cream (Mostefa et al., 2006), gluten-free bread (Sanchez et al., 2004; McCarthy et al., 2005), pork frankfurters (Pappa et al., 2000), walnut oil-in-water beverage emulsion (Gharibzahedi et al., 2012), Iranian white brine cheese (Alizadeh et al., 2005), homogenized infant foods (Martínez et al., 2004) and sweet potato-based products (Singh et al., 2004). Gharibzahedi et al. (2012) studied the effects of Arabic gum content and walnut-oil concentration on properties of prepared emulsions, namely, turbidity loss rate, density, particle size and stability, using RSM. They reported that the R optimum formulation contained 3% (w/w) walnut oil and 9.62% (w/w) Arabic gum. The above studies successfully used RSM to optimize formulation processes revealed by suitable validity criteria (good R^2_{adj} and R^2_{Pred} values, insignificant P values for lack-of fit, and good compatibility of the predicted and experimental values). In these studies, by applying RSM to optimize formulation, different factors, such as independent variables, have been investigated. These are dependent on case to case; for instance, in the work of Gharibzahedi et al. (2012), concentrations of hydrocolloid, oil and sweetener, and conditions of process, such as temperature and pH, were the main independent variables. On the other hand, physicochemical, textural and rheological properties were the main dependent variables.

6.4 Application of RSM in microencapsulation and controlled-release processes

Microencapsulation is a technique in which very small particles or droplets are surrounded by a wall material in order to make small capsules. This technique is widely used in the food and pharmaceutical industry to protect bioactives from environmental deteriorative factors and also to develop and produce improved food and drug delivery systems. However, there are many factors influencing the stability and functionality of the microcapsules, such as the type of wall material, the properties of the core materials (volatility and concentration), the characteristics of the initial emulsion (droplets size, total solids and viscosity), and the conditions of the drying process (temperature, air flow and humidity) (Jafari et al., 2008). Thus, it is important to optimize the microencapsulation process with techniques like RSM. In recent years, some researchers have applied RSM in order to optimize the parameters of the microencapsulation process efficiently for oils, probiotics, vitamins, minerals, and controlled-release of tablets. For instance, Chen et al. (2006) employed the RSM model in order to develop an optimal composition for encapsulating probiotics using sodium alginate blended with peptides, fructooligosaccharides and isomaltooligosaccharides as the wall materials. They reported that the optimal combination of all materials for the microcapsules was found to be 3% sodium alginate mixed with 1% peptides, 3% fructooligosaccharides and 0% isomaltooligosaccharides. In another study, these researchers reported that 1% sodium alginate mixed with 1% peptide and 3% fructooligosaccharides as wall materials would provide the highest protection for the probiotics (*Bifidobacterium bifidum, Bifidobacterium longum,* and *Lactobacillus acidophilus*) (Chen et al., 2005). The results of Kha et al. (2014), who optimized microencapsulation of gac oil using spray drying by RSM, revealed a combination wherein wall concentration of 29.5% and oil load of 0.2 were the optimum conditions for microencapsulation. Huynh et al. (2008) introduced a condition wherein 40% feed concentration (w/w), 18% oil concentration and 65°C outlet air temperature of spray dryer were the optimum conditions for microencapsulation. Similarly, microencapsulation of rice bran oil, sunflower oil and flaxseed oil has been optimized using RSM by Suh et al. (2007), Ahn et al. (2008) and Tonon et al. (2011), respectively. Recently, Assadpour et al. (2016) applied RSM for nano-emulsification and encapsulation of folic acid and Salimi et al. (2014) encapsulated lycopene through emulsification optimization by RSM. Pharmaceutical formulators are also seeking to find the right combination of pellets, which will produce a product with optimum properties. Optimizing the formulation is very important when it is a controlled-release dosage form because many factors can affect the release rate. RSM has been used to optimize the formulation of drugs in the development of a product with controlled-release properties. For example, Kim et al. (2007) studied formulation and optimization of a novel orally-controlled delivery system for tamsulosine hydrochloride using RSM. They found a combination of 10% hydroxypropyl methylcellulose and 10% hydroxypropyl methylcellulose phthalate at a coating level of 25% to be the optimal coating formulation. In another study, Ko et al. (2003) developed and optimized chitosan microparticles for controlled drug release using RSM. They reported that the optimum rate of drug release was observed at 1.8% chitosan concentration, pH 8.7 for the tripolyphosphate solution and 9.7 minutes cross-linking time. In addition to these studies, controlled-release of AZT-loaded microspheres, nicardipine hydrochloride and mucoadhesive tablets of atenolol have been optimized using RSM by Abu-Izza et al. (1996), Huang et al. (2005) and Singh et al. (2006), respectively.

6.5 *Optimization of the enzymatic hydrolysis and clarification*

The enzymatic hydrolysis depends on many factors, such as type and specificity of the enzyme, hydrolysis time, enzyme concentration, incubation temperature, and pH. Since the enzymatic hydrolysis is widely used in the food industry, it is necessary to investigate enzymatic performance during hydrolysis and optimize the process for obtaining a high yield by utilizing low-enzyme contents. Classical techniques are laborious and time-consuming and often do not guarantee the determination of optimal conditions. Recently, RSM has been applied for the optimization of enzymatic hydrolysis processes. For instance, Kurozawa et al. (2008) studied optimization of the enzymatic hydrolysis of chicken meat using RSM, and their results revealed that a condition with 52.5°C, 4.2% (w/w) enzyme/substrate ratio and pH 8.00 had the optimum conditions. In another study, Peričin et al. (2009) reported 0.137% (v/v) enzyme concentration, 0.84% (w/v) NaCl concentration and 32.5 hour as the optimum conditions for enzymatic hydrolysis of protein isolate from hull-less pumpkin oil cake. Results of Kaur et al. (2008) showed that a condition with the enzyme concentration 0.70 mg/100 g guava pulp, 7.27 hour incubation, and 43.3°C is ideal for enzymatic hydrolysis pretreatment of juice recovery from guava fruit. Similarly, RSM has been used to optimize the enzymatic hydrolysis condition of dogfish (*Squalus acanthias*) muscle (Schwabe and Büllesbach, 2013), maize starch (Kunamneni and Singh, 2005), whey protein isolate (Cheison et al., 2007) and pectin (Rodríguez-Nogales et al., 2007). The turbidity and viscosity of fruit juice are caused mainly by polysaccharides, such as pectin, cellulose, hemicellulose, lignin and starch. These colloids may lead to a fouling problem during the filtration process, and an enzymatic treatment is necessary in order to prevent this problem with enzymes such as pectinase and cellulose. Generally, these enzymes increase juice yield, soluble solids content, and clarity in fruit and vegetable juices (Alvarez and Canet, 1999). A number of factors could affect the enzymatic clarification, such as enzyme concentration, time, and temperature incubation of the treatment. Usually, in these studies, the responses are turbidity, clarity, viscosity and antioxidant activity. RSM has been used successfully to optimize enzymatic clarification processes. For instance, Rai et al. (2004) found 99.27 minutes time, 41.89°C temperature, and 0.0004% (w/v) concentration of pectinase as the best conditions for clarification of mosambi juice using CCD. Lee et al. (2006) reported that 0.084% pectinase concentration, incubation temperature of 43.2°C, and an incubation time of 80 minutes are the best conditions for clarifying banana juice. The results of Chen et al. (2012) revealed that the optimal enzymatic clarification of green asparagus juice using CCD is pectinase concentration of 1.45%, incubation temperature of 40.56°C and pH 4.43. In some other published works, RSM has been used for optimizing enzymatic clarification of sapodilla juice (Abu-Izza et al., 1996; Sin et al., 2006a; Sin et al., 2006b), litchis (*Litchi Chinensis Sonn.*) juice (Sanchez et al., 2004), white pitaya juice (Nur 'Aliaa, et al., 2010) and pomegranate juice (Neifar et al., 2009).

6.6 Optimization of the blanching processes

Blanching is a heat treatment which is widely used in the agro-food sector and particularly important in the processing of green vegetables to inactivate the enzymes involved in the spoilage of fresh vegetables (Garrote et al., 2004). There are different techniques for blanching, such as steam or hot water, microwave, and ultrasound. Several factors are important during blanching, including temperature, time, thickness, particle size, ultrasound and microwave power, and concentration of solvents, which should be optimized in order to achieve an appropriate blanching process. RSM has been used to optimize different blanching processes by considering parameters such as enzymatic activity (EA), vitamin C content, color, etc., as dependent variables or responses (Yolmeh et al., 2014). As an example, Jackson et al. (1996) studied optimization of blanching for banana chips using RSM and reported that the crispiest chips could be produced at blanching conditions of 69°C and 22 minutes. Alvarez and Canet (1999) studied optimization of stepwise blanching of frozen-thawed potato tissues and showed that a temperature range of 60–65°C and time of 25–35 minutes were optimum conditions for the blanching. Ismail and Revathi (2006) evaluated the effects of blanching time, evaporation time, temperature, and hydrocolloid on physical properties of chili (*Capsicum annum* var *kulai*) puree by RSM. They reported that complete inactivation of the enzymes was obtained with the blanching temperature of 100°C for 6 minutes. First-order reaction kinetics fitted adequately to predict color loss and pectin in order to improve the viscosity and total soluble solids of the puree. Mestdagh et al. (2008) used a CCD to investigate the effects of blanching time and temperature on the extraction of reducing sugars from potato strips and slices. Their results revealed that blanching at 70°C for 10 minutes was more efficient than lower temperatures that appeared more time-consuming. However, the extraction yield of reducing sugars was over 10% lower when the potato cuts were blanched in water that was previously used for blanching, leading to over 10% less reduction in the final acrylamide content too. Yolmeh et al. (2014) investigated the ultrasound blanching of green bean by RSM and they found that a temperature of 90°C, sonication time of 58.27 s, and duty cycle of 0.79 s were optimum conditions in which residual peroxidase activity and vitamin C loss were measured as 9.64 and 8.92%, respectively. In addition to these examples, other studies have also been carried out to optimize blanching of frozen Jalapeno pepper (*Capsicum annuum*), potato (*Solanum tuberosum* L.), frozen/thawed mashed potatoes, and olive leaf through RSM by Quintero-Ramos et al. (1998), Reyes-Moreno et al. (2001), Fernandez et al. (2006) and Stamatopoulos et al. (2012), respectively.

6.7 Application of RSM in production of microbial enzymes and other metabolites

Production of microbial enzymes and metabolites in microorganisms is highly affected by several factors, such as type of carbon and nitrogen sources, temperature, pH, metal ions, incubation time and inoculum volume (Liew et al., 2005; Açıkel et al., 2010). Therefore, it is significant to optimize the production conditions of microbial enzymes and metabolites, which can be carried out by conventional and statistical techniques. Statistical experimental approaches, such as RSM, are much better since they suggest simultaneous, systematic and efficient variation of all the components for process optimization economically (Abdel-Fattah, 2002). The highest EA of xylanase was measured at 61 U/mL from *Chaetomium thermophilum* by Katapodis et al. (2007). They reported that this EA was observed in 3.9% (w/v) wheat straw and 0.7% (w/v) sodium nitrate. The highest EA of lipase was observed at 6230

IU/ mL from *Candida* sp. 99–125 by He and Tan (2006) at 4.18% (w/v) soybean oil, 5.84% soybean powder, 0.284% K_2HPO_4, 0.1% KH_2PO_4, 0.1% $(NH_4)_2SO_4$, 0.05% $MgSO_4$, and 0.1% Span 60 as the optimum culture medium. The highest EA of α-amylase was measured at 6583 U/g from *Aspergillus oryzae* by Francis et al. (2003) at 30°C, initial moisture of 70%, and an inoculum rate of 1×107 spores/g dry substrate as the optimum conditions. The highest EA of phytase was observed at 2250 U/L from *Escherichia coli* by Sunitha et al. (1999) at 11.75 g/L tryptone, yeast extract 5.88 g/L, and 5.32 g/L NaCl as the optimum condition. The highest EA of protease was measured at 1939 U/mL from *Bacillus* sp. by Puri et al. (2002) at 15 mg/mL starch, 7.5 mg/mL peptone, 144 h incubation time, and 1% inoculums density as the optimum condition.

7. Computer Software for Implementation of RSM

There are now many software packages which can perform regression analysis and generate surface plots using RSM technique. Some of this software is presented in Table 4. It should be noted that there are also other statistical software and developed codes for RSM analysis which have not been included in this table.

8. A Practical Example for Application of RSM

In this section, a practical example is demonstrated in order to show the applicability of RSM in optimizing an extraction process. Picrocrocin, safranal, and crocin are three important bioactive compounds found in saffron which can be extracted using the maceration process in different operating conditions. The data is

Table 4: The most common RSM software.

Software	Description	Company
Design-Expert	• Specifically designed for DOE. • Offers comparative tests, screening, characterization, optimization, robust parameter design, mixture designs and combined designs. • Provides test matrices for screening up to 50 factors. • Graphical tools to identify the impact of each factor on the desired outcomes and reveal abnormalities in the data.	Stat-Ease, Inc., USA
R software	• Provides functions to generate response-surface designs, fit first- and second-order response-surface models. • It compiles and runs on a wide variety of UNIX platforms and similar systems (including FreeBSD and Linux, Windows and Mac). • Free for use.	Bell Laboratories, USA
MATLAB	• RSMdemo (interactive response surface demonstration) statistical toolbox in MATLAB opens a group of three graphical user interfaces for interactively investigating response surface methodology (RSM), nonlinear fitting, and the design of experiments.	MathWorks, USA
Minitab	• Surface design to model curvature in data and identify factor settings that optimize the response. • Predict responses for different factor settings. • Plot the relationships between the factors and the response. • Find settings that optimize one or more responses.	Minitab, Inc., USA
ReliaSoft's DOE++	• Design of experiments (DOE). • Analysis of response data. • Extensive plotting capabilities to present analysis results graphically. • Powerful optimization utility.	ReliaSoft Corporation, USA
JMP	• Powerful in selection of experimental design. • RSM analysis and optimization.	SAS Institute, USA
STATGRAPHICS	• DOE, response surface design and optimization.	Statgraphics Technologies, Inc., USA
STATISTICA	• DOE, response surface design and optimization.	Stat-soft, Inc., USA

derived from the published paper of Sarfarazi et al. (2015). The aim of this example is to find the optimum combination of ethanol concentration, temperature and extraction time in order to optimize the extraction of the mentioned important ingredients. The RMS process includes (1) screening and experimental design, (2) performing extraction process according to the DOE, (3) representing the generated response surface models, (4) optimization, and (5) verifying the optimum operating conditions. This example is solved using Design Expert 10, but any other software, including the software named in Table 4, could be utilized to perform this analysis.

8.1 Screening and experimental design

Picrocrocin, safranal and crocin are responsible for the bitter taste, odor and colour of saffron, respectively. Extraction of these compounds is a critical procedure and should be studied thoroughly. Many authors have suggested that extraction time, type of the filter, extraction solvent, stage of filtration, temperature and some other factors have a critical effect on extraction yield. Among these factors, extraction time, solvent (ethanol) concentration and the temperature were selected as the most influential factors in this study, and experiments were primarily conducted to determine the upper and lower levels of independent variables. The independent variables and their levels are shown in Table 5 (Sarfarazi et al., 2015).

The next step is the design of experiments (DOE). A central composite design (CCD) is the most used design in second-order modeling (Yolmeh and Jafari, 2017), a three-factor full-factorial CCD with five levels was generated. First, we start the design-expert program from the start menu or by clicking on the program icon, then we click on a new design to perform DOE for the new case study (Fig. 4). A new window appears with a toolbar in the left side. The "Response surface" tab should be picked in order to show the available RSM designs. As we want to use central composite design, we click on it, otherwise, we could select other available designs, such as Box-Behnken, etc. (Fig. 4, right).

Click on the menu beside "Numeric Factors" and select the number of independent variables or factors. In the study of Sarfarazi et al. (2015), alpha was selected equal to 2. For determination of alpha level, click on "Options" near the bottom of the CCD screen and set alpha at 2 in the "other" tab. In

Table 5: The upper and lower limits of independent variables in extraction example.

Factor	Unit	Low	High
Ethanol concentration (X_1)	(v/v %)	0	100
Time (X_2)	h	2	7
Temperature (X_3)	°C	5	85

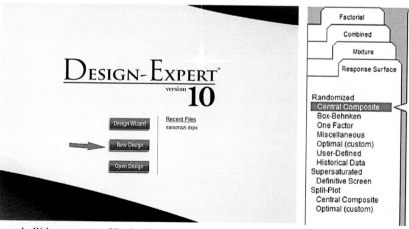

Figure 4: Welcome screen of Design-Expert program (left) and response surface design toolbar (right).

rotatable CCD, alpha is set at 1.68179 for three-factor design by default. In face-centered design, it is set at 1.0. Replicates of factorial, star and center points are determined by the software, but can be changed in this window by the user too. Fill the main Table of CCD window by inserting the name of the factors (ethanol concentration, time and temperature), their units and low and high levels. Note that "Enter factor ranges in terms of alpha" should be checked in order to define the low and high coded values automatically by the software. At the bottom of the CCD window, the type of CCD design can be determined as "Full" or "Small" CCD. Also, the number of blocks could be determined. For example, if you perform the experiments in two days, select 2 in the "Blocks" field. Designing CCD has been completed and now you can press continue at the bottom of the page. In the next page, determine the number of responses (3 for the amount of extracted picrocrocin, safranal and crocin) and their units. In the end, click "Finish" to see the design layout. Save the design and perform the experimental runs. After conducting the experiments according to the experimental design, responses would be entered into the file. The next step is analyzing the results. Under analysis branch, click on one of the responses (e.g., picrocrocin content). In order to complete the analysis, we should use the tabs from left to right. "Transform" tab is for the transformation of the response function into different forms. If we don't want to perform the transformation, we accept the default option "None". In the next tab, linear, two-factor interaction, and higher order polynomials are fitted to the response. If the probability ("prob > F") falls below 0.05 (or any other selected statistical significance level), the source term is significant. The other table in this tab is "Lack-of-Fit Tests". Insignificant lack-of-fit probability value shows the model's suitability for use. Note that, in such tables, the cubic model might be marked as "Aliased" by the software. As the CCD matrix has a few design points to determine the cubic model and it has originally set up for the quadratic model, don't pay attention to this warning by the software. At the lower table, the program automatically marks the best model as "Suggested". After investigating the model statistics, you can see the terms of the model in the "Model" tab. The suggested model in "Fit Summary" tab is set by default in "Process order" in this tab. The next tab is the ANOVA Table, which determines the adequacy of the model. The next table is ANOVA statistics, which shows the values of R square, PRESS, etc., which should be reported in the papers. The following table in this tab shows model coefficients. In the last table, the coded and actual (uncoded) values of model coefficients are shown. This model is also reported as the prediction model for the response in the experimental range. The next tab in the analysis window is the "Diagnostics" tab. Normal probability and some other plots could be verified in this tab. After checking the statistical problems of the model, visualization of the model output could be performed in the "Model Graphs" tab. 2D, 3D and other types of plots are available in this tab. The second and third response would be analyzed the same as the first response, and all of the resulting data should be saved. At the last step, the optimization of the process should be done. There are two options for "Numerical" and "Graphical" optimization in the "Optimization" menu. At the numerical optimization window, the target of optimization should be determined. Here, we want to maximize extraction of three compounds, so set the "Goal" for these responses as "maximize" and the input parameters as "in range". There are other options to set the optimization goal that could be determined by the user according to the case study. Also, the "weight" and upper and lower desired limits for each variable can be determined in this tab. The next tab in the numerical optimization window is "Solutions". If "Report" is selected in a floating "Solution tool". This table shows a different set of variables which yields the desired response. The "Desirability" term should be near 1 at the optimum conditions. The final step in RSM analysis is response prediction and validation. By clicking on "Point prediction", the predicted response for any set of factors is displayed. Some experiments (e.g., at an optimum point) should be performed and the experimental results can be compared to the predicted results in order to validate the RSM results.

9. Conclusion

The complex nature of food processing operations yields various responses at different operating conditions. Today, surviving in the competitive marketplace requires optimization of the process and products. Nowadays, the application of RSM in the optimization of different food industry processes is common due to its advantages compared with conventional methods. In this chapter, the basic steps of

RSM analysis were discussed and the interpretation of the obtained results and graphical illustrations were explained. Finally, it should be noted that RSM also has some limitations, but careful design of the experiment and the selection of model can help to overcome these limitations, although there might be some cases, such as determination of reaction kinetics, for which RSM is not ideal.

References

Abdel-Fattah, Y.R. 2002. Biotechnology Letters 24(14): 1217–1222.

Abu-Izza, K.A., Garcia-Contreras, L. and Lu, D.R. 1996. Preparation and evaluation of sustained release AZT-loaded microspheres: optimization of the release characteristics using response surface methodology. J. Pharm. Sci. 85(2): 144–149.

Açıkel, Ü., Erşan, M. and Sağ Açıkel, Y. 2010. Optimization of critical medium components using response surface methodology for lipase production by Rhizopus delemar. Food and Bioproducts Processing 88(1): 31–39.

Acosta, O., Víquez, F. and Cubero, E. 2008. Optimisation of low calorie mixed fruit jelly by response surface methodology. Food Quality and Preference 19(1): 79–85.

Ahn, J.-H., Kim, Y.-P., Lee, Y.-M., Seo, E.-M., Lee, K.-W. and Kim, H.-S. 2008. Optimization of microencapsulation of seed oil by response surface methodology. Food Chemistry 107(1): 98–105.

Alizadeh, M., Hamedi, M. and Khosroshahi, A. 2005. Optimizing sensorial quality of iranian white brine cheese using response surface methodology. Journal of Food Science 70(4): S299–S303.

Alvarez, M.D. and Canet, W. 1999. Optimization of stepwise blanching of frozen-thawed potato tissues (cv. Monalisa). European Food Research and Technology 210(2): 102–108.

Anandharamakrishnan, C. 2017. Handbook of Drying for Dairy Products, John Wiley & Sons.

Assadpour, E., Maghsoudlou, Y., Jafari, S.-M., Ghorbani, M. and Aalami, M. 2016. Optimization of folic acid nano-emulsification and encapsulation by maltodextrin-whey protein double emulsions. International Journal of Biological Macromolecules 86: 197–207.

Barton, R.R. 2013. Response surface methodology. Encyclopedia of Operations Research and Management Science, Springer 1307–1313.

Baş, D. and Boyacı, İ.H. 2007. Modeling and optimization I: Usability of response surface methodology. Journal of Food Engineering 78(3): 836–845.

Bezerra, M.A., Santelli, R.E., Oliveira, E.P., Villar, L.S. and Escaleira, L.A. 2008. Response surface methodology (RSM) as a tool for optimization in analytical chemistry. Talanta 76(5): 965–977.

Bruns, R.E., Scarminio, I.S. and de Barros Neto, B. 2006. Statistical Design-Chemometrics, Elsevier.

Burande, R.R., Kumbhar, B.K., Ghosh, P.K. and Jayas, D.S. 2008. Optimization of fluidized bed drying process of green peas using response surface methodology. Drying Technology 26(7): 920–930.

Cheison, S.C., Wang, Z. and Xu, S.-Y. 2007. Use of response surface methodology to optimise the hydrolysis of whey protein isolate in a tangential flow filter membrane reactor. Journal of Food Engineering 80(4): 1134–1145.

Chen, K.-N., Chen, M.-J., Liu, J.-R., Lin, C.-W. and Chiu, H.-Y. 2005. Optimization of incorporated prebiotics as coating materials for probiotic microencapsulation. Journal of Food Science 70(5): M260–M266.

Chen, K.-N., Chen, M.-J. and Lin, C.-W. 2006. Optimal combination of the encapsulating materials for probiotic microcapsules and its experimental verification (R1). Journal of Food Engineering 76(3): 313–320.

Chen, X., Xu, F., Qin, W., Ma, L. and Zheng, Y. 2012. Optimization of enzymatic clarification of green asparagus juice using response surface methodology. Journal of Food Science 77(6): C665–C670.

Erdogdu, F. 2008. Optimization in Food Engineering, CRC Press.

Felberg, I., Deliza, R., Farah, A., Calado, E. and Donangelo, C.M. 2010. Formulation of a soy-coffee beverage by response surface methodology and internal preference mapping. Journal of Sensory Studies.

Fernandez, C., Dolores Alvarez, M. and Canet, W. 2006. The effect of low-temperature blanching on the quality of fresh and frozen/thawed mashed potatoes. International Journal of Food Science and Technology 41(5): 577–595.

Francis, F., Sabu, A., Nampoothiri, K.M., Ramachandran, S., Ghosh, S., Szakacs, G. et al. 2003. Use of response surface methodology for optimizing process parameters for the production of α-amylase by Aspergillus oryzae. Biochemical Engineering Journal 15(2): 107–115.

Gallagher, E., O'Brien, C.M., Scannell, A.G.M. and Arendt, E.K. 2003. Use of response surface methodology to produce functional short dough biscuits. Journal of Food Engineering 56(2-3): 269–271.

Gan, C.-Y., Abdul Manaf, N.H. and Latiff, A.A. 2010. Optimization of alcohol insoluble polysaccharides (AIPS) extraction from the Parkia speciosa pod using response surface methodology (RSM). Carbohydrate Polymers 79(4): 825–831.

Gardiner, W.P. and Gettinby, G. 1998. Experimental design techniques in statistical practice: A practical software-based approach, Elsevier.

Garrote, R.L., Silva, E.R., Bertone, R.A. and Roa, R.D. 2004. Predicting the end point of a blanching process. LWT - Food Science and Technology 37(3): 309–315.

Gharibzahedi, S.M.T., Mousavi, S.M., Hamedi, M. and Ghasemlou, M. 2012. Response surface modeling for optimization of formulation variables and physical stability assessment of walnut oil-in-water beverage emulsions. Food Hydrocolloids 26(1): 293–301.

Ghorbannezhad, P., Bay, A., Yolmeh, M., Yadollahi, R. and Moghadam, J.Y. 2016. Optimization of coagulation–flocculation process for medium density fiberboard (MDF) wastewater through response surface methodology. Desalination and Water Treatment 57(56): 26916–26931.

Gunst, R.F., Myers, R.H. and Montgomery, D.C. 1996. Response surface methodology: Process and product optimization using designed experiments. Technometrics 38(3): 285.

He, Y.-Q. and Tan, T.-W. 2006. Use of response surface methodology to optimize culture medium for production of lipase with *Candida* sp. 99-125. Journal of Molecular Catalysis B: Enzymatic 43(1-4): 9–14.

Huang, Y.-B., Tsai, Y.-H., Lee, S.-H., Chang, J.-S. and Wu, P.-C. 2005. Optimization of pH-independent release of nicardipine hydrochloride extended-release matrix tablets using response surface methodology. International Journal of Pharmaceutics 289(1-2): 87–95.

Huynh, T.V., Caffin, N., Dykes, G.A. and Bhandari, B. 2008. Optimization of the microencapsulation of lemon myrtle oil using response surface methodology. Drying Technology 26(3): 357–368.

Ismail, N. and Revathi, R. 2006. Studies on the effects of blanching time, evaporation time, temperature and hydrocolloid on physical properties of chili (*Capsicum annum* var *kulai*) puree. LWT - Food Science and Technology 39(1): 91–97.

Jackson, J.C., Bourne, M.C. and Barnard, J. 1996. Optimization of blanching for crispness of banana chips using response surface methodology. Journal of Food Science 61(1): 165–166.

Karazhiyan, H., Razavi, S.M.A. and Phillips, G.O. 2011. Extraction optimization of a hydrocolloid extract from cress seed (*Lepidium sativum*) using response surface methodology. Food Hydrocolloids 25(5): 915–920.

Katapodis, P., Christakopoulou, V., Kekos, D. and Christakopoulos, P. 2007. Optimization of xylanase production by Chaetomium thermophilum in wheat straw using response surface methodology. Biochemical Engineering Journal 35(2): 136–141.

Kaur, S., Sarkar, B.C., Sharma, H.K. and Singh, C. 2008. Optimization of enzymatic hydrolysis pretreatment conditions for enhanced juice recovery from guava fruit using response surface methodology. Food and Bioprocess Technology 2(1): 96–100.

Kaushik, R., Saran, S., Isar, J. and Saxena, R.K. 2006. Statistical optimization of medium components and growth conditions by response surface methodology to enhance lipase production by Aspergillus carneus. Journal of Molecular Catalysis B: Enzymatic 40(3-4): 121–126.

Kha, T.C., Nguyen, M.H., Roach, P.D. and Stathopoulos, C.E. 2014. Microencapsulation of gac oil by spray drying: Optimization of wall material concentration and oil load using response surface methodology. Drying Technology 32(4): 385–397.

Khazaei, K.M., Jafari, S.M., Ghorbani, M., Kakhki, A.H. and Sarfarazi. M. 2015. Optimization of anthocyanin extraction from saffron petals with response surface methodology. Food Analytical Methods 9(7): 1993–2001.

Khuri, A.I. 2006. Response Surface Methodology and Related Topics. Singapore, World Scientific Publishing.

Khuri, A.I. and Mukhopadhyay, S. 2010. Response surface methodology. Wiley Interdisciplinary Reviews: Computational Statistics 2(2): 128–149.

Kim, M.-S., Kim, J.-S., You, Y.-H., Park, H.J., Lee, S., Park, J.-S. et al. 2007. Development and optimization of a novel oral controlled delivery system for tamsulosin hydrochloride using response surface methodology. International Journal of Pharmaceutics 341(1-2): 97–104.

Ko, J.A., Park, H.J., Park, Y.S., Hwang, S.J. and Park, J.B. 2003. Chitosan microparticle preparation for controlled drug release by response surface methodology. Journal of Microencapsulation 20(6): 791–797.

Kumar, D., Prasad, S. and Murthy, G.S. 2014. Optimization of microwave-assisted hot air drying conditions of okra using response surface methodology. J. Food Sci. Technol. 51(2): 221–232.

Kunamneni, A. and Singh, S. 2005. Response surface optimization of enzymatic hydrolysis of maize starch for higher glucose production. Biochemical Engineering Journal 27(2): 179–190.

Kurozawa, L.E., Park, K.J. and Hubinger, M.D. 2008. Optimization of the enzymatic hydrolysis of chicken meat using response surface methodology. Journal of Food Science 73(5): C405–C412.

Lee, W.C., Yusof, S., Hamid, N.S.A. and Baharin, B.S. 2006. Optimizing conditions for enzymatic clarification of banana juice using response surface methodology (RSM). Journal of Food Engineering 73(1): 55–63.

Liew, S.L., Ariff, A.B., Raha, A.R. and Ho, Y.W. 2005. Optimization of medium composition for the production of a probiotic microorganism, Lactobacillus rhamnosus, using response surface methodology. International Journal of Food Microbiology 102(2): 137–142.

Lundstedt, T., Seifert, E., Abramo, L., Thelin, B., Nyström, Å., Pettersen, J. et al. 1998. Experimental design and optimization. Chemometrics and Intelligent Laboratory Systems 42(1): 3–40.

Madamba, P.S. 2002. The response surface methodology: an application to optimize dehydration operations of selected agricultural crops. LWT - Food Science and Technology 35(7): 584–592.

Mäkelä, M. 2017. Experimental design and response surface methodology in energy applications: A tutorial review. Energy Conversion and Management 151: 630–640.

Maran, P.J. and Manikandan, S. 2012. Response surface modeling and optimization of process parameters for aqueous extraction of pigments from prickly pear (*Opuntia ficus-indica*) fruit. Dyes and Pigments 95(3): 465–472.

Martínez, B., Rincón, F., Ibáñez, M.V. and Bellán, P.A. 2004. Improving the nutritive value of homogenized infant foods using response surface methodology. Journal of Food Science 69(1): SNQ38–SNQ43.

McCarthy, D.F., Gallagher, E., Gormley, T.R., Schober, T.J. and Arendt, E.K. 2005. Application of response surface methodology in the development of gluten-free bread. Cereal Chemistry Journal 82(5): 609–615.

Mestdagh, F., De Wilde, T., Fraselle, S., Govaert, Y., Ooghe, W., Degroodt, J.-M. et al. 2008. Optimization of the blanching process to reduce acrylamide in fried potatoes. LWT - Food Science and Technology 41(9): 1648–1654.

Montgomery, D.C. 2017. Design and Analysis of Experiments, John Wiley & Sons.

Mostefa, N.M., Hadj Sadok, A., Sabri, N. and Hadji, A. 2006. Determination of optimal cream formulation from long-term stability investigation using a surface response modelling. International Journal of Cosmetic Science 28(3): 211–218.

Myers, R.H., Montgomery, D.C. and Anderson-Cook, C.M. 2016. Response Surface Methodology: Process and Product Optimization using Designed Experiments, John Wiley & Sons.

Neifar, M., Ellouze-Ghorbel, R., Kamoun, A., Baklouti, S., Mokni, A., Jaouani, A. et al. 2009. Effective clarification of pomegranate juice using laccase treatment optimized by response surface methodology followed by ultrafiltration. Journal of Food Process Engineering 34(4): 1199–1219.

Nicolai, R. and Dekker, R. 2009. Automated response surface methodology for simulation optimization models with unknown variance. Quality Technology & Quantitative Management 6(3): 325–352.

Nugent, A., Moskowitz, H.R. and Maier, A. 2017. Response surface methodology and consumer-driven product optimization. Accelerating New Food Product Design and Development, John Wiley & Sons Ltd. and the Institute of Food Technologists 323–364.

Nur 'Aliaa, A.R., Siti Mazlina, M.K., Taip, F.S. and Liew Abdullah, A.G. 2010. Response surface optimization for clarification of white pitaya juice using a commercial enzyme. Journal of Food Process Engineering 33(2): 333–347.

Nwabueze, T.U. 2010. Basic steps in adapting response surface methodology as mathematical modelling for bioprocess optimisation in the food systems. International Journal of Food Science & Technology 45(9): 1768–1776.

Pappa, I.C., Bloukas, J.G. and Arvanitoyannis, I.S. 2000. Optimization of salt, olive oil and pectin level for low-fat frankfurters produced by replacing pork backfat with olive oil. Meat Science 56(1): 81–88.

Peričin, D., Radulović-Popović, L., Vaštag, Ž., Mađarev-Popović, S. and Trivić, S. 2009. Enzymatic hydrolysis of protein isolate from hull-less pumpkin oil cake: Application of response surface methodology. Food Chemistry 115(2): 753–757.

Puri, S., Beg, Q.K. and Gupta, R. 2002. Optimization of alkaline protease production from *Bacillus* sp. by response surface methodology. Current Microbiology 44(4): 286–290.

Quintero-Ramos, A., Bourne, M.C., Barnard, J. and Anzaldua-Morales, A. 1998. Optimization of low temperature blanching of frozen jalapeno pepper (*Capsicum annuum*) using response surface methodology. Journal of Food Science 63(3): 519–522.

Rai, P., Majumdar, G.C., DasGupta, S. and De, S. 2004. Optimizing pectinase usage in pretreatment of mosambi juice for clarification by response surface methodology. Journal of Food Engineering 64(3): 397–403.

Reyes-Moreno, C., Parra-Inzunza, M.A., Milán-Carrillo, J. and Zazueta-Niebla, J.A. 2001. A response surface methodology approach to optimise pretreatments to prevent enzymatic browning in potato (*Solanum tuberosum* L.) cubes. Journal of the Science of Food and Agriculture 82(1): 69–79.

Rodríguez-Nogales, J.M., Ortega, N., Perez-Mateos M. and Busto, M.D. 2007. Experimental design and response surface modeling applied for the optimisation of pectin hydrolysis by enzymes from A. niger CECT 2088. Food Chemistry 101(2): 634–642.

Sablani, S.S., Datta, A.K., Rahman, M.S. and Mujumdar, A.S. 2006. Handbook of Food and Bioprocess Modeling Techniques, CRC Press.

Said, K.A.M. and Amin, M.A.M. 2016. Overview on the Response Surface Methodology (RSM) in extraction processes. Journal of Applied Science & Process Engineering 2(1).

Salimi, A., Maghsoudlou, S.M., Jafari, A.S. Mahoonak, Kashaninejad, M. and Ziaiifar, A.M. 2014. Preparation of lycopene emulsions by whey protein concentrate and maltodextrin and optimization by response surface methodology. Journal of Dispersion Science and Technology 36(2): 274–283.

Sanchez, H.D., Osella, C.A. and de la Torre, M.A. 2004. Use of response surface methodology to optimize gluten-free bread fortified with soy flour and dry milk. Food Science and Technology International 10(1): 5–9.

Sarabia, L. and Ortiz, M. 2009. Response surface methodology. In Comprehensive Chemometrics 1: 345–390.

Sarfarazi, M., Jafari, S.M. and Rajabzadeh, G. 2015. Extraction optimization of saffron nutraceuticals through response surface methodology. Food Analytical Methods 8(9): 2273–2285.

Schwabe, C. and Büllesbach, E.E. 2013. Relaxin and the Fine Structure of Proteins, Springer Science & Business Media.

Sevda, S., Singh, A., Joshi, C. and Rodrigues, L. 2012. Extraction and optimization of guava juice by using response surface methodology. American Journal of Food Technology 7: 326–339.

Sin, H.N., Yusof, S., Abdul Hamid, N.S. and Rahman, R.A. 2006a. Optimization of hot water extraction for sapodilla juice using response surface methodology. Journal of Food Engineering 74(3): 352–358.

Sin, H.N., Yusof, S., Sheikh Abdul Hamid, N. and Rahman, R.A. 2006b. Optimization of enzymatic clarification of sapodilla juice using response surface methodology. Journal of Food Engineering 73(4): 313–319.

Singh, B., Chakkal, S.K. and Ahuja, N. 2006. Formulation and optimization of controlled release mucoadhesive tablets of atenolol using response surface methodology. AAPS PharmSciTech. 7(1): E19–E28.

Singh, S., Raina, C.S., Bawa, A.S. and Saxena, D.C. 2004. Sweet potato-based pasta product: Optimization of ingredient levels using response surface methodology. International Journal of Food Science and Technology 39(2): 191–200.

Stamatopoulos, K., Katsoyannos, E., Chatzilazarou, A. and Konteles, S.J. 2012. Improvement of oleuropein extractability by optimising steam blanching process as pre-treatment of olive leaf extraction via response surface methodology. Food Chemistry 133(2): 344–351.

Suh, M.-H., Yoo, S.-H. and Lee, H.G. 2007. Antioxidative activity and structural stability of microencapsulated γ-oryzanol in heat-treated lards. Food Chemistry 100(3): 1065–1070.

Sunitha, K., Lee, J.-K. and Oh, T.-K. 1999. Optimization of medium components for phytase production by *E. coli* using response surface methodology. Bioprocess Engineering 21(6): 477–481.

Thakur, S. and Saxena, D.C. 2000. Formulation of extruded snack food (gum based cereal–pulse blend): Optimization of ingredients levels using response surface methodology. LWT - Food Science and Technology 33(5): 354–361.

Tonon, R.V., Grosso, C.R.F. and Hubinger, M.D. 2011. Influence of emulsion composition and inlet air temperature on the microencapsulation of flaxseed oil by spray drying. Food Research International 44(1): 282–289.

Wang, H., Liu, Y., Wei, S. and Yan, Z. 2012. Application of response surface methodology to optimise supercritical carbon dioxide extraction of essential oil from *Cyperus rotundus* Linn. Food Chemistry 132(1): 582–587.

Witek-Krowiak, A., Chojnacka, K., Podstawczyk, D., Dawiec, A. and Pokomeda, K. 2014. Application of response surface methodology and artificial neural network methods in modelling and optimization of biosorption process. Bioresource Technology 160(Supplement C): 150–160.

Yolmeh, M. and Najafzadeh, M. 2014. Optimisation and modelling green bean's ultrasound blanching. International Journal of Food Science & Technology 49(12): 2678–2684.

Yolmeh, M. and Jafari, S.M. 2017. Applications of response surface methodology in the food industry processes. Food and Bioprocess Technology 1–21.

CHAPTER 14

A Mathematical Approach to the Modelling of the Rheological Properties of Solid Foods

Ryszard Myhan and *Marek Markowski**

1. Introduction to Food Rheology

1.1 Justification for research into the rheological properties of foods

Rheology is a branch of science that deals with the deformation and flow of matter. Rheology describes the flow of synthetic polymers and their solutions under the influence of external load. The deformation of plastics can be observed in many practical situations. Most foods contain polymers, and such products become deformed under load and display rheological properties which are time-dependent. The rheological properties of foods and other biological materials are an important determinant of their suitability for processing and consumption (Bourne, 1982; Abbott, 1999; Rao and Quintero, 2005; Barrett et al., 2010; Diamante and Umemoto, 2015). A knowledge of the rheological properties of food is vital in commerce, food processing and quality control (Rao, 2007). The structure and chemical composition of food are the main determinants of product quality. The structure of food materials can undergo various changes during processing due to the influence of external factors (Damez and Clerion, 2008). A knowledge of the relationships between the rheological properties and the chemical composition (water, protein and fat content) of food products (Myhan et al., 2016) and the influence of rheological properties on the sensory attributes of food products is essential for evaluating their processing suitability and quality (Torley and Young, 1995; Funami et al., 1998; Bayarri et al., 2007; Ahmed and Ramaswamy, 2007; Stokes et al., 2013; Zhiguo et al., 2013; Ma and Boye, 2013). The rheological properties of food also play an important role in different stages of the production process, including portioning and transport of solid, semi-solid and liquid foods, packaging and storage. A sound knowledge of rheological properties is required for designing and building food processing machines and equipment that guarantee optimal product parameters, minimize losses and maximize the effectiveness of the production process (Li et al., 2010; Diamante and Umemoto, 2015).

Mechanical tests play an important role in empirical evaluations of the rheological properties of raw materials and food products. Liquid, semi-solid and solid materials are investigated with the use of different tests. Liquid and semi-solid foods are most often analyzed with the involvement of dynamic oscillatory tests (Yu and Gunasekaran, 2001), whereas solid foods are studied mostly in quasi-static

Faculty of Engineering, University of Warmia and Mazury in Olsztyn, 10-719 Olsztyn, Oczapowskiego 11, Poland.
* Corresponding author: marek@uwm.edu.pl

creep and stress relaxation experiments (Peleg and Calzada, 1976; Rao and Quintero, 2005; Gonzalez-Gutierrez and Scanlon, 2013). Test results provide information about various phenomena that affect food products and their quality, such as date ripeness in various stages of growth (Hassan et al., 2005), fruit hardness (Blahovec, 1996), cheese hardness (Kaya, 2002) and the mechanical properties of meat-based foods (Andrés et al., 2008; Dzadz et al., 2015).

The relationships between stress and strain in a medium during loading and/or unloading are usually analyzed with the use of ideal solids (e.g., ideally elastic solids, ideally plastic solids, elastic-plastic solids) or ideal fluids (e.g., non-viscous fluids, ideally viscous fluids, viscoelastic fluids), and such objects are regarded as homogeneous isotropic materials. The results are described with generalized Maxwell, Kelvin-Voight or Burgers models (Rao and Quintero, 2005; Steffe, 1996).

These models illustrate the relationships between stress, strain and strain rate, and they combine selected models that describe ideally elastic solids, rigid-plastic solids, elastic-plastic solids, reinforced rigid-plastic solids, reinforced elastic-plastic solids, ideal fluids and non-Newtonian fluids. Analyses of compression tests performed on real samples indicate that, in solid foods, the initial stage of compression can be highly dynamic (Myhan et al., 2012) in creep and stress relaxation tests. Despite the above, most experiments are conducted under ideal conditions and disregard the preparatory stage during which the sample undergoes initial compression (Liu, 2001; Sorvari and Malinen, 2007). The experimental compression characteristics of biological materials differ considerably from those of ideal solids. Compression tests of meat and meat products reveal two distinct compression phases (preliminary and final) and demonstrate that the strain rate exerts a significant non-linear influence on the stress values measured in the sample (Myhan et al., 2015).

1.2 A review of the existing rheological models of liquid foods

Raw materials and food products have liquid (e.g., milk, dairy products, juices, purees), semi-solid (combining both solid and liquid properties, such as gels) and solid consistency (e.g., bread, fruit, vegetables and meat). Liquid foods flow under exposure to external load during processing operations (such as pumping). During liquid flow, viscous forces usually dominate over elastic forces, which is why elastic forces are generally disregarded in analyses of the rheological properties of liquid and semi-solid foods (Steffe, 1996). Steady flow tests are most often applied to analyze the rheological properties of liquid foods. These tests are performed to determine the conditions of fully developed flows and material viscosity at different process parameters, including shear displacement rate and material temperature. Various models describing the flow of liquid and semi-solid foods have been proposed in the literature. These models describe the relationships between the shear displacement rate of fluids and tangential force acting upon the liquid. Some of these models can be used to describe the rheological properties of liquid foods (Ibarz and Barbosa-Canovas, 2014; Rao, 2007). The existing rheological flow models can be divided into three categories: (i) empirical models based on the results of experimental data analyses, (ii) theoretical models based on the fundamental laws of science that describe the dependencies between the physical parameters of the model, and (iii) structural models that are deduced based on the structural properties of a material and its flow kinetics (Rao, 2007). Selected rheological flow models are presented in Table 1. Newtonian and Bingham linear rheological models are described with Eqs. (1) and (2), respectively. The shear displacement rate is proportional to tangential strain in the Newtonian model, and it is proportional to the difference between tangential strain τ and yield stress τ_0. Liquid foods generally display the rheological properties of non-Newtonian fluids, and the Newtonian model is rarely used to analyze the flow of liquids other than oil or honey at room temperature (Steffe, 1996) because it does not illustrate the complex relationships between shear displacement rate and tangential strain within a broad range of values. In recent years, the Newtonian model has been used by Rathod and Kokini (2016) to describe the size distribution of bubbles in a liquid in a twin screw co-rotating mixer. The Bingham model has been used by Hamilton et al. (2018) in an analysis of 3D printed edible structures on bread substrates. The Ostwald-de-Waele power law model (3) is often used for evaluations of various raw materials within a wide range of shear displacement rates 10^1–10^4 s^{-1}. Newtonian, Bingham and Ostwald-de-Waele models can be regarded as special cases of the Herschel-Bulkley model (4), which correctly describes the flow of non-Newtonian fluids with yield stress. Depending on the values of the consistency coefficient n, models

Table 1: Selected rheological flow models describing the relationship between shear rate ($\dot{\gamma}$) and shear stress (τ).

$\tau = \eta\dot{\gamma}$	Newton model	(1)
$\tau = \tau_0 + \eta\dot{\gamma}$	Bingham model	(2)
$\tau = K(\dot{\gamma})^n$	Ostwald-de Waele model	(3)
$\tau = \tau_0 + K(\dot{\gamma})^n$	Herschel-Bulkley model	(4)
$\tau = \eta_\infty\dot{\gamma} + K(\dot{\gamma})^n$	Sisko model	(5)
$\eta = \eta_\infty\dot{\gamma} + \dfrac{\eta_0 - \eta_\infty}{1+(\lambda\dot{\gamma})^m}$	Cross model	(6)
$\eta = \eta_\infty\dot{\gamma} + \dfrac{\eta_0 - \eta_\infty}{(1+(\lambda\dot{\gamma})^2)^N}$	Carreau model	(7)
$\tau^{0.5} = \tau_0 + K(\dot{\gamma})^{0.5}$	Casson model	(8)
$\tau^{0.5} = \tau_0 + K(\dot{\gamma})^n$	Mizrahi-Berk model	(9)

(3) and (4) deal with shear thinning ($0 < n < 1$) or shear thickening ($1 < n < \infty$) behavior. Bingham, Ostwald-de Waele and Herschel-Bulkley models have been used to analyze the rheological properties of blueberry puree with additives (Kechinski et al., 2011). The rheological properties of liquid foods can also be described with other models. Sisko (5), Cross (6) and Carreau (7) models are used to describe the rheological properties of pseudo-plastic materials with shear thinning behavior. Parameter η_∞ denotes infinite-shear-viscosity, namely, viscosity (generally very low) at infinitely high shear displacement rate. Sisko, Cross and Carreau have been used in analyses of the rheological properties of blueberry puree (Nindo et al., 2007; Kechinski et al., 2011), fish gelatin-gum (Anvari and Joyner, 2017), emulsion-filled alginate microgel suspensions (Su et al., 2016) and fenugreek gum (Yanxia et al., 2015). The Casson (8) model and the modified Casson (Mizrahi-Berk) model are generally used to investigate the rheological properties of suspensions, in particular chocolate (Glicerina et al., 2016; Kechinski et al., 2011).

1.3 Measurements of the viscoelastic properties of food

Rheological models describing the free flow of liquid and semi-solid foods cannot be used to determine the viscoelastic properties of solid foods and liquids. In liquid foods, elastic strain is generally much lower than tangential strain, and in most cases, it can be disregarded without compromising the accuracy of measurements. This is not the case in solid foods which are deformed under load and where changing viscous stress and elastic strain are of the same order. For this reason, both elastic strain and viscous stress should be taken into account in analyses of the rheological properties of solid foods and, in some cases, also semi-solid foods. Two types of tests are used for this purpose: (i) quasi-static tests, which are relatively long, and (ii) rapid dynamic tests, usually oscillatory tests. Dynamic tests have been increasingly frequently applied since the mid-20th century (Rao, 2007).

Oscillatory tests

Oscillatory tests of viscoelastic properties rely on the Couette flow of a liquid in a narrow groove between two surfaces, parallel plates or a plate and a conical surface, where one surface moves relative to the other surface in sinusoidal rotational motion with frequency ω. The food material inside the groove is subjected to sinusoidal shear deformation according to Eq. (10):

$$\gamma = \gamma_0 \sin(\omega t) \tag{10}$$

The shear displacement rate can be calculated from Eq. (11):

$$\dot{\gamma} = \gamma_0 \omega \cos(\omega t) \tag{11}$$

When strain amplitude γ_0 is low, it can be assumed that the material undergoes linear deformation, and elastic strain and viscous stress can be described with Hookean and Newtonian models, respectively.

Linear elasticity and viscosity are denoted as G' and G'', respectively, and based on the principle of superposition, the resulting stress in the material can be described with Eq. (12):

$$\sigma(t) = G'\gamma_0\sin(\omega t) + G''\gamma_0\omega\cos(\omega t) \tag{12}$$

Equation (12) can be transformed to (13):

$$\sigma(t) = \sigma_0\sin(\omega t + \delta) \tag{13}$$

Equations (11) and (13) indicate that strain γ is phase shifted relative to total stress σ by phase angle δ. An analysis of Eqs. (12) and (13) indicates that, based on the measured strain amplitude γ_0, stress amplitude σ_0 and phase shift angle δ, the values of moduli G' and G'' can be calculated from Eqs. (14)–(16):

$$G' = \frac{\sigma_0}{\gamma_0}\cos\delta \tag{14}$$

$$G'' = \frac{\sigma_0}{\gamma_0}\sin\delta \tag{15}$$

$$\tan\delta = \frac{G''}{G'} \tag{16}$$

where G' is the storage modulus, G'' is the loss modulus, and tan δ is the loss tangent. The viscoelastic properties of a material can be determined by calculating G', G'' and tan δ. The storage modulus G' denotes elastic energy accumulated in the material, and the loss modulus G'' denotes energy dissipation during one deformation cycle. In food materials, moduli G' and G'' and tan δ are generally determined by frequency ω, changes in strain and stress, and material temperature. Dynamic oscillatory tests can be performed for food materials of any consistency, but tangential strain is applied to liquid and semi-solid foods, whereas normal compressive and tensile strain are applied to solid foods. In recent years, the viscoelastic properties of foods have been analyzed with the use of oscillatory tests by Li et al. (2018), Bi et al. (2018), Majid et al. (2018) and others.

Quasi-static creep test

A creep-compliance test can provide valuable information about the viscoelastic properties of foods. In the initial stage of the test, constant load σ_0 is applied to non-deformed material throughout the entire compression phase. When load σ_0 is discontinued, the recovery phase begins and lasts until the end of the test. Sample deformation γ over time t is measured during the test. The creep compliance of the tested material is expressed by the quotient of strain and stress:

$$J(t) = \frac{\gamma(t)}{\sigma_0} \tag{17}$$

The viscoelastic properties of a material are determined by analyzing the phenomena that occur inside the material during the compression phase, based on Eq. (18) (Steffe, 1996; Rao, 2007):

$$J(t) = \frac{1}{G_0} + \sum_i \frac{1}{G_i}(1 - e^{-\frac{t}{\tau_i}}) + \frac{t}{\eta_N} \tag{18}$$

where G_0 is the instantaneous elasticity modulus, G_i is the average time-retarded elasticity modulus within the ith element, t_i is equal to η_i/G_i and denotes the retardation time of the ith element with average viscosity η_i and regarded elasticity G_i, and η_N is Newtonian viscosity. When load is applied, the material undergoes rapid elastic deformation (first component of the sum in Eq. 18). Retarded elastic strain (second component of the sum) and Newtonian viscous strain (third component of the sum) develop in the material over time. Depending on the type of the analyzed material, the middle component of the sum in Eq. (18) can be created based on one or more Kelvin-Voigt elements used in the kinematic stress model.

Quasi-static stress relaxation test

A stress relaxation test is an alternative method for evaluating creep compliance, and it can also provide valuable information about the viscoelastic properties of food. In a relaxation test, the sample is subjected to constant stress, and changes in strain are observed as a function of time. A kinematic model is usually developed based on one or the entire spectrum of Maxwell elements joined in parallel with elasticity G_i and viscosity η_i. The resulting strain decreases over time and can be calculated with the use of Eq. (19):

$$\sigma(t) = \sum_i G_i e^{-\frac{t}{\tau_i}} \tag{19}$$

where G_i is elasticity, η_i is viscosity in the ith Maxwell element, τ_i is stress relaxation time in the ith Maxwell element, and $\tau_i = \eta_i/G_i$.

Similarly to dynamic oscillatory tests, quasi-static creep and relaxation tests can be applied to food materials of any consistency based on tangential or normal strain. Dynamic and quasi-static tests have been described in greater detail by Steffe (1996) and Rao (2007) who presented numerous equations, examples and interpretations.

2. A Rheological Model of Solid Foods

2.1 Rheological properties of solid foods

Three types of strain are identified in ideal homogeneous isotropic materials: Elastic, plastic and viscous (Steffe, 1996). Solid materials are generally analyzed with the use of quasi-static creep and relaxation tests (Rao and Quintero, 2005; Gonzalez-Gutierrez and Scanlon, 2013). However, the behavior of biological materials differs considerably from that of ideal solids; therefore, various rheological models are applied in order to describe different materials subjected to various loads (Peleg, 1984; Holdsworth, 1993; Alvarez et al., 1998; Guerrero-Beltrán et al., 2009). The resulting solutions are mostly semi-empirical equations or multi-parametric models combining simple models joined in series or in parallel connections, where some parameters cannot be interpreted based on the laws of physics applicable to solid bodies (Mohsenin and Mittal, 1977; Mitchell and Blanshard, 1976). Various models are also used to describe different processes in the same body (Yilmaz et al., 2012), and the physical properties of a material are described with different parameters, depending on the type of process. The main weakness of these models is that the values of rheological properties (elasticity, viscosity) are completely different in analyses of the same material. For example, when the Maxwell model was used to describe stress relaxation and when the Kelvin-Voight model was used to describe creep in a biological material (Myhan et al., 2014), the estimates of stress retardation time η/E differed significantly (Fig. 1). The above points to significant differences in the estimates of at least one parameter, elasticity E or viscosity η, when the same material is analyzed under repeatable conditions. A compression test (Fig. 2) indicates that biological materials have complex rheological properties. In the first stage of compression, the sample behaves like a viscoelastic material, and this stage continues until threshold strain is attained, after which, rapid material flow is observed. The sample behaves like reinforced elastic-plastic material in the following stage of the test, and like an ideally plastic material in the final stage. Creep and relaxation tests are usually conducted under idealized conditions. In most published analyses of relaxation tests, the compression stage and the final stage are generally disregarded (Moresi et al., 2012; Knauss and Zhao, 2007; Liu, 2001; Sorvari and Malinen, 2007). In the above approach, valuable information about the analyzed material can be lost, and the developed rheological models adequately describe the behavior of the tested material in the stress relaxation phase and the creep phase, but they have low prognostic value in the initial and final phases of the test. In the creep test presented in Fig. 3, the material is compressed in two distinct stages: The compression stage and the recovery stage, during which the registered strain rates are higher than in the remaining stages. Due to the inertia of the measuring head, the preset constant external load cannot be maintained, which is why compressive forces appear during recovery. These observations can also be used to explain the phenomena at the beginning of recovery when compressive force probably appears because the rate of material recovery is higher than the return speed of the measuring head.

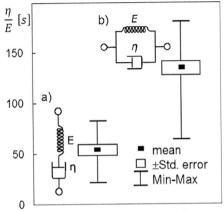

Figure 1: Stress relaxation time: a – stress relaxation test (Maxwell model), b – creep test (Kelvin-Voight model).

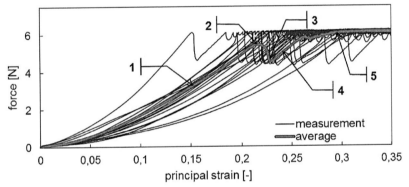

Figure 2: Material compression in a creep test: 1 – viscoelastic strain; 2 – boundary layer; 3 – rapid material flow; 4 – elastic-plastic strain; 5 – plastic strain (Myhan et al., 2012).

Color version at the end of the book

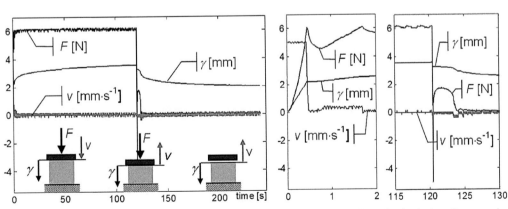

Figure 3: Creep test performed with the use of the TA.XT2i texture analyzer; a – full run, b – beginning of compression, c – creep to recovery (Myhan et al., 2012).

The actual behavior of a material in a relaxation test that adequately represents most measurements is presented in Fig. 4. The real relaxation test is composed of three phases: The compression phase, the stress relaxation phase and the expansion phase during which the measuring head returns to its initial position.

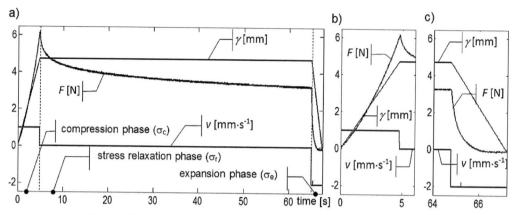

Figure 4: Relaxation test; a – full run, b – compression phase, c – expansion phase.

2.2 A rheological model of homogeneous isotropic solid foods

A rheological model of homogeneous isotropic solid material was developed on the assumption that it should reliably describe the results of the real creep test and the relaxation test, while accounting for the compression phase and the expansion phase. All modeled parameters should have strict physical interpretation and should correspond to the main physical properties of a solid, such as elasticity, and the loss of mechanical energy resulting from internal friction. Various mechanisms can be responsible for the loss of mechanical energy. In fluid mechanics, internal friction is generally associated with viscosity, whereas sliding friction resulting from the mutual displacement of individual structural elements should also be taken into account in analyses of biological materials with solid or semi-liquid consistency. The proposed model would account for unidirectional stress-strain on the assumption that the analyzed material is isotropic in a given direction of strain. A simple model combining the properties of an ideal Hookean elastic solid, an ideal Saint-Venant plastic solid and ideal Newtonian viscous fluid was selected. An analogous representation of the selected model would be a system combining a spring, a hydraulic damper and a slide in a parallel connection (Fig. 5). A distinction between strain rate $\dot{\varepsilon}$ and flow velocity ϑ has to be made for the model to be logically validated in all phases of the real creep test and the stress relaxation test (Fig. 3, Fig. 4).

Material strain appearing in every test phase can be regarded as the vector sum of strain resulting from its elasticity σ_H, viscosity σ_N and sliding friction σ_V:

$$\sigma = \sigma_H + \sigma_N + \sigma_V. \tag{20}$$

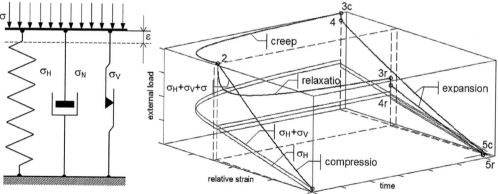

Figure 5: Rheological model: (1-2-3c-4c-5c) – creep test, (1-2-3r-4r-5r) – relaxation test.

The value of stress σ, expressed by the general equation $\sigma = f(\varepsilon, \vartheta, t)$ is determined by relative strain ε, relative flow velocity ϑ and time t.

The compression phase is the first stage of the real creep test and the relaxation test. The compression phase can be divided into two stages: The initial stage and the actual compression stage (Myhan et al., 2015). The initial stage begins when the measuring head is set into motion and ends in point 1 (Fig. 5). It can be assumed that this stage involves only plastic deformation of surface micro-irregularities and that the actual compression phase begins in point 1 and ends in point 2 (Fig. 5). At the beginning of the compression phase, $t_0 = 0 \rightarrow \gamma(t_0) = 0$. The state of the system in time t_0 to t_2 is determined by the ratio of the instantaneous value of external load $\sigma(t)$ to the value of sliding friction $\sigma_V = \mu \cdot sign(-\dot{\varepsilon})$. If $\sigma(t) \leq \sigma_V$, then $\varepsilon(t) = 0$, which implies that the modeled material is ideally rigid under the experimental conditions. If $\sigma(t) > \sigma_V$, then:

$\varepsilon(t) \geq 0 \wedge \sigma(t) = \sigma_H(t) + \sigma_N(t) + \sigma_V$

where: $\sigma_H(t) = E \cdot \varepsilon(t);$ $t \in \langle t_1, t_2 \rangle$ (21)

       $\sigma_N(t) = \eta \cdot \vartheta(t);$ $t \in \langle t_1, t_2 \rangle.$ (22)

The results of the measurements reveal a non-linear relationship between flow velocity ϑ and relative strain ε when the relative strain rate $\dot{\varepsilon}$ is constant ($\dot{\varepsilon} = const$). The above could result from self-reinforcement or radial swelling of the tested sample (Sudhagar et al., 2006). According to the authors (Myhan et al., 2015), stress $\sigma(t)$ during the compression phase can be expressed by the following power function:

$$\sigma(t) = k_c \cdot [\varepsilon(t)]^{n_c}. \tag{23}$$

The material flow velocity determined based on the above equation is also described by Eq. 5:

$$\vartheta = \frac{1}{\eta} \cdot (k_c \cdot [\varepsilon(t)]^{n_c} - \varepsilon(t) \cdot E - \sigma_V), \tag{24}$$

which indicates that flow velocity is directly proportional to time and inversely proportional to viscosity.

The creep phase begins in point 2 and ends in point 3c (Fig. 5). This phase is characterized by constant external load

$$\sigma(t) = \sigma_H(t) + \sigma_N(t) + \sigma_V = \sigma_{max} = const, \tag{25}$$

which is accompanied by an increase in relative strain $\varepsilon(t)$ from the value of $\varepsilon(t_2)$ to $\varepsilon(t_{3c})$. Since the first two elements of the model (Fig. 5) correspond to the Voigt-Kevin model, for every t from t_2 to $t_{3c} \rightarrow \infty$:

$$\varepsilon(t) = \varepsilon(t_2) + \left[\frac{\sigma_{max}}{E} - \varepsilon(t_2) \right] \cdot \left[1 - exp\left(-\frac{E}{\eta}(t - t_2) \right) \right], \tag{26}$$

therefore: $\sigma_N(t) = \sigma(t_2) - \varepsilon(t) \cdot E - \sigma_V$ (27)

The stress relaxation phase, similarly to creep, is preceded by a compression phase; therefore, it begins in point 2 and ends in point 3r (Fig. 5). During that phase, the value of the external load $\sigma(t)$ decreases from the initial value of $\sigma(t_2)$ to the final value of $\sigma(t_{3r})$, which is known as steady-state flow (Ma and Boye, 2013). This phenomenon occurs at constant relative strain $\varepsilon(t)$, which indicates that stress $\sigma_{(H+V)}$ resulting from the material's elastic and plastic properties also remains constant:

$$\varepsilon(t) = \varepsilon(t_2) = const \rightarrow \sigma_{(H+V)} = const, \tag{28}$$

therefore, material flow causes changes in stress $\sigma(t)$.

Significant changes in the material's density, composition and structure do not take place during the stress relaxation phase; therefore, it can be assumed that the value of coefficient η, as one of the parameters that describe the tested material, also remains constant. The above assumption implies that unknown flow velocity ϑ changes in a non-linear manner during that time. This observation is generally described with Peleg's model (Peleg, 1976), but in biological materials, the results are much more reliable when Peleg's model is modified with the use of the below equation (Myhan et al., 2015):

$$\sigma(t) = \sigma(t_2) \cdot \left(1 - \frac{(t-t_2)^n}{k_1 + k_2(t-t_2)^n} \right), \qquad (29)$$

where k_1 is the initial rate of stress relaxation and k_2 is the hypothetical asymptotic value of normalized stress.

Equation (29) can be used to determine the total stress resulting from the material's elastic-plastic properties in point 2:

$$\sigma_{(H+V)}(t_2) = \lim_{t \to \infty} \sigma(t) = \sigma(t_2) \cdot \left(1 - \frac{1}{k_2} \right), \qquad (30)$$

therefore: $\sigma_N(t) = \sigma(t_2) \cdot \left(\dfrac{1}{k_2} - \dfrac{(t-t_2)^n}{k_1 + k_2(t-t_2)^n} \right),$ \hfill (31)

and the calculated relaxation time T_{rel} is expressed by the following equation:

$$T_{rel} = \left(\frac{(e-1)k_1}{k_2} \right)^{\frac{1}{n}} \cong \left(\frac{1.7183 k_1}{k_2} \right)^{\frac{1}{n}}. \qquad (32)$$

The expansion phase is the final phase of the stress relaxation test and the creep test. The expansion phase can be divided into two stages: A transitional stage (Fig. 3c) and the actual expansion stage. The transitional stage begins in point 3r (3c) and ends in point 4r (4c), and it results from a change in strain rate and the resulting change in the sense of vector σ_V; therefore:

$$t_3 = t_4; \quad \varepsilon(t_4) = \varepsilon(t_3); \quad \sigma(t_4) = \sigma(t_3) - 2\sigma_V. \qquad (33)$$

The actual expansion stage begins in point 4r (4c) and ends in point 5r (5c), and the process is determined by the relative return speed of the measuring head. If the head's return speed is higher than the strain rate (Fig. 3a), the material is expanded in the absence of external load:

$$\sigma(t) = \sigma_H(t) + \sigma_N(t) + \sigma_V = 0 \qquad (34)$$

Since the first two elements of the model (Fig. 5) correspond to the Voigt-Kelvin model, for every $t \in \langle t_{4r(c)}, t_{5r(c)} \rangle$ and initial condition $\varepsilon(t) = \varepsilon(t_{4r(c)})$:

$$\varepsilon(t) = \frac{\sigma_{max} - 2_{\sigma V}}{E} \cdot exp\left(\frac{-E}{\eta} \cdot (t-t_2) \right) \cdot \left[exp\left(\frac{-E}{\eta} \cdot (t_4 - t_2) \right) - 1 \right] + \frac{\sigma_V}{E} \qquad (35)$$

When the return speed of the measuring head is lower than the rate of self-expansion of a sample not subjected to an external load, then:

$$\sigma(t) = \sigma_H(t) + \sigma_N(t) + \sigma_V, \qquad (36)$$

and for $t = t_5$, stress $\sigma(t)$ decreases to zero. If the observed process is analogous to compression (Eq. 4), it can be assumed that:

$$\sigma(t) = k_e \cdot (\varepsilon(t) - \varepsilon(t_5))^{n_e}. \qquad (37)$$

Method of estimating the modeled parameters – example

Rheological parameters were estimated based on the averaged result of 20 relaxation tests and creep tests (Fig. 6). The tests were performed on cubes of roasted turkey breast, measuring $10 \times 10 \times 10$ mm (Myhan et al., 2012). The experiment was performed using the Stable Micro Systems TA.XT2i texture analyzer

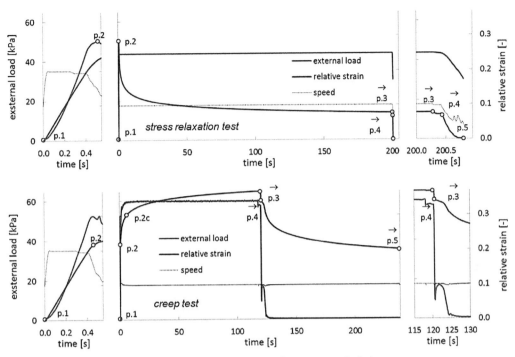

Figure 6: Stress relaxation test and creep test – calculations.

(maximum compressive force – 6 N; maximum head speed –0.0005 ms^{-1}; duration of the creep test – 240 s; duration of the relaxation test – 200 s; time step – 0.02 s).

The rheological properties of the analyzed material were estimated sequentially in successive steps. The parameters of the function (Eq. (23)) describing variations in stress $\sigma(t)$, identical for both compression phase tests, were described in the first step. The results are presented in Table 2.

The form of the function describing changes in stress during the relaxation phase $\sigma(t)$, $t \in \langle t_2, t_{3r} \rangle$ and the function describing variations in relative strain during the creep phase $\varepsilon(t)$, $t \in \langle t_2, t_{3c} \rangle$ is determined in the following step. In both cases, the transition from the compression phase to the relaxation (creep) phase is dynamic (Fig. 2) when the relative strain rate is undefined $\dot\varepsilon \neq const$ (Fig. 6). For this reason, the hypothetical time of transition between those phases has to be determined (points 2r and 2c). In the relaxation test, transition time can be determined on the assumption of strain equality:

$$\overline{\dot\varepsilon_{1-2}} \cdot \overline{t_{2r}} = \overline{\varepsilon_{2-3}} \rightarrow t_{2r} = \frac{\overline{\varepsilon_{2-3}}}{\overline{\dot\varepsilon_{1-2}}} = \frac{0.249}{0.5 \, s^{-1}} \cong 0.498 \, s, \tag{38}$$

where: $\overline{\dot\varepsilon_{1-2}}$ – actual compression rate; $\overline{\varepsilon_{2-3}}$ – actual relative strain in the stress relaxation phase.

Table 2: Estimation of the parameters in Eq. 4.

	Parameter	Estimation	Standard error	Coefficient of determination
Creep test	k_c [kPa]	503.57	9.12	0.9998
	n_c [–]	1.448	0.010	
Relaxation test	k_c [kPa]	465.61	8.77	0.9998
	n_c [–]	1.451	0.010	
Creep and relaxation test	k_c [kPa]	484.96	23.60	0.9975
	n_c [–]	1.450	0.026	

In the creep test, hypothetical transition time can be determined based on the assumption of stress equality:

$$\overline{\sigma}_{2-3} = k_c(\overline{\dot{\varepsilon}}_{1-2} \cdot t_{2p})^{n_c} \rightarrow \ln(t_{2p}) = \frac{1}{1.45}\ln\left(\frac{59.92\ kPa}{484.96\ kPa}\right) - \ln(0.5^{\ s-1}) \rightarrow t_{2p} \cong 0.473\ s, \tag{39}$$

where: $\overline{\sigma}_{2-3}$ – averaged real stress in the creep phase.

Changes in strain in the relaxation phase are described by Eq. 11. The parameters were estimated for the initial conditions given by Eq. 21:

$$\sigma(t) = k_c\ (\overline{\dot{\varepsilon}}_{1-2} \cdot t_{2r})^{n_c} \cdot \left(1 - \frac{(t-t_{2r})^{n_r}}{k_{1r} + k_{2r}(t-t_{2r})^n}\right) = 61.77 \cdot \left(1 - \frac{(t-t_{2r})^{n_r}}{k_{1r} + k_{2r}(t-t_{2r})^n}\right). \tag{40}$$

Changes in relative strain ε in the creep phase are described by Eq. (26), but different conditions should be applied due to the number of parameters that need to be estimated. By analogy with Eqs. (29) and (40) and the initial condition in Eq. (39), these relationships were described in Eq. (41):

$$\varepsilon(t) = (\overline{\dot{\varepsilon}}_{1-2} \cdot t_{2p}) \cdot \left(1 - \frac{(t-t_{2r})^{n_r}}{k_{1r} + k_{2r}(t-t_{2r})^n}\right) = 61.77 \cdot \left(1 - \frac{(t-t_{2r})^{n_p}}{k_{1r} + k_{2r}(t-t_{2r})^n}\right) \tag{41}$$

A preliminary analysis of the estimated values of exponents n_r and n_p indicates that the difference between them does not exceed the value of standard error, and their averaged value corresponds to a fraction of exponent n_c. It can be assumed that they are not only the parameters of Eqs. (40) and (41), but certain values describing the rheological properties of the analyzed material. The differences between parameters k_{2r} and k_{2p} are not significant either. For this reason, a single computational model was used to estimate the parameters in Eqs. (40) and (41) on the assumption that $n_r = n_p = n_{rp} = n_c - 1 = 0.45$ and $k_{2r} = k_{2p} = k_{2rp}$. The results are presented in Table 3.

Since relative strain $\varepsilon(t)$ is constant during stress relaxation and the process is infinitely long, the values of $\sigma_{(H+V)}(t)$, $t \in \langle t_{2r}, t_{3r} \rangle$ can be calculated from Eqs. (31) and (40):

$$\max\left[\sigma_{(H+V)}(t)\right] = \lim_{t \to \infty}\left[\sigma(t_{2r}) \cdot \left(1 + \frac{(t-t_{2r})n_{rp}}{k_{1r} + k_{2rp}(t-t_2)^{n_{rp}}}\right)\right] \tag{42a}$$

$$\max\left[\sigma_{(H+V)}(t)\right] = \sigma(t_{2r}) \cdot \left(1 - \frac{1}{k_{2rp}}\right) = 61.77\ kPa \cdot \left(1 - \frac{1}{1.178}\right) = 9.33\ kPa \tag{42b}$$

and stress relaxation time can be calculated from Eqs. (32) and (40):

$$T_{rel} = \left(\frac{(e-1)\cdot k_{1r}}{k_{2rp}}\right)^{\frac{1}{n_{rp}}} \cong \left(\frac{1.7183 \cdot 1.167}{1.178}\right)^{\frac{1}{0.45}} \cong 3.26\ s, \tag{43}$$

Table 3: Estimation of the parameters in Eqs. (40) and (41).

	Parameter	Estimation	Standard error	Coefficient of determination
Creep & relaxation test	$k_{1r}\ [s^n]$	1.167	0.001	0.9997
	$k_{1p}\ [s^n]$	5.147	0.003	
	$k_{2rp}\ [-]$	1.178	0.001	

Since stress $\sigma(t)$ remains constant during the creep phase and the process is infinitely long, Eq. (41) can be used to determine maximum relative strain:

$$\max[\varepsilon(t)] = \lim_{t \to \infty} \left[\varepsilon(t_{2p}) \cdot \left(1 + \frac{(t - t_{2p})^{n_{rp}}}{k_{1p} + k_{2rp}(t - t_{2p})^{n_{rp}}} \right) \right] \tag{44a}$$

$$\max[\varepsilon(t)] = \varepsilon(t_{2p}) \cdot \left(1 + \frac{1}{k_{2rp}} \right) = 0.2364 \cdot \left(1 + \frac{1}{1.178} \right) \cong 0.437, \tag{44b}$$

and retardation time:

$$T_{ret} = \left(\frac{(e-1)k_{1p}}{k_{2rp}} \right)^{\frac{1}{n_{rp}}} \cong \left(\frac{1.7183 \cdot 5.147}{1.178} \right)^{\frac{1}{0.45}} \cong 88.22 \, s. \tag{45}$$

The final deformation of the sample when expansion time goes into infinity has to be known in order to determine the basic properties of the analyzed material, such as η, E and σ_V. Final deformation can be estimated based on the results of the expansion phase in the creep test, the approximation function similar to Eq. (41) and the following initial conditions:
- expansion begins with maximum relative strain $\max[\varepsilon(t)]$ described in Eq. (44):
- the time t_{e0} when the expansion process begins is unknown

$$\varepsilon(t) = \max[\varepsilon(t)] \cdot \left(1 - \frac{(t - t_{e0})^{(n_{c-1})}}{k_{1e} + k_{2e}(t - t_2)^{(n_{c-1})}} \right) = 0.437 \cdot \left(1 + \frac{(t - t_{e0})^{0.45}}{k_{1e} + k_{2e}(t - t_{e0})^{0.45}} \right), \tag{46}$$

where $t_{e0} = 1.943 \pm 0.012 \, s$, $k_{1e} = 3.178 \pm 0.003$, $k_{2e} = 1.492 \pm 0.001$ with a coefficient of determination $R = 0.9991$. Therefore, when expansion time goes into infinity, sample deformation will equal:

$$\min[\varepsilon(t)] = \lim_{t \to \infty} \left[\max[\varepsilon(t)] \cdot \left(1 - \frac{(t - t_{e0})^{(n_{c-1})}}{k_{1e} + k_{2e}(t - t_2)^{(n_{c-1})}} \right) \right] \tag{47a}$$

$$\min[\varepsilon(t)] = \max[\varepsilon(t)] \cdot \left(1 + \frac{1}{k_{2e}} \right) = 0.437 \cdot \left(1 - \frac{1}{1.492} \right) \cong 0.144, \tag{47b}$$

and the deformed sample does not return to its initial state at the end of the test.

When expansion time goes to infinity ($\sigma = 0$), the strain resulting from the material's viscous properties ($\sigma_N = 0$) is eliminated, and based on Eq. (34):

$$\min[\varepsilon(t)] \cdot E = \sigma_V \to \frac{\sigma_V}{E} = \min[\varepsilon(t)] \cong 0.144 \tag{48}$$

Equations (39) and (42) indicate that when real relative strain $\overline{\varepsilon_{2-3}} = 0.249$ in the relaxation phase, total strain $\max[\sigma_{(H+V)}] = 9.33 \, kPa$, therefore:

$$E = \frac{\max[\sigma_{(H+V)}(t)]}{\overline{\varepsilon_{2-3}} + \min[\varepsilon(t)]} = \frac{9.33 \, kPa}{0.249 + 0.144} \cong 23.74 \, kPa \tag{49}$$

$$\sigma_V = \min[\varepsilon(t)] \cdot E = 0.144 \cdot E \cong 3.42 \, kPa. \tag{50}$$

Time t_1 after which sample deformation begins can be determined based on the values of sliding friction determined in Eq. (49) and based on Eq. (23):

$$\sigma_V = k_c (\overline{\dot{\varepsilon}_{1-2}} \cdot t_1)^{n_c} \to \ln(t_1) = \frac{1}{n_c} \ln \left(\frac{\sigma_V}{k_c} \right) - \ln(\overline{\dot{\varepsilon}_{1-2}}) \to t_1 = 0.064 \, s \tag{51}$$

Parameter η, which describes material viscosity, can be determined based on the empirical data from the creep test and from Eq. (26). When the relationships described in Eqs. (39), (40) and (41) are taken into account, changes in $\varepsilon(t)$ can be described as follows:

$$\varepsilon(t) = \varepsilon(t_{2p}) + \left[\frac{\overline{\sigma_{1-2}}}{E} - \varepsilon(t_{2p}) \right] \cdot \left[1 - exp \left[\left(-\frac{E}{\eta}(t - t_{2p}) \right) \right] \right] \tag{52}$$

The above equation cannot be used to determine viscosity when parameter is constant; therefore, changes in parameter η were described by approximating Eq. (26) to Eq. (41) which well describes the process (coefficient of determination = 0.9997):

$$\eta = -E \cdot (t - t_{2p}) \cdot \left(\ln \left[1 - \frac{\varepsilon(t_{2p}) \cdot (t - t_{2p})^{n_{rp}}}{(k_{1p} + k_{2rp} \cdot (t - t_{2p})^{n_{rp}}) \cdot (\max[\varepsilon(t)] - \varepsilon(t_{2p}))} \right] \right)^{-1} \tag{53}$$

The above equation describes changes in viscosity η over time, but it is also indirectly related to changes in relative strain $\varepsilon(t)$ over time. If viscosity is a characteristic property of the analyzed material, it should not be dependent on time, whereas the influence of strain can be rationally justified by the accompanying changes in material density $\varrho(t) = \varrho_0 \cdot (1 - \varepsilon(t))^{-1}$. In the analyzed time interval $\langle t_{2p}, t_{3p} \rangle$, the values of $\eta(t)$ calculated with the use of Eq. (53) were regarded as the dependent variable, and the values of $\varepsilon(t)$ calculated with the use of Eq. (41)—as the independent variable. These relationships were described with Eq. (54):

$$\eta(t) = \eta_0 + \left(\frac{k_\eta \cdot \varepsilon(t)}{1 - \varepsilon(t)} \right)^{n_\eta}, \tag{54}$$

where: $\eta_0 = 131.88 \pm 0.78 \, kPas$, $k_\eta = 4.304 \pm 0.003$, $n_\eta = 8.456 \pm 0.006$ with a coefficient of determination $R = 0.9998$. Parameter η_0 should be regarded as a value describing the viscous properties of an unloaded sample.

When the above is substituted into Eq. (24), material flow velocity will be described by Eq. (55):

$$\vartheta = \frac{\varepsilon(t)}{\eta_0 + \left(\frac{k_\eta \cdot \varepsilon(t)}{1 - \varepsilon(t)} \right)^{n_\eta}}, \left[k_c \cdot \varepsilon(t)^{(n_c - 1)} - E - \frac{\sigma_V}{\varepsilon(t)} \right], \tag{55}$$

which indicates that material flow velocity is determined by strain and the material's rheological properties.

The linear changes in strain in the expansion phase of the relaxation test suggest that in this case, the return speed of the measuring head is lower than the self-expansion rate of an unloaded sample, and that the process has been adequately described by Eq. (37). The relevant parameters are determined on the following boundary conditions:

- for $t = t_4 \rightarrow \sigma(t) = \overline{\sigma_4} \wedge \varepsilon(t_4) = \overline{\varepsilon_4}$ (Eq. 40),
- for $t = t_5 \rightarrow \sigma(t) = 0 \wedge \varepsilon(t_5) = \min[\varepsilon(t)]$ (Eq. 41),

where $\overline{\sigma_4}$ and $\overline{\varepsilon_4}$ are the actual stress and strain at the beginning of expansion. These parameters are estimated at $k_e = 3.12 \pm 0.04 \, GPa$ and $n_e = 5.47 \pm 0.21$, with a coefficient of determination $R = 0.9963$. The value of the coefficient of determination can be regarded as a measure of adequacy of the model describing the progression of the relaxation test and the creep test.

Conclusions

The values of exponents n_c and n_e of power functions (Eq. (23) and Eq. (37)) indicate that the viscous properties of the analyzed material correspond to the shear thickening behavior of a non-Newtonian fluid ($n_c > 1$, $n_e > 1$). Similar values were obtained at different strain rates (Myhan et al., 2015) for canned pork ham (n_c = 1.45, n_e = 5.43), which could suggest that exponent values are not influenced by the type of biological material or the rate of compression.

The exponential function can also be applied in the described case and in creep tests of similar materials (Takahiko et al., 2005; Chattong et al., 2007), but the exponent will be less than 1 (n_{rp} = 0.47 in the described case). In view of the above, it can be hypothesized that the value of the exponent will be influenced by the material flow velocity described by Eq. (55). In turn, the above hypothesis implies that the viscous properties of a material, subject to its flow velocity, can correspond to the shear thinning behavior of a non-Newtonian fluid ($n < 1$), the shear thickening behavior of a non-Newtonian fluid ($n > 1$) or the behavior of an ideal Newtonian fluid ($n = 1$).

A formal analysis of the structure of Eqs. (40) and (41) indicates that viscosity can be described by the inverse of coefficient :

$$\eta_r = \frac{1}{k_{2rp}} = \frac{1}{1.178} \cong 0.849,$$

or the following formula:

$$\kappa = \frac{k_{1rp}\sigma_2}{k_{2rp}} = \frac{1.178 \cdot 61.77 \, kPa}{5.147 \, s^n} \cong 14.14 \, kPas^{-n} \cdot$$

In the first case, η_r describes the proportion of viscous stress σ_N at maximum sample load σ_2, where coefficient $k_2 \geq 1$ ($k_2 = 1$ – ideally viscous material, $k_2 = \infty$ - completely non-viscous material). Parameter κ corresponds to the flow consistency index K in the Ostwald-de Waele model $\sigma_N = K\dot{\gamma}^n$. The value of K is comparable to the value of η_0^n in Eq. (54).

2.3 A rheological model of solid foods with a cellular structure

The rheological properties of materials, including biological materials, are interpreted based on the estimated values of rheological parameters in phenomenological models. Such models include the classical Maxwell and Kelvin-Voight models (Rao and Quintero, 2005) as well as the model described in the previous chapter. In biological materials, the experimental compression characteristics significantly deviate from the characteristics of homogenous isotropic materials. In meat and meat products (Myhan et al., 2012, 2014, 2015), strain rate $\dot{\varepsilon}$ exerts a non-linear influence on the measured load. The rate of change in load (increase in strain) $\frac{d\sigma}{dt}$ (at $\dot{\varepsilon} = const$) increases with strain across the entire range of compression values. The reference material (rubber with a density of 1.25 g/cm^3 and Shore hardness of 68) is characterized by two compression periods (where $\frac{d\sigma}{dt} \approx const$ in the second period), whereas in samples of fresh beetroots, celery roots and potato tubers, $\frac{d\sigma}{dt}$ increases across the entire range of compression values (Fig. 7).

In these materials, compression is well described by Eq. (23) (coefficient of determination $R > 0.975$). However, this equation has only informative value because it does not explain the differences in the compression of plant tissue and the reference material. Those variations could be attributed to differences in the internal structure of the compared materials. Fresh vegetables are highly hydrated (intracellular water and capillary water), multiphasic materials with a cellular structure, as demonstrated by the images of plant tissues captured under a scanning electron microscope in Fig. 8.

A rheological model of tissue structure

The analyzed materials have cellular tissue structures that resemble parenchymal structures (Zdunek et al., 2004; Gibson, 2012). In these types of structures, fluids can be displaced during compression, which is not observed in homogeneous materials. The analyzed structures were described (Fig. 9) with a wet tissue model (Bruce, 2003) on the following assumptions:

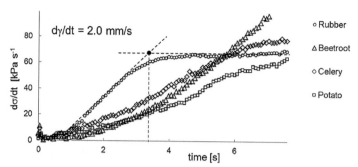

Figure 7: The influence of strain on the rate of increase in strain.

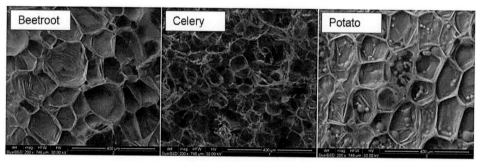

Figure 8: Cellular structure of plant tissues (Quanta 200 environmental scanning electron microscope, FEI Company, Hillsboro, Oregon, USA).

- samples of vegetable roots have a cellular structure, where cells have spherical shape with identical radius ρ_0, and every cell is surrounded by an elastic cell wall with thickness δ,
- every sphere is filled with incompressible fluid with pressure p_0, and intercellular spaces are filled with a mixture of liquid and gas,
- the compression of biological samples leads to spatial deformation of cell and tissue structures, liquid filtration, gas and liquid effusion, displacement of water (cell juice) inside the structure and water extrusion to the surface (loss of cell juice).

The geometric relationships (Fig. 9b) between cell deformation and the displacement of the measuring head are described by Eq. (56):

$$\begin{cases} V_0 = \frac{4}{3}\pi\rho_0^3; V_\gamma = 2\pi\left(\frac{2}{3}\rho_\gamma^3 + \frac{\pi}{2}b\rho_\gamma^2 + b^2\rho_\gamma\right) \\[3mm] V_\gamma = V_0 \to b = \frac{\pi\rho_\gamma}{4}\left(\sqrt{1 + \frac{32}{3\pi^2}\left(\left(\frac{\rho_0}{\rho_\gamma}\right)^3 - 1\right)} - 1\right), \\[3mm] \left(\rho_\gamma = \rho_0 - \frac{h}{2} \wedge \frac{32}{3\pi^2} \cong 1\right) \to b \cong \frac{\pi}{8}\left[2\rho_0\left(\sqrt{\frac{2\rho_0}{2\rho_0 - h}} - 1\right) + h\right] \\[3mm] h = 2\rho_0\frac{\gamma}{H} \to b \cong \frac{\pi}{4}\rho_0\left(\sqrt{\frac{H}{H-\gamma}} - \frac{H-\gamma}{H}\right) \end{cases} \qquad (56)$$

where: b – cell "flattening" radius; h - cell deformation; H – sample height; V – cell volume; γ – displacement of measuring head; ρ – cell radius.

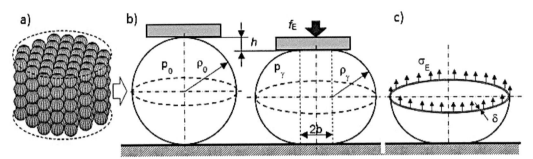

Figure 9: Tissue structure model: (a) geometric model, (b) cell deformation, (c) distribution of stress in the cell wall.

Cell deformation by factor h increases cell surface area and, consequently, relative elongation ε_c of the cell wall, which is accompanied by an increase in stress in the cell wall by σ_E (Fig. 6c). The increase in stress increases the pressure of intracellular fluid by Δp. These relationships are described by Eq. (57):

$$
\begin{cases}
\varepsilon_c = \dfrac{2\pi\rho_\gamma + 4b - 2\pi\rho_0}{2\pi\rho_0} = \dfrac{2\pi\left(\rho_0 - \dfrac{h}{2}\right) + 4b - 2\pi\rho_0}{2\pi\rho_0} = \dfrac{4b - \pi h}{2\pi\rho_0} = \dfrac{1}{2}\left(\sqrt{\dfrac{H}{H-\gamma}} - 1\right) \\[4mm]
\sigma_E = E_c\varepsilon_c \rightarrow \Delta p = E_c\varepsilon_c \dfrac{2\pi(\rho_\gamma + b)\delta}{\pi(\rho_\gamma + b)^2} = \dfrac{16 E_c \delta}{\rho_0} \cdot \dfrac{1 - \sqrt{1 - \dfrac{\gamma}{H}}}{4(4-\pi)\sqrt{\left(1 - \dfrac{\gamma}{H}\right)^3} + \pi}
\end{cases}
\tag{57}
$$

where: ε_c – relative elongation; σ_E – stress; E_c – elastic modulus; Δp – pressure; δ – cell wall thickness.

If the rheological properties of a material are described by the Kelvin-Voigt model, compressive force F can be regarded as the vector sum of the force resulting from elastic strain F_E and frictional resistance F_η, where F_E is the sum of all elastic forces f_E in individual cells of a monolayer. The variation in force F_E as a function of displacement γ is described by Eq. (58):

$$
F_E = \frac{\pi D^2}{4\rho_0^2} f_E = \frac{\pi^2 D^2}{4\rho_0^2}\Delta p b^2 = \frac{\pi^4 E_c D^2 \delta}{4\rho_0} \cdot \frac{1 - \sqrt{1 - \dfrac{\gamma}{H}}}{4(4-\pi)\sqrt{\left(1 - \dfrac{\gamma}{H}\right)^3} + \pi} \cdot \frac{\left(1 - \dfrac{\gamma}{H}\right)^3 - 2\sqrt{\left(1 - \dfrac{\gamma}{H}\right)^3} + 1}{1 - \dfrac{\gamma}{H}}
\tag{58a}
$$

If F_E is the product of the modified elastic modulus E_r, modified relative strain ε_r and the cross-sectional area of the sample, Eq. (58) will take on the following form:

$$
E_r = \frac{\pi^3 E_c \delta}{\rho_0}; \varepsilon_r = \frac{1 - \sqrt{1 - \dfrac{\gamma}{H}}}{4(4-\pi)\sqrt{\left(1 - \dfrac{\gamma}{H}\right)^3} + \pi} \cdot \frac{\left(1 - \dfrac{\gamma}{H}\right)^3 - 2\sqrt{\left(1 - \dfrac{\gamma}{H}\right)^3} + 1}{1 - \dfrac{\gamma}{H}} \rightarrow F_E = E_r \varepsilon_r \frac{\pi D^2}{4}
\tag{58b}
$$

where: ε_r – modified relative elongation; E_r – modified elastic modulus; D – diameter of unloaded sample. According to Eq. (58b), force F_E is determined by the elastic modulus of the cell wall E_c and the ratio of cell wall thickness to cell radius δ/ρ_0.

Successive analyses were carried out on the assumption that viscous friction F_η results from the displacement of intercellular fluids and gases (effusion) (Fig. 10). Effusion was described with the use of basic laws of fluid mechanics:

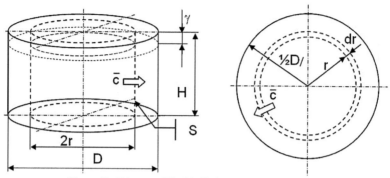

Figure 10: Diagram of liquid effusion in intercellular space.

$$\begin{cases} \sigma_{ef} = \xi \bar{c}^{-2} \rightarrow \dfrac{dF_{\eta}}{dr} = 2\pi r \sigma_{ef} \\[2ex] \bar{c} = \dfrac{\upsilon}{\alpha s}; \quad \upsilon = 2\pi r^2 \gamma \dfrac{\dot{\gamma}}{H} \\[2ex] \alpha = \alpha_0 \dfrac{H-\gamma}{H}; S = 2\pi r(H-\gamma) \end{cases} \tag{59}$$

where: $\bar{c}$ – average rate of discharge; S – area of discharge; α – porosity; $\dot{\gamma}$ – speed of the measuring head; σ_{ef} – effusion stress; υ – volumetric flow rate; ξ – coefficient of local resistance. In view of the above, frictional resistance F_{η} can be described with Eq. (60):

$$\begin{cases} \dfrac{dF_{\eta}}{dr} = 2\pi\xi \left(\dfrac{\gamma\dot{\gamma}}{\alpha_0(H-\gamma)^2} \right)^2 r^3 \rightarrow F_{\eta} = \int_0^{\frac{D}{2}}(dF_{\eta})dr = \dfrac{\pi}{32}\xi\dfrac{D^4}{\alpha_0^2(H-\gamma)^4}\gamma^2\dot{\gamma}^2 \\[3ex] \eta_r = \dfrac{\xi D^2 \gamma^2}{8\alpha_0^2(H-\gamma)^4} \rightarrow F_{\eta} = \eta_r\dot{\gamma}^2\dfrac{\pi D^2}{4} \end{cases} \tag{60}$$

where η_r is the modified coefficient of viscosity.

Averaged stress $\bar{\sigma}$ (expressed by the ratio of compressive force to the surface area of the sample) is the sum of averaged elastic strain $\bar{\sigma}_E$ and averaged viscous stress $\bar{\sigma}_{\eta}$. When Eqs. (58b) and (60) are taken into account, $\bar{\sigma}$ is described by Eq. (61):

$$\bar{\sigma} = \bar{\sigma}_E + \bar{\sigma}_{\eta} = E_r\varepsilon_r + \eta_r\dot{\gamma}^2. \tag{61}$$

In Eq. (61), averaged viscous stress $\bar{\sigma}_{\eta}$ corresponds to the Ostwald-de-Waele model ($\sigma_{\eta} = K\dot{\gamma}^n$), and exponent values ($n > 1$) indicate that compressed cellular materials have the shear thinning behavior of a non-Newtonian fluid.

Method for estimating the modeled parameters – example

Samples of fresh beetroots, celery roots and potato tubers were subjected to a compression test with the use of TA.HDplus texture analyzer (Stable Microsystems Ltd., UK) and the Texture Analysis Screensaver program (version 6.1.1.0). Samples with diameter $D = 19$ *mm* and height $H = 14$ *mm* were compressed to relative strain $\varepsilon_{max} = 0.15$ at different speeds of the measuring head: 0.10, 0.25, 0.50, 1.00, 2.00 mm s^{-1}. Every test was conducted in 10 replications, and compression force F, displacement γ with fixed time step $\Delta t = 0.005$ *s* were registered.

The values of the parameters in Eq. (61) were estimated by the least squares method based on empirical data. The results are presented in Table 4 and Fig. 11. In all analyzed cases, the coefficient of determination was greater than 0.96. The values presented in Table 4 indicate that beetroots and celery roots have similar properties and that they differ significantly from the rheological properties of potato

Table 4: Estimation of the parameters in Eq. (61).

Species	$E_c\dfrac{\delta}{\rho_0}$ [MPa]		$\dfrac{\xi}{\alpha_0^2}$ [MPas^{2}mm^{-2}]		Coefficient of determination
	Estimation	Standard error	Estimation	Standard error	
Beetroot	424.5	1.29	1.18	0.03	0.973
Celery	476.6	1.14	0.38	0.05	0.963
Potato	137.8	0.67	8.75	0.05	0.977

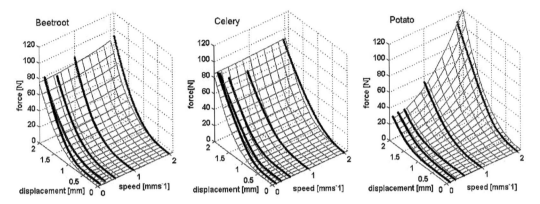

Figure 11: Model fitting (eq. 44) to empirical data.

tubers. In the first case, these structures partly evolved from the top part of the roots and the bottom part of the shoots. In the second case, the analyzed material consisted of underground stem tubers where empty intercellular spaces account for less than 1% of tissue volume (Gao and Pitt, 1991).

3. Summary and Conclusions

The most popular mathematical models for analyzing the rheological properties of liquid, semi-solid and solid foods were presented. The main emphasis was placed on the rheological models of solid foods. The existing rheological models of solids have several weaknesses, in particular when they are used to describe the results of creep and stress relaxation tests of foods with a cellular structure. An original rheological model of a solid with a cellular structure was proposed. The model was developed on the assumption that the analyzed material is composed of identical spherical cells with a known radius, where every sphere is surrounded by an elastic cell wall of known thickness. Every sphere is filled with incompressible fluid with known pressure, and intercellular spaces are filled with a mixture of liquid and gas. The physical processes occurring inside compressed samples were described on the assumption that compression of biological samples leads to spatial deformation of cell and tissue structures, liquid filtration, gas and liquid effusion, displacement of water (cell juice) inside the structure and water extrusion to the surface (loss of cell juice).

The proposed rheological model of a solid with a cellular structure well describes the compression of biological materials. Function $F(\gamma, \dot{\gamma})$ was influenced by the parameters describing tissue structure, such as the elasticity of the cell wall E_c, the dimensionless ratio of cell wall thickness to cell diameter δ/ρ_0, coefficient of local resistance ξ and tissue porosity α. An analysis of Eq. (60) indicates that averaged viscous stress $\bar{\sigma}_\eta$ is also influenced by the shape factor expressed by the ratio of sample height and sample diameter H/D. The averaged stress $\bar{\sigma}$ (Eq. (61)) for different values of displacement rate $\dot{\gamma}$ and shape factor H/D also differed for the same displacement γ. In the future, the proposed rheological model of solid foods with a cellular structure should be evaluated for its sensitivity to the geometric and physical properties of individual cells, including in other plant materials.

References

Abbott, J.A. 1999. Quality measurement of fruits and vegetables. Postharvest Biol. Technol. 15: 207–225.

Ahmed, J. and Ramaswamy, H.S. 2007. Dynamic rheology and thermal transitions in meat-based strained baby foods. J. Food Eng. 78: 1274–1284.

Alvarez, M.D., Canet, W., Cuesta, F. and Lamua, M. 1998. Viscoelastic characterization of solid foods from creep compliance data: Application to potato tissues. Eur. Food Res. Technol. 207(5): 56–362.

Andrés, S.C., Zaritzky, N.E. and Califano, A.N. 2008. Stress relaxation characteristics of low-fat chicken sausages made in Argentina. Meat Sci. 79: 589–594.

Anvari, M. and Joyner, H.S. 2017. Effect of fish gelatin-gum arabic interactions on structural and functional properties of concentrated emulsions. Food Res. Int. 102: 1–7.

Barrett, D.M., Beaulieu, J.C. and Shewfelt, R. 2010. Color, Flavor, texture, and nutritional quality of fresh-cut fruits and vegetables: Desirable levels, instrumental and sensory measurement, and the effects of processing. Crit. Rev. Food Sci. Nutr. 50: 369–389.

Bayarri, S., Smith, T., Hollowood, T. and Hort, J. 2007. The role of rheological behaviour in flavour perception in model oil/water emulsions. Eur. Food Res. Technol. 226: 161–168.

Bi, C., Li, L.T., Zhu, Y.D., Liu, Y.D., Wu, M., Li, G. et al. 2018. Effect of high-speed shear on the non-linear rheological properties of SPI/κ-carrageenan hybrid dispersion and fractal analysis. J. Food Eng. 218: 80–87.

Blahovec, J. 1996. Stress relaxation in cherry fruit. Biorheology 33(6): 451–462.

Bourne, M.C. 1982. Food Texture and Viscosity: Concept and Measurement. Academic Press, New York, USA.

Bruce, D.M. 2003. Mathematical modelling of the cellular mechanics of plants. Philos. Trans. R Soc. B 358(1437): 1437–1444.

Chattong, U., Apichartsrangkoon, A. and Bell, A.E. 2007. Effects of hydrocolloid addition and high-pressure processing on the rheological properties and microstructure of a commercial ostrich meat product "Yor" (Thai sausage). Meat Sci. 76(3): 548–554.

Claracq, J., Sarrazin, J. and Montfort, J.P. 2004. Viscoelastic properties of magnetorheological fluids. Rheol. Acta 43(1): 38–49.

Damez, J.L. and Clerjon, S. 2008. Meat quality assessment using biophysical methods related to meat structure. Meat Sci. 80(1): 132–149.

Diamante, L. and Umemoto, M. 2015. Rheological properties of fruits and vegetables: A Review. Int. J. Food Prop. 18(6): 1191–1210.

Dzadz, Ł., Markowski, M., Sadowski, P., Jakóbczak, A. and Janulin, M. 2015. Creep and recovery characteristics of chicken meat frankfurters. J. Agric. Sci. Technol. 17(4): 827–835.

Funami, T., Yada, H. and Nakao, Y. 1998. Thermal and rheological properties of curdlan gel in minced pork gel. Food Hydrocoll. 12: 55–64.

Gao, Q. and Pitt, R.E. 1991. Mechanics of parenchyma tissue based on cell orientation and microstructure. Trans. ASAE 34: 232–238.

Gibson, L.J. 2012. The hierarchical structure and mechanics of plant materials. J. R. Soc. Interface 9(76): 2749–2766.

Glicerina, V., Balestra, F., Rosa, M.D. and Romani, S. 2016. Microstructural and rheological characteristics of dark, milk and white chocolate: A comparative study. J. Food Eng. 169: 165–171.

Gonzalez-Gutierrez, J. and Scanlon, M.G. 2013. Strain Dependence of the uniaxial compression response of vegetable shortening. J. Am Oil Chem. Soc. 90(9): 1319–1326.

Guerrero-Beltrán, J.A., Welti-Chanes, J. and Barbosa-Cánovas, G.V. 2009. Rheological models and characteristics for selected foods. pp. 35–58. *In*: Sosa-Morales, M.E. and Velez-Ruiz, J.F. (eds.). Food Processing and Engineering Topics. Nova Science Publishers, Inc., New York, USA.

Hamilton, C.A., Alici, G. and Panhuis, M. 2018. 3D printing Vegemite and Marmite: Redefining "breadboards". J. Food Eng. 220: 83–88.

Hassan, B.H., Alhamdan, A.M. and Elansari, A.M. 2005. Stress relaxation of dates at khalal and rutab stages of maturity. J. Food Eng. 66: 439–445.

Holdsworth, S.D. 1993. Rheological models used for the prediction of the flow properties of food products: a literature review. Food Bioprod. Process. Trans. Inst. Chem. Eng. Part. C 71(3): 139–179.

Ibarz, A. and Barbosa-Canovas, G.V. 2014. Introduction to food process engineering. CRC Press, Boca Raton, USA.

Jamroziak, K., Bocian, M. and Kulisiewicz, M. 2010. Examples of applications non-classical visco-elastic models in the process of ballistic impact. Modelowanie Inżynierskie 40: 95–102 (in Polish).

Kaya, S. 2002. Effect of salt on hardness and whiteness of Gaziantep cheese during short-term brining. J. Food Eng. 52: 155–159.

Kechinski, C.P., Schumacher, A.B., Marczak, L.D.F., Tessaro, I.C.F. and Cardozo, N.S.M. 2011. Rheological behavior of blueberry (*Vaccinium ashei*) purees containing xanthan gum and fructose as ingredients. Food Hydrocoll. 25(3): 299–306.

Knauss, W.G. and Zhao, J. 2007. Improved relaxation time coverage in ramp-strain histories. Mech. Time Depend. Mater 11(3-4): 199–216.

Li, X., Wang, Y., Lee, B.H. and Li, D. 2018. Reducing digestibility and viscoelasticity of oat starch after hydrolysis by pullulanase from *Bacillus acidopullulyticus*. Food Hydrocoll. 75: 88–94.

Li, Q., Li, D., Wang, L.J., Özkan, N. and Mao, Z.H. 2010. Dynamic viscoelastic properties of sweet potato studied by dynamic mechanical analyzer. Carbohydr. Polym. 79: 520–525.

Liu, Y. 2001. A direct method for obtaining discrete relaxation spectra from creep data. Rheol. Acta 40: 256–260.

Ma, Z. and Boye, J.I. 2013. Microstructure, physical stability, and rheological properties of salad dressing emulsions supplemented with various pulse flours. J. Food Res. 2(2): 167–181.

Majid, I., Dar, B.N. and Nanda, V. 2018. Rheological, thermal, micro structural and functional properties of freeze-dried onion powders as affected by sprouting. Food Biosci. 22: 105–112.

Mitchell, J.R. and Blanshard, J.M.V. 1976. Rheological properties of nalginate gels. J. Texture Stud. 7: 219–234.

Mohsenin, N.N. and Mittal, J.P. 1977. Use of rheological terms and correlation of compatible measurements in food texture research. J. Texture Stud. 8: 395–408.

Moresi, M. and Pallottino, F., Costa, C. and Menesatti, P. 2012. Viscoelastic properties of Tarocco orange fruit. Food Bioproc. Tech. 5(6): 2360–2369.

Myhan, R., Białobrzewski, I. and Markowski, M. 2012. An approach to modeling the rheological properties of food materials. J. Food Eng. 111: 351–359.

Myhan, R., Białobrzewski, I., Karpińska-Tymoszczyk, M., Danowska-Oziewicz, M., Markowski M. and Majewska, K. 2014. The effect of relative air humidity on the rheological properties of roasted turkey breast. J. Texture Stud. 45(4): 288–294.

Myhan, R., Markowski., M., Daszkiewicz, T., Zapotoczny, P. and Sadowski, P. 2015. Non-linear stress relaxation model as a tool for evaluating the viscoelastic properties of meat products. J. Food Eng. 146: 107–115.

Myhan, R., Markowski, M., Daszkiewicz, T., Korpusik, A. and Zapotoczny, P. 2016. Identification of the chemical composition of meat products based on their rheological properties. J. Texture Stud. 47(6): 504–513.

Nindo, C.I., Tang, J., Powers, J.R. and Takhar, P.S. 2007. Rheological properties of blueberry puree for processing applications. LWT - Food Sci. Technol. 40(2): 292–299.

Peleg, M. 1976. Considerations of a general rheological model for the mechanical behavior of viscoelastic solid food materials. J. Texture Stud. 7(2): 243–255.

Peleg, M. and Calzada, J.F. 1976. Stress relaxation of deformed fruits and vegetables. J. Food Sci. 41: 1325–1329.

Peleg, M. 1984. Application of nonlinear phenomenological rheological models to solid food materials. J. Texture Stud. 15(1): 1–22.

Rao, M.A. 2007. Rheology of fluid and semisolid foods: Principles and applications. Springer, New York, USA.

Rao, V.N.M. and Quintero, X. 2005. Rheological properties of solid foods. pp. 101–148. *In*: Rao, M.A., Rizvi, S.S.H. and Datta, A.K. (eds.). Engineering Properties of Foods. Taylor and Francis, Boca Raton, US.

Rathod, M.L. and Kokini, J.L. 2016. Extension rate distribution and impact on bubble size distribution in Newtonian and non-Newtonian fluid in a twin screw co-rotating mixer. J. Food Eng. 169: 214–227.

Sorvari, J. and Malinen, M. 2007. Numerical interconversion between linear viscoelastic material functions with regularization. Int. J. Solids Struct. 44: 1291–1303.

Steffe, J.F. 1996. Rheological Methods in Food Process Engineering. Freeman Press, East Lasing, USA.

Stokes, J.R., Boehm, M.W. and Baier, S.K. 2013. Oral processing, texture and mouthfeel: From rheology to tribology and beyond. Curr. Opin. Colloid Interface Sci. 18: 349–359.

Su, H.S., Bansal, N. and Bhandari, B. 2016. Rheology of emulsion-filled alginate microgel suspensions. Food Res. Int. 80: 50–60.

Sudhagar, M., Lope, G.T. and Shahab, S. 2006. Effects of compressive force, particle size and moisture content on mechanical properties of biomass pellets from grasses. Biomass Bioenergy 30: 648–654.

Takahiko, S., Takamasa, K. and Akiko, H. 2005. Analysis of creep test on gel-type processed food, mainly sausage. J. Jap. Soc. Food Sci. Technol. 52(11): 517–521.

Torley, P.J. and Young, O.A. 1995. Rheological changes during isothermal holding of salted beef homogenates. Meat Sci. 39: 23–34.

Yanxia, W., Yanbin, L., Rui, X., Yunfei, X., Jian, Y. and Ji, Z. 2015. The flow behavior, thixotropy and dynamical viscoelasticity of fenugreek gum. J. Food Eng. 166: 21–28.

Yilmaz, M.T., Karaman, S., Dogan, M., Yetim, H. and Kayacier, A. 2012. Characterization of O/W model system meat emulsions using shear creep and creep recovery tests based on mechanical simulation models and their correlation with texture profile analysis (TPA) parameters. J. Food Eng. 108: 327–336.

Yu, C. and Gunasekaran, S. 2001. Correlation of dynamic and steady flow viscosities of food materials. Appl. Rheol. 11: 134–140.

Zdunek, A., Umeda, M. and Konstankiewicz, K. 2004. Method of parenchyma cells parametrisation using fluorescence images obtained by confocal scanning laser microscope. Electr. J. Polish Agric. Univ. EJPAU 7(1): http://www.ejpau.media.pl/volume7/issue1/engineering/art-01.html (accessed 02.02.2019).

Zhiguo, L., Pingping, L., Hongling Y. and Jizhan, L. 2013. Internal mechanical damage prediction in tomato compression using multiscale finite element model. J. Food Eng. 116: 639–647.

CHAPTER 15

Mathematical Models for Analyzing the Microbial Growth in Food

Jyoti Singh and *Vishal Mishra**

1. Introduction

One of the basic precepts is that bacterial growth is a function of food available in the environment. The microbial species that will deal better in the environment will thrive. The specific information about growth characteristics of microorganism can be given by a number of factors affecting microbial growth. However, it is virtually impossible to attain this goal as there are a number of different foods eaten globally and many biological variations are present within only a single food. Luckily, a number of primary growth factors are limited for bacterial growth. Thus, the behavior of microbes could be estimated by analyzing the response to those factors. This is the fundamental purpose of predictive microbiology, a branch of food microbiology. The prime objective is to determine the relationship between primary factors and microbial growth in order to overcome the requirement for a vast amount of data. The general method comprises of procurement of data that will help to build mathematical relationships. These mathematical models are then used for the prediction of microbial behaviour in a variety of foods, based on physical measurement of the main factors (Buchanan, 1992).

Most fruitful exploration on modeling affects multiple variables on development of pathogenic microbes. Researchers have developed different models for checking their growth. There is a need to have knowledge about microbiology history in order to describe the interactions between several factors mathematically. Earlier, work related to microbiological modeling had been done, but food was not mentioned at all in those models. Conditions that occur during industrial fermentation were extensively modeled and equations, such as the description of variables' impact on the yield, given by Monad, were developed. Models of fermentation occasionally considered many of the variables related to foodborne microbes. Thus, various models for describing the growth of pathogenic microorganisms in food have been developed. The factors responsible for microbial growth and the associated mathematical model are described in this chapter.

School of Biochemical Engineering, Indian Institute of Technology (BHU) Varanasi, Varanasi - 221005.
  Email: jyotisigh.rs.bce18@itbhu.ac.in
* Corresponding author: vishal.bce@itbhu.ac.in

2. Factors that Influence the Growth of Microorganisms in Food

The various factors that influence the growth of microbes in food are discussed below.

2.1 Intrinsic factors

2.1.1 pH (H⁺ concentration)

Measurement of pH is done by H^+ ions concentration. pH values below 7 are acid and above 7 are base in nature. Each and every food possesses some pH, with which its acidity and alkalinity can be determined (Rosso et al., 1995; Hamad, 2012).

The acidic or alkaline formation of a food in the body has nothing to do with the real pH of the food itself. Lemons, for instance, are very acidic, but the end products they produce after digestion and assimilation are alkaline, so lemons in the body form alkaline. Likewise, meat will test alkaline before digestion, but it leaves acidic residue in the body so meat is categorized as acid-forming like almost all animal products. Change in body's serum pH is a dangerous event and the body constantly attempts to keep pH between 7.35 and 7.45. Systemic pH is maintained by bones as they provide minerals. Osteoporosis may occur if a human consumes an alkaline diet.

2.1.2 Moisture content

The availability of water in a usable form is the key requirement for the development of microbes in food products. The requirement of water for the development of microbes in food is demarcated in terms of water activity (α).

$$\alpha = \frac{P}{P_i}$$

where, P and P_i are vapour pressure of solution and solvent phase, respectively.

The requirement of water to assist the microorganism's growth in food is depicted by the value of water activity. Gram (–ve) microbes are more responsive to the requirement of water than gram (+ve) microorganisms. The table is mentioned in which value of water activity for different pathogenic growth on food (Table 1).

Table 1: Approximate values of water activity for the development of pathogens in food (2001).

Bacteria	Maximum	Optimum	Minimum
Campylobacter sp.		0.99	0.98
C. botulinum type E			0.97
Shigella sp.			0.97
Yersinia enterocolitica			0.97
Salmonella sp.	> 0.99	0.99	0.94
B. cereus			0.93
Listeria monocytogenes			0.92
C. perfringens	0.97	0.95–0.96	0.943
C. botulinum type A and B			0.93

2.1.3 Nutrient content in food

Viable cells of pathogens need nutrients (energy, water, macronutrients and micronutrients) for their maintenance and augmentation. The growth of the microorganism and the product formed depends upon the elemental constitution of the food material (MacDonald and Reitmeier, 2017).

Vitamin B and ascorbic acid are available in plenty of meat and fruits. The microbes that can exploit the nutrients available in food without difficulty are generally the prime ones. The accessibility of vital nutrients affects the growth kinetics of microbes.

2.1.4 Antimicrobial substances

Those substances that possess antimicrobial activity are not affected by the pathogenic effect of microbes (Quintavalla and Vicini, 2002).

A list of antimicrobial substances in different food materials are given in Table 2.

Table 2: List of antimicrobial substances in different food materials.

Food materials	Antimicrobial substances
Clove	Eugenol
Garlic	Allicin
Cow's milk	Lactoferrin
Cow's milk	Casein
Egg	Lysozyme
Egg	Conalbumin

2.1.5 Biological structures

There is a physical barrier on the exterior of the food which inhibits the arrival and development of pathogenic microbes. Biological structures in different food materials are mentioned in Table 3. However, during the processing of food (slicing, crushing, etc.), the physical barrier is smashed and the interior of the food becomes contaminated and, as a result, bacterial growth appears.

Table 3: List of biological structures in different food materials (Morris and Groves, 2013).

Food material	Biological structure
Fruits	Peel
Nut	Shell
Egg	Cuticle
Vegetables	Peel

2.2 Extrinsic factors

2.2.1 Storage temperature of food

Microbial growth depends on the temperature at which the food is stored and can be varied over a wide range (Rosso et al., 1995). Depending on temperature necessities, microbes are divided into four different groups (Table 4).

Table 4: Optimal and growth temperature for corresponding group of microbes.

Group	Growth temperature	Optimal temperature (°C)
Psychrophilic	at or below 5°C	12 to 15
Psychotropic	at or below 7°C	20 to 30
Mesophilic	between 20°C and 45°C	30 to 40
Thermophilic	at 45°C or above	55 to 65

2.2.2 Importance of gases in the environment

Air and oxygen are the two important reasons for food spoilage. As air has no taste, color and odor, it is usually presumed inert and ignored as a means to cause food spoilage. Air consists of 21% O_2, 78% N_2 and 1% contains a mixture of various gases. Oxygen is necessary for all of us in order to sustain life but one more fact about it is that it is responsible for spoilage of food as it deteriorates food constituents in several ways, such as oxidation.

2.2.3 Moisture

Water is the most important resource that keeps us all alive as it is a significant element in all food. It affects the appearance, texture and flavor of food. Dry foods and fresh foods contain various levels of moisture. 90 to 95 percent of water is present in fresh vegetables and fruits. Excess moisture results in spoilage of food by microorganisms and chemical reactions. With the help of water in the food, microbial cells entered and, as a result, food got spoiled. With decrease in temperature, relative humidity increases and vice-versa.

Microbial data can be analyzed mathematically by the help of **"Predictive Microbiology"**. It quantifies food bacteriological ecosystems with the use of mathematical models. It helps in assessing the critical point isolation in the production process and the consequence of environmental variables on pathogenic microbes related to their behaviour. It can also be used to reveal microbial behaviour under several physical and chemical conditions (Zwietering et al., 1990).

3. Growth Phase and Growth Constants

Before understanding the various growth phases, let's be clear about the difference between cell concentration and bacterial density.

> "Cell concentration is described as the number of individual cells per unit volume of a culture, while the bacterial density is the dry weight of cells per unit volume of a culture."

> (Monod, 1949).

Let x_1 be the number of cells at time t_1. After a specific time period, cell-division occurs once.

Then, concentration of cell is,

$$x = x_1.2 \tag{1}$$

After n divisions; it becomes,

$$x = x_1.2^n \tag{2}$$

If the number of divisions becomes 'r' at time t_2, then x will be,

$$x = x_1.2^{r(t_2-t_1)} \tag{3}$$

Using logarithm of base 2

$$r = \frac{\log_2 x_2 - \log_2 x_1}{t_2 - t_1} \tag{4}$$

Here, 'r' = mean division rate with a time interval of $t_2 - t_1$.

The increment of bacterial density with concentration of cells occurs if the mean size of cell doesn't vary with stipulated time.

Growth Phases

Bacterial culture growth can be reprented by a curve that comprises four different phases: Lag, log (exponential), stationary and death phase. This curve signifies the presence of live cells over a period of time.

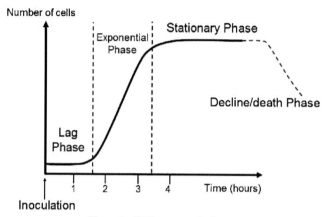

Figure 1: Different growth phases.

The following descriptions of different phases are illustrated in Fig. 1.

1. Lag Phase: Null growth
2. Exponential (Log) Phase: The increment in the growth of a microbial culture
3. Stationary Phase: Null growth
4. Death Phase: Negative growth rate

4. Microbial Growth Kinetics

The cell illustrates logarithmic growth in the log phase. If N_0 denotes the number of cells at $t = 0$, then after a time 't' it becomes N_t (Kovárová-Kovar and Egli, 1998).

Thus,

$$N_t = N_0 \times 2^n \tag{5}$$

where, number of generations is denoted by 'n'.

If 'N' denotes number of cells at a specified time, then the growth rate changes with time as,

$$\frac{dN}{dt} = \mu N \tag{6}$$

If 'X' is the concentration of biomass at a given time, then biomass (growth rate) changes with time as,

$$\frac{dN}{dt} = \mu X \tag{7}$$

The specific growth rate (μ) of bacterial culture is a function of the maximum specific growth rate (μ_{max}), concentration of the limiting substrate (S), and a substrate specific constant (K_s) which is expressed below:

$$\mu = \frac{\mu_{max} \cdot S}{K_s + S} \tag{8}$$

Initially, in a batch culture, the higher amount of substrate is present, i.e., $S > K_S$.

Thus, Eq. (8) becomes,

$$\mu = \mu_{max}$$

At low substrate concentration, it becomes;

$$\mu = \frac{\mu_{max}.S}{K_s} \tag{9}$$

Equation (4) shows resemblance with first order reaction.

Thus,

$$\mu < \mu_{max} \tag{10}$$

Example - Determination of enzyme activity or reaction velocity at different pH.

Generally, a model that contains more parameters should be better fitted. For instance, Table 5 shows maximum enzyme activity of microorganism at different pH ranging from 4.5 to 7. Other than pH, conditions like water activity, temperature and inoculum size should be made alike for batch cultures with the help of which values of μ_{max} can be determined. As seen in Fig. 2 that polynomial with order 2 and 3 were fitted against the data given in Table 5.

Figure 2 shows that order 3 polynomial is best fitted in comparison to ordering 2 as its R^2 value is higher. The order 2 polynomial implies that between pH 6 to 7 there were no data and there μ_{max} firstly shows increment after then decreases with a μ_{max} maximum at 6.5 pH. But, for the same case for order 3 polynomial, μ_{max} first slightly decreases and then increases. Thus, Y1 and Y2 did not give an exact result at 6.5 pH. By introduction of pH' = $(7.pH)^2$, we can rescale the margin in a curve along which μ_{max} increases with increase in pH in the whole experiment. Hence, this curve is the most closely related to the experimental data.

4.1 Mathematical form of the bacterial growth kinetics

The mathematical expression for system kinetics, based on the hypothesis that bacterial population is homogeneous, is given below (Baranyi and Roberts, 1995).

$$\frac{dZ(t)}{dt} = f(Z(t),C(t);D(t)) \tag{11}$$

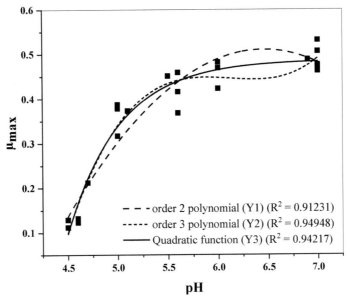

Figure 2: Variation of enzyme activity at different pH.

Table 5: Values of μ_{max} at different pH values (Baranyi and Roberts, 1995).

pH	μ_{max}
4.5	0.112
4.5	0.129
4.6	0.132
4.6	0.123
4.7	0.212
5.0	0.387
5.0	0.317
5.0	0.378
5.1	0.373
5.5	0.451
5.6	0.368
5.6	0.459
5.6	0.416
6.0	0.423
6.0	0.482
6.0	0.469
6.9	0.488
7.0	0.462
7.0	0.531
7.0	0.506
7.0	0.468
7.0	0.476

$$\frac{dC(t)}{dt} = g(Z(t), C(t); D(t)) \tag{12}$$

$$\mu(t) = \phi(Z(t), C(t); D(t)) \tag{13}$$

The functions (f, g, ϕ) given above demonstrate that $D(t)$ is separated from $z(t)$ and $C(t)$ by a semicolon indicates that functions (f, g, ϕ) depend on definite state of growth that doesn't depend on external circumstances, $D(t)$. For example, when the rate constants in functions (f, g, ϕ) depends on exterior temperature.

where, $\mu(t)$ = specific growth rate

$Z(t)$ = referred as physiological or internal state vector

$C(t)$ = external state vector

The concentration of the cell is calculated from the 1st order differential equation. The output is shown below:

$$\frac{dx(t)}{dt} = \mu(t)x(t) \tag{14}$$

4.2 Kinetic models

4.2.1 Monod's model

In the Monod model, $Z(t)$ and $C(t)$ consist of only 1 entry, in which the mass of a cell is measured by $Z_1(t)$ and concentration of substrate around the cell is $C_1(t)$ (Monod, 1949).

The equations, following (11) and (12) are:

$$\frac{dZ_1(t)}{dt} = K_1 \frac{C_1(t)}{K_c + C_1(t)} Z_1(t) \tag{15}$$

$$\frac{dC_1(t)}{dt} = -\frac{1}{Y} \frac{dZ_1(t)}{dt} \tag{16}$$

where K_1, K_c and Y are constants for the respective model.

Assumption: When cell mass reaches a critical value, binary fission takes place. Thus, the specific growth rate of total mass becomes equal to the specific growth rate of cell concentration. It follows Eq. (13) as follows.

$$\mu(t) = \frac{dZ_1(t)}{dt} \tag{17}$$

Two hypothesis was given by Monod (Motta and Pappalardo, 2013)

Hypothesis 1

The growth rate of microbes mainly depends on the concentration of growth-limiting substrate in the medium.

$$\mu(s) = \frac{\mu_m S}{K_h + S} \tag{18}$$

Where,
μ_m = maximal specific growth rate
k_h = half-saturation constant (values of s for $\mu(s) = \mu_m/2$)

Hypothesis II

Instantaneous adjustment of growth rate because of variations in the concentration of substrate.

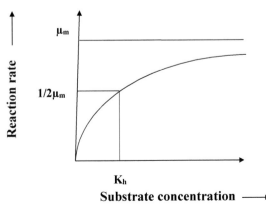

Figure 3: Graphical representation of the Monod equation.

$$\frac{x'(t)}{x(t)} = \mu(s(t)) \tag{19}$$

where,

S(t) = concentration of substrate at time t

x(t) = concentration of microorganism at time t

$\dfrac{x'(t)}{x(t)}$ = specific growth rate of colony

4.2.2 Hills and wright model

For describing the lag phase, along with the growth rate, there is a need to introduce more compartments in the internal state vector, i.e., $Z(t)$. In this model, $Z_1(t)$ and $Z_2(t)$ are two components of internal state vector (Baranyi and Roberts, 1995).

By taking Eq. (11),

$$\frac{dZ_1(t)}{dt} = (K_1 - K_2 Z_1(t))Z_2(t) \tag{20}$$

$$Z_2(t) = Z_1(t) + S_{min} \tag{21}$$

where,

K_1 and K_2 = constants depend on external environment

$Z_1(t)$ = excess biomass of the cell

$Z_2(t)$ = total biomass of the cell

S_{min} = minimum mass

For Eq. (13), in contrast to Monod's model, the equation is applied as below:

$$\mu(t) = \frac{Z_1(t)}{S_{min}} \tag{22}$$

Solving the above equation in a constant external condition,

$$\mu(t) = K_1 \alpha(t) = K_1 \frac{K_2 e^{K_1 t} - K_2 e^{-K_2 t}}{K_2 e^{K_1 t} + K_1 e^{-K_2 t}} \tag{23}$$

Here, value of $\alpha(t) = 0$ to 1.

4.2.3 Baranyi and roberts model

This model is mainly aimed at lag phase computations where the initial number of cells follows Poisson distribution and ANOVA (Baranyi and Roberts, 1995).

$$\frac{dZ_1(t)}{dt} = K_1 Z_1(t) \tag{24}$$

$$\frac{dZ_2(t)}{dt} = \frac{Z_1(t)}{K_2 + Z_1(t)} K_2 \frac{C_1(t)}{K_C + C_1(t)} Z_2(t) \tag{25}$$

$$\frac{dC_1(t)}{dt} = -\frac{1}{Y} \frac{dZ_2(t)}{dt} \tag{26}$$

where, K_1, K_2, K_C, K_Z and Y are model parameters.

Assumption: Cell concentration, x(t) is directly proportional to total biomass.

Thus,

$$\mu(t) = \frac{\frac{dZ_2(t)}{dt}}{Z_2(t)} = \frac{Z_1(t)}{K_2 + Z_1(t)} K_2 \frac{C_1(t)}{K_C + C_1(t)} \tag{27}$$

$\mu(t)$ is the limiting function describing transition to the stationary phase.

Condition: Infinite nutrients results in unlimited growth,

$$\mu(t) \approx 1 - \left(\frac{x(t)}{x_{max}}\right) m$$

where, x_{max} = maximum cell concentration

m = curvature parameter

Model obtained for x(t);

$$\frac{dx(t)}{dt} = \frac{Z_1(t)}{K_2 + Z_1(t)} \mu_{max} (1 - \left(\frac{x(t)}{x_{max}}\right)^m) x(t) \tag{28}$$

where,

$$\frac{dZ_1(t)}{dt} = v Z_1(t) \tag{29}$$

Natural log of cell concentration,

$$y = \ln x(t) \tag{30}$$

The solution for Eq. (30) is,

$$y(t) = y_0 + \mu_{max} A(t) - \frac{1}{m} \ln\left(1 + \frac{e^{m \mu_{max} A(t) - 1}}{e^{m(y_{max} - y_0)}}\right) \tag{31}$$

where,

$y = \ln x(t_0)$ at t = t_0,

$y = \ln x_{max}$

Since, A(t) = delay in time, gradually

$$A(t) = t + \frac{\ln(e^{-\mu_{max} t} + e^{-h_0} - e^{-vt - h_0})}{\mu_{max}} \tag{32}$$

where,

$$h_0 = -\ln \alpha_0$$

$$\alpha_0 = \frac{Z_1(t_0)}{K_2 + Z_1(t_0)}$$

α_0 = physiological state of the cell at t = t_0.

5. Models to Evaluate Microbial Growth

5.1 Square root model

This model evaluates bacterial growth as an element of temperature. It utilizes square root and natural logarithm transformation (Alber and Schaffner, 1992).

This model assesses bacterial development as an element of the temperature. It utilizes square root and regular log change. It is stated as:

$$\sqrt{K} = b(T - T_{min})\{1 - \exp[C(T - T_{max})]\} \tag{33}$$

where,

K = growth rate (Time^{-1})

b = regression coefficient

T = temperature in kelvin

T$_{min}$ = minimum growth temperature

T$_{max}$ = maximum growth temperature

C = regression coefficient

Equation (33) is variate as:

$$K = [b(T - T_{min})]^2 \{1 - \exp[C(T - T_{max})]\} \tag{34}$$

Square root transformation of Eq. (34);

$$\sqrt{K} = b(T - T_{min})\{1 - \exp[C(T - T_{max})]\}^{0.5} \tag{35}$$

Natural log transformation of Eq. (35) gives;

$$\ln(K) = \ln\left([b(T - T_{min})]^2 \{1 - \exp[C(T - T_{max})]\}\right) \tag{36}$$

5.2 Schoolfield model

It is assumed that the microbial growth rate at a specified temperature is regulated by a rate-controlling enzyme (Alber and Schaffner, 1992).

The equation is given as:

$$k = \frac{\rho(298K)\dfrac{T}{298}\exp\left[\dfrac{\Delta H_A^{\neq}}{R}\left(\dfrac{1}{298} - \dfrac{1}{T}\right)\right]}{1 + \exp\left[\dfrac{\Delta H_L}{R}\left\{\dfrac{1}{T_{1/2L}} - \dfrac{1}{T}\right\}\right] + \exp\left[\dfrac{\Delta H_H}{R}\left\{\dfrac{1}{T_{1/2H}} - \dfrac{1}{T}\right\}\right]} \tag{37}$$

where,

k = growth rate (time^{-1})

ρ = growth rate at 298 Kelvin

T = temperature (K)

R = universal gas constant (8.314 JK^{-1}mol^{-1})

$\Delta H_A^{\neq}$ = enthalpy of activation of reaction by rate controlling enzyme (J mol^{-1})

ΔH_L = variation in enthalpy with low-temperature enzyme activation (J mol^{-1})

$T_{1/2L}$ = low temperature when enzyme is 50% inactive

$T_{1/2H}$ = high temperature when enzyme is 50% inactive

ΔH_H = variation in enthalpy with high-temperature enzyme inactivation

Square root transformation for Eq. (37);

$$\sqrt{k} = \sqrt{\frac{\rho(25°C)\dfrac{T}{298}\exp\left[\dfrac{\Delta H_A^{\neq}}{R}\left(\dfrac{1}{298}-\dfrac{1}{T}\right)\right]}{1+\exp\left[\dfrac{\Delta H_L}{R}\left\{\dfrac{1}{T_{1/2L}}-\dfrac{1}{T}\right\}\right]+\exp\left[\dfrac{\Delta H_H}{R}\left\{\dfrac{1}{T_{1/2H}}-\dfrac{1}{T}\right\}\right]}} \tag{38}$$

Natural log transformation;

$$\ln(k) = \ln\left[\frac{\rho(25°C)\dfrac{T}{298}\exp\left[\dfrac{\Delta H_A^{\neq}}{R}\left(\dfrac{1}{298}-\dfrac{1}{T}\right)\right]}{1+\exp\left[\dfrac{\Delta H_L}{R}\left\{\dfrac{1}{T_{1/2L}}-\dfrac{1}{T}\right\}\right]+\exp\left[\dfrac{\Delta H_H}{R}\left\{\dfrac{1}{T_{1/2H}}-\dfrac{1}{T}\right\}\right]}\right] \tag{39}$$

The variant of transformed data is acquired by the expresion

$$(\sigma_i')^2 = \left[\frac{d}{dy_i}\left[f(y_i)\right]\right]^2 \sigma_i^2 \tag{40}$$

where,

($\sigma'i$) = Standard deviation of ith transformed observation

F(yi) = yi transformation

y_i = i^{th} observation

σi^2 = variance of yi

Natural log variance assessed by the specified formula

$$\text{var}\left[\ln(k_i)\right] = \left[\frac{d}{dk_i}\left[\ln(k_i)\right]\right]^2 \text{var}(k_i)$$

$$\text{var}\left[\ln(k_i)\right] = \left[\frac{1}{k_i}\right]^2 \text{var}(k_i) \tag{41}$$

$$\text{var}\left[\ln(k_i)\right] = \text{var}\frac{(k_i)}{k_i^2}$$

where k_i = i^{th} growth rate.

A change in the growth rate alters every residual magnitude that bring about weighted effect on regression. The expression for weighted SSE is given by Eq. (42).

Weighted scheme estimated as natural log transformation can be formulated as;

$$\text{weighted SSE} = \sum\left[\frac{1}{\text{var}(k_i)}(k_i - \hat{k_i})^2\right] \tag{42}$$

where,

SSE = sum of squares due to errors

$1/var(ki)$ = i^{th} weight

k_i = growth rate

$\hat{k}_i$ = expected growth rate

Now, $var(ki)$ in Eq. (42) is substituted by $var[\ln(k)]k^2$ of Eq. (41) as follows:

$$\text{weighted SSE } = \sum\left[\frac{1}{var\left[\ln(k_i)k_i^2\right]}(k_i - \hat{k}_i)^2\right] \tag{43}$$

In natural log transformation regression model,

$$\text{weighted SSE } = \sum\left[\frac{1}{k_i^2}(k_i - \hat{k}_i)^2\right] \tag{44}$$

5.3 Empirical sigmoidal model

This model depicts bacterial growth in food having a single inflection point (Longhi et al., 2017). Here, t_{ifx} (time at inflection point) is the root of Eq. (45).

$$\frac{d^2 y(t)}{dt^2} = 0 \tag{45}$$

Now, the response at the inflection point (y_{ifx}) is attained by replacing (t_{ifx}) in the original model,

$$y(t_{ifx}) = y_{ifx} \tag{46}$$

μ_{max} is obtained by substituting (t_{ifx}) in first derivative of the sigmoid model;

$$\frac{dy(t_{ifx})}{dt} = \mu_{max} \tag{47}$$

While defining μ_{max} and λ,

$$Y_0 = \ln(N_0) \tag{48}$$

where, Y_0 is natural log of initial microbial count (Tsao, 1976; Zwietering et al., 1990).

Equation (48) is obtained (Fig. 4) and can be expressed as:

$$\mu_{max} = \frac{y_{ifx} - y_0}{t_{ifx} - \lambda} \tag{49}$$

Now, from Eq. (49), λ is isolated and we get:

$$\lambda = t_{ifx} - \frac{y_{ifx} - y_0}{\mu_{max}} \tag{50}$$

For a case study, two empirical models, that consist of Shifted Logistics Function (Eq. (51)) and Power Type Growth (Eq. (52)), were chosen.

$$y(t) = y_0 + \frac{A}{1 + \exp\left[k(t_c - t)\right]} - \frac{A}{1 + \exp(kt_c)} \tag{51}$$

$$y(t) = y_0 + \frac{At^n}{b + t^n} \tag{52}$$

Here, A, b, k, n, and t_c are empirical parameters.

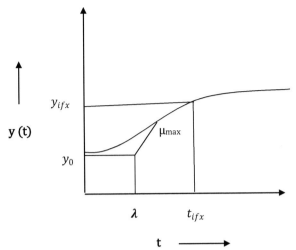

Figure 4: Characteristic sigmoidal curve of bacterial growth.

Now,

$$y(t) = y_0 + A\exp\left\{-\exp\left[\frac{\exp(1)\mu_{max}}{A}(\lambda - t) + 1\right]\right\} \tag{53}$$

$$y(t) = y_0 + \frac{A}{1 + \exp\left[\frac{4\mu_{max}}{A}(\lambda - t) + 2\right]} \tag{54}$$

Here, Eq. (53) and (54) are termed as Gompertz Modified Model and Logistics Modified Model, respectively.

Square Root Secondary Model is applied in order to describe the dependency of μ_{max} and λ with temperature. It is stated as;

$$\sqrt{P} = m(T - T_{min}) \tag{55}$$

where,

P = temperature dependent parameter (μ_{max} or $1/\lambda$)

T = temperature

m = empirical parameter

T_{min} = theatrical temperature for minimal microbial growth

5.4 Richards model

This model, which was not assessed for temperature profile earlier, is now verified and consists of a derivative of two differential equation systems (Teleken et al., 2018).

First equation that describes microbial evolution is expressed as;

$$\frac{dN(t)}{dt} = \mu(N)N(t) \tag{56}$$

where,

N(t) = microbial load

$\mu(N)$ = specific growth rate

Second equation for the dependence of the density of specific growth rate is expressed as;

$$\frac{d\mu(N)}{dN} = -\alpha N^m \tag{57}$$

where,

A = positive parameter, depends on environmental factors

$\mu(N)$ = negative function of population

Isothermal environment

The analytical solution is comprised of equation (1) and (2) at initial condition of $N(0) = N_0$ and $\mu(N_0) = \mu_0$, is given below:

$$N_t = \begin{cases} N_f \left\{ \left(\frac{N_f^{m+1} - N_0^{m+1}}{N_0^{m+1}} \right) e^{-t\left[\frac{\mu_0^{m+1} N_f^{m+1}}{N_f^{m+1} - N_0^{m+1}} \right]} + 1 \right\}^{-(1/(m+1))} & m \neq -1 \\ \ln\left(\frac{N_f}{N_0} \right) \left[1 - e^{\frac{-t\mu_0}{\ln(\frac{N_f}{N_0})}} \right] & m = -1 \\ N_0 e \end{cases} \tag{58}$$

where,

N(t) = microbial load

N_0 = function of initial population

μ_0 = initial specific growth rate

m = shape parameter

N_f = final population

To obtain Eq. (58), that parameter (α) was replaced by the function of some other parameter of the respective model. This can be observed in Eq. (59), as follows

$$\alpha = \begin{cases} \dfrac{\mu_0^{m+1} N_f^{m+1}}{N_f^{m+1} - N_0^{m+1}}, m \neq -1 \\ \dfrac{\mu_0}{\ln(N_f) - \ln(N_0)}, m = -1 \end{cases} \tag{59}$$

The parameters m such as A and μ_{max}, were obtained as follows:

$$A = \log(N_f) - \log(N_0) \tag{60}$$

$$\mu_{max} = \frac{\mu_0^{m+1} N_f^{m+2}}{N_f^{m+1} - N_0^{m+1}} \cdot (m+2)^{\frac{(-m-2)}{(m+1)}} \tag{61}$$

where,

A = net log growth

μ_{max} = maximum specific growth rate

For evaluation of non-isothermal growth;

$$\mu = \mu_0(T) - \frac{\alpha(T)}{m+1} \left(\frac{N^{m+1} - N_0^{m+1}}{N_f^{m+1}} \right) \tag{62}$$

where, $\alpha(T)$ and $\mu_0(T)$ postulate a differential equation for generating non-isothermal growth estimates.

5.5 Logistic model

This model depicts the decrease in the development of microorganisms when the population estimate is extensive and the assets are limited.

The logistic equation starts with $p(t)$ and α but also includes feedback term $\left(1 - \dfrac{P(t)}{k}\right)$ that slows down the growth of bacterial population. The feedback term is negative as it causes decrement in growth (Gibson et al., 1987; Meyer et al., 1999; Fujikawa et al., 2003).

$$\frac{dP(t)}{dt} = \alpha P(t)\left(1 - \frac{P(t)}{k}\right) \tag{63}$$

Here, $P(t)$ is the growth rate of population and α is growth rate constant.

Conditions

1. When $k \gg P(t)$ the feedback term is close to 1.
2. When $P(t) \to k$ the feedback term approaches zero.

The algebraic solution of Eq. (63) is:

$$P(t) = \frac{k}{1 + \exp(-\alpha(t - \beta))} \tag{64}$$

Equation 2 produces an S-shaped sigmoidal curve.

The three parameters (α, β and k) are needed to specify the curve.

i. α postulates the width of a curve. It is helpful to substitute α with Δt, i.e., characteristic duration which equals to $\dfrac{\ln 81}{\alpha}$. Δt is the time period required to grow trajectory from 10% to 90% of the limit k.

ii. β postulates the time at which the curve will arrive at the midpoint of the trajectory, often called t_m.

iii. k is the carrying capacity.

These three parameters (Δt, t_m and k) define the parameterization of the logistic growth model mentioned below:

$$N(t) = \frac{k}{1 + \exp\left[-\dfrac{\ln(81)}{\Delta t}(t - t_m)\right]} \tag{65}$$

Figure 5 demonstrates the fitting of the logistic curve for the growth of bacteria in a closed Petri dish consuming minerals and sugar. The available space inside the petri dish limits k. The growth rate of bacteria slows down as the available nutrition inside the dish becomes exhausted, hence, the S-shaped curve.

5.6 Von bertalanffy model

It is a standout amongst the most broadly-utilized models and it is particularly essential in the study of fisheries (James, 1991; Lester et al., 2004).

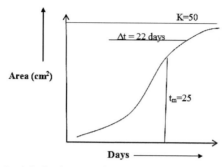

Figure 5: Logistic development of microbial colony (Meyer et al., 1999).

It is expressed as;

$$\frac{dl}{dt} = k(L_\infty - l)$$

(66)

where,

t = time

l = length

k = growth rate

L_∞ = Asymptotic length where growth is zero

Now, integrating the above equation,

$$l_t = L_\infty \left(1 - e^{-k(t-t_0)}\right)$$

(67)

Here,

L_∞, k and t_0 are parameters and,

$L_\infty = 3h_1/g$

k = ln (1 + g/3)

$$t_0 = T + \ln\left(1 - \frac{g(T - t_1)}{3}\right) / \ln\left(1 + \frac{g}{3}\right)$$

(68)

5.7 Weibull model

It is generally used for the fitting of thermal or non-thermal decay of many microbes, expressed as a probabilistic-like function, as follows (Bevilacqua et al., 2015).

$$f(t) = \frac{\beta}{\alpha}\left(\frac{t}{\alpha}\right)^{\beta-1} \exp\left(-\left(\frac{t}{\alpha}\right)^{\beta}\right)$$

(69)

Cumulative distribution of the expression is stated as

$$F(t) = \exp\left(-\left(\frac{t}{\alpha}\right)^{\beta}\right)$$

(70)

For survival kinetics,

$$\ln S(t) = -\left(\frac{t}{\alpha}\right)^{\beta}$$

(71)

where, $S = N/N_0$

β = shape of the curve ($\beta < 1$, concave curve; $\beta = 1$, straight line; $\beta > 1$, convex curve)

α = modification factor of slope

The decimal logarithmic form of this model is stated as

$$\log N = \log N_0 - \left(\frac{t}{\delta}\right)^P \tag{72}$$

where,

δ = reduction time

P = dimensionless parameter

Modified form;

$$\log \frac{N}{N_0} = 1 - \left(\frac{t}{dt}\right)^P \tag{73}$$

Next modification:

$$\log N = \log\left[(N_0 - N_{res})10^{\left(-\left(\frac{t}{dt}\right)^P\right)} + N_{res}\right] \tag{74}$$

Where,

P, N_0, δ, N_{res} are degrees of freedom.

Discussion

Each model discussed here in this chapter has parameters, and identifying the value of these parameters is an important task after making any mathematical model. Predictive food microbiology helps us build computer-based models that can be applied on a variety of food products. This will ultimately help food microbiologists explore the impact of different microorganisms within a food. These models will help us get one step closer to a long-term goal with the ability to design bacteriological safety and quality into a merchandise, instead of endeavoring to derive these characteristics afterward, utilizing final result testing.

References

Available: https://mathinsight.org/bacteria_growth_logistic_model. 2001. Safe Practices for Food Processes. A Report of the Institute of Food Technologists for the Food and Drug Administration of the U.S. Department of Health and Human Services.

Alber, S.A. and Schaffner, D.W. 1992. Evaluation of data transformations used with the square root and school field models for predicting bacterial growth rate. Applied and Environmental Microbiology 58: 3337–3342.

Baranyi, J. and Roberts, T.A. 1995. Mathematics of predictive food microbiology. International Journal of Food Microbiology 26: 199–218.

Bevilacqua, A., Speranza, B., Sinigaglia, M. and Corbo, M.R. 2015. A focus on the death kinetics in predictive microbiology: Benefits and limits of the most important models and some tools dealing with their application in foods. Foods (Basel, Switzerland) 4: 565–580.

Buchanan, R. 1992. Predictive Microbiology: Mathematical Modeling of Microbial Growth in Foods.

Fujikawa, H., Kai, A. and Morozumi, S. 2003. A new logistic model for bacterial growth. Shokuhin Eiseigaku Zasshi 44: 155–60.

Gibson, A.M., Bratchell, N. and Roberts, T.A. 1987. The effect of sodium chloride and temperature on the rate and extent of growth of *Clostridium botulinum* type A in pasteurized pork slurry. J. Appl. Bacteriol. 62: 479–90.

Hamad, S.H. 2012. Factors affecting the growth of microorganisms in food. *In:* Bhat, R., Karim Alias, A. and Paliyath, G. (eds.). Progress in Food Preservation.

James, I.R. 1991. Estimation of von bertalanffy growth curve parameters from recapture data. Biometrics 47: 1519–1530.

Kovárová-Kovar, K. and Egli, T. 1998. Growth kinetics of suspended microbial cells: From single-substrate-controlled growth to mixed-substrate kinetics. Microbiology and Molecular Biology Reviews 62: 646.

Lester, N.P., Shuter, B.J. and Abrams, P.A. 2004. Interpreting the von Bertalanffy model of somatic growth in fishes: The cost of reproduction. Proceedings. Biological Sciences 271: 1625–1631.

Longhi, D., Dalcanton, F., Aragao, G., Carciofi, B. and Laurindo, J. 2017. Microbial growth models: A general mathematical approach to obtain µ max and λ parameters from sigmoidal empirical primary models.

Macdonald, R. and Reitmeier, C. 2017. Chapter 7 - Nutrition and Food Access. *In:* Macdonald, R. and Reitmeier, C. (eds.). Understanding Food Systems. Academic Press.

Monod, J. 1949. The growth of bacterial cultures. Annual Review of Microbiology 3: 371–394.

Morris, V.J. and Groves, K. 2013. Introduction. *In:* Morris, V.J. and Groves, K. (eds.). Food Microstructures. Woodhead Publishing.

Motta, S. and Pappalardo, F. 2013. Mathematical modeling of biological systems. Briefings in Bioinformatics 14: 411–422.

Quintavalla, S. and Vicini, L. 2002. Antimicrobial food packaging in meat industry. Meat Science 62: 373–380.

Rosso, L., Lobry, J.R., Bajard, S. and Flandrois, J.P. 1995. Convenient model to describe the combined effects of temperature and pH on microbial growth. Applied and Environmental Microbiology 61: 610.

Meyer, P.S., Yung, J.W. and Ausubel, J. 1999. A Primer on Logistic Growth and Substitution: The Mathematics of the Loglet Lab Software.

Teleken, J.T., Galvão, A.C. and Robazza, W.D.S. 2018. Use of modified Richards model to predict isothermal and non-isothermal microbial growth. Brazilian Journal of Microbiology 49: 614–620.

Tsao, G.T. 1976. Principles of microbe and cell cultivation, S. John Pirt, Halsted Press, Division of John Wiley and Sons, New York, 274 pages. AIChE Journal 22: 621–621.

Zwietering, M.H., Jongenburger, I., Rombouts, F.M. and van 't Riet, K. 1990. Modeling of the bacterial growth curve. Applied and Environmental Microbiology 56: 1875–1881.

CHAPTER 16

Computational Fluid Dynamics (CFD) Simulations in Food Processing

Abhishek Dutta,[1] *Ferruh Erdoğdu*[2]* and *Fabrizio Sarghini*[3]

1. Introduction

Mathematical modelling approaches are important tools for optimizing and designing food processing operations. The first computational mathematical modelling study in food process engineering was introduced by Texieira et al. (1969), where a finite difference numerical methodology was used to optimize the retention of nutrients in conduction-heated foods. Mathematical modelling in food process engineering relies primarily on physically fundamental mechanisms governing a process which can be helpful in providing a basic definition of the process (Singh and Vijayan, 1998). It might be categorized into two general groups: Experiment-based empirical modelling and physics-based modelling. The first group relies greatly on the availability of the experimental data, and the resulting empirical equation is mostly available under the given range of the experimental conditions. Such an approach is often based on data fitting equations rather than real modelling. The use of the available instruments and lab conditions also affect the generality of these types of models. On the other hand, physics-based models rely on solving the required (partial) differential equations describing the process with valid initial and boundary conditions assumptions. Hence, once the developed mathematical models are (experimentally) validated, their uses for generality like to virtualize the process or for further design and optimization purposes are more adequate. The solution of the partial differential equation set, depending upon system complexity and initial and boundary conditions, might be carried out in three general ways:

- The first approach, if simplifications can be made, is to use the exact solutions of the differential equation, limited to simplified models on simple or regular geometries.
- The second approach is based on using numerical methods, like finite difference, finite element or finite volume schemes, in order to solve simplified models based on partial differential equations governing the process with the required initial and boundary conditions. This is typically referred to as the computational fluid dynamics (CFD) approach.

---

[1] KU Leuven, Campus Groep T Leuven, Faculteit Industriële Ingenieurswetenschappen, Andreas Vesaliusstraat 13, B-3000 Leuven, Belgium.

[2] Ankara University, Department of Food Engineering, Golbasi-Ankara, Turkey.

[3] University of Naples Federico II, Department of Agricultural Sciences, Naples, Italy.

* Corresponding author: ferruherdogdu@ankara.edu.tr; ferruherdogdu@yahoo.com

Most CFD software are based on finite element method or finite volume method (Wang and Sun, 2003), in which finite volume method is particularly preferred since it incorporates the numerical flexibility of finite element method and the computational speed of finite difference method (Nicolai et al., 2001). As a matter of fact, all the three methods, namely, finite difference, finite element and finite volume, can be considered to be derived from a weighted residual approach, where the weight function is either a Dirac function, a function whose value is 1 on each single control volume and 0 elsewhere, or a trial function used to discretize the solution (Finlayson, 1972).

Consistency (difference approximation converges to the differential equation as discretization step sizes reduces), stability (solution to the difference equation does not increase with time at a rate faster than the solution to the differential equation) and convergence (solution to the difference equation approach those of the differential equation under refined grid sizes) are significant issues in any numerical approximation where the solution can be unstable if round-off (consequence of using finite precision floating point numbers), iterative convergence (difference between the exact and iterative solutions of discrete equation) or truncation (difference between differential equation and the corresponding difference equation) errors grow unboundedly in the following iteration steps (Clausing, 1969). Conversely, the solution is stable if these errors damp out. For a consistent solution, truncation errors approach zero as spatial and time increments approach zero. Consistency and stability are necessary conditions for convergence, and a numerical method is convergent if the solution approaches the exact solution as both time and space increments are reduced (Palazoglu and Erdoğdu, 2008). While the finite element and finite volume methods superseded the finite difference method with their flexibility to apply over irregular domains or with moving boundaries, the first two cases have gained a significance in the last decade based on the availability of the increased computer power and the relatively easy-to-use software packages available. The finite volume approach expresses the transport equations in conservation form by converting the volume integrals into surface integrals using Gauss's divergence theorem (Versteeg and Malalasekera, 2007; Norton et al., 2013). In addition, finite volume approach leads to the algebraic equations to promote the solver robustness. The combination of all these complex mathematical backgrounds with a rather easy-to-learn interface and computational geometry-mesh generation abilities is the general approach of the present day CFD software. This increases the number of users and the number of studies applying CFD for various processes in the food engineering domain. However, these easy-to-use software packages, with their user-friendly interface, also led to an increase in the number of unconscious end users, adopting such techniques without knowing the fundamentals of the modelling approach. This created a problem of black-box usage, resulting sometimes in colourful pictures quite far from the reality of the given problem.

CFD is, in fact, a part of fluid dynamics which uses numerical approximations to solve and analyze problems involving fluids. Since the early 1950s, implementation of CFD using a certain computational power has continued to evolve, especially with the advancements in the computer world (Norton and Sun, 2007). A rather broad use of various software now includes all transport-based problems in the CFD category, from the complexity of the fluid-structure interactions in a multi-physics domain to the rather simple conduction problems in regular geometry solids. It must be noted that the fundamental basis for a CFD problem is based on numerically solving the Navier-Stokes equations with appropriate closure conditions. Navier-Stokes equations are developed from conservation principles of mass, momentum and energy for fluid flow (compressible and incompressible, inviscid and creeping, laminar and turbulent and Newtonian and non-Newtonian flows) (Verboven et al., 2004). Coupling of fluid flow with energy equation and internal heat generation through, e.g., dielectric heating, is nowadays a possible capability of the improved CFD approaches (Erdoğdu, 2009).

As a matter of fact, CFD has become a useful tool for analyzing single and multiphase flows in various industrial applications. This is because the rapidly decreasing costs of computational hardware accompanied by faster computing times have made it possible to investigate a thorough numerical study of fluid flows at a much lower cost than before (Dutta et al., 2010). In addition to analyzing the fluid behavior as mentioned, CFD is also used in complex heat transfer problems combined with chemical reactions as well. In food process engineering, a rather broad set of CFD applications have been demonstrated, especially in the last two decades. Among these studies, some are focused on canning and aseptic processing for conventional and innovated thermal processing for possible process improvement; others on improving the performance of refrigerators, refrigerated cabinets and cold stores to determine the air movement and temperature distribution; others on continuous drying and baking processes; others on heat exchangers for shape optimization and modelling of hydrodynamics in order to improve the thermal and fouling analysis; others on impingement studies for heating, cooling and freezing of food products; on extrusion processes for design, scale-up and optimization; on prediction of hygiene in food processing systems; on modelling airflow in vented packages of horticultural commodities for process and package design and on process innovations using microwave and radio frequency. An extensive summary of these studies was summarized in the literature already published.

In this chapter, flow-related problems in general and in food processing will be specifically focused on. As the name suggests, in this method computer-based simulations are performed in order to obtain a solution to the given problem. CFD finds its applications in many industries. Some examples of application areas are the aerospace and automotive industry (analyzing aerodynamic lift and drag), chemical processing industry, biomedical industry (simulating blood flow through arteries and veins), oil & gas industry (simulating oil and gas flow through the pipelines), power generation industry and food processing. CFD analysis carried out in industries mainly finds its applications in redesigning the product for manufacturing so that it has a better performance, developing new product and conceptually analyzing new designs before manufacturing. In other words, the CFD approach is used for virtualization of the process or a given problem.

Scott and Richardson (1997) were some of the first to review the implementation of CFD in the food industry. While there was a certain increase in the CFD use in the food processing area, more comprehensive reviews were also published (Xia and Sun, 2002; Norton and Sun, 2006; Sun, 2007; Norton et al., 2013; Erdoğdu et al., 2017). These studies mostly focused on the background of CFD and various case studies from the literature. In recent years CFD has started gaining popularity against the conventional branch of fluid dynamics namely experimental fluid dynamics to analyze fluid flow. This is because it has certain advantages compared to experimental techniques. Some of these advantages were reported in earlier work by Anderson et al. (1984) as follows:

- It helps in predicting the performance of a system before manufacturing the system, therefore, an efficient system can be manufactured at first hand with the help of CFD analysis.
- It provides detailed and exact information about fluid flow in the system.
- It saves cost and time.
- It is more reliable with increased improvements in numerical schemes and methods employed in CFD.

Besides the advantages involved in CFD, there are also limitations in using these techniques. The use of a numerical method to solve a problem with the CFD approach always introduces some kind of a numerical limitations, like discretization, round-off or interpolation errors associated with the methods themselves, however, constant efforts are being taken to reduce these kinds of errors. Also, considerable computational resources (cost & time) are required to run the large-scale models (Erfoctac report, 2000; Chung, 2002; Ko et al., 2010). Even though the given citations are not very recent, the requirement of the computational resources for larger scale simulations is always a must (Sarghini et al., 2017).

Both computational (CFD) and experimental methods have their own advantages and disadvantages, but generally speaking, they integrate each other. Thus, it should always be noted that CFD cannot and should not replace the real experiments. Both experimental and numerical (through CFD) approaches should complement each other in order to obtain an exact and economic prediction of flow processes, which is expected in planned facilities or to optimize the existing ones. The use of CFD models, validated with experimental studies, is ubiquitously the best approach for further design and optimization purposes. Until recently, certain assumptions and approximations in the computational geometry and boundary conditions of the process were preferred to reduce the computational time, but the recent trend favours defining the whole process by minimizing the simplifications on the cost of computational time and cost. The introduction of powerful computers and work stations are helpful in this regard, but still a certain amount of computational time and cost are still required for complicated problems.

In some cases, numerical simulations complement the real experiments models (Brannock et al., 2010; Cordio et al., 2015). Therefore, the objective of this chapter is to introduce a fundamental background on the CFD approach used in food processing for various processes involved in general CFD analysis and equations without restricting the focus on any specific problem.

2. CFD Analysis

A rather general procedure in a CFD analysis starts with the formulation of the problem from the physics-based analysis. Then, the computation domain modelling with the grid (mesh) generation follows the general formulation. Considering that improper mesh structures might lead to possible non-physical solutions, for example, not correctly resolving the local gradients, mesh generation is a significant part of CFD analysis. After specifying the thermal and physical properties with initial and boundary conditions, numerical solution approaches are to be decided, and simulations are performed in order to analyze the results graphically and numerically. Validation of a simulation study (with the presence of experimental data, if possible) and analysis of the sensitivity of the solution are additionally required steps to conclude the study.

2.1 Steps involved in a CFD analysis

2.1.1 Problem formulation

Once the objectives of the problem are clearly established, the mathematical model needs to be developed in order to define the relevant flow behavior. This involves analyzing the type of fluid (Newtonian/non-Newtonian) based on rheological characterization of the fluid and type of flow under consideration. For Newtonian fluids, the governing equations can be developed with the knowledge of fluid viscosity. For non-Newtonian fluids and other rheological complex fluids, the governing fluid dynamic equations can be developed only when the fluid rheology is known. The governing equations for these types of fluids are not discussed in this chapter. When the viscosity of fluid (as a function of temperature and/or shear rate, e.g., Newtonian or non-Newtonian fluid) is known, then it is required to decide upon the dynamic behavior of the flow. Some commonly observed types of flow are turbulent or laminar flow, unsteady or steady flow, incompressible or compressible flow, multiphase or single flow and non-isothermal or isothermal flow. To formulate a problem, the following needs to be addressed:

- objective of the problem
- geometry to be analyzed
- operating conditions involved in the problem
- deciding on the flow domain (whether there is a flow involved) and flow nature (laminar or turbulent)
- a temporary model suitable for the problem (steady or unsteady)

For the food processing point of view, the incompressible flow approach is often the general assumption due to the fact that mostly liquids are involved (Nicolai et al., 2001); specific problems, i.e., two-phase flows, like the liquid-gas interaction or flow with particles, can induce additional complexities with the requirement of high computational cost. Besides, the presence of turbulence (fluctuations in eddies inside the flow) can enhance the heat transfer and cause additional pressure drops in internal configuration problems. For the solution to a flow problem involving turbulence, a mathematical background to calculate the turbulence quantities is required. General turbulence model approaches, like $\kappa - \varepsilon$, $\kappa - \omega$ (where κ is the turbulence kinetic energy, ε is the turbulence energy dissipation rate, and ω is the turbulence frequency), and a more sophisticated approach, like Reynolds stress models, can be implemented. Olsson et al. (2004) gave a detailed comparison of the turbulent models using experimental data available in the literature.

2.1.2 Preparation of the geometry model and domain

Once the problem in formulated, the geometry of the system should be created. For a complex case, the use of a CAD software package might be preferred. The modelling should then be carried out in such a way that the structure and topology of grid generation should be considered.

2.1.3 Grid generation – numerical mesh

In most of the CFD methods (apart from the meshless methods, described in the following sections), the geometry is to be discretized into a grid forming the computational geometry. Grid generation involves first defining the structure and topology and then generating the grid on that topology. The ability of recent CFD software to create the tetrahedral, hexahedral hybrid and polyhedral meshes allows the numerical mesh to conform to any arbitrary geometry, enhancing the ability of the CFD approach in industrial scale use (Norton et al., 2013). The generated grid should demonstrate a certain minimal grid quality requirement, as defined by orthogonality measures at the boundaries, grid skewness, relative grid spacing, etc.

Structured, unstructured and block-structured grids are the three types of grids. With the simplest structured grid, all nodes have same number of elements around it. This is mostly used in simple geometry domains. If there is a complex domain, then an unstructured grid is generally preferred. Block structure grids are the combination of both structured and unstructured grids. In these methods, the whole domain is split into blocks, and then different grids are used in these blocks (Zuo, 2005; Briggs, 1987). It has to be noted that there are recent advancements in CFD where gridless methods have started gaining momentum to be used in the analysis of various cases. Meshless particle methods as a numerical technique for multiscale analysis are explained by Ho et al. (2013) in the recent food engineering modelling literature, while a comparison between the mesh-free and grid-based approaches were presented by Mudiyanselage et al. (2017) for numerical modelling trends in plant food tissues and morphological changes during drying.

2.1.4 Initial and boundary conditions formulation

Within CFD based simulations, knowledge of initial and boundary conditions are required in order to begin the computations, and the method, depending whether the solution is stationary, applies an iterative approach in order to obtain the final converged solution or to advance to the next iteration step. Finally, an input file is also required, as explained below in detail. On a boundary, for a flow-related problem, a very typical boundary condition is the use of no-slip condition over solid surfaces (full shear stress) while free slip (zero shear stress), a specified wall shear stress or wall velocity might be applied depending upon the physical background of the process (Verboven et al., 2004).

2.1.5 Establishing the simulation strategy

In order to perform the CFD simulation and get the correct result, a proper strategy is required. The simulation strategy involves determining the choice of numerical algorithm, choice of turbulence model (if any) and use of space-marching or time-marching algorithms.

2.1.6 Establishing input parameters and simulation files

All CFD solutions, in general, require an input file containing values for input parameters and the adopted simulation strategy. Also, a grid file should be generated for the containing mesh information, as well as the corresponding initial and boundary conditions.

2.1.7 Performing the simulation

Once the files required for starting the simulation are ready, the simulation can be started. The first thing when starting might be to follow whether a quick convergence on the variables is obtained. The simulation can take place either in serial or parallel computational mode. Usually, serial processing is carried out for simulation in a simple domain where a single processor is used to perform the simulation. In problems involving complex domain, parallel processing is done; two or more processors are used to carry out the simulation process in order to decrease the computational time to complete the simulation.

2.1.8 Monitoring the simulation

The simulation, once started, has to be monitored for completion, and the simulation is considered complete when a converged solution is obtained; this is referred to as "iterative convergence". The general criteria used to ascertain iterative convergence includes monitoring residuals (reduction of errors produced over each iteration) to reach a certain magnitude or monitoring the flow variables values which have to reach a steady (space-marching simulation) or pseudo-steady (time-marching) value, with respect to iteration.

2.1.9 Post-processing the simulation

After the simulation is converged, the simulation is stopped. Then, post-processing is done, wherein desired flow properties are obtained from the computed flow field. There are several ways in which the simulation results can be interpreted, as mentioned below:

a. Calculation of derived variables (vorticity, shear stress)
b. Calculation of integral variables (forces, lift/drag coefficients)
c. Calculation of turbulent quantities (Reynolds stress, energy spectra)
d. MPI (Message-Passing Interface) functions required to combine data from different blocks if the computations are run in parallel
e. Visualization

- XY plots (time/iterative history y of residuals and forces)
- contour plots
- velocity vectors
- iso-surface plots
- streamlines, pathlines, streaklines
- animations

f. Other techniques, like Fast Fourier Transform (FFT) and phase averaging, can also be used, depending on the requirement.

2.1.10 Validation of the simulation

The available results are then compared with the already reliable and analytical, computational or experimental results in order to validate the computed results and reliability of the CFD code (Roache, 1998; Lewis, 1990). Datta (2002) suggests to check the convergence with a basic knowledge of the process physics, compare the improvement in results with a simpler problem or use experimental data. In any case, experimental comparison would be the best-case scenario.

For flow related problems, sophisticated technologies are often required for model validation purposes. Particle image displacement velocimetry (PIV) includes a class of flow measuring techniques (Buchhave, 1992), and it is characterized by following the displacement of small seeded particles embedded in a region of a fluid. Magnetic resonance imaging (MRI) is another possibility for providing the direct measurement of flow field maps (Hills, 1995). Another approach to determine the flow field maps is the use of positron emission particle tracking technique. Cox et al. (2003) and Bakalis et al. (2003) used this approach to evaluate flow velocity trajectory of model fluids in rotating cans.

2.1.11 Sensitivity analysis

To better understand the accuracy of the result and performance of computation, sensitivity studies are carried out. The entire simulation process is repeated in order to check for sensitivity of the results by changing:

- Grid topology and density
- Numerical Algorithm
- Turbulence model
- Boundary condition
- Dimensionality (changing H/D or L/W ratios)

These sensitivity studies are important steps for verifying and validating the already built CFD model.

2.1.12 Documenting the results

It is equally important to document the findings from the analysis as they will be helpful in the future. This step includes describing each of the above-mentioned steps in the process pertaining to the specific fluid flow problem. This document will be very useful when a similar problem is encountered in the future as this document will act as a reference and reduce the modelling time considerably.

2.2 Governing equations

The Navier-Stokes equation governs dynamics of the fluids and consists of conservation of mass, momentum and energy. Nicolai et al. (2001) introduced a concise yet comprehensive summary on Navier-Stokes equations:

$$\frac{\partial \rho}{\partial t} + \frac{\partial \left(\rho u_j \right)}{\partial x_j} = 0 \tag{1}$$

$$\frac{\partial \left(\rho u_i \right)}{\partial t} + \frac{\partial \left(\rho u_i u_j \right)}{\partial x_j} = \frac{\partial}{\partial x_j} \mu \left(\frac{\partial u_j}{\partial x_i} + \frac{\partial u_j}{\partial x_j} \right) - \frac{\partial}{\partial x_i} \left(P + \frac{2}{3} \mu \frac{\partial u_j}{\partial x_j} \right) + f_i \tag{2}$$

$$\frac{\partial \left(\rho H \right)}{\partial t} + \frac{\partial \left(\rho u_j H \right)}{\partial x_j} = \frac{\partial}{\partial x_j} \left(k \frac{\partial T}{\partial x_j} \right) + \frac{\partial P}{\partial t} + Q \tag{3}$$

where u_i (i=1,2,3) is the velocity vector (m/s) in Cartesian component, T and H are the temperature (°C or K) and enthalpy (J/kg), respectively, ρ, k and μ are the density (kg/m³), thermal conductivity (W/m-K) and dynamic viscosity (Pa-s), respectively. f_i is the external body forces (N/m³) and Q is the heat source (W/m³).

Continuity equation (Eq. (1)) states that the mass flow entering into a fluid element must balance with the mass flow exiting the fluid element. The momentum equation (Eq. (2)) is based on Newton's second law of motion. It states that the summation of the forces on a fluid element is equal to the change in momentum. The energy equation (Eq. (3)), on the other hand, is based on the first law of thermodynamics. The energy equation is written as a total enthalpy equation. The viscous dissipation term in the energy equation is mostly not included since it becomes significant for high-speed flows. Furthermore, the time derivative in pressure denotes the rate of change of specific enthalpy.

In CFD approach, to solve the given sets of equations, it is important that the above equations are transferred into discretized form. Some of the discretization methods are:

- Finite element   } More commonly used methods in CFD
- Finite volume
- Finite difference
- Spectral element
- Boundary element methods
- High-resolution Discretization Scheme

For this concept, Datta and Teixeira (1988) were the first to demonstrate the solution of Navier-Stokes equations for natural convection heating in a cylindrical can. Out of all the common methods used in CFD, Finite Volume approach is widely used in most commercial software, like Ansys CFX, Ansys Fluent, CD-Adapco, Star-CCM+ and open source ones like OpenFOAM. This is due to the fact that the finite element method has certain advantages compared to other commonly-used methods:

- It uses integral formulation of conservation laws, which is the native form of conservation laws, thus, no assumptions are required regarding smoothness of solution and it can also handle discontinuities in solution.
- It is based on cell-averaged values, thus, when conservation is satisfied in every control volume, it will be automatically satisfied in the whole domain, unlike other methods, where we have to manually check for conservation of mass, momentum and energy.
- It can be applied on unstructured grids, thus, there would be no requirement for a grid to be structured, unlike finite difference method.

Since the finite volume method is widely used and has many advantages, it is discussed in detail in this chapter.

2.3 Finite volume method

The popularity of Finite Volume Method (FVM) approach in CFD is due to its high flexibility and its offering of a discretization method. Even though it was preceded by finite difference method (FDM) and finite element method (FEM), FVM is assumed to have a particularly prominent role for fluid flow and related transport phenomena problems. In FVM, the discretization is carried out directly within the physical space with no need for any transformation between physical and computational coordinate system. This definitely expands the FVM's applicability to encompass a wide range of applications while still retaining the mathematical formulation simplicity. Another important aspect is that it numerically mirrors the physics and conservation principles of the models, such as the integral property of the governing equations and the characteristics of the terms it discretizes.

In an FVM study, governing equations are integrated over finite volumes where the domain has been subdivided, and the Gauss theorem is used to transform volume integrals of convection and diffusion terms into surface integrals. Following this, surface and volume integrals are transformed into discrete ones and integrated numerically through the use of integration points. To clarify this approach, the following example demonstrated an FVM application for a two-dimensional transport problem. The domain can be seen in Fig. 1.

The conservation equation for a general scalar variable ϕ is:

$$\underbrace{\frac{\partial(\rho\Phi)}{\partial t}}_{\text{transient term}} + \underbrace{\nabla.(\rho u \Phi)}_{\text{convective term}} + \underbrace{\nabla.(\Gamma_\Phi \nabla \Phi)}_{\text{diffusion term}} + \underbrace{Q_\Phi}_{\text{source term}} \tag{4}$$

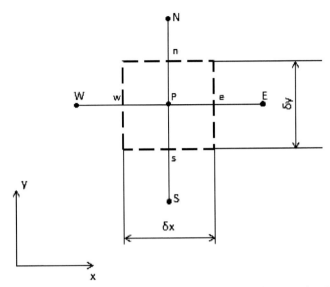

Figure 1: Control volume in 2D computational domain (adapted from Patankar, 1980).

By integrating the above equation over the element P, shown in Fig. 1, the Eq. (4) is transformed into:

$$\int_0^{V_p} \frac{\partial(\rho\Phi)}{\partial t}dV + \int_0^{V_p} \nabla.(\rho u\Phi)dV = \int_0^{V_p} \nabla.(\Gamma_\Phi\nabla\Phi)dV + \int_0^{V_p} Q_\Phi dV \tag{5}$$

Replacing the convection and diffusion terms volume integrals by surface integrals through using the divergence theorem to ensure conservation of scalar variable in the control volume, the above equation becomes:

$$\int_0^{V_p} \frac{\partial(\rho\Phi)}{\partial t}dV + \int_0^{S_p} (\rho u\Phi).dS = \int_0^{S_p} (\Gamma_\Phi\nabla\Phi).dS + \int_0^{V_p} Q_\Phi dV \tag{6}$$

where bold letters indicate vectors, $(\cdot)$ is the dot product operator, Q_Φ represents the source term, **S** the surface vector, **u** the velocity vector, ϕ the conserved quantity, and $\oint_0^{s_p}$ the surface integral over the control volume v_p. To approximate the volume integral in the above equation, one can multiply the volume and the value at the center of the control volume. In order to better understand, one can take an example of discretization of a 2D computational domain, as shown in Fig. 1.

To approximate mass and momentum of control volume P, one can have

$$m = \int_0^V \rho dV \approx \rho_p V, mu = \int_0^{V_i} \rho_i u_i dV \approx \rho_p u_p V \tag{7}$$

and in order to approximate the surface integral, the surface integral is replaced over control volume P by a summation of the flux terms over the faces of control volume P, so the surface integrals of the total fluxes (sum of convection and diffusion fluxes) become

$$\int_0^{S_i} PdS \approx \sum_k P_K S_K \quad k = n, s, e, w \tag{8}$$

where P, N, S, E and W are the nodes located at the center of control volumes, which stores the value of flow variables, while n, s, e and w are the cell faces. The aim of approximation is to transform the above equation into an algebraic equation by expressing surface and volume fluxes in terms of variables at the neighboring cell centers. This transformation has important consequences on FVM properties: One of the most important is rendering the method conservative. In FVM approach, the value of the flow variables is stored at the center of the control volume or in the vertex of the control volume (which is beyond the scope of this text), so a special interpolation scheme is required to get the value of P_k, which are located at the surface of the control volume. There are three widely-used interpolation schemes:

- Central interpolation, where the central differencing scheme is to use consistent expressions to evaluate convective and diffusive fluxes at the control volume faces. One of the major inadequacies of the central interpolation scheme is its inability to identify the flow direction:

$$\Phi_e = \Phi_E \lambda_e + \Phi_p \left(1 - \lambda_e\right) \tag{9a}$$

$$\lambda_e = \frac{x_e - x_P}{x_E - x_P} \tag{9b}$$

- Upwind interpolation: The upwind interpolation or 'donor cell' interpolation scheme considers the flow direction in order to determine the value at a cell face. The convection value of ϕ at a cell face must then be equal to the value at the upstream node. The upwind scheme produces a much more realistic solution with an issue that would not be very close to the exact solution near boundary.

$$\Phi_e = \begin{cases} \Phi_P \text{ if } \left(\vec{\Phi}.\vec{n}\right)e > 0 \\ \Phi_E \text{ if } \left(\vec{\Phi}.\vec{n}\right)e < 0 \end{cases} \tag{10}$$

- Hybrid interpolation: The hybrid difference scheme by Spalding (1970) is a combination of central and upwind difference schemes. It makes use of the central difference scheme as a second order accurate, for small Peclet numbers (|Pe| < 2). For large Peclet numbers (|Pe| > 2), it uses the Upwind difference scheme as a first order accurate, while still considering the fluid convection.

Peclet number (Pe) is a dimensionless parameter demonstrating the measure of the relative strengths of convection and diffusion:

$$Pe = \frac{F}{D} = \frac{\rho u S}{\Gamma S} = \frac{\rho u}{\frac{\Gamma}{\delta x}} \tag{11}$$

where F and D represent convection mass flux and diffusion conductance at cell faces, respectively.

$$\Phi_e = \begin{cases} \left[\left(1 + \frac{2}{Pe_e}\right)\frac{\Phi_E}{2} + \left(1 - \frac{2}{Pe_e}\right)\frac{\Phi_P}{2}\right] \text{ for } -2 < Pe_e < 2 \\ \Phi_E \text{ for } Pe_e \geq 2 \\ \Phi_P \text{ for } Pe_e \leq -2 \end{cases} \tag{12}$$

It uses the favorable properties of the central and upwind difference schemes where it switches to the upwind difference scheme when the central difference scheme produces inaccurate results (for high Peclet numbers). It produces physically realistic solutions and has been proved to be helpful in predicting the practical flows.

3. Case Studies

3.1 Air-impingement in food processing

Boundary condition of heat transfer, applied for computational approaches in conventional thermal processing, is mostly the convective condition. Heat transfer coefficient of the medium is a significant physical property that should be increased in the process to obtain a decrease in process time. Even though the effects are limited for high heat transfer coefficient cases, like canning, a certain significant effect is obtained with lower heat transfer rate processes. In particular, for processes using air as a heat transfer medium, rather low heat transfer coefficients are experienced, and higher values might be observed by only giving a movement to the fluid medium at high velocities. Air impingement systems are typical examples for this case. Impingement is carried out through directing a jet or jets of fluid to result in a certain change. This process cases a higher value of heat transfer coefficient. This value might even approach values similar to a frying process (Ovadia and Walker, 1998). Air impingement has been applied in cooling, cooking and baking-related research and has industrial applications (Li and Walker, 1996; Wahlby et al., 2000). Considering the lower heat transfer coefficient especially obtained at lower temperatures due to the applied temperature and viscosity, air impingement has a certain potential in cooling and thawing processes. For design purposes, the nozzle types and arrangements and use of multiple impinging jets are major components. Significance of nozzle arrangements and multiple jet impingement have been demonstrated in a recent work (Wen et al., 2018).

Due to the evolving flow fields during impingement processes, it is likely that the heat transfer coefficient would be spatially variable and depend on air velocity, air temperature, air humidity and the radius and number of impingement nozzles. Therefore, in order to determine the temperature change of a product undergoing an impingement process, the flow field evolution during the process must be evaluated first. Therefore, in this case study, an air impingement system used for cooling purposes was

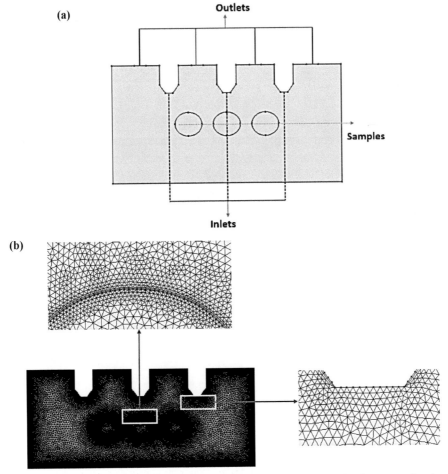

Figure 2: (a) Schematic of a 2-dimensional air-impingement system; and (b) computational geometry showing the mesh structure.

demonstrated. As shown in Fig. 2, the system had 3-impingement nozzles and 3-products for cooling. For simplification purposes, a 2-dimensional system with laminar flow conditions due to the lower velocities applied at the jet inlets (1 m/s) was considered. The following summarizes the boundary conditions applied for the problem, in addition to the governing equations of (1) to (3).

Boundary conditions:

Inlet (constant air velocity):

$$u = -U_0 = 1\,m/s \tag{13}$$

Outlet:

$$p_0 = 0\,Pa \tag{14}$$

At walls and solid surfaces (no-slip condition):

$$u = 0 \tag{15}$$

For the solution of the time-dependent energy equation, the initial condition was assumed to be uniform constant temperature for the samples and the inlet air; 37 and 4°C, respectively, to demonstrate a cooling process. Outflow condition was defined at the outlets of the system. The flow was laminar and incompressible flow, and the gravitational effects were neglected for this particular example.

The computational geometry of the 2-dimensional system was shown in Fig. 2. A finer mesh structure was used along the walls, sample surfaces and air inlets of the system for a faster convergence and to predict the velocity fields precisely. Especially over the sample surface, a mesh boundary layer was

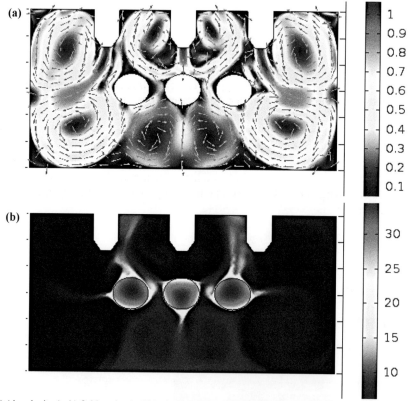

Figure 3: (a) Air velocity (m/s) field evolved within the system (arrows are for the x- and y-components of the velocity); and (b) temperature distribution within the impingement cavity after 1200s.

Color version at the end of the book

applied with a rather fine mesh structure in order to obtain a solution along these boundaries. Figure 2 demonstrates this finer mesh structure. Figure 3(a) shows the air velocity field evolved within the system, while the temperature distribution is shown in Fig. 3(b). As demonstrated in the flow field evolution, based on the given boundary conditions, e.g., presence and location of the outlets, impingement effect on the heat transfer and temperature change of the samples might be observed in Fig. 3(b). This case study illustrates the set-up of a fluid flow problem coupled with heat transfer. Due to the effect of air velocity and spatial variation of heat transfer coefficient, the temperature decrease over the upper surface was more rapid in comparison to the lower surfaces. Based on this temperature variation, it might be concluded that a process modelling using a CFD approach is required for further process design and optimization in order to obtain a uniform and rapid cooling process.

3.2 Screw-drive systems

Screw-drive systems are preferred in thermal food processing where the liquid phases are moved with the rotational effects of a screw. There is a significant liquid and gas interphase interaction during this movement, and the velocity changes inside the liquid, with pressure variations affecting the processing parameters. When these systems are used for pasteurization of particulates carried with the liquid (e.g., water), the residence time distribution becomes a certain issue, and this is highly affected by the design conditions of the screw system. Figure 4 demonstrates the computational geometry of such a system, where a single blade with five pitches is located in a cylindrical casing filled halfway with water. With the externally applied inertial force to the screw system, it creates a mixing effect that leads to a raise in heat transfer. The increased effect of mixing also increases the developed turbulence with the rotation of the screw. All these effects might produce a possible negative pressure inside the system between the pitches, and this might lead to the formation of reverse currents. The particulate product caught up in the reverse currents might spend more time between the pitches with longer residence times and over-processing effect. The geometry of the screw drive systems is governed by external (outer radius and length of the system) and internal (inner radius of the screw, number of pitches and blades, distance between the pitches, etc.) parameters (Rorres, 2000). The internal parameters might be used to optimize the process performance, as suggested by Sarghini et al. (2017). Due to the main rotational effect applied to the screw in order to create the mixing within the system, a case study was simulated to determine the momentum transfer.

For this particular case, the volume of fluid (VOF) approach (Rider et al., 1998) was used, the theory behind this approach was that the fluids (water as liquid and air as gas) inside the system would not be interpenetrating. Since VOF is an advection scheme that allows one to track the geometry and position of the interface, the Navier–Stokes equations describing the motion of the flow were solved separately using the CFD solver of Ansys Fluent v17.2 (Ansys Inc., Canonsburg, PA). The details of the numerical model can be found in Sarghini et al. (2017). The following summarize the physical properties of the computational geometry:

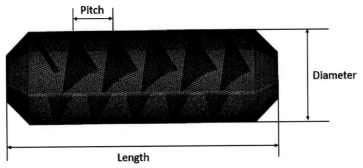

Figure 4: Computational geometry of the screw system with a single blade with five pitches located in a cylindrical casing filled with water in half (Length: 1.8 m – Diameter: 0.6 m – Adapted from Sarghini et al., 2017).

- The length of the rotating screw system (Fig. 4) was 1.8 m with a cylinder casting diameter of 0.6 m.
- Rotation rates were 10 to 40 rpm with an inflow water mass flow rate of 0.17 kg/s at T = 27°C through a circular inlet, and a rectangular to allow the possible outflow.
- No-slip condition was applied over all solid boundaries.
- Initial temperature of the system was 90°C.

Figure 5 demonstrates variation of the liquid and gas interphase at various rotation rates (5 and 40 rpm), while an intense mixing with rotation rates was induced. For the rotational effects, a force analysis is also possible for flow-induced problems. For this purpose, changes in Froude (Fr) and Taylor (Ta) numbers might be used:

$$\text{Fr} = \frac{f_{rf}}{f_{gf}} = \frac{\omega^2 R}{g} \tag{16}$$

$$\text{Ta} = \frac{f_{rotational-inertial}}{f_{viscous}} = \frac{\omega^2 R^4}{v^2} \tag{17}$$

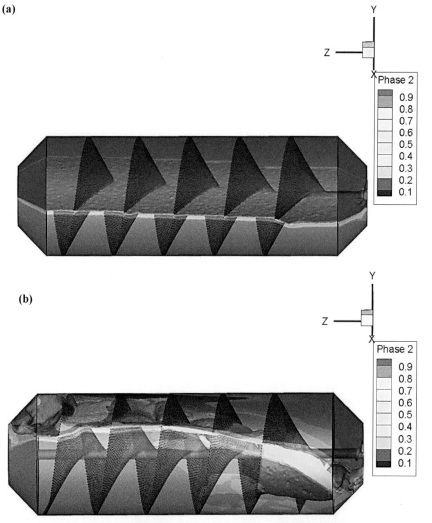

Figure 5: Variation of the liquid–gas interphase at various rotation rates at 4s of the rotational heat transfer process in the screw system (a) 10 rpm; (b) 40 rpm (adapted from Sarghini et al., 2017).

where f_{rf} is the rotational, f_{gf} is the gravitational, $f_{rotational-inertial}$ is the rotational and inertial and $f_{viscous}$ is the viscous forces, $\left(\omega = \dfrac{2\pi f}{60}\right)$ with f as the rotation rate (rpm), and R as the radius of the blades. Based on the resulting dimensionless values, Fr and Ta numbers are defined to be the ratio of rotational and inertial forces to gravitational and viscous forces, respectively. For the given screw rotation problem, Fr number increased 16 times (from 0.032 to 0.519) from 10 to 40 rpm while the Ta number changed from 1.08E11 to 1.73E12 (16 times higher based on the whole fluid volume within the system) at 10 and 40 rpm rotation rates within the initial phase of the process. These increases in the Fr and Ta numbers state that the rotational and rotational inertial forces gain effect over the gravity and viscous forces. This increases the mixing inside the system, leading to rather effective heat transfer which can be easily observed in Fig. 5(b).

4. Future Outlook

The simulation of fluid flow-related problems using CFD has progressed rapidly in the last few decades. Advances in computational technology, a sustained effort by CFD providers to implement comprehensive physical models and advances in numerical methods have been combined to make it possible for engineering and R&D groups to use CFD routinely in many industries (Rubbert, 1994). Advances in simulation capabilities enable reductions in requirement of experiments. In addition to reducing experimental load, physics-based simulation technologies, such as CFD, have offered the added potential of delivering superior understanding and insight into the critical physical phenomena. This has apparently opened new frontiers in design and performance.

Mainstream CFD really has not changed much since it was started with unstructured second order finite volume codes. There has been development of better turbulence models for RANS. These RANS methods have become the high-fidelity method of choice and advances. This was mostly due to using larger meshes, more complex geometries and more numerous runs afforded by decreasing hardware cost. There are also developments of some new resolving techniques (LES, DNS) to better understand turbulent flows for critical design applications, like the development of an aircraft. There has also been progress in developing higher order discretization techniques. However, the workhorse is still either structured finite differences or unstructured finite volume. At present, the CFD development community finds itself poorly positioned to capitalize on the rapidly changing High Performance Computing (HPC) architectures, including massive parallelism and heterogeneous architectures (Davidović et al., 2017). A new paradigm will be required in order to harness the rapidly advancing capabilities, and advancing CFD capabilities must have a larger goal of comprehensive advances in multidisciplinary analysis and optimization capabilities. The recent improvements in CFD analysis which are in the development stage and which will improve significantly in the near future are mentioned in the following sub-sections.

4.1 GPUs-based CFD

While CFD simulations have become computationally demanding to define the whole process without any approximations in the geometry, in many areas of science and industry there is a need for short turnaround times and fast time-to-market. Such goals can be fulfilled only with huge investments in hardware and software licenses. Graphics Processing Units (GPUs) provide completely new possibilities for significant cost savings since the simulation time can be reduced on hardware that is often less expensive than server-class CPUs. Almost every PC contains a graphics card that supports either CUDA or OpenCL.

The computations may be done concurrently on the CPUs and GPUs (Lee et al., 2010). If there are multiple GPUs in the system, independent computing tasks can be solved simultaneously. When cases are solved on the GPU, the CPU resources are free and can be used for other tasks, such as pre- and post-processing. Moreover, the power efficiency per simulation is comparable for a dual-socket multicore CPU and a GPU. Several ongoing projects on Navier-Stokes models and Lattice Boltzmann methods have shown very large speedups using CUDA-enabled GPUs, as seen in Fig. 6. CUDA is a parallel computing

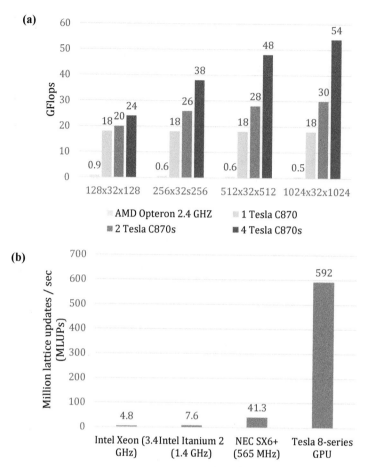

Figure 6: CUDA-enabled GPU performance for CFD analysis (a) Incompressible Navier Stokes equation solver (b) Lattice Boltzmann Method (LBM) solver for 128 × 128 mesh size (adapted from Tompson et al., 2016).

platform and application programming interface (API) model created by NVIDIA. It allows software developers to use a CUDA-enabled graphics processing unit (GPU) for general purpose processing, an approach known as GPGPU. The CUDA platform is a software layer that gives direct access to the GPU's virtual instruction set and parallel computational elements, for the execution of compute kernels.

4.2 o-CFD (CFD-based optimization)

Until now, common use of CFD approach in an optimization study is limited due to the requirement of high computational power and time. However, the detailed CFD simulations open up certain possibilities for further food process optimization. The main idea for this approach is to use the CFD solution of the given problem in an optimization routine, where the explicit and implicit variables are applied in a systematic way in order to further improve the objective function via the use of the decision variables. Foli et al. (2006) presented an approach to combine the CFD analysis with genetic algorithms for shape optimization of a heat exchanger. Heat transfer optimization in heat exchangers were also subject to the use of *o*-CFD (Fabbru, 2000; Rozzi et al., 2007). Shape and geometry optimization is another possible way to use the CFD modelling for optimization purposes. Sarghini et al. (2017) proposed the gradient-based multivariate constrained shape optimization of an extrusion bell shape for pasta production. El-Sayed et al. (2005) presented a review on the shape optimization with CFD to determine the optimum design parameters for product design and improvement using a numerical optimization approach.

Developing new small-scale systems to suit local conditions that can be deployed at low financial risk is an important objective of this method. Here, researchers develop optimization code for the specific application that enables users to determine optimal design and operating conditions before any physical system is constructed. Based on the optimal conditions obtained, the computational design is rapidly transformed into a physical test unit using 3D manufacturing (Ashuri et al., 2017). Researchers use the physical units to validate simulation and design optimization. If discrepancies are found, researchers use this knowledge to refine the simulation optimum, and a new physical system can be created and operated in a day. This method of CFD-based optimization and rapid prototyping using 3D manufacturing is expected to radically change small-scale system development.

4.3 CFD with Artificial Intelligence (AI)

A major role of CFD in the future would be to acquire new knowledge in a number of complex flow problems. It is also expected to play an increasingly effective role in the design and development process. The science of artificial intelligence (AI) can be advantageous for these roles of CFD. For example, neural network approaches have been used to model complex phenomena, such as turbulence (Sarghini et al., 2003). AI is combined along with CFD in order to understand how knowledge-based systems can accelerate the process of acquiring new knowledge in fluid mechanics, how CFD may use expert systems, and how expert systems may speed the design and development process. In this combined model, the two related aspects of CFD, i.e., reasoning and calculating, can be better understood by using AI, through which a substantial portion of reasoning can be achieved. It offers the opportunity of using computers as reasoning machines to set the stage for efficient calculating (Li et al., 2018). In fact, this kind of rule-based CFD system also supports the aforementioned CFD-based optimization by providing the robust simulation model for each design point (Li et al., 2018).

Efficient simulation of the Navier-Stokes equations for fluid flow is a long-standing problem in applied mathematics, for which state-of-the art methods require large computational resources, therefore, machine learning can be used to solve this problem. Machine Learning is a current application of AI based around the idea that we should really just be able to give machines access to data and let them learn for themselves. In this method, a data-driven machine learning approach which approximates inference of the sparse linear system used to enforce the Navier-Stokes incompressibility condition is used (Tompson et al., 2016). This approach is even used to ascertain whether machine learning could be used to predict correction terms in wall function for the pressure spectrum such that the near wall region would not need to be resolved in order to correctly predict the wall pressure (Ling et al., 2017). This method cannot predict the exact solution since it is still in developing stage. In the coming years, however, with increased research activity in this domain, it can be expected that it will give the exact solution to the problem.

Apart from these new developments in CFD, even the existing commercial CFD software are improving their accuracy in predicting results from the simulation. Some improvements which can be seen from this software in near future are:

- Physics-based and predictive modelling: Transition, turbulence, separation, multiphase flows, radiation, heat transfer and constitutive models must reflect the underlying physics more closely than ever before.
- High degree of automation: In all steps of the CFD analysis, including geometry creation, mesh generation and adaptation, creation of large databases of simulation results, understanding the results of a simulation, and ability to computationally steer the process.
- Flexibility: To tackle the capability and capacity of computing tasks in both industrial and research environments.
- Ability: To effectively utilize massively parallel, heterogeneous HPC architectures.
- Management of errors and uncertainties resulting from all possible sources, namely, physical modelling errors and uncertainties, numerical errors from mesh and uncertainties due to lack of knowledge in parameters of particular fluid flow problem.

5. Conclusions

Physics-based mathematical modelling approaches are fundamental tools for food process design and optimization purposes, and CFD is a significant way of using this tool in an effective way. The presence of various software with their easy-to-use and easy-to-learn user interface have created an increasing number of users to apply CFD approaches for mathematical modelling purposes. However, the limited knowledge of most of the users about the mathematical background of CFD have prevented a more efficient use.

Therefore, this chapter has first suggested to use CFD approach with a certain knowledge of the mathematical and physical background of the processes, knowing the fundamental governing differential equations. For this purpose, the steps involved in a CFD analysis were explained in sufficient detail. Governing equations using finite volume method as a mathematical approach to solve the CFD problem were explained. Two case studies from food process engineering were demonstrated. In the first one, a rather simple CFD process was chosen and a step-by-step preparation of the model was demonstrated. The second case was from a rather complex process requiring a comprehensive CFD approach and mathematical analysis on the physical interpretation of the results. A future outlook for CFD studies was also presented, where GPUs-based CFD approaches were explained and data-driven CFD-based optimization studies involving AI were mentioned. In addition, the significance of experimental validation or validation on a mathematical basis were also included in the scope of the chapter.

With the use of more sophisticated CFD software, recent studies in the food engineering literature have focused on grid refinement with moving mesh and multi-physics approaches. For example, a recent study on radio frequency thawing of food products demonstrated the use of moving mesh for the samples moving in radio frequency cavities and the solution of the governing equations for the electromagnetic field distribution and temperature change during the process (Erdoğdu et al., 2017). This particular study also involved phase change operation which increased the complexity of the CFD problem. In another mathematically comprehensive study, Erdoğdu et al. (2017) demonstrated the use of moving mesh in a reciprocally agitated oscillated can during thermal processing, and temperature uniformity, flow field evolution as a function of viscosity effects and reciprocal agitation rate were outlined. With the newly developed features and more sophisticated geometry creation and meshing approaches, CFD studies are expected to dominate the mathematical modelling studies in food processing for further design, scaling up and optimization purposes. In this way, a better process control and process and equipment design would be dominating the food industry in the same way as the automotive industry, which has been using CFD for a long period of time. As also outlined by Olsen (2015), the presence of virtual models might enable the design thinking process to enable further discoveries ahead.

References

Anderson, D.A., Tannehill, J.C. and Pletcher, R.H. 1984. Computational Fluid Mechanics and Heat Transfer. McGraw-Hill Book Company, New York, NY.

Ashuri, T., Martins, J.R., Zaaijer, M.B., van Kuik, G.A. and van Bussel, G.J. 2016. Aeroservoelastic design definition of a 20 MW common research wind turbine model. Wind Energy 19: 2071–2087.

Bakalis, S., Coz, P.W., Wang-Nolan, W., Parker, D.J. and Fryer, P.J. 2003. Use of positron emission particle tracking (PEPT) technique for velocity measurements in model food fluids. Journal of Food Science 68: 2684–2692.

Briggs, W.L. 1987. A Multigrid Tutorial. SIAM, Philidelphia, PA.

Brannock, M., Wang, Y. and Leslie, G. 2010. Mixing characterization of full-scale membrane bioreactors: CFD modelling with experimental validation. Water Res. 44: 3181–3191.

Buchhave, P. 1992. Particle image velocimetry—status and trends. Exp. Therm. Fluid Sci. 5: 586–604.

Chung, T.J. 2002. Computational Fluid Dynamics. Cambridge University Press, Cambridge, U.K.

Clausing, A.M. 1969. Numerical methods in heat transfer. In: B.T. Chao (ed.). Advanced Heat Transfer. University of Illinois Press, Chicago, IL.

Cordioli, M., Rinaldi, M., Copelli, G., Casoli, P. and Barbanti, D. 2015. Computational fluid dynamics (CFD) modelling and experimental validation of thermal processing of canned fruit salad in glass jar. J. Food Eng. 150: 62–69.

Cox, P.W., Bakalis, S., Ismail, H., Forster, R. and Fryer, P.J. 2003. Visualization of three-dimensional flows in rotating cans using positron emission particle tracking (PEPT). J. Food Eng. 60: 229–240.

Datta, A.K. and Teixeira, A.A. 1988. Numerically predicted transient temperature and velocity profiles during natural convection heating of canned liquid foods. J. Food Sci. 53: 191–195.

Datta, A.K. 2002. Simulation-based design of food products and processes. *In*: J. Welti-Chanes et al. (ed.). Engineering and Food for the 21st Century. CRC Press, Boca Raton, FL, Chap. 50.

Davidović, D., Fabregat-Traver, D., Höhnerbach, M. and di Napoli, E. 2017. Accelerating the computation of FLAPW methods on heterogeneous architectures. arXiv: 1712.07206.

Dutta, A., Ekatpure, R., Heynderickx, G., de Broqueville, A. and Marin, G. 2010. Rotating fluidized bed with a static geometry: Guidelines for design and operating conditions. Chem. Eng. Sci. 65: 1678–1693.

El-Sayed, M., Sun, T. and Berry, J. 2005. Shape optimization with computational fluid dynamics. Adv. Eng. Softw. 36: 607–613.

ERCOFTAC. 2000. Best Practices Guidelines for Industrial Computational Fluid Dynamics, Version 1.0.

Erdoğdu, F. 2009. Computational fluid dynamics (CFD) for optimization in food processing. *In*: F. Erdoğdu (ed.). Optimization in Food Engineering, CRC Press, Boca Raton, FL, Chap. 11.

Erdoğdu, F., Sarghini, F. and Marra, F. 2017. Mathematical modeling for virtualization in food processing. Food Eng. Rev. 9: 295–313.

Erdoğdu, F., Tutar, M., Sarghini, F. and Skipnes, D. 2017. Effects of viscosity and agitation rate on temperature and flow field in cans during reciprocal agitation. J. Food Eng. 213: 76–88.

Erdoğdu, F., Altin, O., Marra, F. and Bedane, T.F. 2017. A computational study to design process conditions in industrial radio-frequency tempering/thawing process. J. Food Eng. 213: 99–112.

Fabbri, G. and Chang, M.H. 2000. Heat transfer optimization in corrugated wall channels. Int. J. Heat Mass Tran. 43: 4299–4310.

Finlayson, B.A. 1972. The method of weighted residuals and variational principles with application in fluid mechanics, heat and mass transfer. Volume 87. Academic Press. NY, USA.

Foli, K., Okabe, T., Olhofer, M., Jin, Y. and Sendhoff, B. 2006. Optimization of micro heat exchanger: CFD, analytical approach and multi-objective evolutionary algorithms. Int. J. Heat Mass Tran. 49: 1090–1099.

Hills, B. 1995. Food processing: An MRI perspective. Trends in Food Sci. and Tech. 6: 111–117.

Ho, Q.T., Carmliet, J., Datta, A.K., Defraeye, T., Delele, M.A., Herremans, E. et al. 2013. Multiscale modeling in food engineering. J. Food Eng. 114: 279–291.

Ko, S.-H., Kim, N., Kim, J. Thota, A. and Jha, S. 2010. Efficient runtime environment for coupled multi-physics simulations: Dynamic resource allocation and load-balancing. In: Proceedings of the 10th IEEE/ACM International Conference on Cluster, Cloud and Grid Computing, pp. 349–358.

Lee, V., Kim, C., Chhugani, J., Deisher, M., Kim, D. and Nguyen, A. 2010. Debunking the 100x GPU vs. CPU myth: An evaluation of throughput computing on CPU and GPU. *In*: Proceedings of the 37th Annual International Symposium on Computer Architecture: 19–23 June 2010. Saint-Malo, France: ACM: 2010. p. 451–460.

Lewis, C.H. 1990. Comments on the need for CFD code validation. AIAA J. Spacecraft Rockets 27(2): 97.

Li, A. and Walker, C.E. 1996. Cake baking in conventional, impingement and hybrid ovens. J. Food Sci. 61: 188–191, 197.

Li, L., Lange, C.F. and Ma, Y. 2018. Artificial intelligence aided CFD analysis regime validation and selection in feature-based cyclic CAD/CFD interaction process. Computer-Aided Design and Applications 15: 643–652.

Li, L., Lange, C.F. and Ma, Y. 2018. Association of design and computational fluid dynamics simulation intent in flow control product optimization. Proceedings of the Institution of Mechanical Engineers, Part B: Journal of Engineering Manufacture 232: 2309–2322.

Ling, J., Barone, M.F., Davis, W., Chowdhary, K. and Fike, J. 2017. Development of machine learning models for turbulent wall pressure fluctuations. *In*: 55th AIAA Aerospace Sciences Meeting (p. 0755).

Marvin, J.G. 1995. Perspective on computational fluid dynamics validation. AIAA Journal 33: 1778–1787.

Mudiyanselage, C.M.R., Karunasena, H.C.P., Gu, Y.T., Guan, L. and Senadeeram, W. 2017. Novel trends in numerical modeling of plant food tissues and their morphological changes during drying—a review. J. Food Eng. 194: 24–39.

Nicolai, B.M., Verboven, P. and Scheerlinck, N. 2001. The modeling of heat and mass transfer. *In*: Food Process Modeling. Tijskens, L.M.M. et al., Ed. Woodhead Publishing Ltd., Cambridge, England, Chap. 4.

Norton, T. and Sun, D.-W. 2006. Computational fluid dynamics (CFD) and effective and efficient design and analysis tool for the food industry: A review. Trends in Food Science and Technology 77: 600–620.

Norton, T., Tiwari, B. and Sun, D.-W. 2013. Computational fluid dynamics in the design and analysis of thermal processes: A review of recent advances. Crc. Cr. Rev. Food Sci. 53: 251–275.

Ohlsson, E.E.M., Ahrne, L.M. and Tragardh, A.C. 2004. Heat transfer from a slot air jet impinging on a circular cylinder. J. Food Eng. 63: 393–2004.

Olsen. N.V. 2015. Design thinking and food innovation. Trends in Food Science and Technology 41: 182–187.

Ovadia, D.Z. and Walker, C.E. 1998. Impingement in food processing. Food Technol-Chicago 52(4): 46–50.

Palazoglu, T.K. and Erdoğdu, F. 2009. Numerical solutions-finite difference methods. *In*: F. Erdoğdu (ed.). Optimization in Food Engineering. CRC Press, Boca Raton, FL, Chap. 3.

Patankar, S.V. 1980. Numerical Heat Transfer and Fluid Flow. CRC Press, Boca Raton, FL.

Roache, P.J. 1998. Fundamentals of Computational Fluid Dynamics. Hermosa Publishers, Albuquerque, NM.

Rider, W.J. and Kothe, D.B. 1998. Reconstructing volume tracking. J Comput. Phys. 141: 112–152.

Rorres, C. 2000. The turn of the screw: Optimal design of an Archimedes screw. J. Hydraul. Eng. 126: 72–80.

Rozzi, S., Massini, R., Paciello, G., Pagliarini, G., Rainieri, S. and Trifiro, A. 2007. Heat treatment of fluid foods in a shell and tube heat exchanger: Comparison between smooth and helically corrugated wall tubes. J. Food Eng. 79: 249–254.

Rubbert, P.E. 1994. AIAA Wright Brothers Lecture: CFD and the Changing World of Aircraft Development. ICAS-94-0.2, September.

Sarghini, F., Felice, G. and Santini, S. 2003. Neural networks based sub-grid scale modeling in large eddy simulations. Computers and Fluids 32: 97–108.

Sarghini, F., De Vivo, A. and Marra, F. 2017. Multivariate constrained shape optimization: Application to extrusion bell shape for pasta production. AIP Conference Proceedings: 1896,150007.

Scott, G. and Richardson, P. 1997. The application of computational fluid dynamics in the food industry. Trends in Food Science and Technology 8: 119–124.

Singh, R.P. and Vijayan, J. 1998. Predictive modeling in food process design. Food Sci. Technol. Int. 4: 303–310.

Sarghini, F., De Vivo, A. and Erdoğdu, F. 2017. Analysis of heat and momentum transfer in screw-drive heat transfer systems. Chem Engineer Trans 57: 1729–1734.

Sun, D-W. 2007. Computational Fluid Dynamics in Food Processing. CRC Press, Boca Raton, FL.

Teixeira, A.A., Dixon, J.R., Zahradnik, J.W. and Zinsmeister, G.E. 1969. Computer optimization of nutrient retention in the thermal processing of conduction-heated foods. Food Technol. Chicago 23(6): 137–142.

Tompson, J., Schlachter, K., Sprechmann, P. and Perlin, K. 2016. Accelerating Eulerian Fluid Simulation with Convolutional Networks. CoRR abs/1607.03597.

Verboven, P., De Baerdemaeker, J. and Nicolai, B.M. 2004. Using computational fluid dynamics to optimize thermal processes. *In*: P. Richardson (ed.). Improving the Thermal Processing of Foods. CRC Press, Boca Raton, FL, 2004, Chap. 4.

Versteeg, H.K. and Malalasekera, W. 2007. An Introduction to Computational Fluid Dynamics – The Finite Volume Method. 2nd Ed. Pearson Education Limited, Edinburg, UK.

Wahlby, U., Skjoldebrand, C. and Junker, E. 2000. Impact of impingement on cooking time and food quality. J. Food Eng. 43: 179–187.

Wen, Z.-X., He, Y.-L. and Ma, Z. 2018. Effects of nozzle arrangement on uniformity of multiple impinging jets heat transfer in a fast cooling simulation device. Computers and Fluids. Article in Press.

Xia, B. and Sun, D.-W. 2002. Applications of computational fluid dynamics (CFD) in the food industry: A review. Comput. Electron Agr. 34: 5–24.

Zuo, W. 2005. Introduction to Computational Fluid Dynamics. FAU Erlangen-Nürnberg, JASS, St. Petersburg.

CHAPTER 17

Application of Multivariate Statistical Analysis for Food Safety and Quality Assurance

S Jancy and *R Preetha*[*,†]

1. Introduction

This chapter deals with applications of multivariate ("many variable") analysis methodology to assure food safety and quality in food industries. The multivariate analysis is a powerful tool for analysing more than one variable at a time, which is usually done with the help of statistical software in the modern-day scenario. Availability of different algorithms and computational software enhance the application of multivariate analysis in wide areas of research, such as medicine, chemistry, biology, food science and technology. Multivariate analysis is used in the food industry for analysing quality traits, food adulteration, microbial behavior and other biochemical changes in food samples (Nunes et al., 2015). These methods are widely used as they are not complex compared to other analysis techniques.

Multivariate analysis includes different statistical methods, such as principal component analysis (PCA), discriminate analysis (DA) clustering methods, and so on (Fig. 1) (Mengual-Macenlle et al., 2015). The data generated chemometrically, spectroscopically and sensometrically during the food processing or during quality analysis of food products can be studied using multivariate analysis. After the spectral analysis, the data can be fed into statistical software for multivariate analysis of the food samples, and sometimes the data are auto scaled. Data preprocessing is done using different statistical software, then various models are created based on the type of data collected (Nunes et al., 2015).

At first, the researchers had difficulties in understanding the statistical techniques, but now the task has been made easier, especially interpretation and correlation, through the use of different multivariate analysis software. The multivariate analysis can be used for finding out correlation between large dataset and the data can be interpreted in the form of plots. Different methodologies are available under multivariate analysis (Saona et al., 2004) and have reduced the workload among researchers because they are less time consuming than to conventional techniques.

In general, the objective of this chapter is to explain the multivariate analysis used for food products market research, analysing consumer feedback, and quality assurance in the food/beverages industry. This technique also finds a place in the optimisation of food product preparation, process control, development

Department of Food Process Engineering, SRM Institute of Science and Technology, Kattankulathur, Chennai-603203.
* Corresponding author: preetha.r@ktr.srmuniv.ac.in
† both authors contributed equally

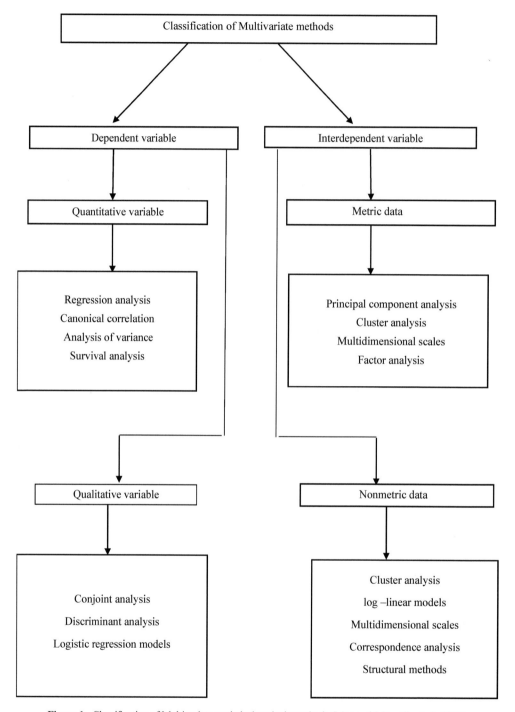

Figure 1: Classification of Multivariate statistical analysis methods (Mengual-Macenlle et al., 2015).

of new products and for the sensory analysis of the products (Nunes et al., 2015). This chapter discusses the statistical techniques based on multivariate analysis in food-based studies and applications.

2. Multivariate Analysis

The multivariate analysis, including chemometrics, are of two types: Modelling and classification. The computational packages (software) follow various models and have reduced the amount of time required to calculate large datasets (Granato and Calado, 2014). The univariate, bivariate and multivariate calibration are used to create complex models based on computational software. The above-mentioned statistical methods have obtained a foothold in most of the common scientific databases (Pubmed, Science Direct, Scopus) and in research areas, viz, food microbiology, nutrition, optimization of products and processes, food chemistry and food technologies. The modeling techniques had been used for studying microbial behavior and the state of chemical compounds during food processing (Corradini and Peleg, 2006; van Boeke, 2008; Tenenhaus-Aziza and Ellouze, 2015). These tools are also widely used in food safety and quality analysis. In the food sector, the modelling tool can be used for the analysis of analytical methods, new food product preparation/optimization, unit operations and production process. The numbers of experiments can be reduced by using this technique because many variables can be varied at a time (Preetha et al., 2015). However, in the traditional method, we can vary only one variable at a time and, moreover, optimum conditions may vary according to changes in other variables, which can be nullified by statistically designed experiments (Preetha et al., 2007a; Preetha et al., 2007b).

Classification is a technique where the data collected is used to build a formula which describes a system behavior by graphical methods. The classification procedures are also known as pattern recognition methods. The classification rules that characterize groups of samples indicate the particular sample measurements (often a small subset of the original measurements), which are very useful in finding out the identity of the group. Sample measurements may be used to classify food samples according to their cultivar, to distinguish normal and abnormal food products or to detect adulteration in food samples (Northstone et al., 2005).

In short, multivariate analysis is a collection of methods which can be used for measuring an individual or object in one or more samples. They are calculated as variables for measurements and for as units for individuals or objects. The PCA (principal component analysis) is a multivariate method in food science which is mainly used to study many foods in combination and for analysis of dietary patterns as a different tool to assess the intakes of individual food items or nutrients in a diet (Northstone et al., 2005). The interrelation between the individual food samples is studied based on the inter-item correlations by analysing data which was collected.

The multivariate analysis was used to reduce the highly prominent independent variables to less than half of the orthogonal factors. In a study, multivariate statistical techniques, such as PCR (Principal Component Analysis) and PLSR (Partial Least Squares regression) models, were used for the data generated based on the spectral reflectance, and the selected spectral indices were tested to predict biochemical parameters of different fruit kinds. The optimisation of data using an optimised model could benefit the study more. Moreover, the above report suggested multivariate methods to analyse food safety and quality control of food by using passive reflectance sensing (Elsayed et al., 2019).

Multivariate analysis was used to predict the origin of the agricultural system of the harvested carrots on the basis of features determined by liquid chromatography and mass spectrometry. The datasets were exported to SIMCA-P v12.0 software (UmetricsAB, Malmö, Sweden) and variables (molecular features) were Pareto scaled. Principal Component Analysis (PCA) was performed for predictive analysis and orthogonal projections to latent structures-discriminant analysis (OPLS-DA) models were built. OPLS-DA models displayed a definite separation between conventional and organic samples, thus, this study proved that multivariate analysis can perform in prediction studies also (Maquet et al., 2018).

A new methodology for the quantification of an artificial dye, sunset yellow (SY), in soft beverages, using image analysis (RGB histograms) and partial least squares regression, was developed and validated using multivariate calibration method. This study suggested multivariate analysis as a very good tool for prediction of various models to analyze different elements in food items (Sena et al., 2014).

Another study using multivariate analysis, such as cluster analysis (CA), principal component analysis (PCA) and discriminant analysis (DA), reported two different clusters (classes) that reflect the different water quality characteristics of the water systems in the Muda River basin, located in northwest

Malayasia. PCA (principal component analysis) was used to describe the importance of variables that helped for groupings and design of the basic properties of the monitoring stations. The multivariate analysis of measurement results also helped Department of Environment, Malayasia to explain water quality status (Azhar et al., 2015).

Chemometrics utilizes mathematical and statistical methods, and has the following applications in food: To design or select the most favorable measurement or procedures or experiments and to supply the maximum amount of information by analyzing chemical data. In food science and technology, chemometrics and multivariate analysis of data is used widely in nutrition studies (Genovese et al., 2013; Vilas Boas et al., 2014; Domingo et al., 2014; Nunes, 2014; Mafra et al., 2014). In particular, principal component analysis (PCA) and soft independent modelling of class analogy (SIMCA) are widely used for analysing a variety of food products.

Principal Component Analysis (PCA) is the most used chemometric method. The PCA scores plot are most often used in order to combine the samples into groups based on their similarities or dissimilarities by using a 2D or 3D projection of the samples. The interpretations of variables are obtained by the loadings plot. In a score plot, the axes scales should be compared in order to avoid misinterpretation of results (Geladi et al., 2003; Kieldahl and Bro, 2010). The above discussed statistical tools were reported for data analysis and explanation in rice plant pathology in order to overcome the yield loss caused by the diseases which affect the rice crop (Nayak et al., 2018). The principal component analysis was also done for studying the diet pattern of children of different age groups by the data collected using questionnaires. The PCA with varimax rotation was done for different food types, such as junk, traditional, health conscious and socio-demographic characteristics at each time interval (Northstone et al., 2005).

3. Multivariate Data Processing and Statistical Software for Multivariate Analysis

The multivariate data analysis follows data preprocessing in order to eradicate the methodical sources of variation in the data set. Methods such as mean-centering and auto scaling the output obtained should be done in order to avoid different interpretation and misuse of the technique. Mean-centering was done by deducting the mean of a variable column from each element in the column. So, the variable has a mean of zero, the data are changed by mean and the center of the data becomes the new origin; So, the knowledge about the origin is not found, but the distances between the data points remains the same (Varmuza and Filzmoser, 2009). Scale differences are not removed in mean-centering and the scaled variables have the same standard deviation.

Auto scaling is nothing but separating each mean-centering component in a post by the standard deviation of the variable. Auto scaling moves the centroid of the data points to the origin and changes the scaling of the axes. Therefore, the comparative distances between the data points are altered, causing a blow-up (highlight) of variables with minor values (Varmuza and Filzmoser, 2009). The dataset was auto-scaled and all variables have an identical variance, then every variable has the same chance of entering the model and if the unceasing data (such as spectra and chromatograms) are auto-scaled, every variable, i.e., real signal or noise from starting point, has the same standard deviation. This blow-up of noise can damage the quality of the model significantly. These effects can be envisioned by using data preprocessing methods (Nunes et al., 2015).

Different multivariate statistical analysis methods were used in food and a good number of software are available for their analysis (Table 1). The statistical software differs based on the data collected. Sigma Plot 11.0 (Systat Software Inc., USA) was used for statistical data analysis and SIMCA 13.0.3.0 software (Umetrics AB, Sweden) for running the multivariate data analysis in a study. In this study, in order to identify veterinary drug abuse, circulating extracellular small RNAs (smexRNAs) in bovine plasma were sequenced and their prospective to serve as biomarkers was evaluated using multivariate data analysis tools. In this case, multivariate methodology such as OPLS-DA (DA, discriminant analysis; OPLS, orthogonal partial least-squares) have proven to possess the best potential to generate discriminative miRNA models, supported by small RNA-Seq data (Melanie et al., 2015).

Table 1: Different multivariate analysis methods used in food product analysis.

S. No.	Type of Analysis/Method	Study Based on	Software/Reference
1	PCA	The diet of children	Windows v.1.1.19 (Northstone et al., 2005)
2	Univariate and multivariate data analysis OPLS-DA	Veterinary drug application based on biomarker signatures using smex RNAs	Sigma plot 11.0 (Systat Software Inc., USA) SIMCA 13.0.3.0 Software (Umetrics AB, Sweden) (Melanie et al., 2015)
3	Multivariate data analysis/ SIMCA and PCA was performed with NIPALS (non-linear iterative partial least squares) algorithm	Bacterial contamination in liquid	Pirouette pattern recognition software (version 3.1 for Windows, Infometrix Inc., Woodinville, Wash) (Saona et al., 2004)
4	Multivariate data analysis/Kernel partial least square (KPLS)	Multivariate KPI for energy management in food Industry	(Corsini et al., 2016)
5	Multivariate data analysis/PCA and OPLS-DA plot	Analysis for food adulteration using AMS (Ambient Mass Spectrometry)	(Black et al., 2016)
6	Multivariate statistical analysis PCA and PLS-DA	Multivariate Analysis for *Oryza sativa* L. based on geographical origin	Web-based metabolomics data processing tool MetaboAnalyst 3.0 (http://www. metaboanalyst.ca/) (Lim et al., 2017)
7	Multivariate chemometric method, PCA and ward method and MCR-ALS (multivariate curve resolution-alternating least squares)	Analysis for PMR (Polygoni Multiflori Radix) using Multivariate chemometric method	MATLAB R2014a (MathWorks, Natick, MA, USA). MCR-ALS GUI 2.0 (the GUI-updated version for the MCR-ALS algorithm) (Sun et al., 2018)
8	Multivariate statistical analysis PCA and 2-way cluster analysis	Analysis on Barnyard millet (*Echinochloa* spp.) for qualitative and quantitative traits	JMP 2009 (JMP, Version 9.0.0. SASInstitute Inc., Cary, NC) and SPSS (Statistical Package for SocialScience, SPSS Inc., Chicago, IL) (Sood et al., 2015)
9	Statistical Analysis GEE (Generalized estimating equations) model	Study based on family food purchase of high-calorie food (HCF) and low-calorie food (LCF) in supermarket	SPSS Statistics 21 (IBM Corporation), For GEE model SAS™ (Version 9.3) (Chrisinger et al., 2018)
10	MANOVA (Multivariate analysis of Variance)	Effect of perceived overeating food addiction attributions and snack choice	(Ruddock and Hardman, 2018)
11	Digital image analysis using MATLAB	Determination of some common food dyes	MATLAB 7.0 (Math works, Natick, MA, USA) (Sorouraddin et al., 2015)
12	Descriptive Analysis and One-way Analysis of variance	Diet quality in children and the role of local food Environment	Stata 12 IC (Stata Corp LP, USA) (Keane et al., 2016)
13	ANOVA	Abrasion of Mulberry fruit	SPSS version 19.0 (Afsharnia et al., 2017)
14	PCA was done by SIMCA	Detection of several common adulterants in raw milk by MID-INFRA RED spectroscopy	MATLAB software, version7.10.0.499–R2010a, The MathWorks, Natick, MA, USA, PLS Toolbox, version 5.2.2, Eigenvector Technologies, Manson, WA, USA (Gondim et al., 2017)
15	ANOVA, PCA, SIMCA and LDA (Linear Discriminant Analysis)	Arabica coffee adulteration with maize and coffee by-products	Chemoface version 1.5 MATLAB program (version 7.5, The Mathworks, Inc.) (Tavares et al., 2016)

Table 1 contd. ...

... Table 1 contd.

S. No.	Type of Analysis/Method	Study Based on	Software/Reference
16	Multivariate statistical analysis/ PCA and cluster analysis	Multivariate statistical analysis to characterize mechanization, structure and energy profile in Italian dairy farms	SPSS Inc., Chicago, IL (Todde et al., 2016)
17	LOCF (Last value-carried-forward) method	To minimize diarrhea in critically ill neurological patients by tube feeding with natural-based food	SAS Version 9.3 (SAS Institute Inc., Cary, NC, USA) (Schmidt et al., 2018)
18	MANOVA and PCA; ANOVA	Whole grains in finishing of culled ewes in pasture or feedlot: Performance, carcass characteristics and meat quality; To study Physico-chemical characteristic of Emu meat after cooking	SAS statistical program (Fruet et al., 2016); ANOVA (Nithyalekshmi and Preetha, 2015)
19	PCA and cluster analysis	Study conducted to identify contamination in water	Microsoft Excel 2007 and IBM SPSS 20 (Barakat et al., 2016)
20	PCA chemometric tools	Detection of adulteration in cumin using electronic sensing coupled with multivariate analysis	(Tahri et al., 2017)
21	Multivariate statistical technique ANOVA	Determination of phenolic compounds from yerba mate-based beverages	Statistica 6.0 (Statsoft Inc., 2001) (Da-Silveira et al., 2016)
22	PCA and PLS-DA	Food fraud in bovine meat	MATLAB Software version 8.4 (The MathWorks, Natick, USA) and PLS Toolbox, Version 7.0 (Eigenvector Technologies Manson, USA) (Nunes et al., 2016)
23	PCA and PLS-DA	Metabolite profiling, antioxidant, and α-glucosidase inhibitory activities of germinated rice	SIMCA-P software version 13.0 (Umetrics, Umea, Sweden) SPSS Inc., Chicago, IL, USA (Pramani et al., 2018)
24	MANOVA (Multivariate Analysis of Variance) and cluster analysis	Analysis in dry cured ham production based on genetic groups in pigs	SAS Windows version 9.1 (SAS® Institute, 2002–2003) (Ventura et al., 2012)
25	PCA and HCA (Hierarchial Cluster Analysis)	Evaluation of nutritional profile of food products	IBM® SPSS® Statistics 20.0 software (IBM Inc., New York, NY, USA) and XLSTAT® 2013.03.30882 (Addinsoft SARL, Paris, France) (Do Prado et al., 2016)
26	ANOVA and PCA	Rheological properties of fermented rice extract based on waxy maize starch	Statsoft, Statistic 7.0, Tulsa, EUA (Costa et al., 2016)
27	Multivariate Statistical Analysis PERMANOVA and PERANOVA	Dam regulation and riverine food-web structure in a Mediterranean River	PERMANOVA and PERANOVA, using 999 permutations on PRIMER-E 6 v.6.1.11 and PERMANOVA+ v.1.0.1 (PRIMER-E Ltd., Plymouth, UK) (Mor et al., 2018)
28	Multivariate analysis Cluster analysis and Pearson correlation	A chemical and sensory acceptance of weaning foods	Statistical 8.0 (Soronikpoho et al., 2017)

Abbreviations: PCA (Principal Component Analysis), DA (Discriminant Analysis), OPLS (Orthogonal Partial Least Squares), SIMCA (Soft Independent Modeling of Class Analogy), CA (Cluster Analysis), ANOVA (Analysis of Variance), PERMANOVA (Permutational Multivariate Analysis of Variance), PERANOVA (Permutational Analysis of Variance), PLS (Partial Least Squares), MANOVA (Multivariate Analysis of Variance).

The accurate statistical techniques are very important for data evaluation in food technology for various fields involving sensory evaluation, development of food products/processes, food microbiology and quality control. In this logic, the concept of statistics applied in food research is commonly used, but anxiety occurs on how the statistical analyses have been developed, particularly when computational software has been used. Several commercial and free statistical software (Fig. 2) are employed in food applications (Nunes et al., 2015).

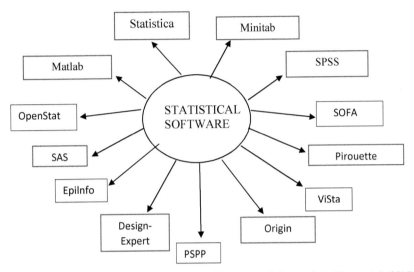

Figure 2: Types of statistical software used for multivariate statistical analysis (Nunes et al., 2015).

4. Application of Multivariate Analysis in Food Quality

In food science and technology, the application of Multivariate analysis has seen a marked rise in the last twenty years (Nunes et al., 2015). The multivariate statistical method is used for plotting the graph with the help of statistical model and software; the data are fed into software programs and the results are interpreted easily for various applications in food (Fig. 3).

4.1 Applications in diet quality

Multivariate analysis has been used in the diet quality analysis of children and people with different socioeconomic backgrounds. In one study, multivariate analysis was used to determine the dietary pattern and quality of children of different age groups, depending on socio-demographic characteristics like sex, maternal age, maternal education and number of siblings. The data collected by questionnaires are analysed by principal component analysis and categorized into three different principal components: Junk-type (sausages, burgers and high fat content foods), traditional-type (meat, poultry and green vegetables) and health-conscious type (vegetables, nuts and pulses). The PCA method helped to conclude that the food intake may differ in each child due to socio-graphic and daily lifestyle changes (Northstone et al., 2005). In another report, the dietary quality of children growing up in Ireland was analysed based on the effect of food availability in the local area. The statistical analysis was done using Stata 12 IC (Stata Corp LP, USA) and the multivariate analysis was presented using quintiles. The descriptive analysis was also performed and the mean Dietary Quality Score (DQS) was calculated for independent predictor variable. The independent sample t-tests and one-way analysis of variance were calculated in order to determine

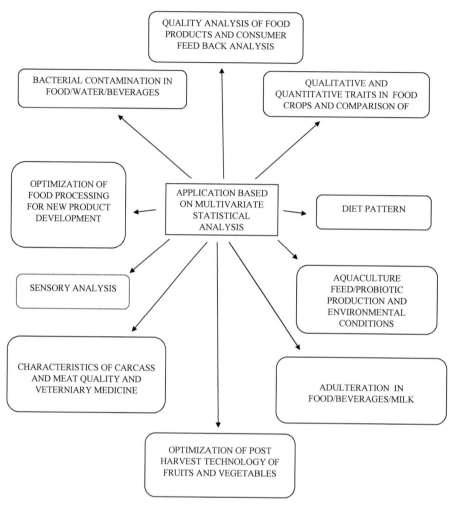

Figure 3: Application based on multivariate statistical analysis (Saona et al., 2004; Preetha et al., 2007a; preetha et al., 2007b; Northstone et al., 2005; Sood et al., 2015; Keane et al., 2016; Afsharnia et al., 2017; Gondim et al., 2017; Tavares et al., 2016; Fruet et al., 2016; Nithyalekshmi and Preetha, 2015; Barakat et al., 2016; Nunes et al., 2016; Mor et al., 2018; Soronikpoho et al., 2017).

the mean difference between the variables. The Electoral division (ED) of each household was used as sample cluster for regression analysis, separate regression models were also used to determine the distance from the participants' households to the nearest supermarkets and convenience stores (Keane et al., 2016).

In another study, multivariate GEE (Generalized estimating equations) were formulated in order to assess predictors for the amount spent by black women on LCF (low-calorie food) and HCF (high-calorie food) while purchasing foods from supermarkets and other retailers in an urban area within the USA. In the abovementioned study, the statistical analyses were done using SPSS statistics 21 (IBM corporation). The descriptive data collection was done to determine HCF and LCF expenditures and participants' characteristics and behavior, and GEE modeling was carried out with a dataset of 450 receipts by SAS™ (Version 9.3) (Chrisinger et al., 2018). A univariate Analysis Of Variance (ANOVA) was conducted in order to evaluate conditions such as overeating, controlled eating and under-eating. The main hypothesis was that the overeating may lead to food addiction based on dietary concern and guilt, and the level of guilt was found to be higher in overeating participants and lower in under-eating participants (Ruddock and Hardman, 2018).

4.1.1 Applications in plant crops and meat quality

A multivariate analysis has been reported for analysis quality and safety of plant crops and meat. The Barnyard millet (*Echinochloa* spp.) core accession was classified for its application in breeding. This qualitative analysis of (Barnyard millet core) was done for germ plasm collection over two years, and the methods performed were PCA (principal component analysis) and cluster analyses based on the trait morphology and quantitative characters of Barnyard millet. The Statistical analysis was done by JMP 2009 (JMP, Version 9.0.0. SAS Institute Inc., Cary, NC) and SPSS (Statistical Package for Social Science, SPSS Inc., Chicago, IL) software. The traits studied included growth habits, culm branching, plant pigmentation, inflorescence shape and compactness for all origins, including India, Japan and other countries (Sood et al., 2015). A Statistical analysis was done for the characteristics of carcass and meat quality. In the abovementioned case, a Shapiro-Wilk test was performed for residual normality and Levene's test for homogeneity, the average was done by Least squares method (LSMEANS). The sample loss was then compared by Tukey's test, the equality of mean vectors was done using MANOVA (Multivariate analysis of Variance), hypotheses evaluation was done by Wilks' test and the principal component analysis was done for graphical assessment of dispersion of standardized variables by association matrix and related treatments (Fruet et al., 2016). In another study, the influence of processing temperature and time on the quality of Emu meat (*Dromaius novaehollandiae*) was studied and the data were analysed by two-way ANOVA using IBM SPSS statistic version 19.0 software (Nithyalakshmi and Preetha, 2016). A study was conducted based on the prospective of circulating extracellular small RNAs (smexRNA) in order to identify misuse of veterinary medicine, because the misuse of drugs leads to the deterioration of milk and meat quality. In the abovementioned study, the identification of biomarker candidates was performed using multivariate methodologies, such as discriminant analysis (DA) and orthogonal partial least-squares analysis (OPLS). Sigma plot 11.0 (Systat Software Inc., USA) and SIMCA 13.0.3.0 Software (Umetrics AB, Sweden) were used for statistical analysis. The multivariate analysis tools helps in revealing the treatment-dependent modifications at miRNA level and the OPLS-DA discriminative models which were used delivered successful results based on high goodness of fit and probability (Melanie et al., 2015).

4.1.2 Applications in evaluation of quality of water and beverages

A study was conducted on the identification of contamination sources affecting the water quality of the Oum Er River and its branch Oued El Abid (Morocco) and the dataset was analysed using Pearson's correlation. In the above case, a computational study was conducted using Microsoft Excel 2007, IBM SPSS 20 and PCA (Principal Component Analysis) in order to find an interrelationship between parameters. In order to identify possible contamination sources of Oum Er River water, cluster analysis was carried out to classify spatial and temporal similarity groups of all monitoring stations, based on their water quality. The multivariate analysis played an important role in the analysis of the large datasets (Barakat et al., 2016). Another study was conducted based on determination of phenolic compounds in yerba mate (native plant from South America) based beverages in which the mathematical methods were obtained by variance analysis and validation of the model was done using ANOVA (Analysis Of Variance). The statistical data was compared by ANOVA and Tukey's test using Statistica 6.0 (Statsoft Inc., 2001). The multivariate statistical technique was proven to be essential in the study due to maximization of resolution and symmetry peak of interest in the analysis (Tahri et al., 2017).

The multivariate analysis technique and Fourier Transform Near-Infrared (FT-NIR) Spectroscopy was used to identify the bacteria in a juice matrix (Saona et al., 2004). In the abovementioned study, the SIMCA (soft independent modelling of class analogy) and PCA (principal component analysis) was performed with NIPALS (non-linear iterative partial least squares) algorithm using Pirouette pattern recognition software (version 3.1 for Windows, Infometrix Inc., Woodinville, Wash.). PCA clustering was used to determine the growth stage of bacterial cells and the discrimination of different *E. coli* strains based on the ethanol treatment, which made differences in spectral readings. The SIMCA model was also helpful in predicting the species of *E. coli* which had contaminated the juice (Saona et al., 2004).

4.1.3 Applications in post-harvest technology

Multivariate analysis technique was also used to assure the quality of product after harvesting. In a report, statistical analysis was done on the abrasion of mulberry fruit while harvesting and postharvest handling using statistical package for social sciences (SPSS version 19.0). The variance analysis of the data was carried out by multivariate factorial design and the data collection was done for maturity stages, dropping height and storage regimes. The ANOVA results obtained showed that C* (chroma), firmness and TAC (Total Anthocyanin Content) had significant impact on the quality of the fruit and the abrasion area on the fruit was influenced by factors such as maturity stages, dropping height and storage conditions (Heidari et al., 2017).

A study was reported on chemical and sensory acceptance of weaning foods (baby food) using multivariate statistical method, in which the cluster analysis was graphically represented in the form of hierarchical clustering procedure, also known as a dendrogram. The abovementioned analysis was performed for commercial and yam-based weaning foods. The Pearson correlation was performed to evaluate samples in order to determine the relationship between sensory properties. ANOVA (Analysis Of Variance) and Turkey HSD test (Soronikpoho et al., 2017) was also done using software Statistical 8.0.

4.2 Applications in food safety and adulteration

Analysis for food adulteration was reported using AMS (Ambient Mass Spectrometry) with the help of PCA and OPLS-DA plots. Both DESI-MS (Desorption Electrospray Ionization Mass Spectrometry) and LESA-MS (Liquid Extraction Surface Analysis Mass Spectrometry) were used to differentiate between five different meat species: Beef, chicken, pork, horse and turkey. The data analysis was done using multivariate statistical software. LESA-MS models were promising; however, DESI-MS models were weaker. In the above mentioned study, OPLS-DA plot was used for discrimination of beef, chicken, pork, horse and turkey using LESA-MS (Black et al., 2016).

A study was conducted to analyse the adulteration of Arabica Coffee with maize and coffee by-products, the principal component analysis (PCA) was done with software chemo face version 1.5 and the estimation of adulteration percentage was done by SIMCA and LDA (Linear Discriminant Analysis) using MATLAB (Version 7.5, The Mathworks, Inc.). The results were obtained based on relative proportion of total tocopherol amounts in pure and adulterated coffee samples by discriminant analysis (Tavares et al., 2016). The detection of adulterants in raw milk was reported using MID-INFRA RED Spectroscopy and data analysis was done using one-class and multi-class multivariate strategies. The result of spectral analysis was introduced to MATLAB software, version 7.10.0.499 – R2010a (The Math Works, Natick, MA, USA), and evaluation was performed using PLS Toolbox, version 5.2.2 (Eigenvector Technologies, Manson, WA, USA). The spectrum baselines and trends were corrected using Multiplicative Scatter Correction (MSC). Principal component analysis was performed as an unsupervised exploratory analysis tool for the visualization of sample distribution in multivariate space in order to determine possible outliers. SIMCA one-class model was also used for prediction of common adulterants in raw milk samples; moreover, the one-class and multi-class multivariate strategies proved to be highly efficient in the analysis of adulterants (Gondim et al., 2017).

5. Merits and Demerits

The multivariate statistical analysis method is used widely because, with the help of software, it is easy to handle lots of data and the results can be interpreted easily, thereby reducing the time of work. However, the multivariate statistical analysis results are difficult to interpret for researchers who do not have in-depth knowledge of mathematics and statistics. Another demerit is that the building of statistical software is expensive (Nunes et al., 2015).

6. Summary and Future Prospective

This chapter discussed the uses of multivariate statistical analysis for analysing the data regarding food quality and safety and also explains the statistical methods and software used for multivariate statistical analysis. This multivariate statistical method is increasingly used in food safety and quality studies because of its merits and usefulness. There are many types and classification under multivariate statistical methods which can be applied in various fields of food science and technology, this is also discussed in short. In most of the studies, statistical analyses were performed and statistical models were constructed for the validation of the variables in order to obtain a better result. This technique is accurate and capable of analysing large datasets, therefore, it reduces the time required and leaves more opportunity for further study.

References

Afsharnia, F., Mehdizadeh, S.A., Ghaseminejad, M. and Heidari, M. 2017. The effect of dynamic loading on abrasion of mulberry fruit using digital image analysis. Information Processing in Agriculture 4: 291–299.

Alezandro, M.R., Granato, D. and Genovese, M.I. 2013. Jaboticaba (Myrciaria jaboticaba (Vell.) Berg), a Brazilian grape-like fruit, improves plasma lipid profile instreptozotocin-mediated oxidative stress in diabetic rats. Food Research International 54: 650–659.

Azhar, S.C., Aris, A.Z., Yusoff, M.K., Ramli, M.F. and Juahir, H. 2015. Classification of river water quality using multivariate analysis. Procedia Environmental Sciences 30: 79–84.

Barakat, A., El Baghdadi, M., Rais, J., Aghezzaf, B. and Slassi, M. 2016. Assessment of spatial and seasonal water quality variation of OumErRbia River (Morocco) using multivariate statistical techniques. International Soil and Water Conservation Research 4: 284–292.

Black, C., Chevallier, O.P. and Elliott, C.T. 2016. The current and potential applications of ambient mass spectrometry in detecting food fraud. Trends in Analytical Chemistry 82: 268–278.

Botelho, B.G., de Assis, L.P. and Sena, M.M. 2014. Development and analytical validation of a simple multivariate calibration method using digital scanner images for sunset yellow determination in soft beverages. Food Chemistry 159: 175–180.

Chrisinger, B.W., DiSantis, K.I., Hillier, A.E. and Kumanyika, S.K. 2018. Family food purchases of high- and low-calorie foods in full-service supermarkets and other food retailers by Black women in an urban US setting. Preventive Medicine Reports 10: 136–143.

Corradini, M.G. and Peleg, M. 2006. On modeling and simulating transitions between microbial growth and inactivation or vice versa. International Journal of Food Microbiology 108(1): 22–35.

Corrêa, S.C., Pinheiro, A.C.M., Siqueira, H.E., Carvalho, E.M., Nunes, C.A. and Vilas Boas, E.V.B. 2014. Prediction of the sensory acceptance of fruits by physical and physical–chemical parameters using multivariate models. LWT — Food Science and Technology 59(2): 666–672. http://dx.doi.org/10.1016/j.lwt.2014.07.042.

Corsini, A., Bonacina, F., Feudo, S., Lucchetta, F. and Marchegiani, A. 2016. Multivariate KPI for energy management of cooling systems in food industry. Energy Procedia 101: 297–304.

Costa, K.K.F.D., Garcia, M.C., de, O., Ribeiro, K., Soares Junior, M.S. and Caliari, M. 2016. Rheological properties of fermented rice extract with probiotic bacteria and different concentrations of waxy maize starch. LWT - Food Science and Technology 72: 71–77.

Cubero-Leon, E., De Rudder, O. and Maquet, A. 2018. Metabolomics for organic food authentication: Results from a long-term field study in carrots. Food Chemistry 239: 760–770.

da Silveira, T.F.F., Meinhart, A.D., de Souza, T.C.L., Filho, J.T. and Godoy, H.T. 2016. Phenolic compounds from yerba mate-based beverages—A multivariate optimization. Food Chemistry 190: 1159–1167.

do Prado, S.B.R., Giuntinia, E.B., Grandea, F. and de Menezesa, E.W. 2016. Techniques to evaluate changes in the nutritional profile of food products. Journal of Food Composition and Analysis 53: 1–6.

Domingo, E., Tirelli, A.A., Nunes, C.A., Guerreiro, M.C. and Pinto, S.M. 2014. Melamine detection in milk using vibrational spectroscopy and chemometrics analysis: A review. Food Research International 60: 131–139.

Elsayed, S., El-Gozayer, K., Allam, A. and Schmidhalter, U. 2019. Passive reflectance sensing using regression and multivariate analysis to estimate biochemical parameters of different fruits kinds. Scientia Horticulturae 243: 21–33.

Fruet, A.P.B., Stefanello, F.S., Rosado Júnior, A.G., de Souza, A.N.M., Tonetto, C.J. and Nörnberg, J.L. 2016. Whole grains in the finishing of culled ewes in pasture or feedlot: Performance, carcass characteristics and meat quality. Meat Science 113: 97–103.

Geladi, P., Manley, M. and Lestander, T. 2003. Scatter plotting in multivariate data analysis. Journal of Chemometrics 17: 503–511.

Gondim, C.S., Junqueira, R.G., Souza, S.V.C., Ruisánchez, I. and Callao, M.P. 2017. Detection of several common adulterants in raw milk by MID-infrared spectroscopy and one-class and multi-class multivariate strategies. Food Chemistry 230: 68–75.

Granato, D., Calado, V.M.A. and Jarvis, B. 2014. Observations on the use of statistical methods in food science and technology. Food Research International 55: 137–149.

Keane, E., Cullinan, J., Perry, C.P., Kearney, P.M., Harrington, J.M., Perry, I.J. et al. 2016. Dietary quality in children and the role of the local food environment. SSM - Population Health 2: 770–777.

Kjeldahl, K. and Bro, R. 2010. Some common misunderstandings in chemometrics. Journal of Chemometrics 24: 558–564.

Lim, D.K., Mo, C., Lee, J.H., Long, N.P., Dong, Z., Li, J. et al. 2017. The integration of multi-platform MS-based metabolomics and multivariate analysis for the geographical origin discrimination of *Oryza sativa* L. Journal of Food and Drug Analysis 1-9 https://doi.org/10.1016/j.jfda.2017.09.004.

Melanie, S., Benedikt, K., Pfaffl, M.W. and Irmgard, R. 2015. The potential of circulating extracellular small RNAs (smexRNA) in veterinary diagnostics—Identifying biomarker signatures by multivariate data analysis. Biomolecular Detection and Quantification 5: 15–22.

Mengual-Macenlle, N., Marcos, P.J., Golpe, R. and González-Rivas, D. 2015. Multivariate analysis in thoracic research. Journal of Thoracic Disease 7(3). doi: 10.3978/j.issn.2072-1439.2015.01.43.

Mor, J.-R., Ruhí, A., Tornés, E., Valcárcel, H., Muñoz, I. and Sabater, S. 2018. Dam regulation and riverine food-web structure in a Mediterranean river. Science of the Total Environment 625: 301–310.

Nayak, P., Mukherjee, A.P., Pandit, E. and Pradhan, S.K. 2018. Application of statistical tools for data analysis and interpretation in rice plant pathology. Rice Science 25(1): 1–18.

Nithyalakshmi, V. and Preetha, R. 2015. Effect of cooking conditions on physico-chemical and textural properties of Emu (*Dromaius novaehollandiae*) meat. International Food Research Journal 22(5): 1924–1930.

Northstone, K., Emmett, P. and The ALSPAC Study Team. 2005. Multivariate analysis of diet in children at four and seven years of age and associations with socio-demographic characteristics. European Journal of Clinical Nutrition 59: 751–760.

Nunes, C.A. 2014. Vibrational spectroscopy and chemometrics to assess authenticity, adulteration and intrinsic quality parameters of edible oils and fats. Food Research International 60: 255–261.

Nunes, C.A., Alvarenga, V.O., Sant'Ana, A.S., Santos, J.S. and Granato, D. 2015. The use of statistical software in food science and technology: Advantages, limitations and misuses. Food Research International 75: 270–280.

Nunes, K.M., Andrade, M.V.O., Santos Filho, A.M.P., Lasmar, M.C. and Sena, M.M. 2016. Detection and characterisation of frauds in bovine meat *in natura* by non-meat ingredient additions using data fusion of chemical parameters and ATR-FTIR spectroscopy. Food Chemistry 205: 14–22.

Pramai, P., Hamid, N.A.A., Mediani, A., Maulidiani, M., Abas, F. and Jiamyangyuen, S. 2018. Metabolite profiling, antioxidant, and α-glucosidase inhibitory activities of germinated rice: Nuclear-magnetic-resonance-based metabolomics study. Journal of Food and Drug Analysis 26: 47–57.

Preetha, R., Jayaprakash, N.S., Philip, R. and Singh, I.S.B. 2007a. Optimization of carbon and nitrogen sources and growth factors for the production of an aquaculture probiotic (*Pseudomonas* MCCB 103) using response surface methodology. Journal of Applied Microbiology 102: 1043–1051.

Preetha, R., Jayaprakash, N.S., Philip, R. and Singh, I.S.B. 2007b. Optimization of medium for the production of a novel aquaculture probiotic, micrococcus MCCB 104 using central composite design. Biotechnology and Bioprocess Engineering 12: 548–555.

Preetha, R., Vijayan, K.K., Jayapraksh, N.S., Alavandi, S.V., Santiago, T.C. and Singh, I.S.B. 2015. Optimization of culture conditions for mass production of the probiotics pseudomonas MCCB 102 and 103 antagonistic to pathogenic vibrios in aquaculture. Probiotics & Antimicro Prot 7 : 137–145. DOI 10.1007/s12602-015-9185-7.

Ruddock, H.K. and Hardman, C.A. 2018. Guilty pleasures: The effect of perceived overeating on food addiction attributions and snack choice. Appetite 121: 9–17.

Saona, L.E.R., Khambaty, F.M., Fry, F.S., Dubois, J. and Calvey, E.M. 2004. Detection and identification of bacteria in a juice matrix with fourier transform near-infrared spectroscopy and multivariate analysis. Journal of Food Protection, 67(11): 2555–2559.

Schmidt, S.B., Kulig, W., Winter, R., Vasold, A.S., Knoll, A.E. and Rollnik, J.D. 2018. The effect of a natural food-based tube feeding in minimizing diarrhea in critically ill neurological patients. Clinical Nutrition: 1–9. https://doi.org/10.1016/j.clnu.2018.01.007.

Sood, S., Khulbe, R.K., Arun Kumar, R., Agrawal, P.K. and Upadhyaya, H.D. 2015. Barnyard millet global core collection evaluation in the submontane Himalayan region of India using multivariate analysis. The Crop Journal 3: 517–525.

Soronikpoho, S., Elleingand, E.F., Fatoumata, C. and Ernest, K. 2017. Multivariate analysis evaluation of technological process on chemical and sensory acceptance of yam and soy-based weaning foods. Food Sci. Nutr. Technol. 2(4): 000132.

Sorouraddin, M.H., Saadati, M. and Mirabi, F. 2015. Simultaneous determination of some common food dyes in commercial products by digital image analysis. Journal of Food and Drug Analysis 23: 447–452.

Sun, L-L., Wang, M., Zhang, H.J., Liu, Y.N., Ren, X.L., Deng, Y.R. et al. 2018. Comprehensive analysis of PolygoniMultifloriRadix of different geographical origins using ultrahigh-performance liquid chromatography fingerprints and multivariate chemometric methods. Journal of Food and Drug Analysis 26: 90–99.

Tahri, K., Tiebeb, C., El Baric, N., Hübertb, T. and Bouchikhia, B. 2017. Geographical classification and adulteration detection of cumin by using electronic sensing coupled to multivariate analysis. Procedia Technology 27: 240–241.

Tavares, K.M., Lima, A.R., Nunes, C.A., Silva, V.A., Mendes, E., Casal, S. et al. 2016. Free tocopherols as chemical markers for Arabica coffee adulteration with maize and coffee by-products. Food Control 70: 318–324.

Tenenhaus-Aziza, F. and Ellouze, M. 2015. Software for predictive microbiology and risk assessment: A description and comparison of tools presented at the ICPMF8 Software Fair. Food Microbiology 45: 290–299.

Todde, G., Murgia, L., Caria, M. and Pazzona, A. 2016. A multivariate statistical analysis approach to characterize mechanization, structural and energy profile in Italian dairy farms. Energy Reports 2: 129–134.

van Boekel, M.A.J.S. 2008. Kinetic modeling of food quality: A critical review. Comprehensive Reviews in Food Science and Food Safety 144: 158.

Varmuza, K. and Filzmoser, P. 2009. Introduction to Multivariate Statistical Analysis in Chemometrics. Boca Raton: CRC Press (336 pp.).

Ventura, H.T., Lopes, P.S., Peloso, J.V., Guimaräes, S.E.F., Carneiro, A.P.S. and Carneiro, P.L.S. 2012. Use of multivariate analysis to evaluate genetic groups of pigs for dry-cured ham production. Livestock Science 148: 214–220.

Zielinski, A.A., Alberti, A., Braga, C.M., Silva, K.M., Canteri, M.H.G., Mafra, L.I. et al. 2014. Effect of mash maceration and ripening stage of apples on phenolic compounds and antioxidant power of cloudy juices: A study using chemometrics. LWT – Food Science and Technology 57: 223–229.

CHAPTER 18

Mathematical Modelling in Food Science through the Paradigm of Eggplant Drying

Alessandra Adrover and *Antonio Brasiello**

1. Introduction

In the field of food process engineering, scientific research and innovative industrial applications were limited by the existence of traditional processes based on consolidated procedures. Only recently, the demand for innovative processed or semi-processed products raised the awareness that a deeper understanding of the chemical and physical phenomena, happening during food transformations, is needed. The consequent requiring of ever more refined experimental and theoretical investigation techniques is the reason for the flourishing research activities.

Mathematical modelling is the essential support for experiments in order to understand the underlying physics, optimize processes (Erdogdu et al., 2017; Collins and Duffy, 2017), improve perceived quality or prolong shelf-life of foods (Shahbaz et al., 2018; Tekin et al., 2017). In particular, mathematical models are also important, in the field of food engineering, for development and analysis of industrial plants or for control systems design. In the first case, for example, they provide useful information in order to increase yields or to improve safety (Gili et al., 2018), while in the second case, for instance, they are a fundamental part of non-linear predictive control systems (Anang et al., 2017).

Depending on the specific objective sets and on the complexity of materials and transformations involved, several modelling approaches are available. Wide differences among each exist in terms of level of details and predicting capabilities, this last issue rarely being emphasized with enough strength in literature.

In this chapter, some useful modelling techniques in the field of food engineering are illustrated. Because the subject, although interesting, is too wide to be treated in a single chapter, the discussion is limited to the case study of eggplants drying, which will be used here, when needed, as a paradigm to illustrate the available techniques and to emphasize the limit of each approach.

Dipartimento di Ingegneria Chimica Materiali Ambiente, Università degli Studi di Roma "La Sapienza" & UdR INSTM di Roma, via Eudossiana 18, 00184, Roma, Italy.
 Email: alessandra.adrover@uniroma1.it
* Corresponding author: antonio.brasiello@uniroma1.it

1.1 The paradigm of eggplant drying

In many foods, the limited shelf-life constitutes a heavy drawback for commercial purpose. Eggplants represent a good example of this since they are common ingredients in food preparations, have a complex structure and high water activity (Wu et al., 2009), the last being responsible for the limited shelf-life (about 10 days at a temperature of about 10–15°C (Hu et al., 2010)). This gave rise to a wide literature about applications to preserve eggplants from deterioration.

In the last decades, several techniques have been developed in order to prolong shelf-life: The use of modified atmosphere packaging (Arvanitoyannis et al., 2005), the application of vacuum-dehydrofreezing for long-term preservation (Wu et al., 2009) and osmotic dehydration (Ahmed et al., 2016) are just few examples. Sometimes, shelf-life improvements are obtained through pre-treatment of the fresh-cut food materials, as described in Hu et al. (2010).

Drying is a well-established technology for improving the shelf-life of foods (Sagar and Suresh Kumar, 2010). It has been widely adopted since fruits and vegetables are common ingredients in dry long-life mixtures. Water removal is one of the best way to preserve foods from deterioration. High water activity can, in fact, sustain the growth of microorganisms (Ansari and Datta, 2003), promote the action of enzymes and activate oxidation reactions (Raitio et al., 2011). Moreover, it is very important that products retain some properties that define food quality, such as colour, texture, global aspect, roughness, taste and nutritional characteristics (Jayaraman and Das Gupta, 1992), after drying. It is worth noting that all these properties depend greatly on the drying conditions.

All these justify the scientific efforts towards the development of even more refined mathematical models in order to achieve the optimal process conditions for extending the shelf-life of foods. Good examples are the works of Basunia and Abe (2001) (rice), Ozdemir and Onur Devres (1999) (hazelnut), Dissa et al. (2008) (Amelie mango), Crisp and Woods (1994) (rapeseed), and Akpinar and Bicer (2005) (potato). The eggplant drying then becomes an easy way to illustrate the available modelling techniques.

1.2 From macro to micro: The spatio-temporal scales

In general, mathematical models consist of a set of algebraic/differential equations that describe the evolution of a set of state variables through which the system under study is defined. The equations have to satisfy precise conditions on the boundary of the domain of the system. A plethora of theories are available in order to take into account different aspects of the problems of interest. Each theory has specific state variables and its own equations.

A useful way to classify such theories, and related models, is from the point of view of the characteristic temporal and length scales of the state variables that they are able to describe. This is just a mere criterion, but it is enough to enlighten the specificity of different approaches. Each theory is, in fact, suitably developed to take into account some specific aspects of the physics of the phenomena happening on very specific and limited time and length scale ranges, relevant (or believed to be so) to the system analysed.

In Fig. 1, a simple scheme taken from the book Brasiello (2009), in which different modelling techniques are classified according to the spatio-temporal timescales, is shown. At the bottom, the equations of quantum mechanics describe the shortest time and length scales, in the range between femtoseconds-picoseconds and angstrom-nanometers, respectively. At the top, the Engineering Design concerns the whole process, also including, for instance, the development and optimization of whole supply chains of foods. The other three approaches bridge the gap between these two extremes and found their main applications in both the characterization of materials and the comprehension of food transformations.

Molecular mechanics deals with the atomistic level of description and, together with the mesoscale modelling (Giona et al., 2017a), has its main applications in the field of property determinations of materials (Frenkel and Smit, 1996). They are both used when one is interested in verifying some assumptions about the motion and the interactions of molecules or atoms in a food matrix. From the mechanics of the microscopic particles, it is possible to derive the values of some macroscopic variables or

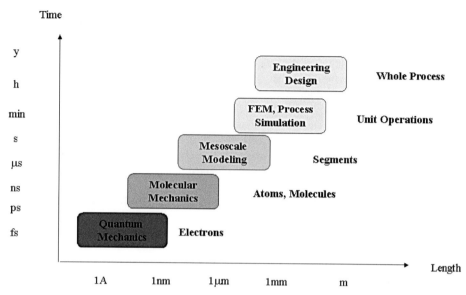

Figure 1: Temporal and spatial scale of mathematical models (Brasiello, 2009).

some exotic behaviour concerning transport mechanisms (Giona et al., 2015; Brasiello et al., 2016; Giona et al., 2016). The validation of the assumptions passes through the experimental verification of the values of such calculated macroscopic variables. The main difference between the two approaches lies in the spatio-temporal scales attainable by each of them: The typical scales of molecular mechanics range from few picoseconds to hundreds of nanoseconds and from few to tens of nanoseconds respectively, whereas, with mesoscale models, one goes from micro-scales (e.g., coarse-grained models, as in Brasiello et al. (2012), or stochastic dynamics, as in Giona et al. (2017b,c)) to the macroscopic ones (e.g., mean fields approaches, bead-spring models). The typical applications are within the study of nucleation processes, phase transitions and polymer dynamics.

When dealing with the macroscopic systems of food engineering, the previous approaches are often useless. Such systems are, in fact, made up of several Avogadro's number of interacting microscopic particles and, hence, there's no interest in modelling the detailed structure of matter. In such a case, the approach thoroughly changes: Matter is considered as a continuum characterized by state variables (i.e., mass, momentum and energy) which are punctual functions of time and space, the evolution of such variables being described by systems of ordinary or partial differential equations.

2. The Microscopic Approach: Atomistic and Coarse-grained Dynamics

Although the macroscopic models are the most commonly used for industrial applications, in the last few years, thanks to the development of computational resources (e.g., parallel and grid computing), the microscopic approaches seem to be very promising for future developments in food science, when some insights into properties of materials are needed. A very promising approach in food science is that followed by Marrink et al. (2007) and De Jong et al. (2013), who developed a model for coarse-grained dynamics of proteins and phospholipids. Analogous approach was used in the papers of Brasiello et al. (2012, 2010) and Pizzirusso et al. (2015) for dynamics and phase-transitions analysis of triglycerides.

Microscopic models are also suitable for the analysis of transport properties in complex materials, as in the paper of Maibaum and Chandler (2003) where a coarse-grained model is used to study the behavior of water confined in a hydrophobic tube.

This section briefly summarizes some basic concepts of molecular mechanics approach. These concepts are very general since they are the same for atomistic, coarse-grained and for higher scale models (e.g., bead-spring models of polymer science). In the last case, stochastic terms, mimic termal fluctuations, are suitably introduced.

2.1 Differential equations and force fields

Molecular mechanics deals with a system (i.e., an ensemble) made up of interacting particles (e.g., atoms), usually called beads, whose motion is ruled by Newton's equations:

$$\frac{d^2x_i}{dt^2} = \frac{F_{x_i}}{m_i} \tag{1}$$

in which i refers to beads. The term on the righthand side of Eq. (1) is due to interactions between the particles and it is, in general, expressed in terms of a potential (also known in literature as *force field*):

$$F_{x_i} = -\frac{\partial \mathbb{V}}{\partial x_i} \tag{2}$$

in which $\mathbb{V}$ is:

$$\mathbb{V}(\mathbf{x}) = \sum_i \mathbb{V}_1(\mathbf{x}_i) + \sum_i \sum_{j>i} \mathbb{V}_2(\mathbf{x}_i, \mathbf{x}_j) + \\ + \sum_i \sum_{j>i} \sum_{k>j>i} \mathbb{V}_3(\mathbf{x}_i, \mathbf{x}_j, \mathbf{x}_k) + ... \tag{3}$$

The first term in Eq. (3) represents the effect of external fields, including container walls effects on the system. The remaining terms properly represent particle interactions; the second term, $\mathbb{V}_2$, being the pair potentials; $\mathbb{V}_3$ the three body potentials, and so on. Moreover, in the second term $\mathbb{V}_2$ are included both bonded and non-bonded interactions.

In molecular dynamics, molecules are considered to be made up of beads which are linked together through potentials defining bonds, bond angles and torsional angles oscillating around equilibrium values. Moreover, suitable non-bonded potentials are also introduced in order to take into account electrostatic interactions as well as dipole-dipole and Van Der Waals interactions.

The most simple choice for a bond potential is to use a Hooke's law:

$$\mathbb{V}(l) = \frac{k}{2}(l - l_0)^2 \tag{4}$$

l_0 being the reference value of the bond.

Sometimes, in coarse-grained simulations, the more sophisticated FENE (Finitely Extendible Non-linear Elastic) potential is used:

$$\mathbb{V}(l) = -\frac{k}{2}l_0 \log\left(1 - \frac{l^2}{l_0^2}\right) \tag{5}$$

At a short distance, the potential asymptotically goes to harmonic potential with force constant k, while it diverges at distance l_0.

The deviation of an angle (three-body bonded interaction) formed by three contiguous atoms from the reference value is also frequently described using Hooke's law:

$$\mathbb{V}(\theta) = \frac{k_\theta}{2}(\theta - \theta_0)^2 \tag{6}$$

where k_θ is the force constant and θ_0 is the reference value. In general, the force constant is smaller than that of bond stretching in Eq. (4). This means that less energy is required to distort an angle away from equilibrium than to stretch or compress a (chemical) bond.

Torsional potentials are instead often expressed as a cosine series expansion. A functional form is, for instance, the following:

$$\mathbb{V}(\psi) = \sum_n \frac{k_{\psi,n}}{2}\left(1 + \cos(n\psi - \psi_0)\right) \tag{7}$$

$k_{\phi,n}$ influences the relative barriers to rotation; n is the *multiplicity*, i.e., the number of minimum points in the function as the bond is rotated through 360°; ϕ_0 is the phase factor which determines when the torsional angle passes through the minimum value.

Non-bonded interactions are *through-space* interactions, i.e., interactions between non-bonded particles (read beads) inside the same molecule or owing to different molecules. The non-bonded terms usually considered are:

- Coulomb potentials to take into account electrostatic interactions:

$$\mathbb{V} = \frac{qQ}{4\pi\varepsilon_0 l} \tag{8}$$

  q and Q being the charges of the two interacting beads.

- Lennard-Jones 12-6 potentials to take into account Van der Waals interactions:

$$\mathbb{V} = 4\varepsilon\left[\left(\frac{\sigma}{l}\right)^{12} - \left(\frac{\sigma}{l}\right)^6\right] \tag{9}$$

  in which σ is the collision diameter (the separation for which the energy is zero) and ε is the well depth.

2.2 Periodic boundary conditions

In a molecular dynamics simulation, great attention is directed to the treatment of boundary conditions since *macroscopic* properties have to be calculated from simulations of relatively small numbers of beads. This means that the number of beads interacting with the boundary is comparable with that in the bulk.

Periodic boundary conditions are used in order to simulate ensemble of beads as it was taken from the bulk of the material. With periodic boundary conditions it is possible to do this by using a relatively small number of beads. To understand how periodic boundary conditions work, one can imagine replicating the box containing the ensemble of beads under study in all directions to give a periodic array. If a bead goes out from the central box during the simulation then it is replaced by its image entering from the opposite side, the number of particles within the central box remaining constant. It is easy to understand that periodic boundary conditions have some drawbacks which have to be carefully taken into account to avoid serious mistakes. In-depth information can be found in Allen and Tildesley (1989) and Frenkel and Smit (1996).

3. The Macroscopic Approach: Kinetic Models and Continuum Mechanics

Food processes often deal with the motion of fluids or/and physicochemical transformations in a macroscopic volume of matter. In practice, for the description of the evolution of state variables, two different kind of macroscopic approaches are possible: The choice between them depends on the availability of information about the physical and chemical laws behind the phenomena involved. When such information lacks at all, as well as when it is very poor, one deals with the empirical models in which the evolution in time (t) of the set of state variables (**A**) is described through kinetic equations, in which the parameters ($\mathbf{P}_A$) of the process also appear, namely:

$$\frac{d\mathbf{A}}{dt} = \mathbf{f}\,(\mathbf{A}, \mathbf{P}_A, t) \tag{10}$$

where **f** is a vectorial function whose expression, for what said, has to be inferred from experimental data.

The empirical models are very common in food engineering due to the complexity of the processes. They are easy to deal with, but do not provide any physical knowledge enhancement. However, their use is not suitable for extrapolation in parameters' regions that are far from experiments, which is, in our opinion, the main drawback.

The most widespread empirical models in literature (e.g., Elhussein and Şahin, 2018; Ramos et al., 2017) are, for instance, those of Newton ($M = exp(-kt)$), Page ($M = exp(-kt)^n$), Henderson and Pabis ($M = a \cdot exp(-kt)^n$), Weibull ($M = a - bexp(-kt)^n$), Wang and Singh ($M = 1 + at + bt^2$) (Akpinar and Bicer, 2005; Ertekin and Firat, 2017). Many applications are found in literature where they are adopted to describe many aspects of drying processes (Khallou et al., 2009, 2010). Being very general, they are easily and successfully used for many foodstuffs (Onwude et al., 2016), such as, for instance, in the works of Doymaz (2011) and Ertekin and Yaldiz (2004) (eggplants), Hernando et al. (2008) (mushrooms), Koc et al. (2008) (quince), Heredia et al. (2007) (cherry tomato), Togrul and Pehlivan (2003) (apricot), and Panchariya et al. (2002) (black tea). Most of these models are modified exponential functions which, as can be seen in the following, is also the functional form of the solution of the diffusion equation (see Eq. 29). Roughly speaking, they mimic the solutions of such equation.

In Fig. 2, for instance, the comparison between curves derived from a kinetic model $dA/dt = k_A A^n$ and experimental Toughness data for green asparagus during cold-storage is reported (Albanese et al., 2007). Model parameters k_A and n have been determined through a non-linear regression procedure.

In Fig. 3, the curve predicting the chlorophyll content evolution is also reported. The form of the kinetic equation is the same for the two cases. It is worth noting that the same functional form is able to predict very different trends by simply tuning its parameters k_A and n. Parameters values are useful also to quantify the effect of different treatments on the observed variables.

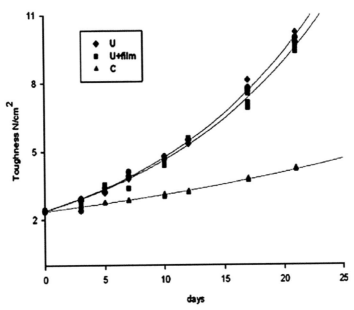

Figure 2: Non-linear regression curves of the kinetic model $dA/dt = k_A A^n$ for the increase in time of toughness for green asparagus samples during cold-storage (U, untreated samples; U + film, packaged samples; C, dipped and packaged samples) (Source: Albanese et al., 2007).

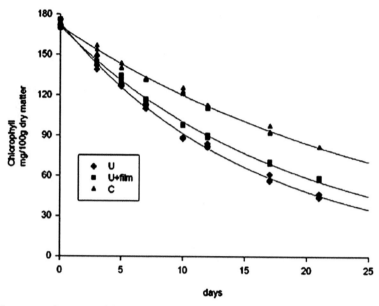

Figure 3: Non-linear regression curves of the kinetic model $dA/dt = k_A A^n$ for the decrease in time of Chlorophyll Content for green asparagus samples during cold-storage (U, untreated samples; U + film, packaged samples; C, dipped and packaged samples) (Source: Albanese et al. (2007)).

When some information about physics is available or when some hypotheses about physics have to be verified, one deals with theoretical models which are based on the balance equations of the continuous mechanics.

In the continuous mechanics approach, the system under study is regarded as a continuous medium, thus disregarding the discrete nature of matter (Landau and Lifshitz, 1959). This means that an element of a system, although infinitesimal, is in any case very large when compared to the molecular distances. In this framework, the variables of the system can be defined as functions of time t and space $\mathbf{P} = (x, y, z)$.

The evolution of the system's variables is governed by the following general balance equations which can refer to the whole system under study as well as to a subsystem (for a complete discussion of the issue see Bird et al., 2002):

$$\frac{\partial \rho}{\partial t} = -(\nabla \cdot \rho \mathbf{v}) \tag{11}$$

$$\frac{\partial \rho \mathbf{v}}{\partial t} = -(\nabla \cdot \rho \mathbf{v}\mathbf{v}) - \nabla p - (\nabla \cdot \tau) + \rho \mathbf{g} \tag{12}$$

$$\rho \frac{DU}{Dt} = -\nabla \cdot \mathbf{q} - p(\nabla \cdot \mathbf{v}) - \tau : \nabla \mathbf{v} + G \tag{13}$$

in which D/Dt is the substantial derivative which is equal to $\partial/\partial t + \mathbf{v} \cdot \nabla$.

Equations (11–13) are mass, momentum and internal energy balances, respectively. They refer to a well-defined volume (V) which represents the system under study. Outside the system, is the environment which is linked to it through suitable equations at the boundary (Ω). The term G is a generic net generation term which takes into account contributions different from those related to conduction, compression or viscous dissipation just taken into account with the terms at the righthand side. For instance, it can be due to electromagnetic fields, such as in microwave oven drying (e.g., Zhu et al., 2015).

In the case of multicomponent systems, Eq. (11) is the sum of the following continuity equations written for each of the component i:

$$\frac{\partial \rho_i}{\partial t} = -(\nabla \cdot \mathbf{n}_i) + \mathcal{R}_i \tag{14}$$

in which $\mathbf{n}_i = \rho_i \mathbf{v} + \mathbf{j}_i$ is the mass flux of the component i ($\mathbf{j}_i$ being the mass flow due to diffusion) and $\mathcal{R}_i$ is the reaction rate for i, where it is obvious that $\sum_i \mathbf{n}_i = \rho \mathbf{v}$ and $\sum_i \mathcal{R}_i = 0$. Moreover, for multicomponent systems Eqs. (12) and (13) can be rewritten as follows:

$$\frac{\partial \rho \mathbf{v}}{\partial t} = -(\nabla \cdot \rho \mathbf{v} \mathbf{v}) - \nabla p - (\nabla \cdot \tau) + \sum_i \rho_i \mathbf{g}_i \tag{15}$$

$$\rho \frac{DU}{Dt} = -\nabla \cdot \mathbf{q} - p(\nabla \cdot \mathbf{v}) - \tau : \nabla \mathbf{v} + \sum_i (\mathbf{j}_i \cdot \mathbf{g}_i) + G \tag{16}$$

Equations (11–13) as well as Eqs. (14–16) are written here in the most general form and were reported here only for scholarly interest, since they are completely useless as they are. They need constitutive equations in order to define internal energy U, conduction heat flux $\mathbf{q}$, mass fluxes $\mathbf{j}_i$ and stress tensor τ as functions of either measurable state variables (e.g., temperature, mass concentration) or kinematic variables (positions and velocities) as well as an equation of state $p = p(\rho, \rho_i, T)$. This means that, in order to be able to solve them, one has to specify thermodynamics as well as to set the proper mechanical properties of the materials. This set of equations will be now particularized for eggplants drying.

The main difficulty when dealing with this approach is the computation of the solutions, since analytical resolution of Eqs. (11–14) is possible only in very few cases. In all the other cases, the computational burden to solve them is often very high and entrusted to calculators. However, in a lot of practical cases, they can be deeply simplified.

3.1 The diffusion equation

In the paradigm of eggplant drying, the system volume represents an eggplant sample.

The simplest model derives from Eqs. (14–16) where the following assumptions are enforced:

- the volume of the system is constant;
- the system is homogeneous and isotropic;
- system velocity $\mathbf{v}$ is null;
- fluid is inviscid;
- thermal transient is negligible;
- conduction heat flux $\mathbf{q}$ is null;
- generation term in energy balance is null;
- diffusion follows Fick's law;
- density is constant;

namely:

$$\frac{\partial \rho_w}{\partial t} = D\nabla^2 \rho_w \tag{17}$$

Equations (12) and (13) being identities. It is worth noting that, although food matrix is, as in this and in many other cases, very complex and, therefore, very far from being homogeneous and isotropic, the two assumptions are often both enforced in order to streamline the model as well as the computations. Equation (17) is known as *Fick's second law of diffusion* or sometimes simply as *Diffusion Equation* and it was solved for many different cases (Two of the most exhaustive books about analytical and numerical resolutions of the diffusion equation in many cases are those of Crank (1975) and Carslaw and Jaeger (1959)).

Sometimes, it is necessary to relax the assumptions of negligible thermal transient as well as of null conduction heat flux (Llave et al., 2016). This happens, for instance, when drying processes suffer changes in environmental conditions (e.g., in the case of natural air drying) or when changes in temperature are imposed in order to improve performance (Stormo et al., 2017). In such cases, the following equation, which can be derived from Eq. (16) (see Bird et al., 2002) is added:

$$\rho c_p \frac{\partial T}{\partial t} = k \nabla^2 T \tag{18}$$

in which the case of energy generation term due, for instance, to a microwave heating is excluded.

It is important to underline that the system of Eqs. (17–18) represents the simplest set of equations for the description of a non-isothermal material transport process in solid foods and, in this sense, is very general.

3.2 A standard method of solution for the diffusion equation when the diffusion coefficient is constant

Equation (17) is formally equal to Eq. (18). Sometimes, in literature this fact is named as *analogy between mass and heat transfer*. Apart from physical implications, this means that the same mathematical methods for solution of mass transport can be also applied for heat transport.

In this section, the standard resolution of the diffusion equation (Eq. (17)) is discussed. The reason that this equation is widely used as the first approximation in a lot of applications is then justified.

In particular here one deals, for simplicity, with the mono-dimensional case which, for instance, describes the water diffusion in the case of a food slice, namely when one dimension of the sample is much lower than the others. In this case, Eq. (17) becomes:

$$\frac{\partial \rho_w}{\partial t} = D \frac{\partial^2 \rho_w}{\partial x^2} \tag{19}$$

Initial and boundary conditions are set, for instance, as follows:

$$t = 0 \quad 0 < x < L \quad \rho_w = \rho_{w,0} \tag{20}$$

$$x = 0 \quad x = L \quad t > 0 \quad \rho w = \rho_{w,eq} \tag{21}$$

The standard method is based on the assumption that the solution of Eq. (19) is of type $\rho_w(x, t) = \mathbb{X}(x)\mathbb{T}(t)$ which, by substitution in Eq. (19), gives:

$$\frac{1}{\mathbb{T}} \frac{\partial \mathbb{T}}{\partial t} = D \frac{1}{\mathbb{X}} \frac{\partial^2 \mathbb{X}}{\partial x^2} \tag{22}$$

the left-hand side depending on t only, while the right-hand side depends on x only. Hence, both have to be equal to the same constant which is conveniently taken as $-\lambda^2 D$, namely:

$$\frac{1}{\mathbb{T}} \frac{\partial \mathbb{T}}{\partial t} = -\lambda^2 D \tag{23}$$

$$\frac{1}{\mathbb{X}} \frac{\partial^2 \mathbb{X}}{\partial x^2} = -\lambda^2 \tag{24}$$

The solutions of Eqs. (23) and (24) are, respectively:

$$\mathbb{T} = e^{-\lambda^2 Dt} \tag{25}$$

$$\mathbb{X} = a_1 sin\lambda x + b_1 cos\lambda x \tag{26}$$

Hence, the most general solution of Eq. (19) is the following:

$$\rho_w = \sum_{m=1}^{\infty}(a_m \, sin\lambda_m x + b_m \, cos\lambda_m x)e^{-\lambda^2 Dt} \tag{27}$$

The boundary conditions (Eq. (21)) impose that $b_m = 0$ and $\lambda_m = m\pi/L$ and, hence, initial condition (Eq. (20)) becomes:

$$\rho_{w,0} = \sum_{m=1}^{\infty} a_m \, sin\left(\frac{m\pi}{L}x\right) \tag{28}$$

By multiplying both sides of Eq. (28) by $sin(p\pi x/L)$ and integrating from 0 to L, one finds that $a_m = 4\rho_{w,0}/(m\pi)$ with $m = 1, 3, 5...$. Then, the solution of the problem Eq. (19) with Eqs. (20) and (21) is:

$$\rho_w = \frac{4\rho_{w,0}}{\pi} \sum_{m=1}^{\infty} \frac{1}{m}\left(sin\frac{m\pi x}{L}\right)exp\left(\frac{-Dm^2\pi^2t}{L^2}\right) \quad m = 1, 3, 5... \tag{29}$$

3.3 Modelling shrinkage by using modified diffusion equations

Air-drying is often followed by physical and chemical modifications of food materials which can deeply modify the mass transport mechanisms (Senadeera, 2008). One of the most important consequences of these changes, which reduces the perceived quality as well as the rehydration capability, is shrinkage (Jayaraman et al., 1990; McMinn and Magee, 1997a,b). It is a volume reduction, coupled with shape and porosity changes and hardness increase. When shrinkage happens, the pure Fickian model (Eq. 17), unable to take into account phenomena of volume changes, no longer holds, as can be seen in Fig. 4. Therefore, more refined model is needed.

Since shrinkage is connected to water fluxes, an effective diffusion coefficient, depending on the water content, can be introduced in order to take into account the effects of shrinkage (e.g., Brasiello et

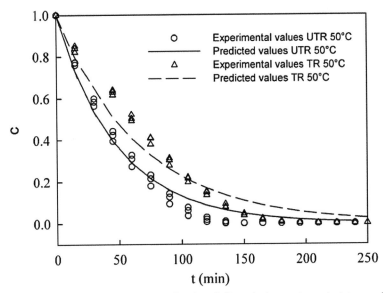

Figure 4: Comparison between experimental and predicted total dimensionless water content temporal profiles during eggplant drying; predicted values are calculated from Eq. (17). The discrepancies between the two sets of data are clearly shown.

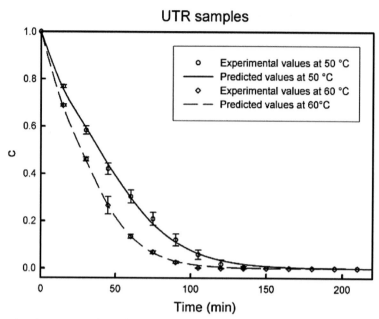

Figure 5: Comparison between experimental and predicted total dimensionless water content temporal profiles during eggplant drying; predicted values are calculated from Eq. (30).

al., 2011). This is one of the easiest choices, derived from the fact that shrinkage modifies food structure and, hence, greatly influences the diffusion coefficient, namely:

$$\frac{\partial \rho_w}{\partial t} = \nabla \cdot (D\,(\rho_w) \nabla \rho_w) \tag{30}$$

Here, the system volume is kept constant; shrinkage is taken into account only through its effect on water transport. This sometimes can be enough for an adequate description of the process (in particular, when the total volume reduction is moderate). In Fig. 5, a comparison between experimental and predicted total dimensionless water content temporal profiles during eggplant drying is reported. Predicted values are obtained through Eq. (30), providing better results with respect to Eq. (17) (see Fig. 4).

A possible alternative choice is the introduction of a fictitious convective term into Eq. (17) in order to mimic the effect of volume reduction on transport phenomena (Carslaw and Jaeger, 1959). Hence, the convection-diffusion equation as follows:

$$\frac{\partial \rho_w}{\partial t} + (\mathbf{v} \cdot \nabla \rho_w) = D_w \nabla^2 \rho_w \tag{31}$$

The fictitious velocity $\mathbf{v}$ in Eq. (31) should be derived from the momentum balance (Eq. (12)) and, hence, from a suitable assumption for the stress tensor. However, for our purposes, it is easiest to either impose the velocity from outside or to introduce it as a function of water content.

In the paper of Brasiello et al. (2013), these two approaches were analyzed. In the first case in particular, a dependence of the diffusion coefficient $D(\rho_w) = A + B * \rho_w$ is introduced, while in the second case the functional form $\mathbf{v} = c \nabla \rho_w$ for the velocity is adopted, both providing good results. It is worth noting that they also demonstrate that, by using suitable boundary conditions, the two approaches are equivalent. Moreover, in the paper by Brasiello et al. (2017), such approaches are used to model the drying process of eggplant samples with quite good results in terms of prediction of the evolution of the water content profiles obtained through Magnetic Resonance Imaging (MRI).

Obviously, the limit of such approaches is from the point of view of the prediction of volume reduction in time because they only allow an overall estimation due to the fixed boundary approximation.

3.4 Modelling shrinkage with diffusion equations with moving boundary

Another possible way to have a precise prediction of both the evolution of water content and the system volume reduction is provided by using a moving boundary approach (Adrover et al., 2019b,a). This means to introduce a further equation for the motion of the boundary, together with a further constitutive equation describing the law governing shrinkage and how this phenomenon is connected to the punctual values of the state variables.

In order to do this, the starting point is the generic convection-diffusion equation (Eq. (31)) which must be rewritten in terms of volume fraction:

$$\frac{\partial \phi}{\partial t} + \nabla \cdot (\phi \mathbf{v}) = D \nabla^2 \phi \tag{32}$$

The convective velocity $\mathbf{v}$ here is a shrinkage velocity. In other words, the local flow of water, in every point $\mathbf{P}$ of the domain, is assumed to be purely diffusive and equal to $-D\nabla\phi$ if observed from a reference frame with the origin at $\mathbf{P}$ and moving with velocity $\mathbf{v}$.

The motion of the boundary is given through the proper choice of a functional form for $\mathbf{v}$, together with a differential equation describing the evolution of the boundary itself.

All the functional forms for $\mathbf{v}$, consistent with the hypothesis made, are obtained by considering the whole balance over the volume V (changing in time). Thus, the integration of both sides of Eq. (32) gives:

$$\int_V \frac{\partial \phi}{\partial t} \, dV = D \int_V \nabla^2 \phi \, dV - \int_V \nabla \cdot (\phi \mathbf{v}) \, dV \tag{33}$$

which, after applying both the Leibniz integral rule and Gauss Theorem, becomes:

$$\frac{\partial}{\partial t} \int_V \phi \, dV = D \int_\Omega \nabla \phi \cdot \mathbf{n} \, d\Omega - \int_\Omega \phi (\mathbf{v} - \mathbf{v}_\Omega) \cdot \mathbf{n} \, d\Omega \tag{34}$$

in which the boundary velocity $\mathbf{v}_\Omega$ is introduced.

The hypothesis of purely diffusive local flow implies that the second term on the right-hand side of Eq. (34) is null, from which the following equation describing the evolution of the boundary derives:

$$\mathbf{v}_\Omega \equiv \frac{d\Omega}{dt} = \mathbf{v}|_\Omega \tag{35}$$

Equation (35) only depends on the shrinkage velocity of the food matrix calculated at the boundary and, in this sense, it is very general. A constitutive equation for $\mathbf{v}$, specific for the class of materials under investigation, is thus needed both in Eq. (32) and in Eq. (35). For example, a very simple choice could be the hypothesis of constant shrinkage velocity $\mathbf{v}$, leading to the decoupling of Eq. (35) from Eq. (32) (see Aprajeeta et al., 2015). However, because in many cases experimental observations suggest a dependence of the boundary evolution from the water content evolution (i.e., flow), in literature (Papanu et al., 1989; Adrover and Nobili, 2015), the shrinkage velocity is often found to be linked to diffusive flow, as follows:

$$\mathbf{v} = -\mathbf{J} \tag{36}$$

By substituting Eq. (36) into Eqs. (32) and (35), one obtains:

$$\frac{\partial \phi}{\partial t} + D \nabla \phi \cdot \nabla \phi = D(1 - \phi) \nabla^2 \phi \tag{37}$$

$$\frac{\partial \Omega}{\partial t} = D \nabla \phi \tag{38}$$

which define the problem.

The required boundary conditions are given by the following relation:

$$-D\nabla\phi|_{\Omega} = h\left(\phi|_{\Omega} - \phi_{eq}\right) \tag{39}$$

The mathematical model can be particularized when dealing with regular geometries. As an example of a cylindrical sample, one can write Eqs. (37–39) as follows:

$$\frac{\partial\phi}{\partial t} + D\left[\left(\frac{\partial\phi}{\partial r}\right)^2 + \left(\frac{\partial\phi}{\partial z}\right)^2\right] = \frac{D}{r}\frac{\partial\phi}{\partial r} + D(1-\phi)\left[\frac{\partial^2\phi}{\partial r^2} + \frac{\partial^2\phi}{\partial z^2}\right] \tag{40}$$

$$\frac{\partial L}{\partial t} = D\frac{\partial\phi}{\partial z}\bigg|_{L} \tag{41}$$

$$\frac{\partial R}{\partial t} = D\frac{\partial\phi}{\partial r}\bigg|_{R} \tag{42}$$

$$-D\frac{\partial\phi}{\partial z}\bigg|_{L} = h\left(\phi|_{L} - \phi_{eq}\right) \tag{43}$$

$$\cdot D\frac{\partial\phi}{\partial z}\bigg|_{0} = 0 \tag{44}$$

$$-D\frac{\partial\phi}{\partial r}\bigg|_{R} = h\left(\phi|_{R} - \phi_{eq}\right) \tag{45}$$

$$D\frac{\partial\phi}{\partial r}\bigg|_{0} = 0 \tag{46}$$

where r and z are the spatial coordinates, L is half of the dimension of the thickness, and R is the radius of the slab whose initial values are R_0 and L_0, respectively. Boundary conditions Eqs. (44) and (46) impose symmetric profiles.

When a dimension of a sample is much lower than the others, a thin layer approximation can be adopted: The sample is approximated to an infinite slab with finite thickness and, hence, mass transport is considered as evolving along thickness only. This heavily reduces the computational burden. Hence, in Eqs. (40–46), when the radius R is much greater than L, the water transport can be considered as evolving along z direction only. The model is simplified as follows:

$$\frac{\partial\phi}{\partial t} + D\left(\frac{\partial\phi}{\partial x}\right)^2 = D(1-\phi)\frac{\partial^2\phi}{\partial x^2} \tag{47}$$

$$-D\frac{\partial\phi}{\partial t}\bigg|_{L} = H(\phi|_{L} - \psi_{eq}) \tag{48}$$

$$-\frac{\partial\phi}{\partial t}\bigg|_{0} = 0 \tag{49}$$

$$\frac{\partial L}{\partial t} = D\frac{\partial\phi}{\partial x}\bigg|_{L} \tag{50}$$

In the case of a moving boundary, the thin layer approximation could not be satisfied for the whole process duration. This happens when, due to volume reduction, at least one dimension becomes comparable to the thickness of the sample during drying. In this case, the thin layer approximation could be source of error whose magnitude depends on the process itself. It could happen that, although the approximation is no longer valid, the evolution of the process and of water content values in the sample are such that it does not cause excessive error. However, this can not be established *a priori* due to computational difficulties and could represent a serious issue.

One way of simplifying the resolution of Eqs. (47–50) is, for instance to fix the moving front through the suitable choice of a new space coordinate. If one defines $\xi = x/L(t)$, boundary conditions Eq. (48) is fixed at $\xi = 1$, while the equation of the evolution of the boundary (Eq. 50) becomes:

$$\frac{\partial L}{\partial t} = \frac{D}{L} \frac{\partial \phi}{\partial \xi} \bigg|_1 \tag{51}$$

Moreover, Eq. (47) can be written as:

$$\frac{\partial \phi}{\partial t} + (1-\xi)\frac{D}{L^2}\left(\frac{\partial \phi}{\partial \xi}\right)^2 = \frac{D}{L^2}(1-\phi)\frac{\partial^2 \phi}{\partial \xi^2} \tag{52}$$

The problem Eqs. (47–50) can be transformed into a problem with fixed boundary with a transport coefficient D/L^2 (Eq. (52)) depending on L which evolves in time according to Eq. (51) which is a fixed boundary equation in ξ coordinate.

The theoretical models discussed in this chapter, apart from the complications of the last cases due to the treatment of moving boundary, are quite simple from the physical point of view.

For all the analyzed cases, solutions of the model can be, in fact, represented by points in a continuum domain, with regular smooth shape, to which a vector of state variables, such as water content, temperature and so on, is associated. It is worth noting that system velocities are imposed from the outside instead of being derived from the momentum equation.

More complex cases, for instance, are those in which materials are highly non-homogeneous or anisotropic, and transformations occur during the process. In these cases, more detailed models could be proposed, thanks to the balance equations of continuum mechanics (Eqs. (14–16)) together with suitable constitutive equations. A wide literature about mathematical modelling of transport phenomena in complex materials exists. However, even though this is a usual and well-established procedure in other fields, in the field of food process, this could be a losing approach in many cases. This because food materials are sometimes too complex to allow the well-refined experiments in order to find out models parameters. This points to the paradox that, in general, too complex cases force one to use very simple models, such as the empirical ones already discussed in a preceding section.

4. Conclusions

In this chapter, a brief overview of the available techniques for mathematical modelling in food processes was shown.

The aim was to provide basic guidelines for the synergistic use of the various available modelling techniques, as is already happening in other research and development areas. This has been very limited so far, in the context of food science, by the complexity of both materials and their transformation/ production processes.

Mathematical models were here divided according to the specific level of detail, namely, their characteristic time and length scale, in order to highlight the fact that each technique refers to a specific field of application. In the case of the microscopic approach (e.g., Molecular Dynamics), in fact, the aim is to relate macroscopic phenomenology of a material to a precise particle mechanics (as simple as possible). This approach, in a special way, is essential in the field of food science in which the complexity of the involved materials causes the occurrence of exotic phenomena.

The general equation of Molecular Dynamics was presented, together with the key concepts of molecular dynamics, namely, force fields and periodic boundary conditions.

Macroscopic modelling was introduced using the case of eggplant drying to illustrate some of the most used and most simple models of the continuum mechanics. Starting from the *Fick's second law of diffusion* (a simple analytical resolution technique is also reported), more complex models were discussed. In particular, the attention is focused on models able to take into account structural modification during process evolution, namely, volume variations and shrinkage that occur during drying.

Nomenclature

Used notations are reported in the following:

Nomenclature	
a_m	integration constants
$\mathbf{A}$	kinetic variable
b_m	integration constants
c_p	specific heat at constant pressure
D	diffusion coefficient
$\mathbf{g}$	gravity acceleration
G	generation term
H	overall transport coefficient
j	mass diffusion flow
k	thermal transport coefficient
l	distance
L	half height of the cylindrical domain
m	mass
n	mass flux
p	pressure
$\mathbf{P}$	parameters
q	conduction energy flow
r	radial coordinate
$\mathcal{R}$	reaction rate
R	radius of the cylindrical domain
t	time
T	temperature
U	internal energy
$\mathbf{v}$	velocity
V	volume of a spatial domain
$\mathbb{V}$	potential
$\mathbf{x}$	spatial vector
x	spatial coordinate
$\mathbf{z}$	spatial coordinate
θ	angle
λ	integration constant
ξ	spatial coordinate
ϕ	volume fraction
ρ	mass concentration
$\boldsymbol{\tau}$	stress tensor
ψ	torsion
ω	boundary of a spatial domain

Subscript	
0	initial value
∞	final value
eq	equilibrium
w	water

References

Adrover, A. and Nobili, M. 2015. Release kinetics from oral thin films: Theory and experiments. Chemical Engineering Research and Design 98: 188–201.

Adrover, A., Brasiello, A. and Ponso, G. 2019a. A moving boundary model for food isothermal drying and shrinkage: A shortcut numerical method for estimating the shrinkage factor. Journal of Food Engineering 244: 212–219.

Adrover, A., Brasiello, A. and Ponso, G. 2019b. A moving boundary model for food isothermal drying and shrinkage: General setting. Journal of Food Engineering 244: 178–191.

Akpinar, E.K. and Bicer, Y. 2005. Modelling of the drying of eggplants in thin-layers. International Journal of Food Science and Technology 40(3): 273–281.

Albanese, D., Russo, L., Cinquanta, L., Brasiello, A. and Di Matteo, M. 2007. Physical and chemical changes in minimally processed green asparagus during cold-storage. Food Chemistry 101(1): 274–280.

Allen, M.P. and Tildesley, D.J. 1989. Oxford University Press. John Wiley & Sons, Inc., Oxford, UK.

Anang, Hadisupadmo, S. and Leksono, E. 2017. Model predictive control design and performance analysis of a pasteurization process plant. Proceedings of the 2016 International Conference on Instrumentation, Control, and Automation, ICA 2016: 81–87.

Ansari, I.A. and Datta, A.K. 2003. An overview of sterilization methods for packaging materials used in aseptic packaging systems. Food and Bioproducts Processing 81(1): 57–65.

Aprajeeta, J., Gopirajah, R. and Anandharamakrishnan, C. 2015. Shrinkage and porosity effects on heat and mass transfer during potato drying. Journal of Food Engineering 144: 119–128.

Arvanitoyannis, I.S., Khah, E.M., Christakou, E.C. and Bietsos, F.A. 2005. Effect of grafting and modified atmosphere packaging on eggplant quality parameters during storage. International Journal of Food Science and Technology 40(3): 311–322.

Basunia, M.A. and Abe, T. 2001. Thin-layer solar drying characteristics of rough rice under natural convection. Journal of Food Engineering 47(4): 295–301.

Bird, R.B., Stewart, W.E. and Lightfoot, E.N. 2002. Transport Phenomena. Second Edition. John Wiley & Sons, Inc., New York, USA.

Brasiello, A. 2009. Molecular dynamics of triglycerides. Atomistic and coarse-grained approaches. Fridericiana Editrice Universitaria, Napoli, Italy.

Brasiello, A., Russo, L., Siettos, C., Milano, G. and Crescitelli, S. 2010. Multi-scale modelling and coarse-grained analysis of triglycerides dynamics. Computer Aided Chemical Engineering 28(C): 625–630.

Brasiello, A., Crescitelli, S., Adiletta, G., Di Matteo, M. and Albanese, D. 2011. Mathematical model with shrinkage of an eggplant drying process. Chemical Engineering Transactions 24: 451–456.

Brasiello, A., Crescitelli, S. and Milano, G. 2012. A multiscale approach to triglycerides simulations: From atomistic to coarse-grained models and back. Faraday Discussions 158: 479–492.

Brasiello, A., Adiletta, G., Russo, P., Crescitelli, S., Albanese, D. and Di Matteo, M. 2013. Mathematical modeling of eggplant drying: Shrinkage effect. Journal of Food Engineering 114(1): 99–105.

Brasiello, A., Crescitelli, S. and Giona, M. 2016. One-dimensional hyperbolic transport: Positivity and admissible boundary conditions derived from the wave formulation. Physica A: Statistical Mechanics and its Applications 449: 176–191.

Brasiello, A., Iannone, G., Adiletta, G., De Pasquale, S., Russo, P. and Di Matteo, M. 2017. Mathematical model for dehydration and shrinkage: Prediction of eggplant's MRI spatial profiles. Journal of Food Engineering 203: 1–5.

Carslaw, H.S. and Jaeger, J.C. 1959. Conduction of Heat in Solids. Second Edition. Oxford University Press, Oxford, UK.

Collins, O.C. and Duffy, K.J. 2017. Optimal control of foliar disease dynamics for multiple maize varieties. Acta Agriculturae Scandinavica Section B: Soil and Plant Science, 1–12.

Crank, J. 1975. The Mathematics of Diffusion. Oxford University Press, Oxford, UK.

Crisp, J. and Woods, J.L. 1994. The drying properties of rapeseed. Journal of Agricultural Engineering Research 57(2): 89–97.

De Jong, D.H., Singh, G., Bennett, W.F.D., Arnarez, C., Wassenaar, T.A., Schäfer, L.V. et al. 2013. Improved parameters for the martini coarse-grained protein force field. Journal of Chemical Theory and Computation 9(1): 687–697.

Dissa, A.O., Desmorieux, H., Bathiebo, J. and Koulidiati, J. 2008. Convective drying characteristics of amelie mango (mangifera indica l. cv. 'amelie') with correction for shrinkage. Journal of Food Engineering 88(4): 429–437.

Doymaz, I. 2011. Drying of eggplant slices in thin layers at different air temperatures. Journal of Food Processing and Preservation 35(2): 280–289.

Elhussein, E.A.A. and Şahin, S. 2018. Drying behaviour, effective diffusivity and energy of activation of olive leaves dried by microwave, vacuum and oven drying methods. Heat and Mass Transfer/Waerme- und Stoffuebertragung, 1–11.

Erdogdu, F., Altin, O., Marra, F. and Bedane, T.F. 2017. A computational study to design process conditions in industrial radio-frequency tempering/thawing process. Journal of Food Engineering 213: 99–112.

Ertekin, C. and Yaldiz, O. 2004. Drying of eggplant and selection of a suitable thin layer drying model. Journal of Food Engineering 63(3): 349–359.

Ertekin, C. and Firat, M.Z. 2017. A comprehensive review of thin-layer drying models used in agricultural products. Critical Reviews in Food Science and Nutrition 57(4): 701–717.

Frenkel, D. and Smit, B. 1996. Understanding Molecular Simulation. From Algorithm to Applications. Academic Press, San Diego, CA, USA.

Gili, R.D., Torrez Irigoyen, R.M., Penci, M.C., Giner, S.A. and Ribotta, P.D. 2018. Wheat germ thermal treatment in fluidised bed experimental study and mathematical modelling of the heat and mass transfer. Journal of Food Engineering 221: 11–19.

Giona, M., Brasiello, A. and Crescitelli, S. 2015. Ergodicity-breaking bifurcations and tunneling in hyperbolic transport models. EPL 112(3).

Giona, M., Brasiello, A. and Crescitelli, S. 2016. On the influence of reflective boundary conditions on the statistics of poisson-kac diffusion processes. Physica A: Statistical Mechanics and its Applications 450: 148–164.

Giona, M., Brasiello, A. and Crescitelli, S. 2017a. Stochastic foundations of undulatory transport phenomena: Generalized poisson-kac processes—Part I basic theory. Journal of Physics A: Mathematical and Theoretical 50(33).

Giona, M., Brasiello, A. and Crescitelli, S. 2017b. Stochastic foundations of undulatory transport phenomena: Generalized poisson-kac processes—Part II irreversibility, norms and entropies. Journal of Physics A: Mathematical and Theoretical 50(33).

Giona, M., Brasiello, A. and Crescitelli, S. 2017c. Stochastic foundations of undulatory transport phenomena: Generalized poisson-kac processes—Part III extensions and applications to kinetic theory and transport. Journal of Physics A: Mathematical and Theoretical 50(33).

Heredia, A., Barrera, C. and Andrès, A. 2007. Drying of cherry tomato by a combination of different dehydration techniques. Comparison of kinetics and other related properties. Journal of Food Engineering 80(1): 111–118.

Hernando, I., Sanjuàn, N., Pèrez-Munuera, I. and Mulet, A. 2008. Rehydration of freeze-dried and convective dried boletus edulis mushrooms: Effect on some quality parameters. Journal of Food Science 73(8): E356–E362.

Hu, W., Jiang, A., Tian, M., Liu, C. and Wang, Y. 2010. Effect of ethanol treatment on physiological and quality attributes of fresh-cut eggplant. Journal of the Science of Food and Agriculture 90(8): 1323–1326.

Jayaraman, K.S., Das Gupta, D.K. and Rao, N.B. 1990. Effect of pretreatment with salt and sucrose on the quality and stability of dehydrated cauliflower. International Journal of Food Science and Technology 25(1): 47–60.

Jayaraman, K.S. and Das Gupta, D.K. 1992. Dehydration of fruits and vegetables-recent developments in principles and techniques. Drying Technology 10(1): 1–50.

Khalloufi, S., Almeida-Rivera, C. and Bongers, P. 2009. A theoretical model and its experimental validation to predict the porosity as a function of shrinkage and collapse phenomena during drying. Food Research International 42(8): 1122–1130.

Khalloufi, S., Almeida-Rivera, C. and Bongers, P. 2010. A fundamental approach and its experimental validation to simulate density as a function of moisture content during drying processes. Journal of Food Engineering 97(2): 177–187.

Koc, B., Eren, I. and Kaymak Ertekin, F. 2008. Modelling bulk density, porosity and shrinkage of quince during drying: The effect of drying method. Journal of Food Engineering 85(3): 340–349.

Landau, L.D. and Lifshitz, E.M. 1959. Fluid Mechanics. Vol. 6 of the Course of Theoretical Physics. Pergamon Press Ltd., Oxford, UK.

Llave, Y., Takemori, K., Fukuoka, M., Takemori, T., Tomita, H. and Sakai, N. 2016. Mathematical modeling of shrinkage deformation in eggplant under-going simultaneous heat and mass transfer during convection-oven roasting. Journal of Food Engineering 178: 124–136.

Maibaum, L. and Chandler, D. 2003. A coarse-grained model of water confined in a hydrophobic tube. Journal of Physical Chemistry B 107(5): 1189–1193.

Marrink, S.J., Risselada, H.J., Yefimov, S., Tieleman, D.P. and De Vries, A.H. 2007. The martini force field: Coarse grained model for biomolecular simulations. Journal of Physical Chemistry B 111(27): 7812–7824.

McMinn, W.A.M. and Magee, T.R.A. 1997a. Diffusional analysis during air drying of a starch food system. Developments in Chemical Engineering and Mineral Processing 5(1-2): 61–77.

McMinn, W.A.M. and Magee, T.R.A. 1997b. Quality and physical structure of a dehydrated starch-based system. Drying Technology 15(6-8): 1961–1971.

Onwude, D.I., Hashim, N., Janius, R.B., Nawi, N.M. and Abdan, K. 2016. Modeling the thin-layer drying of fruits and vegetables: A review. Comprehensive Reviews in Food Science and Food Safety 15(3): 599–618.

Ozdemir, M. and Onur Devres, Y. 1999. Thin layer drying characteristics of hazelnuts during roasting. Journal of Food Engineering 42(4): 225–233.

Panchariya, P.C., Popovic, D. and Sharma, A.L. 2002. Thin-layer modelling of black tea drying process. Journal of Food Engineering 52(4): 349–357.

Papanu, J.S., Soane (Soong), D.S., Bell, A.T. and Hess, D.W. 1989. Transport models for swelling and dissolution of thin polymer films. Journal of Applied Polymer Science 38(5): 859–885.

Pizzirusso, A., Brasiello, A., De Nicola, A., Marangoni, A. and Milano, G. 2015. Coarse-grained modelling of triglyceride crystallisation: A molecular insight into tripalmitin tristearin binary mixtures by molecular dynamics simulations. Journal of Physics D: Applied Physics 48(49).

Raitio, R., Orlien, V. and Skibsted, L.H. 2011. Free radical interactions between raw materials in dry soup powder. Food Chemistry 129(3): 951–956.

Ramos, K.K., Lessio, B.C., Mece, A.L.B. and Efraim, P. 2017. Mathematical modeling of uvaia byproduct drying and evaluation of quality parameters. Food Science and Biotechnology 26(3): 643–651.

Sagar, V.R. and Suresh Kumar, P. 2010. Recent advances in drying and dehydration of fruits and vegetables: A review. Journal of Food Science and Technology 47(1): 15–26.

Senadeera, W. 2008. The drying constant and its effect on the shrinkage constant of different-shaped food particulates. International Journal of Food Engineering 4(8).

Shahbaz, H.M., Jeong, B., Kim, J.U., Ha, N., Lee, H., Ha, S.-D. et al. 2018. Application of high pressure processing for prevention of greenish-gray yolks and improvement of safety and shelf-life of hard-cooked peeled eggs. Innovative Food Science and Emerging Technologies 45: 10–17.

Stormo, S.K., Skipnes, D., Sone, I., Skuland, A., Heia, K. and Skåra, T. 2017. Modeling-assisted minimal heat processing of atlantic cod (gadus morhua). Journal of Food Process Engineering 40(6).

Tekin, Z.H., Başlar, M., Karasu, S. and Kilicli, M. 2017. Dehydration of green beans using ultrasound-assisted vacuum drying as a novel technique: drying kinetics and quality parameters. Journal of Food Processing and Preservation 41(6).

Togrul, I.T. and Pehlivan, D. 2003. Modelling of drying kinetics of single apricot. Journal of Food Engineering 58(1): 23–32.

Wu, L., Orikasa, T., Tokuyasu, K., Shiina, T. and Tagawa, A. 2009. Applicability of vacuum-dehydrofreezing technique for the long-term preservation of fresh-cut eggplant: Effects of process conditions on the quality attributes of the samples. Journal of Food Engineering 91(4): 560–565.

Zhu, H., Gulati, T., Datta, A. and Huang, K. 2015. Microwave drying of spheres: Coupled electromagnetics-multiphase transport modeling with experimentation. Part I: Model development and experimental methodology. Food and Bioproducts Processing 96: 314–325.

CHAPTER 19

Use of Mathematical Modelling of Dough Biscuits Baking Behaviour

Noemi Baldino,* Francesca R Lupi, Domenico Gabriele and *Bruno de Cindio*

1. Introduction

The texture of the biscuits is recognized as a principal factor influencing the consumer (Zoulias et al., 2002) and its final appearance and taste are mainly affected by the type of ingredients used, but also by processing and particularly by the baking step, that plays an important role in the final structure of these systems, causing important changes in the structure.

The baking process step is the principal element responsible for the chemical and physical transformations of each piece of dough before they can be technically defined "biscuits". Generally speaking, a biscuit is a piece of dough that, after proper baking, shows a firm structure, often brittle, particular textural properties and low or absent deformation under shear stress (Ferrari et al., 2012).

Biscuit baking is a complex process involving many transformations in the structure of the dough due to contemporary heat and mass transfer phenomena, as well as biochemical and physicochemical reactions that also occur. Biscuits are essentially made from flour, fat and sugar, and, according to the ratio between them, it is possible to obtain semi-sweet or short dough biscuits (Baldino et al., 2014; Chevallier et al., 2000; Lene et al., 2004). As a rule, the biscuit dough has a low water content and, for this reason, is less elastic and extensible in comparison to bread dough (e.g., short dough has less than 20% w.b. of water) and can break under tension, but before the dough reaches the final characteristics, others steps are necessary. As said before, the baking step is one of the most important steps, because during this, owing to the thermal flux inside the oven increasing the temperature of the dough, all the phenomena concerning the increment of temperature begin. The most significant phenomenon involves the water evaporation from the dough to the oven, and because of this, the gas cells, previously created during leavening, expand and, as a macroscopic effect, it is possible to observe an increase in the biscuit volume. Then, during the baking, the viscoelastic properties of the medium change and these changes involve denaturation of proteins, starch gelatinization, fat melting and, after a crust formation, a possible browning reaction is expected (Ferrari et al., 2012; Sablani et al., 1998; Chevalier et al., 2002; Mirade et al., 2004).

Department of Information, Modelling, Electronics and System Engineering (D.I.M.E.S.), University of Calabria, Via P. Bucci, Cubo 39C, I-87036 Rende (CS), Italy.
* Corresponding author: noemi.baldino@unical.it

Even today, several research groups continue to study the baking modelling for systems like cookies, approaching the study from a different point of view, because of the complex rheological properties of the initial dough and their evolution during baking, as well as all the variables involved during the heat and mass transfer in the ovens.

Many works are present in the literature on this subject and it is possible to observe that the baking modelling was developed from a mathematical point of view in an attempt to correlate the key oven-design factors and variables with the final texture and geometrical characteristics, with increasing computational complexity in the last few years.

Generally, there are mathematical models, dealing with three different aspects of baking, present in the literature. The first type of mathematical model deals with the properties and simulation of the transport phenomena in the fluid phase inside the oven and considers the temperature and humidity profile inside the biscuit uniform. The second model analyses only the transport phenomena inside the biscuit, and the fluid phase characteristics appear only in boundary conditions; finally, the last aspect is only related to the changes in the product quality parameters like colour or luminosity (Broyart and Trystram, 2003).

In recent years, there has been an increasing amount of attention focused on the modelling of transport phenomena concerning biscuit baking. Thorvaldsson et al. (1999) wrote a 1-D mathematical model, describing the principal transport phenomena concerning heat, water and vapour transport. The equations were solved by the finite difference method, Implicit Euler.

Fan et al. (1999) modelled the evolution of the dough volume within the cooking time in the oven. They considered the increment of the volume of the bubble caused by CO_2 release and water vapour from the aqueous dough phase. The initial radius was fixed and a certain quantity of dough, considered as a viscous material and having a constant mass, was supposed surrounding the bubble. Moreover, the cells number are constant for the whole baking period and a power law was used to model the rheological behaviour. The temperature was considered constant in the dough shell but time-dependent. No coalescence between the bubbles was considered and the 4th Runge-Kutta method was used to solve the model equations.

Broyart and Trystram (2002) developed a model to simulate the baking period, taking into account the heat transfer due to the transport phenomena (radiation, convection, conduction). In this model, the mass transfer fluxes are controlled by the partial vapour pressure difference at the air-biscuit interface, hypothesized at equilibrium. All the material parameters are calculated by literature data and, in order to solve the equations, Euler's method was adopted with the proper initial condition coming from a pilot plant oven. Subsequently, Broyart and Trystram (2003) computed variables like colour and width.

Sosa-Morales et al. (2004) studied the heat and mass transfer modelling for a biscuit baking process. The model allowed studying the transport phenomena during the precooking and freezing of biscuits, and checking of the differences between two different cooking methods: with a conventional and a microwave oven. The comparison was made by the quality attributes of the final product. The biscuit was modelled as a cylinder.

Wong et al. (2006) developed a two-dimensional model for the dough in continuous movement. The dough was considered as solid material with heat parameter only determined by the temperature. It was modelled in unsteady state conditions, writing the equation of momentum, mass and energy conservation, using commercial software and, moreover, using periodic boundary conditions to simulate the movement in the 4 zones of the oven.

Hadiyanto et al. (2007) developed a generic baking model, which can be divided into the heat and mass transfer model, and in the product quality model. The mass balance equations were written for the water in the liquid and vapour states and for the CO_2. The gases were considered as ideal in equilibrium with liquid water, while the energy balance took into account conduction, evaporation-condensation, and the water vapour and CO_2 fluxes. The change of height was caused by the increasing pressure inside the gas cells in the dough due to the release of water vapour and CO_2 from baking powder or from the yeast. Moreover, the dough was considered as a Maxwell fluid and the deformation, caused by expansion, was caused by the pressure difference between the inside and the outside of the bubble. Thanks to the extension calculation of the fluid around the bubble, the height of the biscuit was calculated.

More recently, Ferrari et al. (2012) developed an FEM model of a biscuit baking, taking into account the thermo-physical properties during the whole baking time, but not the mechanical properties. The model considers the heat and mass transfer equations governing the system as a biscuit as isotropic. Unluckily, the moisture profile is not so accurate as the temperature profiles, which instead are well predicted.

Many of the works reported in this brief review consider the dough as a solid material, not taking into account the system heterogeneity and then the connected volume expansion. Only a few of them analyse the dough and impute the expansion to the overpressure inside the bubble, but do not consider the eventuality of the mechanical criterion for the coalescence, which causes the break of the layer around the bubble, or an eventual stabilization criterion.

The biscuit industry is an important sector in the food industry and, generally, the improvement of the existing biscuits version, owing to lifestyle change for example, or the design of a new one, is a complex industrial problem.

Specifically speaking, biscuit production involves several steps, including the measuring of the ingredients, mixing, leavening, and so on. All of the steps are very important for the correct success of the final cookie. The eventual substitution or change in one of the ingredients can interfere in the rheological properties of the dough and with the final texture as well as in the final biscuit weight, thickness, diameter or other dimensions and moisture, which potentially can stop the production (Cronin and Preis, 2000). Then, the slightest change in raw materials or in the processing conditions can yield a different product that, potentially, allows a new share of the market to be gained.

Therefore, in the light of the above, many research groups are still involved in the optimization, from a rheological and textural point of view, to improve the first steps of the process. While others are involved in the possibility of obtaining a good baking model, which allows for the prediction of the final product in terms of temperature and moisture profiles as well as the geometrical characteristics.

The aim of the work described in this chapter was to give the possibility of predicting the final geometrical parameters, as well as the temperature, water distribution and porosity, as a function of the initial rheological characteristics and operating conditions in the oven during the baking time. In fact, the mechanical properties' evolution with temperature and time is very important to understanding the evolution of the materials' characteristics during the process and for the final texture (Dobraszczyk et al., 2003) and the rheological principles and theory that can be used.

In this work, the biscuit system is considered as pseudo-homogeneous, caused by the bubbles put inside the system during leavening, and solid-like.

The proposed model considers the dough as a three-dimensional protein network of gluten, incorporating all the other components like hydrated starch and gas bubbles, formed during mixing and the principal element responsible for the biscuit volume expansion during the rising and/or baking phase, stabilized by the protein membrane that avoids gas leakage (Lene et al., 2004). The model considers micro- and macro-systems, connected to each other. In particular, a single bubble surrounded by a proper amount of viscoelastic dough constitutes the microsystem and the macrosystem is the whole biscuit considered as pseudo-homogeneous and solid-like. In other words, all the phenomena refer to an equivalent homogeneous system and, to do this, the bubbles presence modifies the properties of the system. Consequently, it is necessary to define effective material properties, obtaining in this way a calculative advantage and easy use, because the equations will be simplified, even if the heterogeneous element is lost.

2. Heating Methods

Biscuit baking ovens generally work in continuous operation and are configured as tunnel ovens with lengths varying between 15–100 m (Mirade et al., 2004) and widths varying from about 1 to 4 m (Williamson and Williamson, 2009). Two main types of tunnel oven are conventionally used in cereal biscuits baking and cooking industrial processes: direct-fired ovens and indirect-fired ovens. In the first case, gas burners (working with natural gas, LPG, propane) or electric heating elements (wire, rods, tubes, plates) are placed all along the conveyor band in order to produce the heat energy necessary for cooking. On the contrary, in the indirect-fired ovens, the heat energy is produced in a chamber separated by a steel wall from the tunnel where biscuits are placed (Williamson and Williamson, 2009). Usually, the baking chamber consists of around three to five different zones along the oven (Khatir et al., 2012), where extraction chimneys promote air movement and allow exhausted air to be removed (Yanniotis and Sundén, 2007). Fans of variable speed are also placed along the tunnel with the aim of moving air so as to improve the transport phenomena which normally occur in cooking processes. The oven chamber can be controlled in terms of the temperature and humidity of the drying fluid, as well as the velocity of the fluid, which can be differentiated in order to tune the process conditions to different recipes, making the oven adaptable to different products.

In addition, the speed of the long conveyors can be regulated and, normally, belts are made of wire-mesh or carbon steel sheet. Thanks to the variable speed, the baking time can be adjusted within a limited range. In general, despite all the technical devices described above, the control of the process conditions is complex, owing to the short baking time and the lack of control points along the oven tunnel. Moreover, a small change in the operational conditions can change the texture of the biscuit dramatically.

Nowadays, the operators apply a trial and error approach instead of scientific control of the oven to change the oven conditions. Unfortunately, the influence of unpredictable variations in uncontrolled process conditions during baking can generate drastic changes in the rheological behaviour during the process and in the final texture of the biscuits.

Therefore, the final thickness and moisture of the biscuit should be normally tuned by controlling the heat and mass transfer mechanism (Cronin and Preis, 2000), even if the second phenomenon can be considered as a consequence of the first one. In fact, during baking, the heat in the oven is transferred mainly from air to the biscuit pieces by convection and from the oven walls to the biscuit surfaces by radiation. Finally, heat is transferred by conduction within the biscuit from the surface to the core and, simultaneously, the water diffuses from the core to the surface where it evaporates and is removed by the air flow (Williamson and Wilson, 2009).

3. Biscuit Baking Model Description

As mentioned before, the quality of bakery products relies both on formulation and processing conditions and their optimisation can yield improvements of some quality parameters.

The main idea is to model a dough biscuit by splitting the problem into two parts, called micro- and macro-systems, according to the approach originally proposed by de Cindio and Correra (1995), the first one being heterogeneous, while the latter is pseudo-homogeneous.

Each gas bubble surrounded by a proper amount of paste, constitutes the micro-system operating in an equilibrium condition. The bubbles, inserted during the mixing phase into the cereal paste, are supposed to grow during baking, thanks to the decomposition of raising agents, water evaporation and being at thermodynamic equilibrium.

Then, all the microscopic variables and information, computed at the equilibrium condition (radius, components in the vapour and dough phase) are transferred to the macro-system, operating with pseudo-homogenous properties, where macroscopic heat and mass balance equations are written.

The balance equations give all the variable trend during the baking time (like temperature, humidity, water concentrations, etc.), at any point of the spatial grid into which the dough system can be subdivided and from which the volume of the cookie may be computed at any time.

Here, we consider a general system having a cylindrical shape (Fig. 1) and because the dimension of the initial height of the dough biscuits is usually much lower (one order of magnitude) when compared to the diameter, the mass and heat fluxes from the lateral surface of the biscuit can be neglected with respect to the upper surface. Therefore, only the axial profiles of temperature and concentration have to be considered and it is possible to discretize the system in a 1-D spatial grid.

Thanks to this approach, the heterogeneous character of the dough can be taken into account, even if the system is treated as pseudo-homogeneous, and this has many computational advantages. At any time, the iterative procedure gives back the volume, the temperature profile and water profiles of the biscuit.

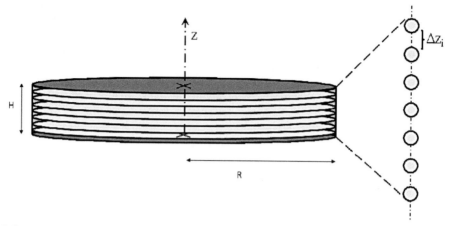

Figure 1: Schematic representation of 1-D grid equivalence of biscuit model (Reprinted with permission from Baldino, N. et al. 2014. Modeling of baking behavior of semi-sweet short dough biscuits. Innov. Food Sci. Emerg. Technol. 25: 40–52).

3.1 Bubble growth microsystem

The present model assumes that the bubbles, formed in the initial mixing phase, grow over time as a consequence of the raising agent, water evaporation and temperature increase. As a consequence, the volume of the bubbles inside the dough will grow over time, then, at the beginning, the bubbles can be considered in an infinite medium, the paste phase. According to the suggestion of dividing the problem of bubble growth into kinetics, thermodynamic and mechanical equilibrium, the system was assumed to be constituted of a single closed spherical gas bubble surrounded by an infinite mass of a viscoelastic paste (Baldino et al., 2014). At any time, this system is subjected to both mechanical and thermodynamic equilibrium. In this way, once a finite integration time interval is fixed, from the reaction kinetics, the amount of either new products or components present at any time in the gas bubble is determined by a thermodynamic flash, while the bubble volume is computed by the mechanical equilibrium. Thus, from a theoretical point of view, the process has been considered to change as a quasi-static transformation, allowing passing through equilibrium stages.

According to various scientific papers (Baldino et al., 2014; Gabriele et al., 2012; de Cindio and Correra, 1995), the bubble expansion modelling is developed, considering the mechanical behaviour of a single bubble expanding in a viscoelastic fluid assumed to be an infinite medium (Fig. 2).

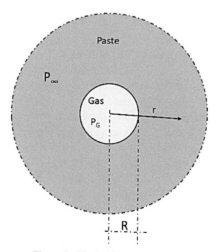

Figure 2: The bubble-paste system.

It can be considered as a homogeneously-mixed medium and it can be assumed that the bubbles of air formed during the mixing phase are uniformly distributed and all have a radius equal to R_0 and then with a uni-modal distribution (Carlson and Bohlin, 1978). Moreover, it can be considered that the system is not concentrated and that the bubbles inside the medium are in such number as not to interact among themselves. Then the spherical bubbles of radius R_0 are considered and a system of spherical coordinates is defined with the origin in the centre of the bubble.

The volumetric gas fraction can be expressed as the ratio between the volume of gas and the total volume of the system:

$$\varphi = volumetric\ fraction = \frac{V_{gas}}{V_{tot}} \tag{1}$$

where the total volume of the bubbles (V_{gas}) is:

$$V_{gas} = n_{bubble} \cdot V_{bubble} \tag{2}$$

If it is supposed that the bubbles contain only air, the pressure p_0, working initially on the inside surface in the bubble, will be given by:

$$P_0 = P_g\,(t=0) = P_{atm} + \frac{2\gamma}{R_0} \tag{3}$$

where γ represents the interfacial tension at the air/paste interface and R_0 is the initial radius of a single bubble.

Bubble unity is defined as the air bubble and the part of the continuous phase, that interacts with it. This mass will be given by:

$$m_{ub} = \rho_p\,(\frac{1-\varphi}{\varphi} \cdot \frac{4}{3}\pi R_0^3) \tag{4}$$

where ρ_p is the density of the continuous phase. It is possible reasonably to assume for the air an ideal gas behaviour, therefore, supposing that n_a^0 moles of air are present, we will have:

$$P_0 \cdot V_{bubble}^0 = n_a \cdot R_g \cdot T_0 \tag{5}$$

where R_g is the ideal gas constant.

$$P_{atm} \cdot \frac{4}{3}\pi R_0^3 = (n_a^0) R_g \cdot T_0 \tag{6}$$

The kinematics analysis of the bubble growth was performed separately considering the continuous side and the phase gas according to Bird et al. (1977) and, as reported in other works (Baldino et al., 2014; Gabriele et al., 2012), assuming spherical coordinates and centred reference frame and a purely radial flow, the momentum balance equation can be written as:

$$R\ddot{R} + \frac{1}{2}\dot{R}^2 = \frac{P_G - P_\infty}{\rho_p} - \frac{2\gamma}{\rho_p R} - \frac{3}{\rho_p}\int_R^\infty \frac{\tau_{rr}^{(b)}}{r_b}dr_b \tag{7}$$

where the surface tension can be neglected due to the low value in comparison to the higher value of dough consistency. In Eq. (7), P_∞ and P_G are, respectively, the pressure at infinite distance from the bubble and the bubble pressure described by ideal gas equation, while R is the bubble radius during time, and $\tau_{rr}^{(b)}$ is the only non-zero component of the stress tensor estimated at the interface. Moreover, the contribution of the gravity forces was neglected.

Equation (7) expresses the functionality between the pressure that is realized inside the gas bubble, the interfacial tension, the radius of the bubble and the stress conditions that are created in the shell of continuous phase that surrounds the bubble. To obtain the solution, it is necessary to compute the integral term, which represents the rheological contribution and ties the value of the stress component to the kinematics and, therefore, to the deformation. Obviously, this contribution will be different according to the models assumed for describing the material behaviour.

It is necessary to specify the term $\tau_{rr}^{(b)}$, and this can be done by postulating a constitutive equation. The material in question is time-dependent and the almost rheological linear equation of Goddard and Miller was used according to Baldino et al. (2014).

Hence, if the initial radius is known together with the material parameters, it is possible to obtain the bubble evolution solving Eq. (7).

This procedure needs to know the behaviour of the dough surrounding the bubble as a function of external conditions, i.e., thermal fluxes and pressure, P_G, which can be supposed to follow the equation of a mixture of ideal gases:

$$P_G = \sum_i P_i = \frac{R_g T}{V_{bubble}} \cdot \sum_i n_i \tag{8}$$

where n_i are the moles of any i-component in the gas bubble.

During the baking, the raising agents decompose and the water evaporates, as a consequence the pressure inside the bubbles increases.

The decomposition reactions occur only in the liquid phase and reaction products can be assumed to be soluble only in water and not in the paste. In addition, CO_2 exhibits a rather low solubility described by Henry's Law.

Reaction products are dissolved, therefore, in the liquid phase, but according to the thermodynamic conditions, they tend to go into the gas phase, according to the liquid/vapour equilibrium and are assumed to be at thermodynamic equilibrium with the gas phase within the bubble. The water is assumed to be partially bonded to the flour paste and the raising agent remains in the liquid phase because of its low volatility. A quasi-static condition can be assumed, if it is possible to assume a fast decomposition rate of the raising agents. Then, assuming a proper interval time, the quantity of the new products can be computed, with respect to the mass balance, and a thermodynamic flash computation can be done to know the gas in the bubbles and the substances in the paste, always at equilibrium state (Smith and Van Ness, 1987, Chap. 12). The following thermodynamic equilibrium equations apply:

$$P_{H_2O} = a_w \cdot P_{H_2O}^0 = P_G \frac{n_{H_2O}}{n_{tot}} \tag{9a}$$

$$P_{CO_2} = H_{CO_2} \cdot x_{CO_2} = P_G \frac{n_{CO_2}}{n_{tot}} \tag{9b}$$

$$P_{NH_3} = P_{NH_3}^0 \cdot x_{NH_3} = P_G \frac{n_{NH_3}}{n_{tot}} \tag{9c}$$

$$P_G = P_{H_2O} + P_{CO_2} + P_{NH_3} + P_{Air} \tag{9d}$$

The CO_2 and NH_3 are water-soluble substances and the amount can be calculated according to Henry's theory for the first species and the ideal Raoult's law for the second one. Regarding the H_2O, the quantity comes from the initial amount and from the decomposition reactions. The air in the bubbles is accepted to be insoluble in the water.

The thermodynamic equilibrium is solved at constant temperature thanks to the set of Eqs. (9a)–(9d) (Smith and Van Ness, 1987) and the gas and liquid compositions represent the important output, being the gas the quantity reaming in the bubble, which contributes to the volume, and the liquid part the moles in the paste phase.

3.2 Biscuit macroscopic baking model

In Baldino et al. (2014), a 1-D macroscopic model was simulated assuming a grid point as shown in Fig. 1.

In this study, the authors analysed the evolution of the baking period for a crunchy biscuit type, which is made by an industrial process that can be schematized as a series of steps: Mixing, standing, forming and baking.

During mixing, a certain amount of air bubble nuclei are inserted into a mixture essentially constituted of flour, water and other minor ingredients. After a proper time of standing, during which the raising agents start to react to produce some substances capable of blowing the dough, the biscuit is subjected to a forming step, usually on a band belt. Finally, the belt passes through a heating oven where the biscuit baking is carried out for few minutes, according to the type and size. The heating is achieved by applying either radiative or convective heat fluxes. Gas bubbles grow due to both the thermal expansion and the thermodynamic equilibrium between the dense phase and the gas, allowing the increase of the total amount of moles of the gas phase. Whilst at the beginning it is possible to admit that the bubbles behave as single closed cells when increasing their diameter, interactions between them occur and bubble coalescence is observed or, owing to different mechanisms like thermal aggregation, a stabilisation of the bubbles is reached. Thus, the biscuit reaches a maximum height with only a small loss of water, thereafter a more or less pronounced mechanical collapse of the structure is observed, evidenced by a sharp decrease in both the biscuit height and its water content. Thus, as a consequence of raising agents and thermal treatment during the entire process, the main phenomena involved in biscuit production are:

1. bubble expansion
2. bubble coalescence/stabilisation
3. biscuit drying

The overall phenomenon is determined by several factors: Dough rheological properties, transport properties, void fraction, heat treatment, process features and so on. In addition, during this stage, the biscuit shrinks as a consequence of water loss that is ascribed to evaporation from both the external surfaces and the new surfaces made available for evaporation by the air cell opening towards the external ambient. For these reasons, transport phenomena of both energy and mass must play a crucial role in determining the final biscuit characteristics.

In this model only the baking step is considered and the other steps, like mixing and standing, are contemplated in the initial conditions. Then, the biscuits are considered with the only change being their dimensions only in height and the deformation on the belt was ignored. Thus, a 1-D model was assumed to be sufficient to describe the biscuit behaviour during baking, according to the steps enumerated before. Every step was modelled according to some specific physical hypotheses, as reported in the following, so generating in such a way a series of microsystems, every one characterised by the assumption that they must represent both thermodynamic and mechanical equilibrium conditions. The microsystem is treated essentially as a sink for the macrosystem and represents the link between them. The thermodynamic and mechanical variables computed by the macrosystem, are put in equilibrium condition by applying a single microsystem according to the considered step. Next, the information obtained by solving the equilibrium equations set (the bubble radius and the components distribution), are sent to the macrosystem, which solves the transport phenomena.

Then the heat and mass balance equations are solved with reference to a discretised 1-D spatial grid and, at any time, by using the proper initial and boundary conditions, the new values of the thermodynamic and mechanical variables are obtained and sent again to the microsystem. This procedure can be iterated at any grid point and at any time. By storing and then plotting all the data coming out of the algorithm, it is possible to obtain the evolution of all the variables during baking, like biscuit volume, temperature, water concentration, etc.

3.2.1 Mass transport (Macrosystem)

The mass transport inside the biscuit must be subdivided into two parts: Transport into the dense phase and transport into the gas cells. Owing to the assumption that the variables must be continuous, a pseudo-homogeneous system is built with effective properties. This is made possible because the transport into the gas is neglected and pure diffusive transport is considered. This agrees with the assumption that the microsystem is in equilibrium and consequently no transport must be considered. On the contrary, the macrosystem is a pseudo-homogenous material, because the presence of the bubbles is considered in the properties, and the water is solubilized inside this. Moreover, it was supposed that the raising agents' reaction products are soluble just in water and that the water can diffuse throughout the paste, whereas the others in the water. This has some effect not only on the transport properties but even on the gas-water solution equilibrium of those species. It should also be noted that the concentration of all the species has to be referred to the system where they are effectively contained. The raising agent products diffuse in water, and then the mass balance equations are written in reference to this mechanism.

Then, according to Baldino et al. (2014), the mass transport for H_2O, NH_3 and CO_2 can be solved with proper boundary conditions starting from the following balance equations, the details of which are reported extensively in the cited paper:

Water transport

$$\frac{\partial(c_W^*(1-\varphi))}{\partial t} = \frac{\partial}{\partial z}\left(D_{eff}\frac{\partial c_W^*}{\partial z}\right) - F_W 4\pi R^2 n_b^0 + R_W(1-\varphi)\cdot\rho_p \tag{10}$$

Carbon dioxide and Ammonia Transport

$$\frac{\partial(c_W^*(1-\varphi)x_i)}{\partial t} = \frac{\partial}{\partial z}\left(-D_{iw}\frac{\partial}{\partial z}[c_W^*(1-\varphi)x_i]+x_i N_W\right) - F_i 4\pi R^2 n_b^0 + R_i(1-\varphi)\cdot\rho_p \tag{11}$$

where $c_W^* = \dfrac{\text{moles of water}}{\text{Paste volume}}$, $x_i = \dfrac{\text{moles of i-component}}{\text{Moles of water}}$, F_W and F_i are, respectively, the water molar flux or *i*-component molar flux that evaporates/condenses in any bubbles, n_b^0 is the number of bubbles per unit of volume, ε the void fraction, R_w and R_i are the water and CO_2 or NH_3 generated by the raising agent decomposition per unit of mass paste, N_w is the water molar flux and D_{eff} and $D_{i,w}$ are the diffusion coefficient of water and of the i-component in water, respectively.

3.2.2 Energy transport (Macrosystem)

The energy transport inside the biscuit is assumed to be purely conductive and only the convective and radiative mechanisms are taken into account. It is possible to model the biscuit and the tray on which the dough is baked, given a flux Φ_{Oven} as an input data, which changes according to the profile imposed during baking, while convection is assumed and, as a consequence, it is possible to use a constant value for h_{ext}, convective heat transfer coefficient.

The energy balance equations for the tray and the biscuit are the following:

$$\frac{\partial}{\partial t}\left(\rho_{Steel}C_{PSteel}T\right) = \frac{\partial}{\partial z}K_{Tray}\frac{\partial T}{\partial z} + S_B^L\cdot h_{ext}\left(T_\infty - T\right) + S_B^L\cdot\left(E_{Oven}\cdot\Phi_{Oven} - E_{Tray}\cdot\sigma\cdot T^4\right) \tag{12}$$

$$\frac{\partial}{\partial t}\left(\rho_P C_{PDough}T\right) = \frac{\partial}{\partial z}K_{Dough}\frac{\partial T}{\partial z} - \sum_i \lambda_i S_e^i + S_B^L\cdot h_{ext}\left(T_\infty - T\right) + S_B^L\cdot\left(E_{Oven}\Phi_{Oven} - E_B\sigma T^4\right) \tag{13}$$

where ρ is the density, σ the Stefan-Boltzmann constant, E is the emissivity, K is the thermal heat conductivity, and the subscript indicates what it refers (tray or dough). S_e^i is molar flux of the i-component moving to/from the gas bubbles, S_B^L lateral biscuit surface, and T_∞ is the bulk temperature in the oven.

All the details of the balance equations are reported in the paper of Baldino et al. (2014) together with the adopted boundary conditions (according to the specific process and operating conditions).

4. Bubbles Interaction and Stabilization Criteria

Many baked goods, like cookies products, are recognized by a typical porous texture. To obtain this, the air has to be put inside during the mixing stage, as stated above. Then, the gas produced during leavening, baking or incorporated in mixing must be retained by the gas cells in order to have a good expansion during baking. Therefore, the volume expansion is directly related to the extent to which gas cells can expand and to the viscoelastic dough properties. The expansion is indeed controlled by the dough film around the bubbles, which can resist collapsing, have a breakdown or be stabilized by physiochemical changes (Baldino et al., 2014; Chevallier et al., 2002).

It is well known that the dough around the bubble has to resist in order to prevent the layer rupture around the cells before the optimum time and, in this case, the elongational properties have to be adequate to resist under baking conditions and to allow the best volume expansion.

Then, Van Vliet et al. (1992) proposed an additional dough property that can be used to understand whether a specific material is able to retain a gas: Strain hardening. This last property, measured by extensional tests, gives the capacity to the dough to resist during the biaxial extension taking place in the oven during the bubbles volume expansion.

The strain hardening ability of the dough could be used as a stabilisation criterion. The previously proposed criterion for bubble expansion in an infinite visco-elastic medium does not consider any interaction of a bubble with surrounding ones. This hypothesis is acceptable for the mass transfer model, but an interaction mechanism should be considered from a mechanical point of view.

In fact, when the bubble radius increases, the bubbles start interacting and then they can either coalesce, yielding the cell "opening", or stabilize yielding closed cells no longer able to expand. A criterion, able to discriminate between these opposite situations and to give information about potential cell stabilisation is, therefore, necessary. The literature provides many examples. Bubble growth can stop if both the reactants are completely consumed and temperature reaches a constant value and then the water is at the equilibrium conditions. Moreover, it is well-known that bubble stabilisation happens earlier than that stated above and, therefore, another mechanical stabilisation mechanism should be taken into account.

From a physical point of view, when a bubble expands, the mass around it becomes thinner and thinner, and at a certain time, it cannot be any longer assumed as an infinite medium as proposed above. Some authors (Gan et al., 1995) state that coalescence occurs when the uniform starch/protein layer surrounding the bubble becomes thin enough to break, therefore, a critical dough thickness value should be evaluated. This theory has been successfully applied to bread dough mainly composed of flour and water, but the same authors report that stabilisation is affected by the dough recipe, especially when plasticizer ingredients are involved, such as the fat and sugar in biscuit recipes. In addition, the evaluation of a critical thickness implies that the layer should be uniform, i.e., a spatial uniform interaction between bubbles should be assumed. On the contrary, physical evidence indicates that the material surrounding bubbles does not form a layer of constant thickness and a region of layer reduction can be individuated, according to the spatial distribution of the bubbles.

To overcome this problem, a new modelling approach can be proposed: The bubble rupture/coalescence phenomenon is linked to the elastic deformation energy accumulated in the dough layer as a result of bubble expansion. From a mass balance, if the bubble volume distribution is known and the mass of paste associated with any bubble is constant during the process, if the water content does not change, the initial layer around the bubble can be computed. The material contained in this layer is subjected to a variable force that tends to squeeze it out from the gap when two contiguous bubbles interact with each other. As a consequence, the material distribution around the bubble is not a uniform layer nor time independent. This effect starts to be important only when the radius becomes sufficiently large so that the infinite paste hypothesis does not apply any longer. During this phenomenon, a certain amount of elastic energy is accumulated in the material, according to the time-dependent rheological properties of the system. When a maximum energy load is reached the dough layer can break and the bubble eventually coalesces.

From a modelling point of view, the problem can be split into two parts: the calculation of the dough layer thickness profile and, secondly, the calculation of the elastic energy associated with the deformation process owing to the smallness of the interaction zone. The local layer thickness evolution can be modelled referring to a planar equivalent squeezing flow whilst the elastic energy can be calculated using the mechanical theory of shell and membranes subjected to an external load (Williams, 1973).

Of course, when the bubbles open towards the ambient, the internal pressure suddenly drops from the actual to the external value. This implies a new mechanical equilibrium with an elastic recovery under constant pressure. Therefore, in order to overcome the problem, a recovery mechanism based on the amount of elastic energy stored during the closed cell behaviour can be computed, and that can be recovered at the opening. This, in turn, means that a material memory function must be inserted to take into account the recovery mechanism in the mechanical properties of the biscuit system.

In other cases, a different stabilisation mechanism can be used, based on Chevallier et al. (2002) and Baldino et al. (2014), which clearly depend on the biscuit texture and recipe. During heating into the oven, a protein cross-linking can happen in a temperature range of about 80–85°C for dough with an initial water percentage of 20–25% w.b. Then, a temperature criterion can be used to stop the bubble growth. Since the protein cross-linking normally depends on the dough recipe, according to Baldino et al. (2014), the value of protein cross-linking criteria can be obtained using rheological methods, performing a dynamic temperature ramp test, which is a test used to observe the evolution of the dynamic moduli in temperature. Generally, the moduli of the dough system decrease while increasing the temperature up to a certain value, owing to kinetics effects, and then start to increase. The dough changes, becoming a "solid-like" material as a consequence of induced aggregation phenomena and network formation. The critical temperature then can be chosen as the value at which the complex modulus stops decreasing.

5. Shrinkage Computation

Typically, when a biscuit is baked it is possible to observe three zones, as reported in Fig. 3. An initial flat zone, a second one where the height of the biscuit increases, reaching a maximum, and a third zone where the height gradually decreases due to the drying process that is coupled with biscuit baking. This last phenomenon implies the onset of shrinkage, which represents the volumetric dimensional reduction mainly due to water loss. It should be noticed that this phenomenon is always present due to the oven conditions that are far from water equilibrium, but the experimental evidence shows that, during bubble growing, water tends to go preferably towards the gas bubble, also owing to the smallness of the evaporating surfaces.

The shrinkage phenomenon was widely studied in the past by Minshkin et al. (1984) and Lang et al. (1994), who described the volumetric variation of dough as a function of the water loss and with a linear equation as follows:

$$V_{bisc}(t) = V_{bisc\_p}\left(1 + K_{shr}\left(U - U_0\right)\right) \tag{14}$$

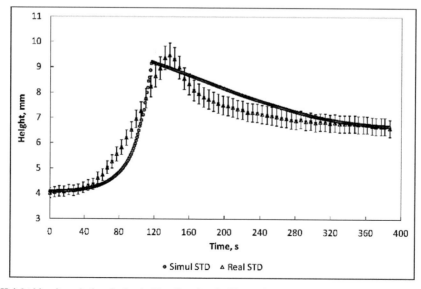

Figure 3: **Height biscuit evolution during baking** (Reprinted with permission from Baldino, N. et al. 2014. Modeling of baking behavior of semi-sweet short dough biscuits. Innov. Food Sci. Emerg. Technol. 25: 40–52).

where $V_{bisc}(t)$ is the time variation of the biscuit volume, V_{bisc_p} is the biscuit volume when the system reaches the maximum height, K_{sh} is a constant that can be obtained by experimental tests and the initial humidity value (U_0) can be assumed as the value of the biscuit humidity at the beginning of the drying zone (3th zone). Therefore, the biscuit volume diminishes linearly according to the water loss during the baking period.

6. Conclusions

Biscuit quality is affected in a significant way by process conditions that should be properly tuned, according to adopted raw materials, in order to guarantee a standard final product with the expected properties.

With the aim of avoiding expensive and time-consuming "trial and error" procedures, a more fundamental approach, based on physical and mathematical modelling of the process, is becoming more used.

This approach is based on the physical analysis of transport phenomena, occurring within the biscuit and the oven, and the mathematical formulation of conservation principles (momentum, mass and energy). Attention has been focused mainly on mass and heat transfer within the biscuit, where the main changes occur, whereas, often, transport phenomena within the surrounding environment (i.e., the oven) are not investigated in detail and are considered only within boundary conditions.

It is worth noticing that biscuit dough is a heterogeneous material (made of a paste surrounding gas cells), therefore, in order to simplify the mathematical description of the problems, different approaches have been proposed in the literature, often based on a pseudo-homogenous materials with "effective" properties. Anyway, these models are not able to describe properly what is really happening within the dough at the air-paste interface where the main phenomena (such as gas and vapour exchange, bubble growth, etc.), occur.

A different and very interesting approach is based on a double-scale system: At "micro" level the paste-cell interface is studied, describing the local mass transfer and the mechanical equilibrium controlling the radius evolution. At "macro" level, mass and heat transfer balance equations, within the biscuit, are written considering a homogenous system with properties computed, at the local level, according to the "micro" balances (such as the local void fraction depending on the bubble evolution). The two systems of equations, working at two different length scales, are strictly interconnected because the bubble evolution and the local mass transfer depend on the macroscopic temperature and concentration profiles, whereas the macroscopic "effective" properties depend on the local void fraction.

This approach maintains the simplicity of the pseudo-homogenous models and is also able to describe the physical changes occurring at the local scale within the biscuit.

It was used in the literature, with very interesting results, to describe the biscuit behaviour during baking.

In addition, it has to be considered that final biscuit properties are determined, also, by the potential bubble coalescence (i.e., cell opening) or stabilization (i.e., halt in bubble growth); therefore, a criterion to evaluate the potential cell stabilization has to be inserted in the model. Different criteria have been proposed, based on different physical parameters, such as a minimum critical thickness, a maximum critical stress, a limiting critical energy, a threshold temperature corresponding to critical changes in paste microstructure, etc.

As a final result, a comprehensive physical model describing the different steps of baking can be obtained, this gives an accurate description of evolution in biscuit properties (such as void fraction, height, moisture, etc.), that can be used to optimise the process conditions as a function of formulation and expected biscuit quality.

References

Baldino, N., Gabriele, D., Lupi, F.R., de Cindio, B. and Cicerelli, L. 2014. Modelling of baking behavior of semi-sweet short dough biscuits. Innov. Food Sci. Emerg. Technol. 25: 40–52.

Bird, R.B., Armstrong, R.C. and Hassager, O. 1977. Dynamics of Polymeric Liquids vol. 1, first ed. New York: Wiley.

Broyart, B. and Trystram, G. 2003. Modelling of heat and mass transfer phenomena and quality changes during continuous biscuit baking using both deductive and inductive (neural network) modelling principles. Chem. Eng. Res. Des. 81: 316–326.

Carlson, T. and Bohlin, L. 1978. Free surface energy in the elasticity of weat flour dough. Cereal Chem. 55(4): 539.

Chevallier, S., Colonna, P. and Lourdin, D. 2000. Contribution of major ingredients during baking of biscuit dough systems. J. Cereal Sci. 31: 241–252.

Chevallier, S., Della Valle, G., Colonna, P., Broyart, B. and Trystram, G. 2002. Structural and chemical modifications of short dough during baking. J. Cereal Sci. 35: 1–10.

Cronin, K. and Preis, C. 2000. A statistical analysis of biscuit physical properties as affected by baking. J. Food Eng. 46: 217–225.

de Cindio, B. and Correra, S. 1995. Mathematical modeling of leavened cereal goods. J. Food Eng. 24: 379–403.

Dobraszczyk, B.J. and Morgenstern, M.P. 2003. Rheology and breadmaking process. J. Cereal Sci. 38: 229–245.

Ferrari, E., Marai, S.V., Guidetti, R. and Piazza, L. 2012. Modelling of heat and Moisture transfer phenomena during dry biscuit baking by using finite element method. Int. J. Food Eng. 8(3): 394–417.

Gabriele, D., Baldino, N., Migliori, M., de Cindio, B. and Tricarico, C. 2012. Modelling flow behaviour of dairy foams through a nozzle. J. Food Eng. 109: 218–229.

Gan, Z., Ellis, P.R. and Schofield, J.D. 1995. Mini Review: Gas cell stabilisation and gas retention in wheat bread dough. J. Cereal Sci. 21: 215–230.

Hadiyanto, A., Asselman, G., van Straten, R.M., Boom, D.C., Esveld, A.J. and van Boxtel, B. 2007. Quality prediction of bakery products in the initial phase of process design. Innov. Food Sci. Emerg. Technol. 8(2): 285–298.

Khatir, Z., Paton, J., Thompson, H., Kapur, N., Toropov, V., Lawes, M. et al. 2012. Computational fluid dynamics (CFD) investigation of air flow and temperature distribution in a small-scale bread-baking oven. Applied Energy 89: 89–96.

Lang, W., Sokhansanj, S. and Rohani, S. 1994. Dynamic shrinkage and variable parameters in Bakker–Artema's mathematical simulations of wheat and canola drying. Dry Technol. 13(8/9): 2181–2190.

Lene, P.A., Kaack, K., Bergsoe, M.N. and Adler-Nissen, J. 2004. Rheological properties of biscuit dough from different cultivars, and relationship to baking characteristics. J. Cereal Sci. 39: 37–46.

Minshkin, M., Saguy, I. and Karel, M. 1984. Optimization of nutrient retention during processing: Ascorbic acid in potato dehydration. J. Food Sci. 49: 1262–1266.

Mirade, P.S., Daudin, J.D., Ducept, F., Trystram, G. and Clément, J. 2004. Characterization and CFD modelling of air temperature and velocity profiles in an industrial biscuit baking tunnel oven. Food Res. Int. 37: 1031–1039.

Sablani, S.S., Marcotte, M., Baik, O.D. and Castaigne, F. 1998. Modeling of simultaneous heat and water transport in the baking process. Lebensm-Wiss u-Technol. 31: 201–209.

Smith, J.M. and Van Ness, H.C. 1987. Introduction to Chemical Engineering Thermodynamics (5th ed.). New York: McGraw-Hill.

Sosa-Morales, M.E., Guerrero-Cruz, G., Gonzalez-Loo, H. and Velez-Ruiz, J.F. 2004. Modelling of heat and mass transfer during baking of biscuits. J. Food Process Preserv. 28: 417–432.

Thorvadsson, K. and Janestad, H. 1999. A model for simultaneous heat, water and vapour diffusion. J Food Eng 55: 167–172.

van Vliet, T., Janssen, A.M., Bloksma, A.H. and Walstra, P. 1992. Strain hardening of dough as a requirement for gas retention. J. Texture Stud. 23(4): 439–460.

Williams, J.G. 1973. Stress Analysis of Polymers. London: Longman Ltd.

Williamson, M.E. and Wilson, D.I. 2009. Development of an improved heating system for industrial tunnel baking ovens. J. Food Eng. 91: 64–71.

Wong, S.Y., Zhou, W. and Hua, J. 2006. Robustness analysis of a CFD model to the uncertainties in its physical properties for a bread baking process. J. Food Eng. 77(4): 784–791.

Yanniotis, S. and Sundén, B. 2007. Heat Transfer in Food Processing. Recent Developments and Applications. Boston: WIT Press.

Zoulias, E.I., Oreopoulou, V. and Tzia, C. 2002. Textural properties of low-fat cookies containing carbohydrate- or protein-based fat replacers. J. Food Eng. 55: 337–342.

CHAPTER **20**

Applications of Principal Component Analysis (PCA) for Fruit Juice Recovery and Quality Analysis

Debabrata Bera,[1,]* *Lakshmishri Roy*[2] and *Tanmoy Bhattacharya*[3]

1. Introduction

The application of mathematical and statistical tools/algorithms and chemometrics in the food sector has been on the rise in recent times. The popularity of their usage can be attributed to their potential in terms of the processing and analyzing of large amounts of complex experimental data. Mathematicians and statisticians have developed advanced methods of solving problems within various sectors of the food and health industries. These methods or models, when implemented in conjunction with computational packages (software), have proved to be useful to scientists in replacement of the manual calculation methods that are time-consuming and usually imprecise (for large datasets). Accurate statistical techniques for handling of data in food industries has benefitted the various aspects of the food sector, including sensory evaluation, development of food products/processes and food microbiology, as well as the quality control in food processes. The concept of statistics applied in food research is extensive, but there is an increasing concern as to the right approach adapted by the scientists and researchers in the interpretation and reporting of the results.

Principal component analysis (PCA) is advantageously applied for the evaluation in food analysis (Pokorny et al., 1995; Velesek et al., 1995; Frau et al., 1999). PCA can contribute both to the quality control of foodstuffs (manufacturing processes, contents, adulterations, contamination, etc.), and to the differentiation of their type characteristics (food authentication) (Frau et al., 1999). Principal component analysis (PCA) is a pattern recognition technique that can be used in order to discover latent associations or structures among the original variables and associations among samples. Cluster analysis is used to detect similarities and associations. The resulting data from PCA can then be applied to the following: Profiling specific product characteristics, comparing and contrasting similar products based on attributes important to consumers, and altering product characteristics with the goal of increasing market share for

[1] Department of Food Technology & Biochemical Engineering, Jadavpur University, Kolkata-700032, West Bengal, India.
[2] Department of Food Technology, Techno India Salt Lake, Kolkata 700091, West Bengal, India.
[3] Department of Computer Science, Techno India, Salt Lake, Kolkata 700032, West Bengal, India.
* Corresponding author: debabrata.bera@jadavpuruniversity.in

a given set of products. In optimizing technological processes, PCA can be a very useful tool to indicate the variables that have a significant impact on the examined process and those which are less important. Since patterns in data can be hard to find in data of high dimension, where the luxury of graphical representation is not available, PCA is a powerful tool for analyzing data.

The main goals of **principal component analysis** are:

- To identify hidden patterns in a data set.
- To reduce the dimensionality of the data by removing the noise and redundancy in the data.
- To identify correlated variables.

PCA method is particularly useful when the variables within the data set are highly correlated.

Correlation indicates that there is redundancy in the data. Due to this redundancy, PCA can be used to reduce the original variables into a smaller number of new variables (i.e., principal components) explaining most of the variance in the original variables.

2. Basic Structure of Principal Component Analysis

Principal Component Analysis is a dimension-reduction tool that can be used advantageously in cases demanding multivariate analysis and those with a substantial number of correlated variables. With Principal component analysis, a large set of variables can be reduced to a small set that still retain most of the information in the large set. The technique enables us to create and use a reduced set of variables, which are called *principal factors*. A reduced set is much easier to analyze and interpret. To study a data set that results in the estimation of roughly 500–700 parameters may be difficult, but if we could reduce these to 5–8, it would certainly make it easier and simpler. These new variables correspond to a linear combination of the originals and are called principal components.

Various steps of PCA method include:

Step 1: Prepare the data

The data obtained from either experiments or survey is tabulated in the form of a matrix.

Center the data: Subtract the mean from each of the variables. This produces a data set whose mean is zero.

For PCA to work properly, we have to subtract the mean from each of the data dimensions. The mean subtracted is the average across each dimension. So, all the x values have the average of x (the mean of the values of all the data points) subtracted, and all the y values have average of y values subtracted from them. This produces a data set whose mean is zero.

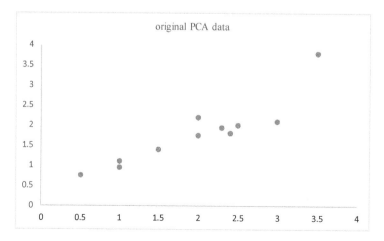

Figure 1: PCA example data, original data on the left, data with the means subtracted on the right and a plot of the data.

If the non-diagonal elements in this covariance matrix are positive, then it is expected that both the x and y variable increase together.

Scale the data: If the variances of the variables in your data are significantly different, it is a good idea to scale the data to unit variance. This is achieved by dividing each of the variables by its standard deviation.

Step 2: Calculate the covariance/correlation matrix

Step 3: Calculation of the eigenvectors and eigenvalues of the covariance Matrix

Eigenvectors are perpendicular to each other. But, more importantly, they provide us with information about the patterns in the data. It may be observed that the eigenvectors go through the middle of the points, like drawing a line of best fit. The first eigenvector is indicative of how the two data sets are related along that line. The second eigenvector gives us the other, less important, pattern in the data, that all the points follow the main line, but are off to the side of the main line by some amount. So, by this process of taking the eigenvectors of the covariance matrix, one is able to extract lines that characterize the data. The rest of the steps involve transforming the data so that it is expressed in terms of thin lines.

In the figure cited below, it may be observed that the data has quite a strong pattern. As expected from the covariance matrix, the two variables do indeed increase together. Both the eigenvectors have been plotted as well. They appear as diagonal dotted lines on the plot. Computing the eigenvectors and eigenvalues of the covariance matrix, wherein it is noticed that these eigenvectors are unit eigenvectors, i.e., their lengths are both 1. This is very important for PCA.

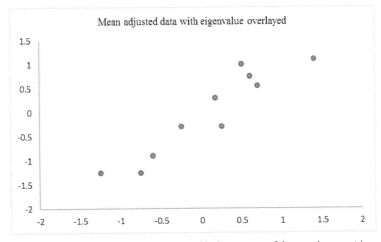

Figure 2: A plot of the normalized data (mean subtracted) with eigenvectors of the covariance matrix overlayed on top.

Step 4: Choosing components and forming a feature vector (Principle Component)

In this step, data compression and reduced dimensionality comes into play. On observing the eigenvectors and eigenvalues, it may be stated that the eigenvalues are quite different values. In fact, it turns out that the eigenvector with the highest eigenvalue is the principle component of the data set. In general, once eigenvectors are found from the covariance matrix, the next step is to arrange them by eigenvalue, highest to lowest. This gives the components in order of significance. Now, if required, the components of lesser significance may be ignored. On doing this, some information may be lost, but if the eigenvalues are small, then not much is lost. Thus, on leaving out some components, the final data set will have less dimensions than the original. What needs to be done subsequently is to form a feature vector, which is just a matrix of vectors constructed by taking the eigenvectors that one wants to keep from the list of eigenvectors, and forming a matrix with these eigenvectors in the columns.

Step 5: Deriving the new data set

This is the final step in PCA, and is also the easiest. Once we have chosen the components (eigenvectors) that we wish to keep in our data and formed a feature vector, we simply take the transpose of the vector and multiply it on the left of the original data set, transposed, where the matrix with the eigenvectors in the columns is transposed so that the eigenvectors are now in the rows, with the most significant eigenvector at the top, and the mean-adjusted data transposed, i.e., the data items are in each column, with each row holding a separate dimension. It will give us the original data solely in terms of the vectors we chose. So, what has been done here is basically that we have transformed our data, expressed in terms of the patterns between them, where the patterns are the lines that most closely describe the relationships between the data. This is helpful because we have now classified our data point as a combination of the contributions from each of those lines. Initially we had the simple values and axis that don't really tell us exactly how that point relates to the rest of the data. Now, the values of the data points tell us exactly where (i.e., above/below) the trend lines the data point sets. In the case of the transformation using both eigenvectors, we have simply altered the data so that it is in terms of those eigenvectors instead of the usual axes. However, the single-eigenvector decomposition has removed the contribution due to the smaller eigenvector and left us with data that is only in terms of the other.

3. Fruit Juice Recovery Methods

There are a number of unit operations involved in converting whole fruit to the desired juice. Operations such as cooling, washing, sorting and inspecting require attention to be paid to mass and heat transfer.

Fruits are collected from market or farm for fruit juice production in the food industry. These fruits contain unwanted materials, i.e., dust, sand, microorganisms, etc., on their surface (skin). These unwanted materials should be removed from the surface, by washing with fresh water or blowing air, before use as raw material for juice production in order to avoid contamination and to control the quality of the produced juice. After washing, the peels are removed manually or by machine. The core materials are processed to pulp by a suitable pulper. Pulp is the mixture of fruit juice and insoluble fibers. Fruits juice is entrapped inside the fiber and cell walls. After pulping, the ruptured cell walls release the juice, but a certain amount of juice remains inside the fiber and cell walls after pulping. To get the maximum juice yield, the pulp is treated with different enzymes, like hemi-cellulase, pectin or a mixture of these. These enzymes hydrolyze the cell wall components and release the juice. After juice extraction, it is treated thermally in order to destruct naturally occurring and treated enzymes present in the pulp, to destroy microorganisms present in it. Then, the pressing operation is carried out in order to separate the insoluble fibrous material from the juice. After filtration or centrifugation, the juice is obtained.

The amount of juice extracted depends on different operating parameters, like the type of enzyme, concentration of enzyme, temperature and operation time, etc. The quality of the extracted juice also depends on these operating parameters.

General flow sheet for fruit juice extraction

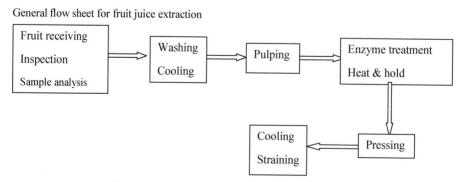

Figure 3: Process flow sheet of the fruit juice production.

4. Quality of Fruit Juice with Recovery Methods

Commonly processed fruit juice is gaining popularity in the world due to its nutritional value and typical flavour. Unit operations employed during industrial processing simplify juice production, but also tend to affect the physicochemical properties of the juice. To date, thermal treatment, enzyme treatment, etc., are still the most important and widely-used unit operations in the food processing industry (Rawson et al., 2011). Phenolic and flavonoid compounds are major bioactive substances present in juice (Hyson, 2011) and these treatments modify the total phenolic contents (TPCs) and total flavonoid content (TFC) of juice. Besides, heat treatment has been reported to increase the total soluble solid (TSS) of fruit pulp (Branko et al., 2016), which has been reported to reduce the pectin methyl-esterase activity (Aghajanzade et al., 2016), affecting the juice stability. Moreover, heat treatment can vaporize different organic acids, making it change the titratable acidity (TA) and pH values of apple juice (Kubo et al., 2013). Physical stability is an important index regarding particle sedimentation and serum separation (Augusto et al., 2012). Homogenization is a commonly used unit operation in juice preparation not only for the preservation of foodstuffs but also as a valuable tool to promote desirable changes in the physical properties of the products (Augusto et al., 2012). It has been reported that homogenization increases the TSS value of strawberry juice (Karacam et al., 2015). Similarly, it also increases the TPC of mango (Guan et al., 2016) and strawberry juices (Karacam et al., 2015). Moreover, homogenization can also affect the turbidity and physical stability of juices (Augusto et al., 2012). Apart from the thermal treatment and homogenization, other common unit operations, like pulping and juicing, also affect the juice quality (Heinmaa et al., 2017). Although the effect of certain unit operations has been widely studied, there is still much work to be done regarding the comparison of the quality parameters of juices based on different technologies. Until now, many production technologies of juices have been designed. Multivariate data analysis was widely used in food analysis and showed advantages in characterizing processing technologies used in food production (Dong et al., 2007; Kaya et al., 2015). Therefore, studies investigating the influence of several industrial production technologies on the physicochemical properties of juice become crucial to identify the most significant parameters. Multivariate statistical analyses, including principal component analysis (PCA), linear discriminant analysis (LDA) and cluster analysis (CA), are employed to investigate the similarities and variations of juices.

5. Why PCA is used in Recovery Methods of Fruit Juice

Principal component analysis (PCA) is a statistical tool that can describe these sets of multivariate data of possibly correlated variables by relatively few numbers of linearly uncorrelated variables. In this process, the uncorrelated variables are formulated as a linear combination of the original variables and in decreasing order of importance so that the first component explains most of the original variation in the data. Each created component is orthogonal to the other to explain the variation that is not already explained by other components. PCA successfully summarizes several different characteristics in order to define fewer characteristics to explain the data well. Again, it removes the redundancy or correlation that can exist when explaining data using a large number of variables and creates a list of uncorrelated variables. In PCA, the variables are transformed in such a manner that they explain the variance of the dataset in a decreasing manner and their covariance is zero. And variance-covariance matrix can be drawn using those variables where the diagonal of the matrix defines the variance explained by those variables and non-diagonal elements are the pair wise covariance in such case. This matrix is transformed in such a way that all non-diagonal elements are zero. PCA is, thus, a statistical tool for exploratory data analysis and is applicable as a predictive model.

Method of Principle component construction as applied in the present study

Step 1: All variables are presented in a single scale.

Step 2: Variance-covariance matrix of those variables is constructed.

Step 3: Eigenvalues and eigenfactors of the covariance matrix are calculated.

Step 4: Eigenvalues are arranged from highest to lowest.

Step 5: First few components are taken, which together satisfy 80% variability of the data set.

There are three methods to find the number of Principle components to be taken for analysis.

- Cumulative 70–80% proportion of variance be satisfied.
- Eigenvalues be greater than unity.
- Scree plot of eigenvalues.

6. Case Study: Pineapple Juice Recovery

Pine apple juice is very popular largely because of its attractive flavour and refreshing sugar acid balance (Bartolome and Ruperez, 1995). More recently, the market has been equally receptive to fruit juice either obtained by traditional clarification process using gelatin and bentonite (Carvalho et al., 2008), as well as by processes using Ultrafilration (UF) and Microfiltration (MF) membranes. Enzymes are used to ensure optimal juice yield and a quality product that ensures consumer appeal. The use of pectinase and hemicellulase preparations in pineapple pulp maceration not only facilitates easy pressing and increased juice recovery but also ensures the highest possible quality of the end products (Kilara, 1982; Kashyap et al., 2001) by facilitating the easy with which both UF and MF are carried out.

Pectin is the essential structural component of fruits and vegetables (plant sources), in combination with hemi-cellulose and it form cell wall of fruit tissues. Pectins are a polymeric chain of D-galacturonic acid, partially esterified with methanol. Fruits are in the best condition for processing before their fully ripened state, as solubilized pectin and fruit tissues will badly affect the efficiency of separation of the juice and the result will be low yields. Different enzymes solubilize the substances and increase the extraction efficiency. This is as a result of their ability to degrade the cell walls, significantly lowering the viscosity of the recovered juices, hence, minimizing membrane fouling (Carvalho et al., 2008).

The collected pineapples were stored in 5°C for seven days before juice production. After washing, the pineapple was cut into small pieces by a stainless-steel knife and then processed in a blender for two minutes. The enzyme treatment of the pineapple pulp was conveniently carried out in three batches, each with 36 samples distributed evenly within the temperature ranges used. All the samples were subjected to two steps of juice extraction. In batch I, 500 g of fresh pineapple pulp was pressed in order to extract the juice in the initial step after being incubated at 35, 37.5 or 40°C for 30 minutes. Subsequently, in the second step, the pulp residue was mixed with water in the ratio of 1:3 (w/w) before treatment with 0.03% (w/w) pectinase and hemicellulase at 35, 37.5 40°C for 30 minutes with constant stirring. The reaction was terminated by immersion of the sample in boiling water for 5 min to inactivate enzymes before juice extraction. In batch II, 500 g of fresh pineapple pulp was treated with an enzyme mixture under similar conditions to batch I. The enzyme treated pulp residue was mixed with water in the ratio of 1:3 (w/w) and the juice was further extracted. In batch III, the above enzymes were employed under similar reaction conditions in both steps of juice extracted from pineapple pulp and residue. The control followed similar process trends as in batch I and II, but without enzyme treatment. The percentage of juice recovery and total phenolic, Brix, titrable acidity, turbidity, viscosity and sensory analysis of the products were done.

Statistical analysis

The amount of pineapple juice extraction was dependant on the type of treatment process, variation of enzyme applied and temperature of enzymatic reaction. In all cases, due to enzymatic reaction, the amount of juice extraction was increased. The results were given in Table 1.

The quality of the extracted pineapple juice depends on its physico-chemical characteristics as well as sensory analysis by an expert panel. In this case, physico-chemical properties were analyzed in terms of total soluble solid (°Brix), titrable acidity, viscosity, turbidity and total phenolics. These parameters determine the taste, colour, appearance, mouth feel, etc., of the juice. Table 2 shows that all of these values depend on the type of extraction method, enzymatic treatment and operational temperature.

Table 1: The effect of enzyme treatment on percent juice recovery from batches of pine apple pulp.

Batches	Enzyme treatment	35°C	37.5°C	40°C	Average
		% Mean Yield at different temperature ± SD			
I	Pectinase	78.69 ± 2.24	78.69 ± 1.58	79.02 ± 2.14	78.8
	Hemicellulase	80.2 ± 2.89	79.18 ± 1.40	78.75 ± 5.42	79.38
	Mixture	77.69 ± 0.80	79.23 ± 1.85	80.4 ± 2.55	79.11
	Control	71.86 ± 6.88	71.72 ± 5.18	65.95 ± 4.71	69.84
II	Pectinase	82.07 ± 1.64	80.36 ± 1.27	85.42 ± 4.7	82.62
	Hemicellulase	83.82 ± 1.4	82.56 ± 3.63	91.28 ± 1.92	85.89
	Mixture	86.28 ± 1.61	83.81 ± 3.229	93.33 ± 3.02	87.81
	Control	68.57 ± 5.95	70.02 ± 3.82	66.75 ± 4.69	68.45
III	Pectinase	81.99 ± 2.37	84.56 ± 2.36	89.33 ± 1.21	85.29
	Hemicellulase	83.55 ± 4.08	91.3 ± 2.80	93.17 ± 3.73	89.34
	Mixture	89.46 ± 4.17	88.83 ± 7.22	94.93 ± 1.14	91.07
	Control	69.43 ± 1.03	71.36 ± 3.29	69.89 ± 1.62	70.23

Table 2: Effect of enzyme treatment of pineapple pulp/residue on brix, titrable acidity, viscosity, turbidity and total phenolic.

Batch	Treatment	Parameter	35°C		37.5°C		40°C	
			Step-I	Step-2	Step-I	Step-2	Step-I	Step-2
I	Control	°Brix	14.0	4.8	14.6	4.2	14.5	4.6
		TA	1.128	0.433	1.139	0.0318	1.130	0.280
		Viscosity	3.9	2.6	4.1	3.0	4.1	3.4
		Turbidity	1.518	1.538	1.501	1.522	1.504	1.524
		TP	437		400		450	
	Pectinase	°Brix	14.5	4.6	15.2	5.4	15.0	5.1
		TA	1.152	0.398	1.140	0.342	1.143	0.336
		Viscosity	4.1	2.3	3.8	2.0	3.7	2.4
		Turbidity	0.828	1.128	0.816	1.424	0.845	1.548
		TP	594		560		595	
	Hemicellulase	°Brix	14.5	5.6	15.4	6.0	15.0	4.7
		TA	1.468	0.486	1.507	0.501	1.102	0.320
		Viscosity	3.8	2.0	3.6	1.9	3.0	2.4
		Turbidity	0.834	1.145	0.986	1.345	1.002	1.098
		TP	537		560		595	
	Mixture	°Brix	14.9	4.6	15.8	4.2	14.6	4.8
		TA	1.548	0.523	1.345	0.464	1.1520	0.468
		Viscosity	3.1	2.0	3.8	2.1	4.1	2.4
		Turbidity	0.846	1.234	0.845	1.402	0.946	1.365
		TP	596		580		575	

Table 2 contd. ...

... Table 2 contd.

Batch	Treatment	Parameter	35°C		37.5°C		40°C	
			Step-I	Step-2	Step-I	Step-2	Step-I	Step-2
II	Control	°Brix	14.8	4.0	14.9	4.5	14.1	4.8
		TA	1.130	0.318	1.128	0.433	1.139	0.315
		Viscosity	4.1	3.0	3.9	3.1	4.2	2.9
		Turbidity	1.423	1.556	1.545	1.654	1.496	1.502
		TP	425		440		435	
	Pectinase	°Brix	15.0	5.4	15.2	4.8	14.9	5.3
		TA	1.150	0.402	1.152	0.335	1.149	0.398
		Viscosity	2.9	2.1	3.0	2.5	2.8	2.4
		Turbidity	0.845	1.326	0.946	1.502	0.875	1.524
		TP	597		601		597	
	Hemicellulase	°Brix	15.0	4.8	14.9	4.1	14.6	5.0
		TA	1.334	0.364	1.402	0.432	1.156	0.338
		Viscosity	2.6	2.2	2.6	2.4	2.9	2.3
		Turbidity	0.845	1.153	0.966	1.075	0.927	1.112
		TP	600		595		610	
	Mixture	°Brix	15.1	5.0	14.9	4.6	14.5	5.1
		TA	1.546	0.467	1.438	0.344	1.152	0.467
		Viscosity	2.8	2.0	2.9	2.1	3.0	2.6
		Turbidity	0.846	1.323	0.845	1.024	0.892	1.009
		TP	600		612		615	
III	Control	°Brix	14.5	4.0	14.1	4.4	14.3	4.2
		TA	1.342	0.456	1.167	0.334	1.153	0.289
		Viscosity	4.1	3.0	3.7	2.6	4.0	2.8
		Turbidity	1.235	1.645	1.436	1.626	1.507	1.562
		TP	435		437		497	
	Pectinase	°Brix	15.0	5.1	14.8	4.9	14.5	5.2
		TA	1.153	0.364	1.150	0.398	1.402	0.335
		Viscosity	3.0	2.4	2.7	2.3	2.9	2.5
		Turbidity	0.867	1.143	0.976	1.075	0.872	1.098
		TP	605		600		630	
	Hemicellulase	°Brix	15.0	5.2	15.1	5.0	14.9	4.8
		TA	1.152	0.280	1.156	0.345	1.238	0.404
		Viscosity	2.4	2.0	2.7	2.2	2.9	2.3
		Turbidity	0.895	1.486	0.852	1.503	0.879	1.152
			590		584		623	
	Mixture	°Brix	15.0	5.2	15.2	4.6	14.7	4.6
		TA	1.554	0.543	1.522	0.436	1.465	0.476
		Viscosity	3.1	2.6	2.6	2.3	2.9	2.7
		Turbidity	0.867	1.456	0.896	1.290	0.982	1.109
		TP	602		598		635	

The collected fruit juices from different treatment processes were analyzed by a taste panel. The sweetness, tartness, colour, odour and overall acceptability were judged and rated on the five-point Hedonic scale. Where 1.00–1.80 is not acceptable; 1.81–2.61 is fair acceptable; 2.62–3.42 is acceptable; 3.43–4.23 is very acceptable; 4.24–5.00 is highly acceptable. The average sensory values are given in Table 3.

Table 3: Average score on a five-point Hedonic scale of the organoleptic properties of the extracted juice.

Batch	Treatment	Sweetness	Tartness	Colour	Odour	Overall acceptability
I	Control	4.7	3.2	4.7	2.8	4.2
	Pectinase	4.0	4.1	3.6	2.4	3.6
	Hemicellulase	4.4	4.5	4.0	2.7	4.0
	Mixture	3.8	4.6	3.4	2.4	3.9
II	Control	4.6	3.4	4.5	2.9	4.6
	Pectinase	4.2	4.3	4.0	2.5	3.4
	Hemicellulase	4.6	4.4	4.4	2.8	4.1
	Mixture	3.4	4.5	3.6	2.3	3.8
III	Control	4.2	4.2	4.6	2.8	4.1
	Pectinase	3.7	4.4	3.8	2.2	3.4
	Hemicellulase	4.4	4.0	4.6	2.9	3.8
	Mixture	3.2	4.7	3.8	2.2	3.7

The recovery percentage of fruit juice is dependent on the employment of enzyme(s) (single/mixture), temperature of the enzymatic action and time of reaction. Brix, titrable acidity, viscosity, turbidity and total phenolics are considered as quality attributes of the final product. Final quality of the extracted juice was judged on the basis of sensory evaluation in terms of sweetness, tartness, colour, odour and overall acceptability and ranked on the Hedonic scale (1–5). The main objective of commercial juice extraction is the production of large quantities with good physico-chemical properties and remarkable sensory output.

Eleven variables are involved in the juice production. Final quality and volume of the product are regulated by these variables. To determine the correlation and interaction between these, multivariable data Principal Component Analysis (PCA) is applied.

The box whisker plot simultaneously shows different important features of the data, such as trends, variability and identification of the outliers. This plot displays the three quartiles, the minimum and the maximum of the data on a rectangular box and it is aligned either horizontally or vertically. The horizontal line in the box represents the median and the vertical bars display the range of the data. The box plot study is made in order to observe differences among juices extracted by different methods.

Figure 4 shows the Brix results of different types of pineapple juice obtained through different extraction processes. Brix of different samples are more consistant, dispersion is very low, whereas data for total phenolic contents are more dispersed.

Then the variables were plotted for Scree plot which contributes the significance of principal components. Which is indicated by eigenvalues. This plot indicates that there is a significant contribution by the first four components, which contribute 85.8% cumulative correlation. Hence, we consider four principle components. Here, all of these eigenvalues are greater than 1.0.

Again the data plotted in score plot based on first two components, it is observed that the majority of the group I components present in the left hand side. The first component is very effective in separating the components of groups II and III. A few components of group I are misclassified in groups II and III, but components of group II are unable to separate components from group III. Some of the components of groups II and III are misclassified on the right-hand side. Principle components against variable

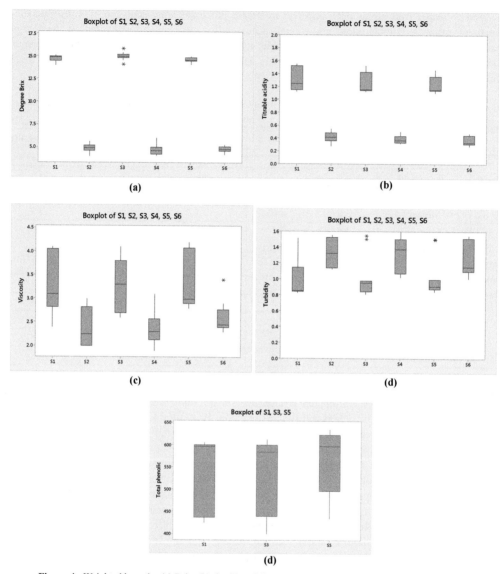

Figure 4: Weighted box plot (a) Brix, (b) titrable acidity, (c) viscosity, (d) turbidity, (e) total phenolics.

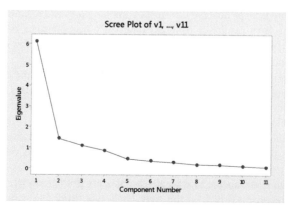

Figure 5: Scree plot using Eigenvalues.

coefficients indicate that principal components are linearly correlated with different variables. This will provide important inputs during cluster analysis.

Outlier plot of variables indicated that the result of observation 31 is out of the range, hence, this experimental observation should be excluded when determining the interaction and correlation of different results.

The loading plot graphs the coefficients of each variable for the first component versus the coefficients for the second component. Use the loading plot to identify which variables have the largest effect on each component. Loadings can range from –1 to 1. Loadings close to –1 or 1 indicate that the variable strongly influences the component. Loadings close to 0 indicate that the variable has a weak influence on the component.

Loading plot of the variables indicates that v1, v4 and v7 are well correlated with other variables. On the other hand, v10 and v6 are correlated with other variables and explained to some extent. However, v2, v3, v5, v9, v11, v8 and v6 are not properly correlated with other variables. The feature should be properly explained by PCA.

This graph indicates the similarity of different variables. Variables v1, v6, v8, v2, v3 are similar more than 70.07% and variables v4, v11, v5, v9, v7 and v10 are similar upto 50%. And indicates the separate two clusters. Variable v1 is highly similar to variable v6, v4 is highly similar to v11, v5 is to v9 and v7 to v10. From previous study, some variables are not properly explained or correlated with others, but Dendrogram results indicate that highly correlated data are similar to non-correlated data.

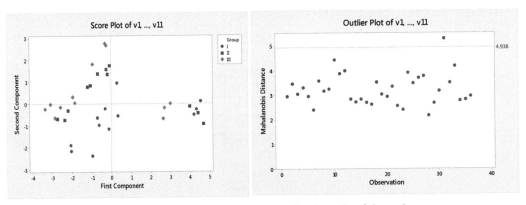

Figure 6: Score plot of two components and out layer plot of observations.

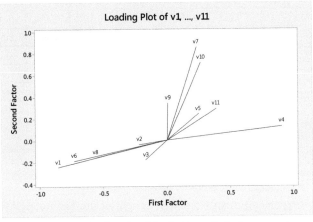

Figure 7: PCA loading plot of the first two principal components.

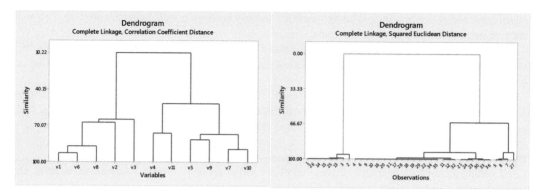

Figure 8: Dentogram linkage and correlations.

Table 4: Unrotated and rotated factor matrices obtained using PCA.

variable	F1	F2	F3	F4	F5	F6	F7	F8	F9	F10
Unrotated										
Brix	−0.869	0.392	−0.023	0.150	−0.084	−0.163	0.099	0.145	0.022	0.031
TA	−0.548	−0.049	0.405	−0.632	0.330	−0.072	0.106	0.067	0.064	0.007
Viscosity	−0.508	−0.264	0.635	0.426	0.142	0.133	−0.138	0.114	−0.125	0.042
Turbidity	0.740	−0.541	−0.077	−0.047	−0.136	−0.152	−0.119	0.255	0.147	0.080
TP	0.755	0.363	−0.115	0.120	0.316	−0.363	−0.130	−0.057	−0.105	0.076
Sweetness	−0.936	0.172	−0.041	−0.128	−0.170	0.016	0.007	−0.003	−0.089	−0.013
Tartness	0.737	0.402	0.172	−0.364	−0.240	0.142	−0.049	0.044	−0.148	0.165
Colour	−0.858	0.092	0.242	−0.041	−0.141	−0.097	−0.290	−0.209	0.178	0.073
Odour	0.583	0.686	0.028	0.161	0.172	0.290	−0.021	0.072	0.203	0.011
Overall acceptability	0.765	0.327	0.424	−0.082	−0.179	−0.154	−0.122	0.064	−0.014	0.066
Rotated										
Brix	0.941	0.206	−0.124	0.101	0.065	0.155	0.109	−0.073	−0.002	−0.012
TA	0.208	0.037	−0.117	0.939	0.148	0.102	0.134	−0.080	0.050	−0.007
Viscosity	0.181	0.171	−0.938	0.117	0.106	−0.051	0.148	−0.084	0.034	−0.003
Turbidity	−0.752	−0.111	0.123	−0.193	0.137	−0.157	−0.084	0.097	−0.555	0.027
TP	−0.264	−0.241	0.225	−0.191	−0.288	−0.150	−0.817	0.103	−0.033	0.021
Sweetness	0.739	0.167	−0.080	0.219	0.239	0.311	0.323	−0.158	0.125	−0.041
Tartness	−0.291	−0.839	0.277	−0.026	−0.274	−0.135	−0.111	0.089	−0.001	−0.144
Colour	0.585	0.138	−0.272	0.242	0.207	0.183	0.212	−0.616	0.075	−0.001
Odour	−0.057	−0.337	0.132	−0.186	−0.866	−0.108	−0.245	0.086	0.042	0.020
Overall acceptability	−0.278	−0.723	0.036	−0.067	−0.237	−0.356	−0.285	0.009	−0.099	0.346

The second Dendrogram plot indicates the first cluster of different observations, covering about 90% similarity, and the 2nd cluster covers up to 66% similarity. Out of 36 observations, 34 are assigned correctly with their corresponding group, but one is wrongly assigned as group III instead of group II and another assigned as II instead of I that means more than 94% are correctly assigned in the right group.

A rotated varimax matrix draw figures about the magnification of the data information. The projected data of rotated factor 1 versus rotated factor 2 shows the clustering of the data according to their juice types. A degree of dispersion is also observed. Dispersed data are placed another way in the form of cluster. Outlier data is placed totally outside of the cluster. From this cluster figure, we can predict the type of juice, interaction between different data and ambiguity of data if there is any.

7. Conclusion

PCA is one of the most popular chemometric methods in food research. PCA score plots are commonly used in order to group samples based on their similarities or dissimilarities, usually using a 2D or 3D projection of the samples. In addition, an interpretation about how the variables influence this pattern is achieved through a loadings plot. However, some practices can contribute to misinterpreted results, such as:

- Non-equivalent axes scales in score plots that lead to misinterpretation, i.e., dissimilar samples may appear similar.
- Excessive stretching of axes yielding to plots with filled plot area which otherwise would not have been fitted in (Geladi et al., 2003; Kjeldahl and Bro, 2010).
- Amount of variance described by each PC must be noted as different values may lead to different interpretations.
- PCA and PCA-like techniques are gaining increased application in food science and technology and its allied fields (Beebe et al., 1998).

These, along with certain toolboxes, aid in pattern recognition, classification and multivariate calibration. However, in order to obtain meaningful results, meaningful data, logical analysis and realization of the purpose of analysis are vital. Application of sophisticated tools without conceptual understanding of the tools, misuse of data preprocessing, pattern characterization, classification of models, etc., will yield erroneous results and thereby cause limitations to the utility of these methods. In this regard, we strongly discourage the use of computational software to 'click and go' without the appropriate knowledge on what was obtained or how to interpret holistically the outputs.

References

Aghajanzadeh, S., Ziaiifar, A.M., Kashaninejad, M., Maghsoudlou, Y. and Esmailzadeh, E. 2016. Thermal inactivation kinetic of pectin methylesterase and cloud stability in sour orange juice. Journal of Food Engineering 185: 72–77.

Augusto, P.E.D., Ibarz, A. and Cristianini, M. 2012. Effect of high-pressure homogenization (HPH) on the rheological properties of tomato juice: Time-dependent and steady-state shear. Journal of Food Engineering 111(4): 570–579.

Bartolome, A.P. and Ruperez, P. 1995. Dietary fiber in pineapple fruit. Journal of Clinical Nutrition 49(S2): 61–S263.

Beebe, K., Pell, R. and Seasholtz, M.B. 1998. Chemometrics: A Practical Guide, Wiley-Interscience, New York.

Branco, I.G., Moraes, I.C.F., Argandoña, E.J.S., Madrona, G.S., Dos Santos, C., Ruiz, A.L.T.G. et al. 2016. Influence of pasteurization on antioxidant and *in vitro* anti-proliferative effects of jambolan (*Syzygium cumini* (L.) Skeels) fruit pulp. Industrial Crops and Products 89: 225–230.

De Carvalho, L.M.J., de Castro, I.M. and da Silva, C.A.B. 2008. A study of retention of sugars in the process of clarification of pineapple juice (*Ananas comosus*, L. Merril) by micro- and ultra-filtration. Journal of Food Engineering 87: 447–454.

Dong, W., Hu, R., Chu, Z., Zhao, J. and Tan, L. 2017. Effect of different drying techniques on bioactive components, fatty acid composition, and volatile profile of robusta coffee beans. Food Chemistry 234: 121–130.

Frau, M., Simal, S., Femenia, A., Sanjuan, E. and Rossel, O.C. 1999. Use of principal component analysis to evaluate the physical properties of Mahon cheese. Eur Food Research Technology 210: 73–76.

Geladi, P., Manley, M. and Lestander, T.A. 2003. Scatter plotting in multivariate data analysis. Journal of Chemometrics 17(8-9): 503–511.

Guan, Y., Zhou, L., Bi, J., Yi, J., Liu, X., Chen, Q. et al. 2016. Change of microbial and quality attributes of mango juice treated by high pressure homogenization combined with moderate inlet temperatures during storage. Innovative Food Science & Emerging Technologies 36: 320–329.

Heinmaa, L., Moor, U., Põldma, P., Raudsepp, P., Kidmose, U. and Lo Scalzo, R. 2017. Content of health-beneficial compounds and sensory properties of organic apple juice as affected by processing technology. LWT - Food Science and Technology 85: 372–379.

Hyson, D. A. 2011. A comprehensive review of apples and apple components and their relationship to human health. Advances in Nutrition (Bethesda, Md.) 2(5): 408–420.

Karacam, C.H., Sahin, S. and Oztop, M.H. 2015. Effect of high-pressure homogenization (microfluidization) on the quality of Ottoman Strawberry (*F. Ananassa*) juice. LWT - Food Science and Technology 64(2): 932–937.

Kashyap, D.R., Vohra, P.K., Chopra, S. and Tewari, R. 2001. Applications of pectinases in the commercial sector: A review. Bioresource Technol. 77: 215–227.

Kaya, Z., Yıldız, S. and Ünlütürk, S. 2015. Effect of UV-C irradiation and heat treatment on the shelf life stability of a lemon–melon juice blend: Multivariate statistical approach. Innovative Food Science & Emerging Technologies 29: 230–239.

Kilara, A. 1982. Enzymes and their uses in the processed apple industry: A review. Process Biochemistry 17: 35–41.

Kjeldahl, K. and Bro, R. 2010. Some common misunderstandings in chemometrics. Journal of Chemometrics 24: 558–564.

Kubo, M.T.K., Augusto, P.E.D. and Cristianini, M. 2013. Effect of high pressure homogenization (HPH) on the physical stability of tomato juice. Food Research International 51(1): 170–179.

Pokorny, J., Kalinova, L. and Velisek, J. 1995. Time intensity bitterness evaluation of bitter liquers. Potrav Vědy 13: 409–417.

Rawson, A., Patras, A., Tiwari, B.K., Noci, F., Koutchma, T. and Brunton, N. 2011. Effect of thermal and non-thermal processing technologies on the bioactive content of exotic fruits and their products: Review of recent advances. Food Research International 44(7): 1875–1887.

Velisek, J., Mikulcova, R., Mikova, K., Kassahun, B.E., Link, J. and Davidek, J. 1995. Chemometric investigation of mustard seed. Potrav Vědy 13: 1–12.

Wlodarska, K., Pawlak-Lemanska, K., Khmelinskii, I. and Sikorska, E. 2016. Explorative study of apple juice fluorescence in relation to antioxidant properties. Food Chemistry 210: 593–599.

CHAPTER 21

Use of Artificial Neural Networks in Optimizing Food Processes

RA Conde-Gutiérrez, U Cruz-Jacobo and *JA Hernández**

1. Introduction

In the sustainable food processes, the transformation of raw material and energy is carried out in order to obtain a desired product. However, to perform the transformation of raw material and energy, various interacting factors are involved. Also, these transformations are increasingly coupled with energy saving. The complexity in the processes lies in the fact that some perform chemical reactions, where the input variables interact with each other and undergo changes due to their physical and chemical properties. In these processes, it is essential to have a model capable of predicting their behavior, contemplating the conditions of entry, in order to simulate the process and obtain benefits such as: Improvement of the operation, economic savings and control of the process.

The simplest way to model a process is through a linear correlation, but in most processes, due to the complexity of using multiple input variables and the chemical reactions involved, the behavior of the process is not linear. An artificial neural network (ANN) is considered as a model of mathematical-computational type that allows one to simulate a desired output of a process, independently of whether the behavior is linear or non-linear. This feature provides great strength to the model, which is demonstrated by applying it in various investigations, such as those presented by Mohanraj et al. (2012), Kalogirou (2001), Dande and Samant (2018) and Jani et al. (2016a,b, 2017). ANNs are able to model behaviors with acceptable accuracy, even in complex systems, such as those presented by the researchers Jani et al. (2016a,b), namely, simulating the performance of a rotary dehumidifier and the cooling capacity for a solid desiccant, both in air conditioning systems.

An artificial neural network is founded on emulating the behavior of human reasoning, by carrying out learning through experience and obtaining knowledge through a set of data. The ANN performs an imitation of the brain's neuronal structure through a computer with the ability to perform parallel calculation.

In food engineering, some processes, such as classification, drying and cooling, are focuses of interest. In these processes, several challenges which imply a degree of complexity during the production

Centro de Investigación en Ingeniería y Ciencias Aplicadas (CIICAp-IICBA), Universidad Autónoma del Estado de Morelos (UAEM), Av. Universidad 1001, Col. Chamilpa, 62209, Cuernavaca, Morelos, México.
  Emails: roberto.conde@uaem.mx; ulises.cruzjac@uaem.edu.mx
* Corresponding author: alfredo@uaem.mx

of food are posed. Some challenges consist of classifying raw material based on shape, size, colour, etc., in order to carry out a certain function, determining the exact time of drying of any fruit or vegetable, taking into account its state of maturity, colour, taste, etc., and specifying the temperature necessary to preserve a finished product, preventing it from decomposing. These brief examples of food problems will be described in detail throughout the chapter, where the application of the ANN model is justified. At present, human oversight is still used in various food processes due to the complexity of the processes or the challenges that arise. For this reason, an intelligent model such as ANN, which can solve or improve the process, is implemented, thus reducing errors caused by the human factor.

2. Basic Principles of Artificial Neural Networks

In nature, there are organisms that perform well-defined functions and that have been a focus of interest to imitate and take advantage of the function. For example, the sunflowers that follow the passage of the sun throughout the day and the colonies of bees and ants that organize and keep the queen alive in order to continue with the progeny, among others. However, these organisms are not alone in providing functions worthy of imitation; human beings themselves have organs that perform very specific functions and the organ of greatest interest is the human brain. The basic principle of artificial neural networks consists of learning from acquired experiences to perform a function.

2.1 Biological foundation of the artificial network

The brain is the main organ of the human body and is very to analyse, in comparison to other organs, and this is basically due to its unique capacity to carry out certain functions. According to the first researches conducted on the human brain, there were about 100 billion cells, called neurons, but in the investigation of Herculano-Houzel (2009), an average of 14 billion was counted. Neurons perform the function of communicating with each other accurately and rapidly over long distances and with other cells. Electrical signals, known as nerve impulses, are transmitted through the neurons. The body of a neuron is the source of synthesis of an organic molecule and these molecules are transported as a nerve impulse through the axon until reaching the terminals, where they are stored to be applied in the stimulation, as described by Williams (2003). Nodes of Ranvier are interruptions between the axon that enable the nervous impulse to travel at a higher speed. The nucleus of the neuron is related to the synthesis of ribonucleic acid (RNA) and dendrites are the extensions of neurons that appear to be branches or points that extend from the cell body with the function of receiving chemical messages from other neurons (Fig. 1).

When physically connecting a population of neurons to perform a function, it is called a neural network. According to the work of researchers Andreev et al. (2018), a network consists of the neurons receiving signals to process together, which implies an electrochemical process. In this process, once a neuron has been excited, it will depolarize, transmitting a signal through the axon in order to excite the neurons nearby. Finally, from the biological functioning of brain neurons, artificial neural networks were developed to emulate the learning process.

3. General Conformation of an Artificial Neural Network

An artificial neural network is a programmable model that involves the interaction of neurons in different layers to carry out a learning process and simulate a desired behavior. As happens in the human brain, in ANN, the neurons are interconnected and pass a signal, each connection carrying a numerical value, called weight, plus a factor known as bias. Contemplating the research of Yegnanarayana (1999) and Sablani et al. (2007), some of the most important advantages when applying the ANNs are:

1. Fault tolerance: Unlike the human brain, where the decomposition of nerve cells affects the function of neurons, the ANN is able to tolerate incomplete data. Having a smaller amount of data to carry out the learning does not prevent the ANN from performing its simulation function.

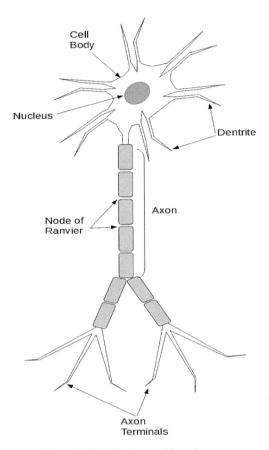

Figure 1: Biological composition of a neuron.

2. Flexibility and adaptability: The ANN automatically adjusts to the environment to be simulated and has the capacity to adapt its weights for the changes that are carried out.
3. Nonlinearity: The ANN can simulate linear behaviors as non-linear.
4. High computational speed and collective computation: The ANN performs simulation quickly, since it operates in parallel and, in turn, carries out distribution tasks.

3.1 Model of an artificial neural network

The origin of the ANN model begins with the researchers McCulloch and Pitts (1943), who proposed for the first time a model that implements an artificial neuron. The combination of the two researchers allowed us to take advantage of the biological processes of the human brain and the application of artificial neural networks in artificial intelligence.

In the inputs of the model ANN *In* for (1, 2 .... *k*), a corresponding weight factor (*Wi*) is assigned to it (Fig. 2). In addition, an external factor of bias is included. The input weights when added with the bias (*b1*), an input occurs (n_s) which are described as:

$$n_s = Wi_1 \times In + Wi_2 \times In_2 + \cdots + Wi_k \times In_2 + b1 \tag{1}$$

Subsequently, the input (n_s) passes through a transfer or activation process, which will produce an output. The weighted sum of the patterns obtained by the transfer function (*Wo* and *b2*) is carried out, where they are coupled into an equation, which can be written as:

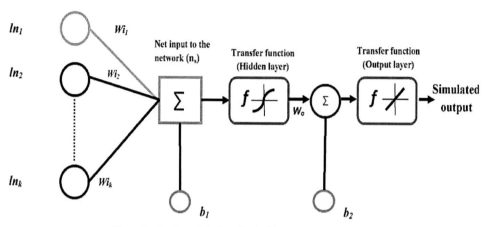

Figure 2: Basic model of an ANN with two transfer function.

$$Out = Wo \times \left[f\left(Wi_k \times In_k + b1\right)\right] + b2 \qquad (2)$$

The transfer function f can be linear or non-linear and is used to calculate the response output of a neuron. In neurons that are in the same layer, the same transfer functions are used, as mentioned by Sivanandam et al. (2006). Some of the functions most used today are given in Table 1.

Table 1: Transfer functions used by ANN.

Transfer function	Equation	Plot
Linear	$f(x) = x \quad (3)$	Linear Function
Binary Step	$f(x) = \begin{cases} 0, x < 0 \\ 1, x \geq 0 \end{cases} \quad (4)$	Binary Step

Table 1 contd. ...

... Table 1 contd.

Transfer function	Equation	Plot
Logistic function	$f(x) = \log sig(x) = \dfrac{1}{1+e^{-x}}$ (5)	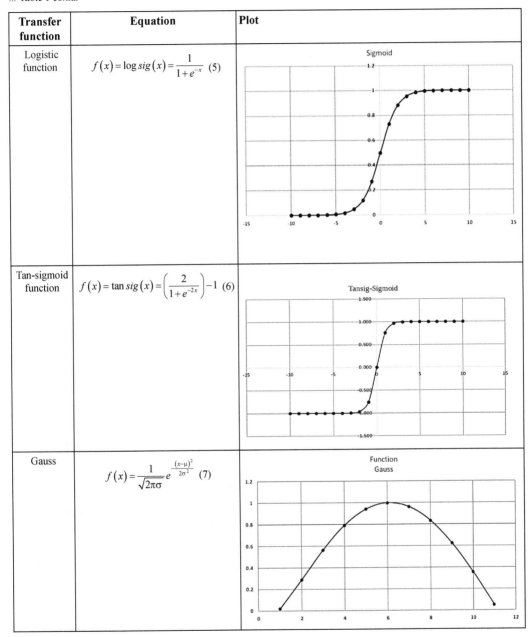
Tan-sigmoid function	$f(x) = \tan sig(x) = \left(\dfrac{2}{1+e^{-2x}}\right) - 1$ (6)	
Gauss	$f(x) = \dfrac{1}{\sqrt{2\pi}\sigma} e^{-\frac{(x-\mu)^2}{2\sigma^2}}$ (7)	

3.2 Learning procedure during the formation of an artificial network

The main feature of the ANN is to carry out a learning process which can be supervised or unsupervised. The difference between both types of learning is summarized in that supervised learning provides a correct answer to learn during training but in unsupervised learning the objective is not present and is adjusted to the observations. In the ANN, the most used learning is supervised, where an algorithm known as "Backpropagation" is applied. According to Rumelhart et al. (1985), the "Backpropagation" algorithm consists of reducing the error between the desired and the simulated output using mathematical methods, such as the descending gradient (Rumelhart et al., 1985). The operation of the algorithm consists of applying a pattern to the input as a stimulus, later, it is propagated through the layers of the network until an output is obtained. The result is equated with the required output and the difference between the two is calculated (error). The errors propagate backward, from the exit to the intermediate neurons, in order to improve learning and minimize the error as best as possible, so that the simulated value is very similar to the desired one. Currently, multiple training algorithms have been developed, with the aim of not falling into overfitting, and are used in ANN to cover various applications (Fig. 3 and Table 2).

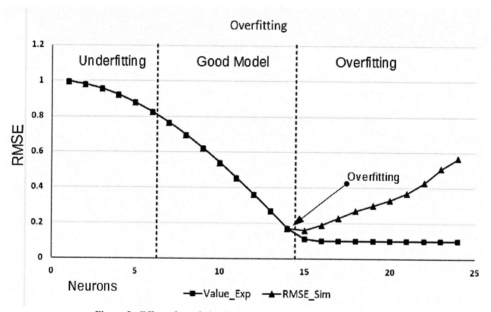

Figure 3: Effect of overfitting in the development of the ANN model.

Table 2: Training algorithms Backpropagation used in ANN proposals in various applications.

Training Algorithm	Application	Reference
Radial basis functions neural networks (RBFNNs)	Microbial growth on limit conditions	Fernández-Navarro et al., 2010
Recurrent Neural Network (RNN)	Efficient language modelling	Grachv et al., 2019
Broyden, Fletcher, Goldfard and Shanno (BFGS)	Osmotic Treatment of fish	Ćurčić et al., 2015
Convolutional neural networks based	Food recognition	Ciocca et al., 2017
Wavelet Neural Network	Optimization	Masoumi et al., 2014
Sequential Monte Carlo	Optimization strategy	Freitas et al., 2000
Feedforward - Bird Mating optimizer (BMO)	Wisconsin breast cancer classifier	Askarzadeh and Rezazadeh, 2013
Artificial neural network-genetic algorithm (ANN-GA)	Solid-phase micro extraction	Dil et al., 2016
Seasonal (SANN)	Seasonal time series	Hamzaçebi, 2018

4. Types of Structuring in Artificial Networks

The general structure of an ANN is made up of several layers, where the neurons, when interacting with each other, allow one to obtain a simulated output. The layers of the ANN are the following:

a) The input layer. In this layer, signals from the environment are received through data acquirer or sensors that monitor the system.
b) The hidden layer. In this layer, upon receiving the stimuli from the input layer, an output is emitted without maintaining contact with the outside. An internal representation of the output to be simulated is made by processing the information and obtaining learning patterns.
c) The output layer. In this layer, a direct signal is sent to the outside, where the signal is the output resulting from the interaction of the previous layers, obtaining a simulated output.

4.1 Single layer structure

The simplest design of an artificial network is formed by a single layer. In this structure, the outputs are fed directly from the inputs and are considered as the simplest form of Feedforward. The information passes through the single-layer in one direction (Fig. 4). This type of network is usually used to classify linear patterns and filtrations (Da Silva et al., 2017).

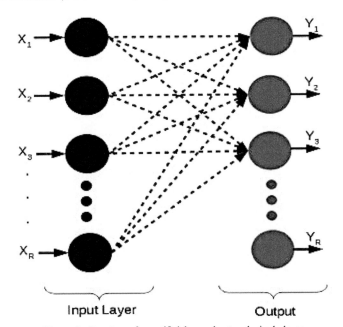

Input Layer   Output

Figure 4: Structure of an artificial neural network single-layer.

4.2 Multilayer structure

Multilayer networks are composed of multiple layers, where each layer has one or more hidden neurons and are also considered as a form of Feedforward (Fig. 5). This type of ANN consists of sequentially grouping several layers from the input to the output. The Multilayer networks are the most applied design to solve problems of classification, as well as problems of choice. According to Cantarella et al. (2005), in order to carry out a Multilayer network, it is necessary to make the following specifications:

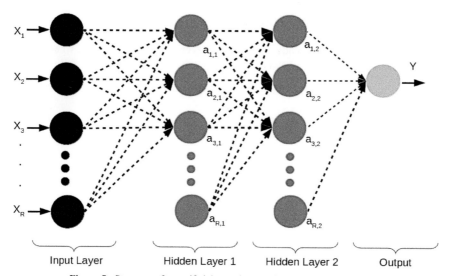

Figure 5: Structure of an artificial neural network multilayer feedforward.

1. Number of layers in the input, in the hidden and output.
2. Number of processing units (weights and bias), obtained in each layer.
3. Activation functions. Generally, in the same layer, the same activation function is used to calculate the processing units.

4.3 *Structure of recurrent type*

This type of network is characterized by having recurrent connections where the information flows forward (Feedforward) and backwards (Feedbackward), through connections that communicate with neurons in previous layers, which allows one to obtain a dynamic temporal behavior for a certain time sequence (Fig. 6). Because of the feedback characteristics, recurrent networks are used for identification and optimization systems and control processes, among others (Da Silva et al., 2017).

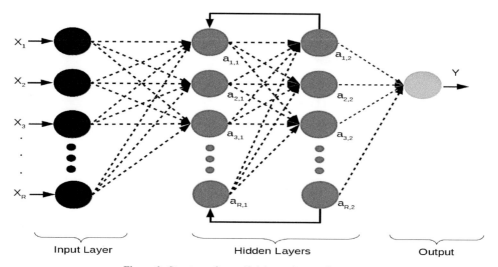

Figure 6: Structure of an artificial neural network recurrent.

5. General Steps for the Development of the ANN Model

In order to carry out the development of artificial neural networks, it is necessary perform a series of steps. These steps are included in three main tasks: Generation of data, Training-learning process and Selection of the best architecture. Each of these tasks is related to the others, so if a failure occurs in any of them, it directly affects the simulated output, preventing one from obtaining a desired precision. Among various works, the investigations by Bassam et al. (2014), Kumar and Singh (2018), Tang et al. (2018) and Leśniak et al. (2018), highlight when proposing a general algorithm on the development of an ANN (Fig. 7), which encompasses the three main tasks and shows how their interaction leads to obtaining a satisfactory simulated output.

5.1 Generation of data

In order to begin the development of the artificial neural network, it is necessary to provide a set of data, with the purpose of obtaining the appropriate information to carry out the learning. The data to train the network can come directly from experimental equipment, simulations of mathematical methods or compiled in the literature. Before entering the ANN model, the data must be homogenized (*Hom*), with the objective that the data is a numerical interval of [0.1- 0.9], since during the training of the ANN it is easier to handle the information in this form. The equation most used to perform the normalization of data can be written as:

$$y_{i,Hom} = \left(\frac{Y_{i,Real} - Y_{min}}{Y_{max} - Y_{min}} \right) \times 0.8 + 0.1 \qquad (8)$$

5.2 Training of ANN

Network training consists of randomly dividing the normalized data into three groups. The training and test group is used in order for the ANN model to perform the learning and finally compare it with the validation data. The learning is performing in an iterative way, where the objective is to minimize the difference between the simulated value and the desired value. Different algorithms of Backpropagation have been used for their efficiency to reduce the error and the speed of adjustment of the weights and bias in order to obtain a better precision of the output to simulate (Table 2).

5.3 Selection of the optimal network

During the training of the network, an internal comparison is made between the data obtained by the simulated value (Sim) and the real value (Real), in order to detect if it is necessary to increase the number of transfer functions through the hidden layer. To determine the optimal architecture of the artificial network, there are several statistical criteria. However, the most used are: Root Mean Square Error (RMSE), to compare the error obtained during the training, Mean Percentage Error (MPE) to compare the values directly (simulated vs real) and obtain a percentage error, and, to determine the correlation between the values, the Coefficient of Determination (R^2) is applied. These statistical criteria can be described as:

$$RMSE = \sqrt{ \frac{ \sum_{i=1}^{n} \left(P_{Sim(i)} - P_{Real(i)} \right)^2 }{ n } } \qquad (9)$$

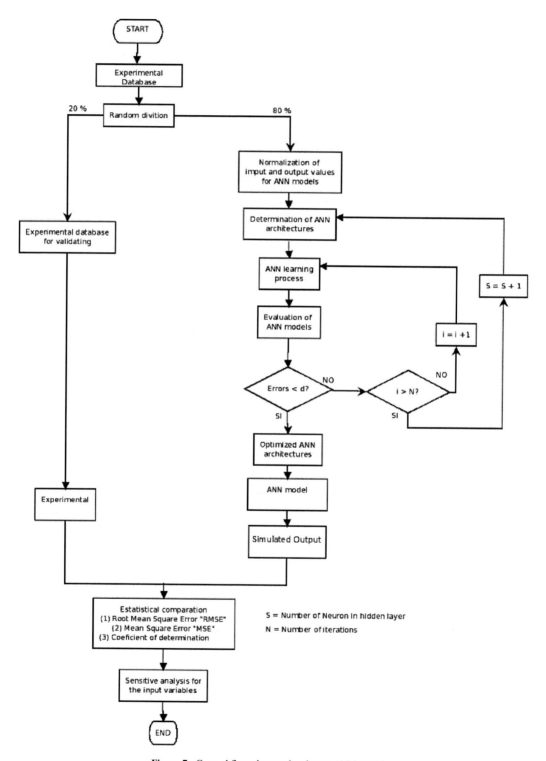

Figure 7: General flow chart to develop an ANN model.

$$MPE = \frac{\sum\limits_{i=1}^{n}\left(\dfrac{P_{Real(i)} - P_{Sim(i)}}{P_{Real(i)}}\right)}{n} \times 100(\%) \tag{10}$$

$$R^2 = 1 - \frac{\sum\limits_{i=1}^{n}\left(P_{Real(i)} - P_{Sim(i)}\right)^2}{\sum\limits_{i=1}^{n}\left(P_{Real(i)} - \overline{P}_{Real(i)}\right)^2} \tag{11}$$

where: $\overline{P} = \dfrac{1}{n}\sum\limits_{I=1}^{N} p_i$; $P_{Sim(i)}$ value taken from the population predicted by the ANN model; $P_{Real(i)}$ set of desired values.

Once the best network architecture with the best statistical values has been chosen, a general validation is carried out. One way to perform this validation is through the use of various statistical criteria, such as the Fischer test and the T-student test, as applied by Márquez-Nolasco et al. (2017). With these tests, it is demonstrated that the population simulated and the population real can be considered as the same or similar through the variance and mean. Finally, after all these steps, it is concluded that the ANN obtained is able to model a linear or non-linear behavior with an acceptable precision and, therefore, reliable to perform control, identification, classification and other functions.

6. Applications of the ANN Model in Food Processes

Artificial neural networks and their ability to model nonlinear behaviors are ideal for modelling food processes. Within the food processes, the three fundamentals highlighted here are classification, drying and cooling. Each of these processes is important to the food industry, as they are necessary during production. For this reason, several researchers have focused on modelling using ANN to improve each of these processes and achieve a global benefit in the food industry.

6.1 Food classification

This process is of great importance to the food industry, basically due to the fact that with the classification of food it is possible to determine the quality of the raw material, control its storage, etc., and therefore, obtain a satisfactory quality of the final product.

6.1.1 Introduction

Currently, the acquisition of images is used to classify the raw material in food, since with this, the classification process can be carried out more quickly and effectively. The analysis and classification of images is very useful to automatic operation in industries. At present the systems that use this technique are capable of solving few problems, in sectors like security, industrial, scientific and commercial. In the industries of food, it has a particular relevance, since price quality and useful life is very important for clients in the market. The automation of classifier system with artificial intelligence can decrease higher degree of error in manual selection.

In the automation selection process of food, it is necessary to recognize the characteristics of the product, such as: color, texture, shape, weight, etc. However, some characteristics are difficult to classify, for example, the different colors that a product can present. In order to avoid errors in the automated selection process, images of three-dimensional objects are digitized. This digitization is represented in a matrix, where each element in this matrix is assigned a characteristic numerical value, called pixel. Figure 8 shows a block diagram that describes a classification system applied to food, composed as follows: Light source, image capture equipment, computer and processing software. In this way, the data with which the Neural Network is trained are acquired.

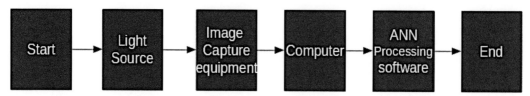

Figure 8: General scheme of data acquisition in a food classification process to train an ANN model.

6.1.2 Description of problem

In the process of classification in the food industry, the need to solve various problems arises, depending on the type of product desired, such as those presented by the following researchers. Debska and Guzowska-Świder (2011) show the importance of quality control and authenticity of products in the brewery industry. The procedure for evaluating the quality and potential alterations in the industry depends on manual inspections and on the results of laboratory tests. However, manual inspections are prone to human errors and laboratory tests are time-consuming. For this reason, it is necessary to build automated decision-making systems for product evaluation, to classify brand origin as authentic or altered. Wang et al. (2017) investigated the food security incidents that occurred in China. Due to the fact that, when moving food from the place of origin to the consumer, it can lose quality and can even break down, it is necessary to develop a system to manage the traceability of food. Kılıç et al. (2007) looked into the problems that arise when consumers choose products according to their visual properties, when purchasing them in the food section of a market. Inspection of the products, such as vegetables, cereals, and fruits, is based on visual properties, i.e., colour, shape and size. These properties are used to distinguish between high- and poor-quality items, in consequence, this job is very important to the consumer. Nevertheless, inspection and classification by humans is based on perception of environment, so human error is common and an automated system would be preferable.

6.1.3 Solve problem

For the solution to the problems raised in the previous section, the researchers implemented the artificial neural networks model, obtaining satisfactory results.

Debska and Guzowska-Świder (2011) solved the problem of quality control in the brewing industry by developing an ANN model to classify the quality of the beer of a particular brand, from two different types of quality. The ANN model considers the following as input variables: Transparency, colour, lack of sweetness of beer, extract in original work, apparent extract, real extract, amount of alcohol in beer, real degree of fermentation, apparent fermentability, pH, dissolved carbon dioxide in beer and level of acidity in beer. The input parameters were used to train the ANN and determine whether the quality of the beer was good or not. In the development of the ANN model, a multilayer perceptron (MLP) and a radial base function (RBF) were applied to solve the classification problem. In order to achieve satisfactory precision in the learning process, the intelligent problem solver (IPS) was used to search for an optimal network. This work also carried out the classification with the automated network designer (AND), with the purpose of generating neural networks that rank beer samples by the quality being good or not. In the results, it is shown that the two best networks correctly classified the quality of the food with 100% of totality, using a training algorithm of Broyden-Fletcher-Goldfarb-Shannona (BFGS) (Table 3). Finally, this work demonstrates how artificial neural networks designed by statistical modules (IPS, AND) are able to achieve 100% accuracy in the classification of the quality of a product.

Wang et al. (2017) proposed an improved system of food traceability, in order to evaluate the quality of food and provide consumers with information about the evaluation. The proposed system starts with a fuzzy classification method, to classify the quality of the food as very bad, bad, medium, good and very good. Subsequently, with data collected training a model of artificial neural networks that determines the final quality of the food and classifies it as high quality, medium quality and low quality. The proposed architecture of the ANN has five inputs and three outputs (Fig. 9). The ANN was developed through a feedforward, using a sigmoid activation function, and the backpropagation algorithm was employed to

Table 3: Network architecture with the best results obtained by AND (Modified from Debska and Guzowska-Świder (2011)).

ANN No.	Architecture	Type of neuron	Classification level Q%			Training algorithm	Transfer function		Error function
			Training	Test	Validation		Hidden neurons	Output neurons	
1	12-13-2	MLP	100.00	100.0	100.0	BFGS	Exponential	Linear	SOS
2	12-10-2	MLP	100.00	100.0	100.0	BFGS	Tangent	Softmax	CE
3	12-14-2	MLP	96.0	100.0	100.0	BFGS	Exponential	Exponential	SOS
4	12-15-2	RBF	94.0	100.0	100.0	RBFT	Gaussa	Softmax	CE
5	12-15-2	RBF	92.0	100.0	100.0	RBFT	Gaussa	Linear	SOS
6	12-11-2	MLP	92.0	100.0	100.0	BFGS	Logistic	Softmax	CE
7	12-12-2	MLP	90.0	100.0	100.0	BFGS	Tangent	Tangent	SOS
8	12-8-2	MLP	90.0	100.0	100.0	BFGS	Tangent	Tangent	SOS
9	12-7-2	MLP	90.0	100.0	100.0	BFGS	Exponential	Softmax	CE
10	12-6-2	MLP	90.0	100.0	100.0	BFGS	Tangent	Exponential	SOS

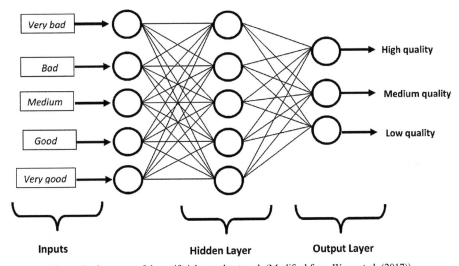

Figure 9: Structure of the artificial neural network (Modified from Wang et al. (2017)).

carry out the training of the ANN. Finally, the results reflected that not only is an effective traceability to guarantee food safety and quality obtained, but also, when applying the ANN, a way of evaluating them is obtained in order to visualize the quality and meet the consumer expectations.

Kılıc et al. (2007) developed a system of computer vision (CVS) for the inspection of beans, centered on the size and colorimetry of the samples. To identify the quantification of the color of the samples, an artificial neural network was developed. The architecture of the ANN model consists of using twelve neurons in the input layer, twelve neurons in the hidden layer and two neurons in the output layer. The activation function used in the hidden layer was hyperbolic tangent (TANSIG) and a positive linear (POSLIN) in the output layer. To train the ANN model, fourteen images containing 511 bean samples obtained from CVS were used, of which 69 were used for training; 71 for validation and 371 for testing. Once trained, another sample of 371 grains was tested, making a comparison of results with the results of human inspectors, where the system overall performance for classification of beans was 90.56% (Table 4). Finally, through the ANN, the system achieves quantification of the colour of the beans, displaying advantages over human inspection, such as the erroneous classifications by the human factor and the high performance when making the classification quickly.

Table 4: Comparison of the classification process between the computerized way and that performed by inspectors (Modified from Kılıç et al. (2007)).

Colour			Percentage		
Types*	No sample	Percentage of acceptance (%)	Types	No sample	Percentage of acceptance (%)
White beans	272	99.3	Undamaged	272	98.9
Yellow–green damaged	45	93.3	Low damaged	51	74.5
Black damaged	54	69.1	Highly damaged	48	93.8
Total	371	95.1	Total	371	94.9
Overall	90.6%				

* As decided by the inspectors.

6.1.4 Conclusion

The classification process turns out to be one of the most complex processes within the food industry, since it is necessary to carry it out with a certain degree of accuracy. During this section, three different investigations were highlighted, where the common problem consisted in carrying out a classification process. Due to the importance of this process, researchers implement an intelligent system, with the ability to classify specific aspects of food, such as the quality of the product and the evaluation of the product during its traceability, as well as differentiate between raw materials by their colour, texture, etc. Finally, the researchers emphasize that, when modelling the process with artificial neural networks, more than 90% acceptance is obtained when classifying the products, thus obtaining a better way to carry out the process.

6.2 Drying food

In general, the process of drying food is used in the food industry to preserve food and its characteristics for longer. Some products to which it is feasible to apply drying are fruits, vegetables and meats.

6.2.1 Introduction

The first method of drying food was to use the sun, but this drying induces changes in the biochemical aspects of matter. Dehydration, or drying, is the act of removing most of the water from something, in others words, to apply a method that removes almost all the moisture contained in a body, either plant or animal products. Drying is the most useful transformation process to preserve food because it is easy to apply and the subsequent low weight of the product is too easier to handle. When drying a product, several phenomena interfere, such as heat and mass transfer.

The break-even point of a thermal drying process is given by water activity; another factor that is very important, in order to know the end point on the drying of food, is shrinkage. Figure 10 shows in a general manner how the drying process is carried out. Some of the factors involved in this method are temperature, humidity, product geometry, air velocity and food shrinkage. A basic example is to circulate

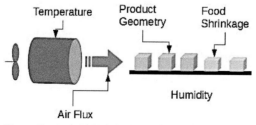

Figure 10: General diagram of the drying process in food.

hot air over food horizontally or vertically. With higher air temperature will come faster drying, in the same way, the speed with which hot air flows affects the rate of drying.

6.2.2 Description of problem

In the drying process, challenges and inconveniences that make the system complex arise, basically because they depend on the moisture conditions of the raw material and the necessary characteristics of the output product. For this reason, researchers approach this process through the following problems.

Hernández et al. (2000) raised an interest in knowing the temperatures of drying and humidification in mango and cassava using empirical equations. Later, Hernández-Pérez et al. (2004) realized the importance of controlling input variables and the need to apply a computational model. The researchers noted that, when applying the existing empirical equations, it was difficult to carry out the modelling of non-linear behaviors, such as those that occur during the drying process. Finally, Hernández (2009), using his past research, proposed an optimization technique in order to know how to optimize the input variables.

Tripathy and Kumar (2009) denoted the difficulty of removing moisture in food products when using solar dryers. Various efforts have been made to produce dehydrated foods, reducing the energy used and minimizing the degradation of the nutrients of the food. For this case, potatoes with cylinder geometry and slices were used.

6.2.3 Solve problem

Hernández-Pérez et al. (2004) proposed an ANN model as a solution to the problems posed during the drying of mango and cassava. Figure 11 specifies the input and output variables used to carry out the development of the ANN model.

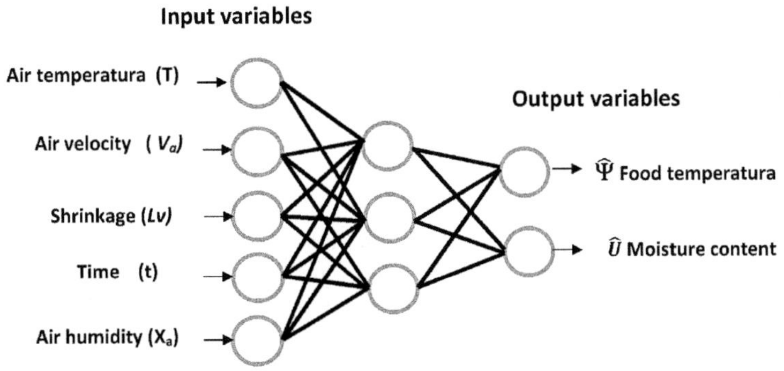

Figure 11: Conformation of neural network to simulate the drying process in mango and cassava (Modified from Hernández-Pérez et al. (2004)).

For the training and learning process, a database that consisted of 27 experimental tests for mango and 13 tests for cassava was used. Once the training and learning process was completed, the ANN model with three neurons in the hidden layer was established using an optimization algorithm of Levenberg-Marquardt type. During the process to determine the best network architecture in the hidden layer, it was highlighted that, when using more than four neurons, an overfitting was presented, in spite of continuing to decrease the error. The uncertainty of the ANN model between the learning and test data began to increase after the third neuron, as shown in Fig. 12.

The following equations were used to simulate the two output variables chosen, where the weight and bias coefficients can be found in Hernández-Pérez et al. (2004).

$$\hat{\psi} = \sum_{j=1}^{S} \left[W_{o(j)} \cdot \left(\frac{2}{1 + \exp\left(-2 \cdot \left(\sum_{k=1}^{K} \left(W_{i(j,k)} \cdot In_k \right) + b1_{(j)} \right) \right)} - 1 \right) \right] + b2 \tag{12}$$

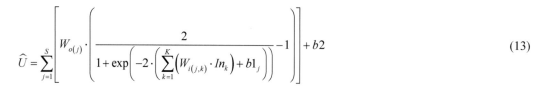

$$\widehat{U} = \sum_{j=1}^{S} \left[W_{o(j)} \cdot \left(\frac{2}{1+\exp\left(-2 \cdot \left(\sum_{k=1}^{K} \left(W_{i(j,k)} \cdot In_k\right) + b1_j \right)\right)} - 1 \right) \right] + b2 \qquad (13)$$

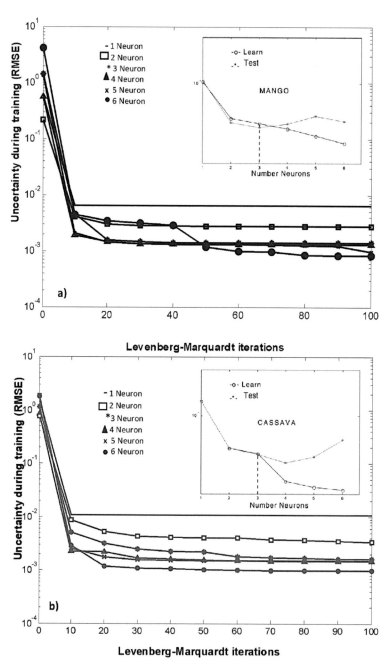

Figure 12: Confrontation of ANN model Uncertainty (RMSE) with respect to the number of neurons for (a) Mango and (b) Cassava (Modified from Hernández-Pérez et al. (2004)).

When the Eqs. (12) and (13) are integrated with their respective coefficients, a direct confrontation can be made between the experimental and simulated data, obtaining an excellent concordance, with values of $r^2 = 1$. Figure 13 shows an example is this comparison, applied in the prediction of moisture during the drying of the mango.

Finally, a validation is made with the experimental data in order to demonstrate how well the ANN model is capable of simulating the desired outputs. Figure 14 shows how the ANN model from various input conditions predicts the desired output (Moisture content).

The researcher Hernández (2009), using part of his last research in 2004, proposes the inverse artificial neural network model (ANNi). The ANNi model has as a novelty to optimize variables of inputs from a desired result. The ANNi model was structured to optimize the drying time in mango and cassava, obtaining as a result an excellent agreement when comparing the experimental data and those optimized by ANNi. The errors between the required value and the simulated to find the optimal drying time in mango and cassava were zero, respectively (error = $\hat{U}_{(required)} - \hat{U}_{(simulated)} = 0$). Finally, the ANNi model was able to obtain the optimum value in 0.3 s making it feasible to apply the model online and control the drying process in the food industry.

Tripathy and Kumar (2009), discussed the problem that arises in the process of elimination of humidity in potatoes through thermal drying. An ANN model was proposed for the prediction of the transient temperature of drying in mixed solar mode. The experimental variables to train the ANN model were the solar radiation, intensity and temperature of the ambient air, easy to obtain with the appropriate instrumentation. For the training of the ANN, 79 and 64 randomly selected data were used for the prediction of potatoes depending on its geometry (cylinders and slices), while 15 and 22 data were used for validation. To obtain the best prediction, various transfer functions and training algorithms were applied (Tables 5 and 6).

To classify potatoes according to their morphology, it was found that a neural network can be used to obtain a better performance with greater precision, furthermore, this is a tool that can be used for other applications in the food process.

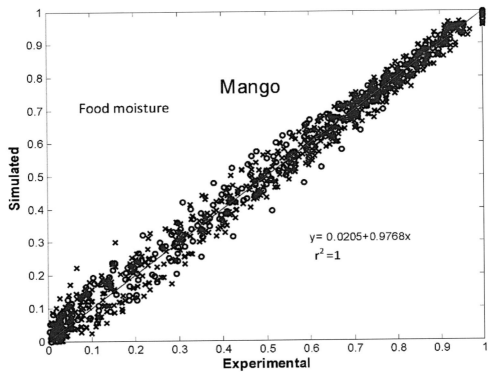

Figure 13: Direct confrontations between the real data and those obtained by the ANN model (Modified from Hernández-Pérez et al. (2004)).

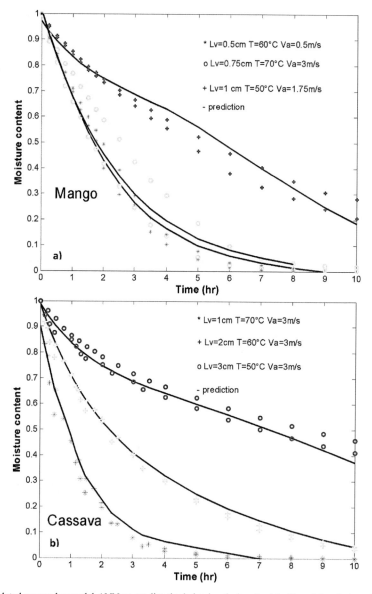

Figure 14: Simulated curves by model ANN to predict the behavior during the kinetics of the drying of the mango and cassava (Modified from Hernández-Pérez et al. (2004)).

6.2.4 Conclusion

Throughout this section, the complexity that exists during the food drying process has been denoted. The aforementioned researchers propose using the ANN model, since the existing empirical formulas do not allow them to model the process with precision. Once the ANN model was applied, it was possible to model the process, including complex variables, such as the drying time and humidity of the product, as an output to be simulated. It is worth mentioning that researchers apply this computational tool because of its ability to use experimental data, unlike other computational tools that need to be programmed based on theoretical models, such as Computational Fluid Dynamics (CFD) and ASPEN, among others. These models and computational programs are designed with theoretical criteria of operation of the

Table 5: Transfer functions and optimization algorithms (back propagation) used during the ANN model training (Modified from Tripathy and Kumar (2009)).

Sl. No.	Transfer function	Abbreviation	Training algorithms (Back propagation)	Abbreviation
1	Logarithmic	Logsig	Scaled conjugate gradient	SCG
2	Hyperbolic tangent	Tansig	Polak–Ribiere conjugate gradient	CGP
3	Positive linear	Poslin	Quasi-Newton	BFG
4	Saturating linear	Satlin	Levenberg–Marquardt	LM
5	-----------		Resilient	RPROP

Table 6: Results of error measurements in food temperature prediction using the ANN model with various transfer functions (Modified from Tripathy and Kumar (2009)).

Error	Type of geometry							
	Cylinder				Slice			
	Transfer functions							
	log*	tan*	Sat+	Pos+	log*	tan*	Sat+	Pos+
Mean absolute error	0.633	0.642	0.675	0.967	0.475	0.524	0.606	0.632
Root mean square error	0.769	0.798	0.848	1.161	0.670	0.741	0.744	0.860
Standard error	0.213	0.221	0.235	0.321	0.146	0.161	0.162	0.188
Correlation coefficient	0.950	0.947	0.940	0.889	0.975	0.970	0.969	0.960

* Sigmoid type, +Linear transfer function type.

processes and, therefore, have to consider some "assumptions", which make their degree of accuracy less significant. While the ANN models, for their learning capacity, become interesting as an alternative for modelling processes in the food industry.

6.3 Food cooling

In order to preserve food, several techniques, such as cooling, have been applied. This process is widely used because it can reduce microorganism growth, and chemical and cellular reactions are delayed. In addition, cold storage conditions are ideal for the properties of certain foods, such as flavor, texture and nutritional benefits.

6.3.1 Introduction

Cooling food extends the shelf life of a wide variety of foods. This process, along with drying, is one of the most frequently and widely used by humans in order to preserve food. This methodology decreases without completely stopping the action of microorganisms, preventing decomposition. When a low temperature is applied slowly, the ice crystals form on the outside of the cell, if it is administered with enough time, the water of the cells manages to leave due to the osmotic pressure. Controlling the freezing process allows a high degree of quality in food products in this process, including prior freezing and storage after freezing. A rapid freeze is a determining factor for the formation of ice crystals, dehydration of the cell and damage to cell walls. Changes in the colouring of fruits and vegetables are related to three physicochemical mechanisms and along with taste are the factors that indicate the quality of frozen foods.

The distribution and storage affect the quality of frozen foods. This is a slow and irreversible process, so in order to maintain good quality it is necessary to be careful in each step of the process of freezing and thawing.

In the process of freezing the food, great care must be taken when carrying out the pre-cooling in the preparation and the cooling for its storage. During pre-cooling, in order for the freezing to preserve the characteristics of food, such as colour, smell, taste, etc., it is necessary to apply some other techniques, namely, hot water treatment and cooking.

The selection of a cooling method is based on the type of food that needs to be preserved. There is a wide range of methods and techniques, among them are contact metal plates, submersion, blow of air, flow bed and cryogenization. Figure 15 shows a flow diagram to determine the steps to follow when preserving animal meat in pre-cooling and cooling process.

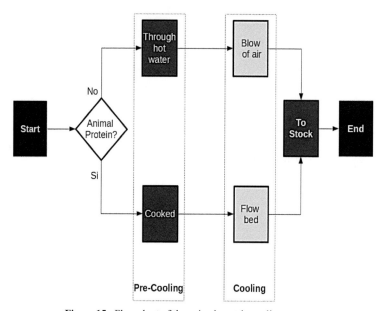

Figure 15: Flow chart of the animal protein cooling process.

6.3.2 Description of problem

The human consumption of frozen food products is on a massive scale. For this reason, some researchers have applied artificial neural networks to make the food cooling process more efficient and improve the quality of the desired product.

Kong et al. (2016) mentioned that one of the most widely commercialized varieties of fish is common carp, since it is found in most rivers and freshwater bodies. For this reason, their preservation is very useful. However, the need to find the right concentration of salt to preserve the quality and its natural characteristics is presented.

Goyal and Goyal (2011) denoted the problem that exists in food, once exposed for consumption. The priority is to maintain food quality to obtain client satisfaction. For this reason, a mathematical model that predicts the moment in which the food meets the adequate quality for consumption should be designed.

Ahmad et al. (2014) state that in the frozen food market, the shrimp maintain an annual demand growth of 20%, which has created the need to monitor the quality of the product once it has been packed and frozen, taking into consideration temperature as a preponderant factor. The changes in the temperature of storage or transport are reflected in variations in quality, this as a function of time.

6.3.3 Solve problem

Kong et al. (2016) developed an ANN model by a radial basis function neural networks (RBFNNs). The ANN model chosen is of type "feedforward", for its ability to reach an accurate prediction. The architecture of the model is designed through an input layer, a hidden layer and an output layer. In the input layer, control variables are entered. In the hidden layer, the input variables are transformed through activation functions. The output layer has a linear function, that give a response. This architecture of network was applied to study and quantifies the quality of common carp fillets during frozen storage and has four experimental variables: Free fatty acids (FAA), salt extractable protein (SEP), total sulfhydryl (SH) and Ca^{2+}-ATPase activity; to be predicted on 1 to 17 wk and 261, 253 and 245 k. The training of the network was a supervised two-stage method. First, a self-organized learning stage and then a supervised learning stage. The performance of the network used the mean square error (MSE) and regression coefficient (R^2), between the experimental and predicted values (Table 7).

The low MSE in the network with 28 neurons is the better architecture for development. In the results, the performance of the model ANN is considered acceptable, predicting with a relative error of ± 5%, for case of 245 k (Table 8). The application of RBFNNs provides the optimal time and temperature for the frozen storage of carp fillets.

Goyal and Goyal (2011) developed an ANN model with the function of predicting the shelf-life of a milky white dessert jeweled with pistachio. This work is interesting for dessert manufacturers, since analyzing the shelf life is essential for the preservation of the product. For the development of the ANN model, a generalized regression (GR) function was used to perform the approximation. Additionally, a linear layer train model (LLT) was applied. The experimental data used to train the model are: Tyrosine, moisture, free fatty acids, titratable acidity, peroxide value. The number of neurons applied in the hidden layer varied from 1 to 30, performing the training of the network with 100 epochs. As a result of training with LLT, a neural network with 22 was obtained as the best architecture (Table 9). For the radial basis network (GR), the best result was spread constant = 15, with RMSE 0.03076 and $R^2 = 0.98864$ (Table 10).

Table 7: Evaluation of the performance of the Network (Modified from Kong et al. (2016)).

No. Neurons	Error	No. Neurons	Error	Spread	Error	R^2
2	0.07557	18	0.00553	0.50	0.00027	1.000
4	0.37460	20	0.00476	1.00	0.00026	1.000
6	0.02076	22	0.00357	1.50	0.00007	1.000
8	0.01372	24	0.00225	2.00	0.00170	0.999
10	0.01220	26	0.00175	3.00	0.00431	0.992
12	0.00884	28	0.00007			

Table 8: Confrontation of errors in predicted values with respect to those experimental (Modified from Kong et al. (2016)).

Indicators		Storage time (wk)		
		1	**7**	**17**
FFA	Predicted value	1.96	3.02	6.79
	Experimental value	1.97 ± 0.08	3.09 ± 0.45	6.79 ± 0.80
	Relative error (%)	0.43	2.23	0.02
SEP	Predicted value	72.95	62.05	51.23
	Experimental value	72.96 ± 6.03	62.18 ± 8.97	51.23 ± 3.68
	Relative error (%)	0.02	0.22	0.00
SH	Predicted value	5.51	5.31	4.73
	Experimental value	5.51 ± 0.06	5.33 ± 0.22	4.73 ± 0.04
	Relative error (%)	0.13	0.47	−0.03
Ca^{2+}-ATPase	Predicted value	0.22	0.17	0.12
	Experimental value	0.22 ± 0.01	0.17 ± 0.01	0.12 ± 0.02
	Relative error (%)	−2.17	0.92	0.26

Table 9: RMSE and R² that outcome training of network LLT (Modified from Goyal and Goyal (2011)).

Neurons	RMSE	R²
4	0.0709	0.9047
8	0.0441	0.9766
14	0.0503	0.9697
20	0.0202	0.9951
22	**0.0169**	**0.9966**
26	0.0731	0.9359

Table 10: RMSE and R² training of network GR (Modified from Goyal and Goyal (2011)).

Spread constant	RMSE	R²
3	0.0696	0.9419
7	0.0402	0.9806
15	**0.0308**	**0.9886**
18	0.0851	0.9130
20	0.0949	0.8920

In conclusion, the ANN model can be considered as suitable tool to determine the life time of food products.

Also, Ahmad et al. (2014) developed an ANN model to predict the time limit of food storage for manage to preserve its quality. In this work, freezing rate, thawing rate, storage time, were considered as input variables. In addition, the thickness, width and length of the samples were included. The training was optimized with genetic algorithms, the results were classified in L *, a *, b *, with which cohesiveness hardness is evaluated and how chewy the food is. In the proposed ANN architecture, the hyperbolic tangent activation function and the Levenberg-Marquart training algorithm were used. The number of neurons applied in the hidden layer varied from 1 to 25 and changing the number of layers between 1 and 3. In this training a total of 500 data was used (Table 11).

Table 11: Interval of input and output used for training of ANN model (Modified from Ahmad et al. (2014)).

	Min	Max	Mean
Input			
Freezing rate °C	0.1	0.6	0.35
Thawing rate °C	3.9	5.8	4.85
Storage time	12	90	51
Sample Thickness (m)	0.01	0.019	0.0145
Sample width (m)	0.008	0.015	0.012
Sample length (m)	0.1202	0.154	0.1371
Output			
L*	37.89	61.655	49.7725
a*	0.32	3.03	1.675
b*	17.37	27.4	22.285
Hardness	17.37	95.2	56.285
Cohesiveness	0.054	3.94	1.997
chewiness	10.66	261.77	136.215

The genetic algorithm (GA) was capable of predicting values that have a strong relationship between experimental and predicted data. In consequence, this model of ANN optimized with GA had the generalization ability and could be used in real scenarios. The model can plot a dataset of 50 points that prove a high correlation $R^2 => 0.9$. In conclusion, ANN is a tool capable of handling large data sets, very accurately in predicting color and texture values.

6.3.4 Conclusion

The ANN models developed to solve the problems raised during the cooling process demonstrate that this tool can predict the storage-time and aspects of food, like colour and texture. This tool is an efficient auxiliary in the food cooling process, due to its ability to model nonlinear behaviors, such as those displayed by various foods at the time of preservation.

7. Optimizing Food Processes

In accordance with previous applications, artificial neural networks have been shown to be able to model fundamental food processes but also perform optimization functions. The following table summarizes articles where artificial neural networks were applied in order to optimize food processes (Table 12).

Table 12: Application of ANN in food optimization.

Optimization	Problem to solve	Database	Type of ANN Models	Learning process	Reference
Optimization of the manufacture of oligosaccharides to detect the ideal sugar beet	Determine the yield and conversion of the solid during beet from the sugar plant	Experimental data of a local factory	Feedforward, sigmoid function	Backpropagation	Astray et al., 2006
Optimization in the extraction of Stevia	Estimate the total yield of the extract, stevioside and Reb-A	Data obtained from a microwave-assisted extractor	Multilayer perceptron, feedforward	Backpropagation	Ameer et al., 2017
Optimum operating conditions in foodstuffs drying	Determine the optimal time	Coefficients obtained during the training of an ANN model	Feedforward, sigmoid function	Backpropagation	Hernández, 2009
Optimization in feed production	Prediction of the production rate for a poultry feed recipe	Data obtained directly from the raw material	Feedforward, sigmoid function	Backpropagation	Sudha et al., 2016
Optimization pressure food processing	Prediction of temperature and time	The data is based on a simulation model	Multilayer perceptron, feedforward	Backpropagation	Torrecilla et al., 2007
Optimization in the making of a cake	Determine the best conditions in the moisture content, time, height and temperature of the mixture	Obtained directly from food preparation	Feedforward, sigmoid function	Backpropagation	Mukhopadhyay et al., 2015
Optimization of thermal processing in food manufacturing	Determine the sterilization temperature, process time and quality degradation of the food	The data was obtained from a mathematical model	Feedforward, sigmoid function	Backpropagation	Kseibat et al., 1999
Optimization of heating in food	Estimate the optimal amount of retention and retort temperature	Database is obtained from an orthogonal experimental	Feedforward	–	Chen and Ramaswamy, 2002

8. Concluding Remarks

In this chapter, reference is made to a mathematical-computational model capable of solving the various conflicts that arise in the food industry. The articles and books cited are a strong basis for how ANN can model nonlinear processes and, therefore, be able to predict behaviors in food processes. The ANN model has been applied in processes such as classification, drying and cooling of food, demonstrating that it provides an excellent alternative for carrying out the control of variables. The purpose of modelling and controlling food processes by ANN is to obtain the following benefits: To save time and money and render the operation of the processes more efficient. Finally, the application of the ANN extends to optimizing food processes by obtaining optimal input variables. By determining specific optimal input variables that improve the performance of the process, it is possible to save a significant amount of money in the food industries, which makes the ANN methodology more valuable and of greater interest to apply in more food processes such as packaging, washing, fermentation, preservation and sales.

References

Ahmad, I., Jeenanunta, C., Chanvarasuth, P. and Komolavanij, S. 2014. Prediction of physical quality parameters of frozen shrimp (*Litopenaeus vannamei*): An artificial neural networks and genetic algorithm approach. Food Bioprocess Tech. 7(5): 1433–1444.

Ameer, K., Bae, S.W., Jo, Y., Lee, H.G., Ameer, A. and Kwon, J.H. 2017. Optimization of microwave-assisted extraction of total extract, stevioside and rebaudioside-A from *Stevia rebaudiana* (Bertoni) leaves, using response surface methodology (RSM) and artificial neural network (ANN) modelling. Food Chem. 229: 198–207.

Andreev, A.V., Makarov, V.V., Runnova, A.E., Pisarchik, A.N. and Hramov, A.E. 2018. Coherence resonance in stimulated neuronal network. Chaos Soliton Fract. 106: 80–85.

Askarzadeh, A. and Rezazadeh, A. 2013. Artificial neural network training using a new efficient optimization algorithm. Appl. Soft. Comput. 13(2): 1206–1213.

Astray, G., Gullón, B., Labidi, J. and Gullón, P. 2016. Comparison between developed models using response surface methodology (RSM) and artificial neural networks (ANNs) with the purpose to optimize oligosaccharide mixtures production from sugar beet pulp. Ind. Crop Prod. 92: 290–299.

Bassam, A., Conde-Gutiérrez, R.A., Castillo, J., Laredo, G. and Hernández, J.A. 2014. Direct neural network modeling for separation of linear and branched paraffins by adsorption process for gasoline octane number improvement. Fuel 124: 158–167.

Cantarella, G.E. and de Luca, S. 2005. Multilayer feedforward networks for transportation mode choice analysis: An analysis and a comparison with random utility models. Transportation Research Part C: Emerging Technologies 13(2): 121–155.

Chen, C.R. and Ramaswamy, H.S. 2002. Prediction and optimization of constant retort temperature (CRT) processing using neural network and genetic algorithms. J. Food Process. Eng. 25(5): 351–380.

Ciocca, G., Napoletano, P. and Schettini, R. 2017. Food recognition: A new dataset, experiments, and results. IEEE J. Biomed. Health 21(3): 588–598.

Ćurčić, B.L., Pezo, L.L., Filipović, V.S., Nićetin, M.R. and Knežević, V. 2015. Osmotic treatment of fish in two different solutions-artificial neural network model. J. Food Process. Preserv. 39(6): 671–680.

Da Silva, I.N., Spatti, D.H., Flauzino, R.A., Liboni, L.H.B. and dos Reis Alves, S.F. 2017. Artificial Neural Networks: A Practical Course. Springer International Publishing, Switzerland.

Dande, P. and Samant, P. 2018. Acquaintance to Artificial Neural Networks and use of artificial intelligence as a diagnostic tool for tuberculosis: A review. Tuberculosis 108: 1–9.

Dębska, B. and Guzowska-Świder, B. 2011. Application of artificial neural network in food classification. Anal. Chim. Acta. 705(1-2): 283–291.

Dil, E.A., Ghaedi, M., Asfaram, A., Mehrabi, F., Bazrafshan, A.A. and Ghaedi, A.M. 2016. Trace determination of safranin O dye using ultrasound assisted dispersive solid-phase micro extraction: Artificial neural network-genetic algorithm and response surface methodology. Ultrason Sonochem. 33: 129–140.

Fernández-Navarro, F., Valero, A., Hervás-Martínez, C., Gutiérrez, P.A., García-Gimeno, R.M. and Zurera-Cosano, G. 2010. Development of a multi-classification neural network model to determine the microbial growth/no growth interface. Int. J. Food Microbiol. 141(3): 203–212.

Freitas, J.F.G.d., Niranjan, M., Gee, A.H. and Doucet, A. 2000. Sequential monte carlo methods to train neural network models. Neural Comput. 12(4): 955–993.

Goyal, S. and Goyal, G.K. 2011. A new scientific approach of intelligent artificial neural network engineering for predicting shelf life of milky white dessert jeweled with pistachio. Int. J. Sci. Eng. Res. 2(9): 1–4.

Grachv, A.M., Ignatov, D.I. and Savchenko, A.V. 2019. Compression of recurrent neural networks for efficient language modeling. Appl. Soft. Comput. 79: 354–362.

Hamzaçebi, C. 2018. Improving artificial neural networks' performance in seasonal time series forecasting. Inform. Sciences 178(23): 4550–4559.

Herculano-Houzel, S. 2009. The human brain in numbers: A linearly scaled-up primate brain. Front Hum. Neuro-Sci. 3: 31.

Hernández, J.A., Pavon, G. and Garcia, M.A. 2000. Analytical solution of mass transfer equation considering shrinkage for modeling food-drying kinetics. J. Food Eng. 45: 1–10.

Hernández, J.A. 2009. Optimum operating conditions for heat and mass transfer in foodstuffs drying by means of neural network inverse. Food Control 20(4): 435–438.

Hernández-Pérez, J.A., García-Alvarado, M.A., Trystram, G. and Heyd, B. 2004. Neural networks for the heat and mass transfer prediction during drying of cassava and mango. Innov. Food Sci. Emerg. Technol. 5: 56–64.

Jani, D.B., Mishra, M. and Sahoo, P.K. 2016a. Performance prediction of rotary solid desiccant dehumidifier in hybrid air-conditioning system using artificial neural network. Appl. Therm. Eng. 98: 1091–1103.

Jani, D.B., Mishra, M. and Sahoo, P.K. 2016b. Performance prediction of solid desiccant - vapor compression hybrid air-conditioning system using artificial neural network. Energy 103: 618–629.

Jani, D.B., Mishra, M. and Sahoo, P.K. 2017. Application of artificial neural network for predicting performance of solid desiccant cooling systems—A review. Renew Sust. Energ. Rev. 80: 352–366.

Kalogirou, S.A. 2001. Artificial neural networks in renewable energy systems applications: A review. Renew. Sust. Energ. Rev. 5(4): 373–401.

Kılıç, K., Boyacı, I.H., Köksel, H. and Küsmenoğlu, I. 2007. A classification system for beans using computer vision system and artificial neural networks. J. Food Eng. 78(3): 897–904.

Kong, C., Wang, H., Li, D., Zhang, Y., Pan, J., Zhu, B. et al. 2016. Quality changes and predictive models of radial basis function neural networks for brined common carp (*Cyprinus carpio*) fillets during frozen storage. Food Chem. 201: 327–333.

Kseibat, D., Basir, O.A. and Mittal, G.S. 1999. An artificial neural network for optimizing safety and quality in thermal food processing. In Intelligent Control/Intelligent Systems and Semiotics, 1999. Proceedings of the 1999 IEEE International Symposium 393–398.

Kumar, J. and Singh, A.K. 2018. Workload prediction in cloud using artificial neural network and adaptive differential evolution. Future Gener. Comput. Syst. 81: 41–52.

Leśniak, A. and Juszczyk, M. 2018. Prediction of site overhead costs with the use of artificial neural network based model. Arch. Civ. Mech. Eng. 18(3): 973–982.

Márquez-Nolasco, A., Conde-Gutiérrez, R.A., Hernández, J.A., Huicochea, A., Siqueiros, J. and Perez, O.R. 2017. Optimization and estimation of the thermal energy of an absorber with graphite disks by using direct and inverse neural network. J Energ Resour-ASME 140(2): 020906-020906-13.

Masoumi, H.R.F., Basri, M., Kassim, A., Abdullah, D.K., Abdollahi, Y., Gani, S.S.A. et al. 2014. Optimization of process parameters for lipase-catalyzed synthesis of esteramines-based esterquats using wavelet neural network (WNN) in 2-liter bioreactor. J. Ind. Eng. Chem. 20(4): 1973–1976.

McCulloch, W.S. and Pitts, W. 1943. A logical calculus of the ideas immanent in neurons activity. Bull Math Biophys. 5: 115–133.

Mohanraj, M., Jayaraj, S. and Muraleedharan, C. 2012. Applications of artificial neural networks for refrigeration, air-conditioning and heat pump systems—A review. Renew Sust. Energ. Rev. 16(2): 1340–1358.

Mukhopadhyay, S., Mishra, H.N., Goswami, T.K. and Majumdar, G.C. 2015. Neural network modeling and optimization of process parameters for production of chhana cake using genetic algorithm. Int. Food Res. J. 22(2): 465–475.

Rumelhart, D.E., Hinton, G.E. and Williams, R.J. 1985. Learning internal representations by error propagation. University San Diego La Jolla Institute for Cognitive Science, California.

Sablani, S.S., Datta, A.K., Rahman, M.S. and Mujumdar, A.S. 2007. Handbook of Food and Bioprocess Modeling Techniques. CRC Press/Taylor & Francis, New York.

Sivanandam, S.N., Sumathi, S. and Deepa, S.N. 2006. Introduction to Neural Networks using Matlab 6.0. Tata McGraw-Hill Publishing Company Limited, New Delhi.

Sudha, L., Dillibabu, R., Srinivas, S.S. and Annamalai, A. 2016. Optimization of process parameters in feed manufacturing using artificial neural network. Comput. Electron. Agr. 120: 1–6.

Tang, J., Liu, F., Zhang, W., Ke, R. and Zou, Y. 2018. Lane-changes prediction based on adaptive fuzzy neural network. Expert Syst. Appl. 91: 452–463.

Torrecilla, J.S., Otero, L. and Sanz, P.D. 2007. Optimization of an artificial neural network for thermal/pressure food processing: Evaluation of training algorithms. Comput. Electron. Agr. 56(2): 101–110.

Tripathy, P.P. and Kumar, S. 2009. Neural network approach for food temperature prediction during solar drying. Int. J. Therm. Sci. 48(7): 1452–1459.

Wang, J., Yue, H. and Zhou, Z. 2017. An improved traceability system for food quality assurance and evaluation based on fuzzy classification and neural network. Food Control 79: 363–370.

Williams, R.J.P. 2003. The biological chemistry of the brain and its possible evolution. Inorg. Chim. Acta 356: 27–40.

Yegnanarayana, B. 1999. Artificial Neural Networks. Prentice-Hall, India.

CHAPTER 22

Application of Neural Networks in Optimizing Different Food Processes

Case Study

KK Dash, GVS Bhagya Raj and MA Gayary*

1. Introduction

An Artificial Neural Network (ANN) is an information processing paradigm that is inspired by the way biological nervous systems, such as the brain, process information. The key element of this paradigm is the novel structure of the information processing system. It is composed of a large number of highly interconnected processing elements (neurons) working in unison to solve specific problems. In the human brain, there are billions of cells called neurons, which processes information in the form of electric signals. External information or stimuli are received by the dendrites of the neuron, processed in the neuron cell body, converted to output and passed through the Axon to the next neuron. The next neuron can choose to either accept it or reject it depending on the strength of the signal. ANN is a very simplistic representation of how neurons work in the brain. It performs various tasks, such as pattern-matching, classification, optimization function, approximation, vector quantization and data clustering. Neural networks have the ability to derive meaning from complicated or imprecise data and, therefore, can be used to extract patterns and detect trends that are too complex to be noticed by either humans or other computer techniques. Unlike conventional computers, ANNs do not require programming. A trained neural network learns from experience, generalizes from examples, and is able to extract essential characteristics from noisy data which can be thought of as an "expert" in the category of information it has been given to analyze. The advantages of ANN are

1. Storing information on the entire network: In ANN, information is not stored on a database but in the entire network, which helps in not preventing the network from functioning even if a few pieces disappeared.
2. They require significantly less time to develop and can respond to conditions unspecified or not previously assigned.

Department of Food Engineering and Technology, Tezpur University, Tezpur, Assam – 784028, India.
 Emails: gvsbraj@gmail.com; mainao.alina@gmail.com
* Corresponding author: kshirod@tezu.ernet.in

3. Adaptive learning: An ability to learn how to do tasks based on the data given for training or initial experience. ANNs have the ability to learn and model non-linear and complex relationships, which is very important because in real life, many of the relationships between inputs and outputs are non-linear as well as complex.

4. Ability to work with incomplete knowledge: The input data may produce output with incomplete information even after training. The performance of the ANN depends on the missing information.

5. Self-Organization: An ANN can create its own organization or representation of the information it receives during learning time.

6. Real-Time Operation: ANN computations may be carried out in parallel, and special hardware devices are being designed and manufactured which take advantage of this capability. It has numerical strength that can perform more than one job at the same time.

7. Fault Tolerance via Redundant Information Coding: ANN does not prevent the model from generating output even if one or more cells are corrupted. This feature makes the networks fault tolerant. Partial destruction of a network leads to the corresponding degradation of performance. However, some network capabilities may be retained even with major network damage.

8. Ability to make machine learning: ANNs learn events and make decisions by commenting on similar events.

As a novel approach offering advantages over mathematical modelling, ANN has been recognized as a powerful tool to model complex, dynamic, highly nonlinear and ill-defined scientific and engineering problems (Chen et al., 2011; Aghbashlo et al., 2015). ANNs have emerged as one of the most used methods for modeling and optimization, especially for nonlinear problems (Desai et al., 2008). The prominent advantage of ANN is its ability to efficiently deal with nonlinear relationships, even when the exact nature of the relationship is unknown. The ability of the ANNs to recognize and reproduce the cause-effect relationships through training for the multiple input-output systems makes them efficient in representing even the most complex systems. ANN is an efficient computing system whose central theme is borrowed from the analogy of biological neural networks. In food technology, ANNs are useful for food safety and quality analyses, predicting chemical, functional and sensory properties of various food products during processing and distribution (Baykal and Yildirim, 2013). ANN modeling methods do not require parameters of physical models and have the ability to learn from experimental data (Di Scala et al., 2013).

The progress of neurobiology has allowed researchers to build mathematical models of neurons in order to simulate neural behavior. Neural networks are recognized as good tools for dynamic modeling and have been extensively studied since the publication of the perceptron identification method (Rumelhart and Zipser, 1985). The interest of such models includes modeling without any assumptions about the nature of underlying mechanisms and their ability to take into account non-linearity and interactions between variables (Bishop, 1994). An outstanding feature of neural networks is the ability to learn the solution of the problem from a set of examples and to provide a smooth and reasonable interpolation for new data. Also, in the field of food process engineering, it is a good alternative for conventional empirical modeling based on polynomial and linear regressions. For food processes, the application of neural networks keeps on expanding.

ANN is an emerging non-linear computational modeling method consisting of three layers, namely, input, hidden and output layers. Each layer comprises several processing elements or operating neurons, which are related to each other by parallel connections and organised in patterns similar to biological networks. The principle behind ANNs is to mimic the functioning and learning process of a human brain using an artificial neuron. Links between these neurons are known as weights and biases, the strength of neuron interconnections is determined by the weight associated with them. For every ANN, the input layers constitute independent variables which send data via synapses to the hidden layer neurons and the output layers constitute dependent variables and get data via more synapses from hidden layer neurons. Weights are known as the synapses and store parameters that help in manipulating the data during calculations. The hidden layers exist between input and output layers with multiple neurons as an interface to fully interconnect input and output layers. The pattern of hidden layers to be applied in the hierarchical network can be either multiple layers or a single layer. Each neuron sums all of the inputs that it receives and converts the sum into an output value based on a predefined activation, or transfer function (Soleimani et al., 2013). The connection between neurons in separate layers is defined in terms of weights and biases. More complex systems may have more hidden layers of neurons, the universal approximation theory suggests that a network with a single hidden layer and with a sufficiently large number of neurons can interpret any input-output structure effectively (Prinderre et al., 1998; Pareek et al., 2002; Khani and Nemati, 2013).

ANN can learn similarly to the way that human brain works, and presents some benefits for the learning process:

    i. ANN can learn from incomplete data.
   ii. ANN can use databases with noisy data.
  iii. ANN can model nonlinear behaviors.

The most used kind of ANN model is the Multi-Layer Perceptron (MLP) (Desai et al., 2008). MLP models allow for the modeling of complex and highly non-linear processes through different neural network layers. An ANN is a massively distributed interconnection of adaptive nonlinear processing elements (PE) or neurons, which resembles the human brain in two respects, i.e., a learning process is needed for the network to acquire knowledge from its environment, and interneuron connection strength or synaptic weights are used to store the acquired knowledge (Haykin et al., 2009). These properties of ANN provide higher flexibility and capability in data fitting, prediction, and modeling of nonlinear relationships compared to the RSM.

This book chapter provide basic information about ANNs, Different learning methods of ANN and its application in different food processing techniques, i.e., extrusion, fermentation, drying, baking and thermal processing in food products. ANN has also been applied for food safety and quality analysis, and the prediction of physical, chemical, functional and sensory properties of foods. In addition to these, a case study on the application of ANN for osmotic dehydration of culinary banana slice was illustrated.

2. Applications of ANNs in Various Food Processing Operations

ANNs used by various researchers for modeling and predictions in the field of food process engineering are listed in Table 1.

3. Process for Modeling by ANN

3.1 Neural network type

There are a variety of network types that can be used when creating neural networks. The network type can determine various network parameters, such as the type of neurons that are present in each layer and the method by which network layers are interconnected.

Single Layer Feedforward Network: This type of network consists of two layers, namely, the input layer and the output layer. The output of each layer is simply fed into the next layer; hence, the name feedforward networks. Each layer can have a different transfer function and size. The input layer neurons receive the input signals and the output layer neurons receive the output signals. The synaptic links carrying the weights connect every input neuron to the output neuron, but not vice-versa. Such a network is said to be feedforward. Despite the two layers, the network is termed single layer since it is the output layer alone that performs the computation. The input layer merely transmits the signals to the output layer. Hence, the name "single layer feedforward network".

Multilayer Feedforward Network: This network is made of multiple layers. It possesses an input and an output layer and also has one or more intermediary layers, called hidden layers. The computational units of the hidden layer are known as the hidden neurons or hidden units. The hidden layer aids in performing useful intermediary computations before directing the input to the output layer. The input layer neurons

Table 1: Application of ANN in various food processing operations.

Food process Operation	Type of process	Application of ANN	Author
Extrusion	Neural networks in extrusion process identification and control.	Dynamic changes of torque, specific mechanical energy and pressure in flat bread extrusion were modelled and controlled using the multi-input and multi-output approach and two independently taught feedforward ANNs.	(Eerikäinen et al., 1994)
Extrusion	Modelling and control of a food extrusion process using ANN and an expert system.	Modelling the start-up of a food extrusion process using ANNs.	(Popescu et al., 2001)
Extrusion	Optimization of extrusion process variables using a genetic algorithm.	Optimum process conditions separately for each (individual) and commonly for all (common) extrudate properties extruder barrel temperature, extruder screw speed and fish content and moisture content of feed, for maximum expansion ratio (ER), minimum bulk density (BD), minimum hardness (H) and maximum water solubility index (WSI).	(Tumuluru and Bandyopadhyay, 2004)
Microwave heating	Finite Element Modeling (FEM) of Heat and Mass Transfer in Food Materials During Microwave Heating.	Developed a three-dimensional FEM to simulate coupled heat and mass transfer in food materials during microwave heating and verified the FEM using experimental time-dependent temperature and moisture results for cylinder-shaped and slab-shaped potato specimens.	(Zhou et al., 1995)
Baking	Neural Networks (NN) vs Principal Component Regression (PCR) for Prediction of Wheat Flour Loaf Volume in Baking Tests.	Used NN and PCR for prediction of a baking test of wheat flour dough and their comparison was done.	(Horimoto et al., 1995)
Rheological Property	Prediction of Dough Rheological Properties Using Neural Networks.	Development of a neural network to predict dough rheology (farinograph peak, extensibility, and maximum resistance of dough) using the work input (mixer torque) during mixing.	(Ruan et al., 1995)
Rheological Property	Prediction of rheological properties of Iranian bread dough from chemical composition of wheat flour by using ANN.	The dependence of rheological parameters of dough (farinographical parameters), on the chemical composition of wheat flour and developed a neural network model for the dynamic prediction of frinographical factors from chemical composition of wheat flour.	(Razmi-Rad et al., 2007)
Drying	Dynamic Models for Drying and Wet-Milling Quality Degradation of Corn Using Neural Networks.	Neural network model of drying and wet-milling quality degradation of corn was developed.	(Trelea et al., 1997)
Drying	Dynamic optimal control of batch rice drying process.	The optimization algorithm used a dynamic compartment model of the drying process coupled with a quality degradation kinetic. Constraints on total drying time, final moisture content and allowed range of operating conditions were imposed.	(Olmos et al., 2002)
Drying	Experimental and theoretical investigation of shelled corn drying in a microwave-assisted fluidized bed dryer using ANN.	Drying behavior of shelled corn in a microwave-assisted fluidized bed dryer at different microwave energy levels and drying air temperatures and development and evaluation of ANN model used to predict the shelled corn drying time.	(Momenzadeh et al., 2011)

Table 1 contd. ...

...Table 1 contd.

Food process operation	Type of process	Application of ANN	Author
Microbial growth prediction	Improving ANNs with a pruning methodology and genetic algorithms for their application in microbial growth prediction in food.	Growth or inhibition of microorganisms on food predicted using ANN model.	(García-Gimeno et al., 2002)
Thermal processing	Modelling and optimization of variable retort temperature (VRT) thermal processing using coupled neural networks and genetic algorithms.	Development of ANN prediction models for variable (process time, average quality retention, surface cook value) related to VRT function parameters.	(Chen and Ramaswamy, 2002)
Thermal processing	Complex method for nonlinear constrained multi-criteria (multi-objective function) optimization of thermal processing.	Development of an easy methodology for the multi-objective optimization of thermal processing for food systems consisting of two different geometries.	(Erdogdu and Balaban, 2003)
Storage study	Dynamic optimization using neural networks and genetic algorithms for tomato cool storage to minimize water loss.	Variable temperature profile (optimal 1–step setpoints of temperature) of the storage unit to minimize water losses of tomatoes.	(Morimoto et al., 2003)
Cheese Production	Modeling of pH and acidity for industrial cheese production.	Neural Network was used for modelling and predicting the pH or acidity of curd and cheese.	(Doganis et al., 2006)
Fermentation Technology	Application of ANN to food and fermentation technology.	The intelligent modeling approach of models employing ANN in combination with other data analysis systems is able to solve a very important problem—processing of scarce, uncertainty and incomplete numerical and linguistic information.	(Bhotmange and Shastri, 2011)
Quality	Quality of beer was modelled using ANN.	Good quality beer and the unsatisfactory quality of beer were selected for the model.	(Swider, 2011)
DPPH free radical scavenging activity	Determination of DPPH free radical scavenging activity by using ANN.	The experimental results and the predicted values are in good correlation.	(Musa et al., 2016)
Sensory analysis	Two ANN models applied and studied for sensory quality of instant coffee flavored sterilized drink.	From the two ANN models, feedforward backpropagation model is better than cascade forward artificial intelligence model in terms of prediction of sensory quality of instant coffee flavored Sterilized Drink.	(Goyal and Goyal, 2011)
Honey characterization	Honey characterization was modelled using ANN.	The relationships between ash content, antioxidant activity, total phenolic content, and honey colour studied using ANN and compared with computer vision system.	(Shafiee et al., 2014)
Thermal conductivity prediction	Thermal conductivity prediction of foods by Neural Network and Fuzzy.	A neuro fuzzy model was developed for the prediction of thermal conductivity of food product under different conditions and it was found to be a very powerful tool in predicting thermal conductivity.	(Shafiur et al., 2011)
Stability of olive oil	Stability of extra virgin olive oil was studied by ANN.	ANN model with 11 input neurons, 18 hidden neurons and 5 output neurons showed high classification performance for tinplate cans, amber and transparent PET bottles stored in the light and dark.	(Faria et al., 2015)

are linked to the hidden layer neurons and the weights on these links are referred to as input-hidden layer weights. Similarly, the hidden layer neurons are linked to the output layer neurons and the corresponding weights are referred to as hidden-output layer weights.

Recurrent Networks: These networks differ from feedforward network architectures in the sense that there is at least one feedback loop. Thus, in these networks, there could exist one layer with feedback connections. There could also be neurons with self-feedback links, i.e., the output of a neuron is fed back into itself as input.

3.2 Learning methods in a neural network

The Learning methods in NNs can be grouped into 3 basic types.

   i) *Supervised Learning*: In this, every input pattern that is used to train the network is associated with a target or the desired output pattern. A teacher is assumed to be present during the learning process when a comparison is made between the network's computed output and the correct expected output, to determine the error. Tasks that fall under this category are Pattern Recognition and Regression.
  ii) *Unsupervised Learning*: In this learning method, the target output is not presented to the network. The system learns on its own by discovering and adapting to structural features in the input patterns as if there is no teacher to present the desired patterns. Tasks that fall under this category include clustering, compression and filtering.
 iii) *Reinforced Learning*: In this method, the teacher, though available, does not present the expected answer, it only indicates if the computed output is correct or incorrect. The information provided helps the network in its learning process. However, reinforced learning is not one of the popular forms of learning.

3.3 Experimental designs for ANN

3.3.1 Full factorial design (FF): This design is most common and intuitive strategy of experimental design which includes all possible combinations of all independent variables with two or more levels. The experiment grows exponentially with the number of parameters and the number of levels. This design used the data effectively and do not confound the effects of the parameters. The main and the interaction effect can be evaluated. Full factorial design was implemented in ANN modelling of improving the quality of canned foods (Chen and Ramaswamy, 2002), residence time distribution of carrot cubes (Chen and Ramaswamy, 2000) and microwave-assisted extraction of natural dye from seeds of Bixa orellana (Sinha et al., 2013).

3.3.2 Fractional factorial designs: A full factorial design may become very difficult to be used as the number of parameters increases, so the most important combinations of the variables are included in Fractional factorial designs and significance of effects is expressed using statistical methods. Fractional factorial design uses a subset of the full factorial experiments and provides good information on the main effects and some information about interaction effects. Fractional factorial design used with ANN model to remove zinc ions from wastewater by the application of micellar-enhanced ultrafiltration (Rahmanian et al., 2011).

3.3.3 Central composite design (CCD): The central composite design (CCD) used to allocate the operation variables into a range of evaluation. It is composed of a factorial design, a set of central points, and axial points equidistant to the center point. Each variable in the factorial design component of CCD has a low and high value, represented with a coded value of -1 and $+1$, respectively. The significance of the factorial component is determined by a cube which represents the variables of the model. In the cube, each corner represents the interaction of the factors. For example, with a 3-independent variable, a total of 8 interactions can be evaluated. The axial component of CCD are the points that are equidistant from the center of the cube formed for the earlier component. For the axial component, a sphere is represented with a positive axial value $(+\alpha)$ and a negative axial value $(-\alpha)$, which add two more levels to the independent variables. The α value can be calculated by using $2^{n/4}$ where n represents number of variables. The average of the high and the low value of the factorial component are the values for central point component of CCD, represented with numerical value 0, which may be defined as the region where the optimal conditions are supposedly met. There are three types of designs in CCD, namely, central composite circumscribed (CCC), central composite inscribed (CCI), and central composite face-centered (CCF). For CCC and CCI, a total of 5 levels taken for each variable and for CCF 3 levels taken for each variable. CCD was implemented for ANN modelling for extracellular protease production from a newly isolated *Pseudomonas* sp. (Ray et al., 2004), Spray-Dried Pomegranate Juice (Youssefi et al., 2009) and air drying process for Artemisia absinthium leaves (Karimi et al., 2012).

3.3.4 Box-Behnken design (BBD): The variables in this design have three levels and the experimental design is built using the combination of two variables with two levels, high and low, as factorial design and other variables at mean level throughout the block. As the number of parameters grows, this design helps to limit the sample size, which is kept to a value sufficient for the estimation of the coefficients. At the end of the design, central points are added. BBD was used with ANN modelling for microwave-assisted extraction in order to determine Zinc in Fish Muscles (Moghaddam and Khajeh, 2011) and Nicotinic Acid and Nicotinamide in Food (Mu et al., 2013).

3.3.5 Mixture design: Mixture design can be implemented when the factors are in proportions to formulate a specific blend. The fact that the proportions must add up to one is the key attribute of mixture designs. Mixture design is fundamentally different from screening because screening experiments are orthogonal. Which means that, over a period of time, the setting of one factor varies independently of any other factor. The interpretation is simple, as the effect of the factors on the response are separable in the screening experiment. With mixtures, it is impossible to vary one factor independently of all the others. When you change the proportion of one ingredient, the proportion of one or more other ingredients must also change in order to compensate. This simple fact has a profound effect on every aspect of experimentation with mixtures such as the factor space, the design properties and the interpretation of the results. General mixture designs used are optimal, simplex centroid, simplex lattice, extreme vertices and ABCD design.

3.4 Coding of independent and dependent variables for neural networking

In neural network modelling, the input and output data set were coded for developing the model. The coding of input dataset was carried out within the range -1 and $+1$ and the output data set between 0 and 1.

If X_{max} and X_{min} represent the maximum and minimum values of input variables X, respectively, x is the dimensionless coded form between the coded value $+1$ and -1. The relationship between X and x is expressed as follows,

$$X_M = (X_{max} + X_{min})/2 \tag{1}$$

$$X_D = (X_{max} - X_M) \tag{2}$$

$$x = (X - X_M)/X_D \tag{3}$$

$$X = xX_D + X_M \tag{4}$$

The values of the dependent variable in real (Y) and coded (y) form lies between $+1$ and 0. Y_{minp} represents the pseudo-minimum of the actual minimum Y_{min} expressed as follows,

$$Y_{minp} = 2Y_{min} - Y_{max} \tag{5}$$

The relationship between the real (Y) and coded (y) values of response can be expressed in the following equations,

$$Y_M = (Y_{max} + X_{minp})/2 \tag{6}$$

$$Y_D = (Y_{max} - Y_M) \tag{7}$$

$$y = (Y - Y_M)/Y_D \qquad (8)$$

$$Y = yY_D + Y_M \qquad (9)$$

4. Modelling by Neural Network

Neural network is a simplified model which imitates the living animal nervous system. A layer of neurons acts as input and another as output. It consists of computational units organized into three kinds of layers: Input, hidden, and output (Fig. 1). A hidden layer is present between the input and output layers. The relationship between a set of independent variables X and dependent variables Y can be obtained by using the neural network. The neural network learns directly from the pairs of input X and output Y data and develops a relationship through a set of mathematical equations. After the learning, the network is able to predict the outputs from an input data set that has not been previously used for the learning.

Figure 1 shows the basic structure of a feedforward backpropagation neural network. Backpropagation is a method of training multilayer ANNs which uses the procedure of supervised learning. Supervised algorithms are error-based learning algorithms that utilize an external reference signal and generate an error signal by comparing the reference with the obtained output. Based on the error signal, the neural network modifies its synaptic connection weights in order to improve the system performance.

Connecting lines between the neurons represent the synaptic joints, and the weights associated with these lines represent the electrochemical potential conductance of the joints. The figure shows weight 'u' as the synaptic joints between the input and hidden layer neurons and 'w' as the synaptic joints between hidden and output layer neurons. Each of the neurons in the hidden and output layer has been shown with a 'threshold' value, 'Th' and 'To' show the threshold values for the hidden and output layer neuron, respectively. A neuron in the hidden layer would produce output only when the computed value of input Hi at this neuron exceeds its 'threshold' value. Similarly, a neuron at the output layer would produce output only when the computed value of input yi at this neuron exceeds its 'threshold' value.

In the feedforward backpropagation network, computational sequence for the prediction of dependent variable Y follows from input to hidden and then from hidden to output layer. Error value (computed from the actual and predicted values of the dependent variable) is propagated backward, from output to hidden layer and then from the hidden to input layer, in such a way that the new values of weights of connecting links (i.e., u and w) and the threshold values of the output and hidden layers (i.e., Th and To) are established. When this forward and backward computation is repeatedly carried out on all the input-output data pairs for a long time, final values of the weights of connecting links and the threshold values

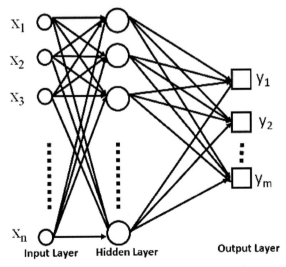

Figure 1: Basic structure of a feedforward backpropagation neural network with one input, one hidden and one output layer.

of the output and hidden layer neurons are obtained. The network, thus established, would then be able to predict the values of output (i.e., Y_1, Y_2, Y_3 .....) from a set of input data (i.e., X_1, X_2, X_3, X_4 .....). The values of input must lie within the maximum and minimum values of input data set that was used for training the networks.

Initially, values of *u, Th, w,* and *To* are assumed to lie between 0 and +1. A specific input-output data pair is chosen and new values of *u, Th, w,* and *To* are computed against their previous values. Using the new values of *u, Th, w,* and *To* for the next input-output data pair, another new values of weights and biases are calculated. This sequence is repeated several times for all the input-output data pairs until the relative deviation percent or the mean square of error between the actual and calculated values of output comes out to a very small value.

In neural network modelling input-output data, weights of line joining the neurons and the biases or threshold value of hidden and output layer neuron is designated as follows. The size of the matrix which represents the input data X (e.g., X_1, X_2, X_3, X_4) is (n∗nx) where, n is the number of experiments conducted and nx is the number of independent variables. Coded values of these parameters are allowed to range between +1 and −1. A single input data set is represented by a row matrix x of size (1 ∗ nx). Size of matrix representing the output parameters Y (e.g., Y_1, Y_2 and Y_3) is (*n* ∗ *ny*), where ny is the number of independent variables. Only one out of n data sets are used at a time and the size of matrix Y representing the single data set is (1 ∗ *ny*). The number of input layer (*nx*), output layer (*ny*) and hidden layer (*nh*) depends upon the number of independent and dependent variables.

Weight of synaptic joints between the input and hidden layer neuron have been designated by matrix u and its size is (*nx* ∗ *nh*), where '*nx*' is the number of independent variables and '*nh*' is the number of hidden layer neurons. As a first approximation, the elements of matrix u are assumed to lie between +1 and −1 and they are formed by way of random number generation. 'Th' is a vector or column matrix of size (*nx* ∗ 1). It represents the biases or threshold values of hidden layer neurons. Initially the elements of matrix are assumed to lie between −1 and +1 and formed by method of random number generation. '*w*' is a matrix of size (*nh* ∗ *ny*) where, '*ny*' is the number of neurons in the output layer. It represents the weights of synaptic joint between hidden and output layer neuron. Initially the values of '*w*' matrix are assumed to lie between +1 and −1 and formed by method of random number generation. '*To*' is a vector or column matrix of size (*ny* ∗ 1). It represents the bias or threshold values of the output layer neurons. An initial value of '*To*' matrix is fixed between +1 and −1. '*yh*' is the computed output values of hidden layer neurons. '*yo*' is computed output values of output layer neurons. '*eh*' is the 'backpropagation error' at the hidden layer neuron. '*eo*' is the backpropagation error at the output layer neurons. The backpropagation error '*eo*' and '*eh*' are used for updating the values *u, w, Th* and *To* in the next computation cycle. The process is called 'training' where the network 'learns' to relate a given input output data pair. The learning method given here is known as the 'gradient descent method' of learning. From the initial assumed values of *u, w, Th* and *To* computation of values of *yh, yo, eh* and *eo* are carried out by application of neural network method.

Output yh of hidden layer neuron is given by,

$$yh = \frac{1}{1+\exp[-(u'*x'+Th)]} \tag{10}$$

where, *u'* and *x'* are the transpose of matrices u and x, respectively. From sizes of matrices *u* and *x*, the size of yh matrix was evaluated as (*nh* ∗ 1).

Output yo at the output layer neuron is given by,

$$yo = \frac{1}{1+\exp[-(w'*yh+To)]} \tag{11}$$

Computed value of error *eo* at the output layer neuron is given by,

$$eo = (y'-yo)_\circ * yo_\circ * (1-yo) \tag{12}$$

where, *y'* is the transpose of the matrix y, which represents a single set of data containing the dependent variable. The dots ($_\circ$) in Eq. (12) represent that '*eo*' is a scalar product of matrices (*y'*–yo), yo and (1–yo).

Computed value of error eh at the hidden layer neuron is given by,

$$eh = yh_o * (1-yh)_o * (w * eo) \tag{13}$$

In Eq. (13), matrixes yh and $(1-yh)$ form scalar product, whereas $(w * eo)$ is a vector product. Size of the matrix 'eh' comes out to be $(nh * 1)$.

New values of bias or threshold parameter 'To_{new}' of output layer neurons for the next computation cycle will be,

$$To_{new} = L * eo + To \tag{14}$$

where L is the learning rate the value of which should lie between 0.6 and 0.9.

New values bias or threshold parameter 'Th_{new}' of hidden layer neurons for next computation cycle is,

$$Th_{new} = L * eh + Th \tag{15}$$

Values of weights 'u_{new}' of lines joining the input and hidden layer neurons can be computed from the following equation.

$$u_{new} = x' * L * eh' + u \tag{16}$$

Values of weights 'w_{new}' of lines joining the hidden and output layer neuron for next computation cycle are given by,

$$w_{new} = L * yh * eo' + w \tag{17}$$

In this method, the initially assumed values of u, w, Th and To were obtained through random number generation. A single computation cycle carries out this procedure for all the data pairs. After a large number of computation cycles, final values of u, w, Th and To established would give the values of yo for all input data. Here, yo is the coded value, which lies in the range of 0 and +1. These coded values are then converted to real values. Relative deviation percent between the actual and estimated values of dependent variable will reduce gradually with the completion of computation cycles in an automated fashion. The effective number of network parameters remains approximately constant when the network has been sufficiently trained (i.e., Relative deviation is relatively constant).

The new values of u, w, Th and To (i.e., u_{new}, w_{new}, Th_{new} and To_{new}) obtained (Eqs. (14)–(17)) are used for the second set of input-output data pairs. The training process was repeated several times in order to get the best performance of the ANN, due to a high degree of variability. Likewise, a training process was carried out repeatedly for 2000–5000 times on the first set of observed data and the final values of the weights and biases were obtained. Towards the end of the training process the relative deviation values become constant. The validity of the model is determined by mean relative percent between the actual and predicted values of independent variables:

$$E = \frac{100}{N} \sum_{i=1}^{N} \frac{|X_\alpha - X_p|}{X_p} \tag{18}$$

where, N is the number of experiments, X_a and X_p are the actual and predicted values of independent variables, respectively.

The performance of the NN model was validated by Mean Squared Error (MSE) and regression coefficient obtained as follows (Izadirfar and Jahromi, 2007);

$$MSE = \frac{1}{N} \sum_{i=1}^{N} (y_i - y_{di})^2 \tag{19}$$

$$R^2 = 1 - \frac{\sum_{i=1}^{N} (y_i - y_{di})^2}{\sum_{i=1}^{N} (y_{di} - y_m)^2} \tag{20}$$

where, N is the number of data, y_i is the predicted value by ANN model, y_{di} the actual or experimental value, y_m is the average of actual values, and R^2 is regression coefficient.

Relative importance of independent variables X (i.e., X_1, X_2, X_3 and X_4) for predicting the response Y can be found out by using the final values (viz. u, w, Th and To) of the network parameters. In order to find out the effect of X_n on Y_n (where n is the dependent variable number, i.e., Y_1, Y_2 and Y_3), the coded values x_1, x_2, x_3, x_4 of independent variables were set to zero, and found out the values of yo_n at $x_1 = +1$, $x_2 = 0$, $x_3 = 0$, $x_4 = 0$ and at $x_1 = -1$, $x_2 = 0$, $x_3 = 0$, $x_4 = 0$. The difference, $(\Delta yon)_{x1}$ between the two yo_n values would show the effect of the independent variable X_n on the dependent variable Y_n

i.e.,

$(\Delta yo_n)_{x1} = (yon)+1,0,0,0-(yon)-1,0,0,0$ (21)

$(\Delta yo_n)_{x2} = (yon)0,+1,0,0-(yon)0,-1,0,0$ (22)

$(\Delta yo_n)_{x3} = (yon)0,0,+1,0-(yon)0,0,-1,0$ (23)

$(\Delta yo_n)_{x4} = (yon)0,0,0+1-(yon)0,0,0-1$ (24)

Comparing the magnitude of $(\Delta yo_n)_{x1}$, $(\Delta yo_n)_{x2}$, $(\Delta yo_n)_{x3}$ and $(\Delta yo_n)_{x4}$ the relative importance of independent variables X_1, X_2, X_3 and X_4 on a dependent variable Y_n can be predicted. The increase in the value of independent variable X will cause the change in the value of the dependent variable Y which can be found out by observing the sign of (Δyo_n); a negative sign would indicate that the dependent variable would decrease with an increase in the magnitude of X value.

5. Case Study on Application of ANN for Osmotic Dehydration of Culinary Banana Slice

The factors influencing the osmotic dehydration are temperature of osmotic dehydration (x_1), processing time (x_2), salt concentration (x_3) and solution to sample ratio (x_4), with three responses being water loss (y_1), solid gain (y_2) and weight reduction (y_3). The levels of independent and dependent variables in coded and uncoded forms for osmotic dehydration are presented in Table 2.

Table 2: The levels of process variables in coded and uncoded forms for osmotic dehydration.

Coded levels	Uncoded values of process variables			
	Temperature (°C)	Processing time (min)	Salt concentration (%, w/w)	Solution to sample ratio (w/w)
−2	25	30	5	5:1
−1	30	60	10	10:1
0	35	90	15	15:1
+1	40	120	20	20:1
+2	45	150	25	25:1

5.1 Experimental Procedure for osmotic dehydration process

The osmotic dehydration process was conducted in a 250 ml beaker placed in an incubator shaker maintained at constant temperature. The experiments were carried out in the temperature range of 25–45°C. The culinary banana samples were weighed and then placed into dehydrating beaker of varying salt concentrations 5–25% (w/v) when the solution rises to desired temperature. The ratio of the weight of the medium to pieces of sample was maintained over a range of 5:1–25:1 (w/w). After each sampling time (30–150 min), the culinary banana slices were taken out of the osmotic solution and gently blotted with tissue paper. Slices were weighed and dry matter was determined by hot air drying at 105°C for 24 hr. The experiments were conducted in duplicate and for each experiment fresh salt solutions were used. Constant agitation condition of 150 rpm was used for the experiment. Agitation is given during osmosis to reduce the mass transfer resistance at the surface of the solid and to provide homogeneous mixing of the osmotic solution. It also aids in regulating the temperature of the osmotic medium. The water loss solid gain and weight loss were evaluated under each experimental condition.

Water loss from the sample at any time of osmotic dehydration (WL) was determined as the net loss of water from the fresh sample after osmotic dehydration process on the basis of the initial sample weight and is calculated by the following equation.

$$WL = \frac{W_o - (W - S)}{S_o + W_o} \times 100 \tag{25}$$

where, W_o denotes the initial weight of water, W is the final weight of culinary banana slice after dehydration; S is the final weight of solids at the end of the process and S_o represents the initial weight of solids.

Samples are withdrawn at each predetermined time interval and oven dried to determine the dry mass. Salt gain or solid gain (SG) by the sample was found out as net uptake of solids by the osmotically treated sample, based on the initial sample weight given by the following mass balance equation.

$$SG = \frac{S - S_o}{S_o + W_o} \times 100 \tag{26}$$

Weight reduction in the sample was determined by Eq. (27).

Weight reduction $(WR) = WL - SG$ \hfill (27)

5.2 Modelling of osmotic dehydration by ANN

The values of the experimentally designed independent variables and measured dependent variables are implemented in order to obtain the model parameters of ANN. The weight of synaptic joints between input and hidden layer neurons 'u' and between hidden and output layer neuron 'w' were generated randomly by random number generation (between 0 and 1). The threshold values of hidden (Th) and output layer (To) neuron were also generated by the same process.

The cross-validation data set was used to test the performance of the network while training was in progress as an indicator of the level of generalization and the time at which the network has begun to over train. Relative deviation between the experimental and predicted values of the responses, R^2 and MSE values are evaluated. Testing data set was used to examine the network generalization capability. The final values of matrices u, w, Th and To are as follows,

u =

3.3900	1.5571	0.5205	5.4298	0.5738	−1.5598	1.3937	4.1664	2.0940	−0.3703	0.6403
3.5517	−2.6525	0.6286	4.0051	9.7950	−0.6425	4.1918	−2.4251	−9.9996	−0.6894	−0.6976
0.1887	−3.3897	−3.5946	5.4354	3.2153	3.2080	3.0335	2.4760	3.5437	7.1845	4.6168
4.9217	0.5169	4.2410	3.1674	0.4176	5.8632	3.0022	−1.7229	0.8352	0.3622	0.9226

w =			Th =	To =
−0.1980	−1.8222	0.6769	−0.3072	−2.6834
−0.1895	3.8544	−1.8725	−2.8724	−2.7799
−2.0286	−1.1852	−2.3797	0.7099	−1.9065
−0.5711	3.5197	−1.9969	−2.0736	
2.3838	2.4067	2.2406	2.2177	
1.3858	−0.0140	1.6751	0.2166	
1.0472	−1.7529	1.7205	−1.5339	
−0.0334	−3.1186	0.6687	−2.4429	
2.2205	1.3016	2.8305	−1.3664	
0.4324	2.1678	−0.5881	0.9821	
1.3285	3.3325	1.7582	−4.1892	

5.3 Processing elements (PE) in hidden layer

In the next step, to assess the generalization of the ANN model, the experimental and predicted values of ANN model were calculated from the final values of u, w, Th and To, which were compared statistically. In the experimental range of independent variables, the optimum number of hidden layers for minimizing the difference between the predicted ANN values and the experimental values using Relative deviation during testing as performance indicator was 11.

The mean relative deviation for prediction by ANN model was 2.59%, which is lower than 10%, indicating that ANN adequately fit to the agreement for optimization process of the osmotic dehydration of culinary banana slices. The obtained R^2 value and RMSE between the actual and predicted for the three responses are also analyzed. The maximum R^2 (> 0.983) and minimum RMSE (< 0.554) values for WL and SG showed a good agreement of the NN model and the experimental data confirm a very good generalization of the neural network for the three responses. Figures 2, 3 and 4 show the plots of the predicted and the experimental data for WL, SG and WR, respectively. The predicted values were close to the actual values in case of water loss, solid gain and weight reduction in terms of the percentage relative deviation.

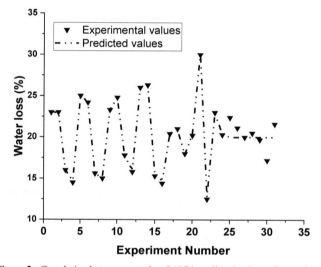

Figure 2: Correlation between actual and ANN predicted values of water loss.

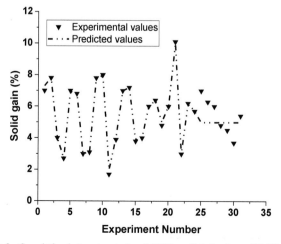

Figure 3: Correlation between actual and ANN predicted values of Solid gain.

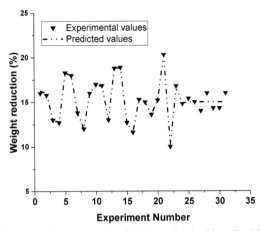

Figure 4: Correlation between actual and ANN predicted values of weight reduction.

5.4 Effects of independent variables on the responses

It was observed that the independent variables affected the osmotic dehydration of culinary banana in terms of responses, water loss (Y_1), solid gain (Y_2) and weight reduction (Y_3). Table 3 shows the influence of independent variables on water loss. Water loss is positively correlated with salt concentration of the osmotic solution (Table 3). A small incremental change in salt concentration resulted in significant change in osmotic pressure because of its lower molecular weight. The relative influence of solution temperature (x_1) and processing time (x_2) is negative, while concentration (x_3) and solution to sample ratio (x_4) is positive. Out of the four independent variables, salt concentration showed the highest effect on water loss, followed by the solution to sample ratio. On the other hand, temperature and processing time showed negative impact on water loss of culinary banana slices, highlighting the insignificance of these two independent variables on water loss.

A similar trend of the effect of variables was observed in terms of solid gain (SG) by the culinary banana slices during OD treatment (Table 4). With increase in the value of temperature and processing time, solid gain increases, showing positive impact for solid gain.

Table 5 shows the relative influence of the independent variables on the weight reduction. The result depicts the positive effect of temperature, salt concentration and solution to sample ratio and negative effect of time on weight reduction. Increasing the values of independent variables (temperature, salt concentration and ratio) increases the weight reduction rate while increase in processing time resulted in decrease of weight reduction.

It was observed that the final values of the network parameters depend on the initially noted values of *u*, *Th*, *w* and *To*, which were obtained though random number generation. A single computation cycle carries out this procedure on all the input–output data pairs. After a large number of computation cycles

Table 3: Relative influence of the coded values of independent parameters on water loss.

x_1	x_2	x_3	x_4	yo_1	Δyo_1
+1	0	0	0	0.443	$(\Delta yo_1)_{x1} = -0.021$
−1	0	0	0	0.464	
0	+1	0	0	0.307	$(\Delta yo_1)_{x2} = -0.132$
0	−1	0	0	0.439	
0	0	+1	0	0.987	$(\Delta yo_1)_{x3} = 0.972$
0	0	−1	0	0.016	
0	0	0	+1	0.583	$(\Delta yo_1)_{x4} = 0.130$
0	0	0	−1	0.453	

Table 4: Relative influence of the coded values of independent parameters on solid gain.

x_1	x_2	x_3	x_4	yo_2	Δyo_2
+1	0	0	0	0.489	$(\Delta yo_2)_{x1} = -0.080$
−1	0	0	0	0.568	
0	+1	0	0	0.407	$(\Delta yo_2)_{x2} = -0.082$
0	−1	0	0	0.489	
0	0	+1	0	0.988	$(\Delta yo_2)_{x3} = 0.820$
0	0	−1	0	0.168	
0	0	0	+1	0.573	$(\Delta yo_2)_{x4} = 0.096$
0	0	0	−1	0.477	

Table 5: Relative influence of the coded values of independent parameters on weight reduction.

x_1	x_2	x_3	x_4	Yo_3	Δyo_3
+1	0	0	0	0.494	$(\Delta yo_3)_{x1} = 0.040$
−1	0	0	0	0.454	
0	+1	0	0	0.332	$(\Delta yo_3)_{x2} = -0.151$
0	−1	0	0	0.483	
0	0	+1	0	0.993	$(\Delta yo_3)_{x3} = 0.981$
0	0	−1	0	0.012	
0	0	0	+1	0.636	$(\Delta yo_3)_{x4} = 0.128$
0	0	0	−1	0.509	

(5000), final values of To, Th, w and u will give the values of yo Eq. (11) for all the input data. The yo's (yo_1, yo_2 and yo_3) are the coded values of response, which lie between 0 and +1. These values are then decoded and compared with the actual values of the dependent parameters.

6. Future Prospective of ANN in the Food Industry

Basically, the emerging technologies of food processing are more nonlinear and complex in modelling. ANN can be regarded as a promising tool for modelling of different applications of emerging technologies in the food processing sector because it has the capacity to correlate any forms of nonlinear relationships and it does not require prior specification of suitable fitting function.

High-pressure processing (HPP) is a very effective food processing technology in the preservation of food products which retains the nutritional quality and enhances the value of the product. The modelling of HPP is very difficult due to the thermal behavior of the foodstuffs during high-pressure treatment. ANN can be the best solution for modelling of the HPP processing by considering temperature, pressure and time as the independent variables to control the microbial or enzymatic activity of the product. Pulsed electric field (PEF) is another emerging non-thermal technique, where mass transfer rates can be improved at very low energy input compared to other conventional techniques. Applying ANN for the modelling can accurately predict the output parameters like colour, microbial or enzymatic activity and the texture of food products treated with PEF by considering the treatment parameters like PEF temperature and electric field strength and, hence, improve the process of PEF treatment. Ultrasound-assisted extraction (UAE) has higher efficiency than other extraction processes and is more popular due to its low solvent consumption, low cost and low pollution impact on the environment. ANN modelling can be a promising tool for the nonlinear multivariate modelling of the UAE parameters, i.e., ultrasound intensity, ultrasound duration and temperature for effective ultrasound treatment. Cold plasma treatment can improve the physical, nutritional and sensory characteristics of the treated products and can be used for preserving unstable bioactive compounds and modulating enzyme activity. ANN modelling also can be applied for the cold plasma treatment of foodstuffs. Electrical input (voltage, frequency, power), treatment time and food material properties can be considered as independent variables for the ANN modelling.

7. Conclusion

ANN modelling is a newly developed strategy and an alternative to conventional modeling techniques. ANNs are a powerful tool for simulating various nonlinear systems and can be applied to many complex problems associated with food processing. The application of the backpropagation technique to a food process modelling problem requires three simple steps: Network design, learning or training and usage. ANNs can be successfully applied in food processing unit operations for the prediction and modelling of the responses or dependent variables. The ANN is better than other modelling techniques because it implements experimental data to develop the model with high accuracy and solve problems related to the food processing industry.

References

Aghbashlo, M., Hosseinpour, S. and Mujumdar, A.S. 2015. Application of artificial neural networks (ANNs) in drying technology: A comprehensive review. Drying Technology, pp. 1397–1462. doi: 10.1080/07373937.2015.1036288.

Baykal, H. and Yildirim, H.K. 2013. Application of Artificial Neural Networks (ANNs) in wine technology. Critical Reviews in Food Science and Nutrition. pp. 415–421. doi: 10.1080/10408398.2010.540359.

Bhotmange, M. and Shastri, P. 2011. Application of artificial neural networks to food and fermentation technology. Artificial Neural Networks - Industrial and Control Engineering Applications, pp. 201–222. doi: 10.5772/16067.

Bishop, C.M. 1994. Neural networks and their applications. Review of Scientific Instruments. AIP, 65(6): 1803–1832.

Chen, C. and Ramaswamy, H. 2002. Modeling and optimization of variable retort temperature (VRT) thermal processing using coupled neural networks and genetic algorithms. Journal of Food Engineering 53: 209–220.

Chen, C., Duan, S., Cai, T. and Liu, B. 2011. Online 24-h solar power forecasting based on weather type classification using artificial neural network. Solar Energy 85(11): 2856–2870. doi: 10.1016/j.solener.2011.08.027.

Chen, C.R. and Ramaswamy, H.S. 2000. A neuro-computing approach for modeling of residence time distribution (RTD) of carrot cubes in a vertical scraped surface heat exchanger (SSHE) 33: 549–556.

Desai, K.M., Survase, S.A., Saudagar, P.S., Lele, S.S. and Singhal, R.S. 2008. Comparison of artificial neural network (ANN) and response surface methodology (RSM) in fermentation media optimization: Case study of fermentative production of scleroglucan. Biochemical Engineering Journal 41(3): 266–273. doi: 10.1016/j.bej.2008.05.009.

Doganis, P., Alexandridis, A., Patrinos, P. and Sarimveis, H. 2006. Time series sales forecasting for short shelf-life food products based on artificial neural networks and evolutionary computing. Journal of Food Engineering 75(2): 196–204. doi: 10.1016/j.jfoodeng.2005.03.056.

Eerikäinen, T., Zhu, Y.-H. and Linko, P. 1994. Neural networks in extrusion process identification and control. Food Control 5(2): 111–119. doi: 10.1016/0956-7135(94)90096-5.

García-Gimeno, R.M., Hervás-Martínez, C. and De Silóniz, M.I. 2002. Improving artificial neural networks with a pruning methodology and genetic algorithms for their application in microbial growth prediction in food. International Journal of Food Microbiology 72(1-2): 19–30. doi: 10.1016/S0168-1605(01)00608-0.

Goyal, S. and Goyal, G.K. 2011. Cascade and feedforward backpropagation artificial neural networks models for prediction of sensory quality of instant coffee flavoured sterilized drink. Canadian Journal on Artificial Intelligence, Machine Learning and Pattern Recognition 2(6): 78-82.

Haykin, S., Haykin, S.S., Haykin, S.S. and Haykin, S.S. 2009. Neural Networks and Learning Machines. Pearson Upper Saddle River, NJ, USA.

Horimoto, Y., Durance, T., Nakai, S. and Lukow, O.M. 1995. Neural networks vs principal component regression for prediction of wheat flour loaf volume in baking tests. Journal of Food Science 60(3): 429–433. doi: 10.1111/j.1365-2621.1995.tb09796.x.

Karimi, F., Rafiee, S., Taheri-Garavand, A. and Karimi, M. 2012. Optimization of an air-drying process for Artemisia absinthium leaves using response surface and artificial neural network models. Journal of the Taiwan Institute of Chemical Engineers 43(1): 29-39. doi: 10.1016/j.jtice.2011.04.005.

Khani, A. and Nemati, A. 2013. Modeling of nitrate removal by nanosized iron oxide immobilized on perlite using artificial neural network. Asian Journal of Chemistry 25(8): 4340–4346.

Moghaddam, M.G. and Khajeh, M. 2011. Comparison of Response Surface Methodology and Artificial Neural Network in Predicting the Microwave-Assisted Extraction Procedure to Determine Zinc in Fish Muscles. 2011 (October), pp. 803–808. doi: 10.4236/fns.2011.28110.

Momenzadeh, L., Zomorodian, A. and Mowla, D. 2011. Experimental and theoretical investigation of shelled corn drying in a microwave-assisted fluidized bed dryer using Artificial Neural Network. Food and Bioproducts Processing. Institution of Chemical Engineers 89(1): 15–21. doi: 10.1016/j.fbp.2010.03.007.

Morimoto, T., Tu, K., Hatou, K. and Hashimoto, Y. 2003. Dynamic optimization using neural networks and genetic algorithms for tomato cool storage to minimize water loss. Transactions of the ASAE. American Society of Agricultural and Biological Engineers 46(4): 1151.

Mu, G., Luan, F., Liu, H. and Gao, Y. 2013. Use of Experimental Design and Artificial Neural Network in Optimization of Capillary Electrophoresis for the Determination of Nicotinic Acid and Nicotinamide in Food Compared with High-Performance Liquid Chromatography, pp. 191–200. doi: 10.1007/s12161-012-9429-z.

Musa, K.H., Abdullah, A. and Al-Haiqi, A. 2016. Determination of DPPH free radical scavenging activity: Application of artificial neural networks. Food Chemistry. Elsevier 194: 705–711.

Olmos, A., Trelea, I.C., Courtois, F., Bonazzi, C. and Trystram, G. 2002. Dynamic optimal control of batch rice drying process. Drying Technology 20(7): 1319–1345. doi: 10.1081/DRT-120005855.

Pareek, V.K., Brungs, M.P., Adesina, A.A. and Sharma, R. 2002. Artificial neural network modeling of a multiphase photodegradation system. Journal of Photochemistry and Photobiology A: Chemistry 149(1-3). doi: 10.1016/S1010-6030(01)00640-2.

Popescu, O., Popescu, D.C., Wilder, J. and Karwe, M.V. 2001. A new approach to modeling and control of a food extrusion process using artificial neural network and an expert system. Journal of Food Process Engineering 24(1): 17–36. doi: 10.1111/j.1745-4530.2001.tb00529.x.

Prinderre, P., Piccerelle, P.H., Cauture, E., Kalantzis, G., Reynier, J.P. and Joachim, J. 1998. Formulation and evaluation of o/w emulsions using experimental design. International Journal of Pharmaceutics. Elsevier 163(1-2): 73–79.

Rahmanian, B., Pakizeh, M., Mansoori, S.A.A. and Abedini, R. 2011. Application of experimental design approach and artificial neural network (ANN) for the determination of potential micellar-enhanced ultrafiltration process. Journal of Hazardous Materials. Elsevier B.V., 187(1-3): 67–74. doi: 10.1016/j.jhazmat.2010.11.135.

Ray, J., Kumar, P. and Banerjee, R. 2004. Optimization of culture parameters for extracellular protease production from a newly isolated *Pseudomonas* sp. using response surface and artificial neural network models. 39: 2193–2198. doi: 10.1016/j.procbio.2003.11.009.

Razmi-Rad, E., Ghanbarzadeh, B., Mousavi, S.M., Emam-Djomeh, Z. and Khazaei, J. 2007. Prediction of rheological properties of Iranian bread dough from chemical composition of wheat flour by using artificial neural networks. Journal of Food Engineering 81(4): 728–734. doi: 10.1016/j.jfoodeng.2007.01.009.

Ruan, R., Almaer, S. and Zhang, J. 1995. Prediction of dough rheological properties using neural networks. Cereal Chemistry pp. 1–4. Available at: http://www.aaccnet.org/publications/cc/backissues/1995/Documents/72_308.pdf.

Rumelhart, D.E. and Zipser, D. 1985. Feature discovery by competitive learning. Cognitive Science. 9(1): 75–112. doi: 10.1016/S0364-0213(85)80010-0.

Scala, K.D., Meschino, G., Vega-Galvez, A., Lemus-Mondaca, R., Roura, S. and Mascheroni, R. 2013. An artificial neural network model for prediction of quality characteristics of apples during convective dehydration. Food Science and Technology (Campinas) 33(3): 411–416. doi: http://dx.doi.org/10.1590/S0101-20612013005000064.

Shafiee, S., Minaei, S., Moghaddam-Charkari, N. and Barzegar, M. 2014. Honey characterization using computer vision system and artificial neural networks. Food Chemistry Elsevier Ltd. 159: 143–150. doi: 10.1016/j.foodchem.2014.02.136.

Shafiur, M., Rashid, M.M. and Hussain, M.A. 2011. Food and bioproducts processing thermal conductivity prediction of foods by Neural Network and Fuzzy (ANFIS) modeling techniques. Food and Bioproducts Processing. Institution of Chemical Engineers 90(2): 333–340. doi: 10.1016/j.fbp.2011.07.001.

Silva, S.F., Anjos, C.A.R., Cavalcanti, R.N. and dos Santos Celeghini, R.M. 2015. Evaluation of extra virgin olive oil stability by artificial neural network. Food Chemistry. Elsevier Ltd, 179: 35–43. doi: 10.1016/j.foodchem.2015.01.100.

Sinha, K., Chowdhury, S., Saha, P.D. and Datta, S. 2013. Modeling of microwave-assisted extraction of natural dye from seeds of Bixa orellana (Annatto) using response surface methodology (RSM) and artificial neural network (ANN). Industrial Crops and Products 41(1): 165–171. doi: 10.1016/j.indcrop.2012.04.004.

Soleimani, R., Shoushtari, N.A., Mirza, B. and Salahi, A. 2013. Experimental investigation, modeling and optimization of membrane separation using artificial neural network and multi-objective optimization using genetic algorithm. Chemical Engineering Research and Design 91(5): 883–903. doi: 10.1016/j.cherd.2012.08.004.

Swider, B.G. 2011. Analytica Chimica Acta 705: 283–291. doi: 10.1016/j.aca.2011.06.033.

Trelea, I.C., Courtois, F. and Trystram, G. 1997. Dynamic models for drying and wet-milling quality degradation of corn using neural networks. Drying Technology 15(3-4): 1095–1102. doi: 10.1080/07373939708917280.

Tumuluru, J.S. and Bandyopadhyay, S. 2004. Optimization of extrusion process variables using a genetic algorithm. Food and Bioproducts Processing 82(2): 143–150.

Youssefi, S., Emam-Djomeh, Z. and Mousavi, S.M. 2009. Comparison of artificial neural network (ANN) and response surface methodology (RSM) in the prediction of quality parameters of spray-dried pomegranate juice. 3937. doi: 10.1080/07373930902988247.

Zhou, L., Puri, V.M., Anantheswaran, R.C. and Yeh, G. 1995. Finite element modeling of heat and mass transfer in food materials during microwave heating—Model development and validation. Journal of Food Engineering 25(4): 509–529. doi: 10.1016/0260-8774(94)00032-5.

CHAPTER 23

Mathematical Modelling for Predicting the Temperatures During Microwave Heating of Solid Foods

A Case Study

Coskan Ilicali,[1,*] *Filiz Icier*[2] *and Ömer Faruk Cokgezme*[2]

1. Introduction

Microwave heating is one of the most important heating methods for foods. It has been proposed as an alternative electrical heating method to traditional heating methods for foods. Unlike conventional heating, where heat flows from the outer surface to the center of the body, heat is generated within the body in microwave heating due to the transmission of an electromagnetic wave, and heat can propagate in all directions depending on the temperature (Knoerzer et al., 2009; Hossan et al., 2010; Bhattacharya and Basak, 2017).

Microwave heating has some advantages, such as highspeed startup, short process time, better internal heating, high energy efficiency and very little pollution. Thus, it is in high demand for industrial and household applications (Navarro and Burgos, 2017). It is used for different purposes in food processing, such as heating, thawing, blanching, pasteurization, sterilization, baking, cooking, extraction, freezing and dehydration (Metaxas and Meredith, 1983; Decaureau, 1985; Buffler, 1993; Roussy and Pearce, 1995; Datta and Anantheswaran, 2001; Schubert and Regier, 2005; Campañone and Zaritzky, 2005, 2010; Turabi et al., 2008; Knoerzer et al., 2009; Cha-um et al., 2011; Wray and Ramaswamy, 2015; James et al., 2015; Zhong et al., 2015; Bhattacharya and Basak, 2017). In summary, an increase in product temperature is a common subject (Knoerzer et al., 2009).

Microwave frequencies range between 10^6 Hz and 10^{12} Hz, although the allocated frequencies are 0.915, 2.45, 5.8 and 22 GHz in commercial applications (Tang et al., 2008; Chiavaro et al., 2009; Mutyala et al., 2010; Baysal et al., 2011). Since the electromagnetic field in dielectric materials polarizes the molecules, they rotate due to dipole moments created. The frictions between the molecules generates heat

[1] Kyrgyz-Turkish Manas University, Faculty of Engineering, Department of Food Engineering, Bishkek, Kyrgyzstan.
[2] Ege University, Faculty of Engineering, Department of Food Engineering, Izmir, Turkey.
* Corresponding author: coskan.ilicali@manas.edu.kg

within the material. Thus, microwave heating can cause a very quick increase in the temperature even if the food has low thermal conductivity (Hossan et al., 2010).

However, the occurrence of non-uniformity inside the food is the main disadvantage for the wide use of microwave heating. Several researchers reported that hot and cold spots inside food material could occur during microwave treatments (Zhou et al., 1995; Sakai et al., 2004; Hossan et al., 2010; Liu et al., 2014; Bhattacharya and Basak, 2017). There are also a number of microwave specific factors that may cause non-uniform heating patterns. Examples of these are focusing effects, corner and edge heating, inhomogeneous electromagnetic field distributions or variation in dielectric properties in heterogeneous food materials (Knoerzer et al., 2009).

The main reasons for non-uniform heating are (1) interference of electromagnetic waves, and (2) uneven absorption of microwave energy and subsequent heat dissipation due to variation in dielectric and thermo-physical properties of foods, depending on changes in temperature and composition during heating (Pitchai et al., 2012). The transmission and reflection from interfaces result in the existence of a standing wave causing non-uniform heating (Hossan et al., 2010).

Several studies have been made to determine the temperature distribution during microwave heating (Wäppling-Raaholt et al., 2006; Knoerzer et al., 2009; Hossan et al., 2010; Liu et al., 2013).

The shape of the sample is one of the major factors affecting the generation of non-uniformity, particularly inside the cylinder and sphere and at the corners in parallelepiped shape (Oliveira and Franca, 2002; Basak and Ayappa, 2002; Campañone et al., 2012, 2014).

The dielectric properties of food materials have an important role in the uniformity of the heating process. Dielectric properties are electrical properties that govern the interactions between MW radiation and food (Llave et al., 2016). The main parameter describing the level of heat generation inside a non-magnetic material, such as food, is the complex permittivity ε^* (Eq. (1)) (İçier and Baysal, 2004; Farag et al., 2012);

$$\varepsilon^* = \varepsilon' - j\varepsilon'' \tag{1}$$

The real part of the complex permittivity is named as dielectric constant ε' and the imaginary part is called "loss factor" ε''. Dielectric constant represents the ability of the product to store electromagnetic energy while loss factor represents the ability of the product to dissipate electromagnetic energy. The ratio between these parts is referred to as the "loss tangent". It is known as the conversion factor of microwave energy to thermal energy within a material (İçier and Baysal, 2004; Farag et al., 2012; Llave et al., 2016), as represented in Eq. (2).

$$tan\delta = \frac{\varepsilon''}{\varepsilon'} \tag{2}$$

Penetration depth is a measure of the penetration capability of electromagnetic radiation inside the product (Mao et al., 2005). The penetration depth (D) is defined as the depth where the magnitude of the electric field drops by a factor 1/e (1/2.718) with respect to the surface value (Farag et al., 2012; Liu et al., 2013b; Llave et al., 2016). Similarly, the power penetration depth (D_p) is the distance in meters where the power density is reduced by a factor 1/e of the surface (Eq. (3)), and dependent on the attenuation factor (Eq. (4)). For materials having high loss ($\varepsilon''_{eff} \geq \varepsilon'$) and low loss ($\varepsilon''_{eff} \leq \varepsilon'$) factors, Eqs. (5) and (6) apply, respectively (Campañone and Zaritzky, 2010; Farag et al., 2012);

$$D_p = \frac{D}{2} = \frac{1}{2\alpha} \tag{3}$$

$$\alpha = f\sqrt{\frac{[\mu'\mu_o\varepsilon'\varepsilon_o]\left[\left(1+\left(\varepsilon''_{eff}/\varepsilon'\right)^2\right)^{1/2}-1\right]}{2}} \tag{4}$$

$$\alpha = \frac{f}{2}\sqrt{\frac{[\mu'\mu_o\varepsilon\varepsilon_o]}{\varepsilon'}}\varepsilon''_{eff} \tag{5}$$

$$\alpha = \frac{\pi \varepsilon'' eff}{\lambda_0 \sqrt{\varepsilon}}, \qquad (6)$$

where f is the frequency (Hz), α is attenuation coefficient (1/m), μ_o is permeability of free space, μ' relative permeability and λ_0 wave length (m).

For materials having penetration depth characteristics that vary with temperature, the heat generation will also change significantly with heating time during microwave heating (Liu et al., 2013a).

In addition, the interaction of transmitted and reflected waves within the material could cause some spatial resonances (Campañone et al., 2014). During penetration of microwaves through the food sample, series parts of transmitted and reflected waves are formed. Thus, the localized heating of the sample at the locations of the constructive interferences could occur (Bhattacharya and Basak, 2017). For this reason, the resonance driven non-uniform heating occurs only if the dimension of the sample is similar to the penetration depth of the microwave. On the other hand, if the dimension of the sample was larger than the penetration depth, the surface temperature would be much higher than the center temperature (Bhattacharya and Basak, 2017).

Several authors have discussed ways to minimize the non-uniform temperature distribution, such as the use of power cycle during the microwave heating (Taher and Farid, 2001; Gunasekaran and Yang, 2007; Campañone et al., 2014), rotation of food inside the cavity (Geedipalli et al., 2007), movement of the material (conveyor belts, rotary systems, fluidized beds, agitation, etc.) (Li et al., 2010), assisting of jet impingement (Walker and Li, 1993; Bows et al., 1999; Datta and Anantheswaran, 2005; Antonio and Deam, 2005), applying variable-frequency technique (Bows et al., 1999; Antonio and Deam, 2005) and on/off power applications (Campañone et al., 2014).

2. Basics of Microwave Heating

2.1 Electromagnetics and boundary conditions

Electromagnetic waves consist of simultaneous electric and magnetic fields. The dipole molecules reorient themselves in order to be in phase with the alternating electric field, and thus, molecular kinetic energy increases. Depending on the electrical and physical properties of the material, its temperature increases, since kinetic energy is converted to heat energy within the material (Yang and Gunasekaran, 2001; Durka et al., 2009; Farag et al., 2012).

The distribution of the electromagnetic field within the material during microwave heating is governed by Maxwell's equations (Pitchai et al., 2012; Navarro and Burgos, 2017);

$$\vec{\nabla} \times \vec{E} = -\frac{\partial \vec{B}}{\partial t} \qquad (7)$$

$$\vec{\nabla} \times \vec{H} = \vec{J} + \frac{\partial \vec{D}}{\partial t} \qquad (8)$$

$$\vec{\nabla} \cdot \vec{D} = \rho_v \qquad (9)$$

$$\vec{\nabla} \cdot \vec{B} = 0 \qquad (10)$$

where $\vec{E}$ is the electric field, $\vec{B}$ is the magnetic induction, $\vec{H}$ is the magnetic field, $\vec{J}$ is the current density, $\vec{D}$ is the electric displacement, and ρ_v is the electric charge density.

$$j = \sigma E \qquad (11)$$

$$D = \epsilon E \qquad (12)$$

$$B = \mu H \qquad (13)$$

where σ is the electric conductivity, μ is the magnetic permeability and ϵ is the electric permittivity (Navarro and Burgos, 2017).

Equations (7) and (8) show the coupling between the electric and magnetic fields while Eqs. (9) and (10) describe the sources of electric and magnetic fields (Knoerzer et al., 2008).

The inner walls of the microwave oven cavity are generally considered to be perfect electrical conductors (Liu et al., 2013a).

2.2 Heat generation

Poynting theorem and Lambert's law are used to calculate the microwave power generation based on the size of the food material. Poynting theorem is generally used to calculate the heat generation due to the electromagnetic wave (Hossan et al., 2010);

$$\vec{q} = \frac{1}{2}\vec{E} \times \vec{H} \tag{14}$$

$$Q(z) = -Re(\vec{\nabla}.\vec{q}) \tag{15}$$

where $\vec{q}$ is the Poynting vector and $\vec{H}$ is the complex conjugate of the magnetic field.

For non-magnetic materials, such as food, the relative permeability can be set to unity, and the permittivity tensor is written as a complex constant (Eq. (16)). Complex constant has real (ε') and imaginary (ε'') parts. σ is the electrical conductivity, and the subscript "0" describes the behavior in a vacuum, thus, ε is a relative value (Knoerzer et al., 2008);

$$\varepsilon''_r = \frac{\sigma}{2.\pi.f.\varepsilon_0} \tag{16}$$

Applying Poynting power theorem, the dissipated power inside a microwave cavity could be represented as energy generated inside a heated material (Farag et al., 2012; Navarro and Burgos, 2017);

$$\Phi(r) = \frac{1}{2}f\varepsilon_0\varepsilon'' EE^* \tag{17}$$

where E* is the complex conjugate of E.

For nonmagnetic materials, the volumetric absorbed power can be calculated as given in Eqs. (18)–(20) (Knoerzer et al., 2008; Farag et al., 2012; Pitchai et al., 2012);

$$\Phi = f\varepsilon_0\varepsilon''_{eff} E^2_{rms} \tag{18}$$

$$\Phi = 2\pi f\varepsilon' tan\delta\, E^2_{rms} \tag{19}$$

$$\Phi = 2\pi f\varepsilon_0\varepsilon'' E^2 \tag{20}$$

where Φ is the volumetric absorbed power (W/m³), f is frequency (1/s), ε_0 permittivity of free space, ε''_{eff} is complex permittivity, ε' is dielectric constant, ε'' is dielectric loss, E_{rms} is the root mean square of the electric field (V/m) and E is the electrical field intensity. The internal volumetric heat generation term (Φ, W/m³) is proportional to the square of the electric field intensity, the dielectric loss factor and the frequency (Eq. (20)) (Curet et al., 2008; Liu et al., 2013a).

On the other hand, Lambert's law is a result of a series of simplifications of Maxwell's equations (Barringer et al., 1995; Navarro and Burgos, 2017). For example, the power penetration in one dimension is described by Lambert's law, as shown in Eq. (21) (Zhou et al., 1995; Bail et al., 2000; Farag et al., 2012).

$$P = P_0 e^{(-2\alpha y)} \tag{21}$$

Equation (19) represents volumetric absorbed power while Eq. (21) represents local power at a certain distance from the surface of the heated material (Farag et al., 2012).

For practical calculations of electric field strength inside a microwave cavity, an energy balance for water used as the material was applied to estimate the power term using an empirical approach (Padua, 1993; Farag et al., 2012);

$$\frac{\sum mCp\Delta T}{Vt} = 2\pi f\varepsilon' \tan\delta\, E_{rms}^2 \tag{22}$$

where m is the mass of the water used (kg), Cp is the average specific heat of the water (J/kg K), ΔT is the temperature difference in the water, V is the volume of the water used (m³), and t is the heating application time(s).

2.3 Heat transfer

The governing equations for simultaneous heat and mass transfer are the continuity equation, the thermal energy equation and Fick's law. In the case of neglecting mass transfer, a general heat transfer equation can be written as (Knoerzer et al., 2008):

$$\rho C_p \frac{\partial T}{\partial t} - \nabla \cdot (k \cdot \nabla T) = -\nabla_{qR} + \Phi \tag{23}$$

where ρ is the density (kg/m³), C_p is the specific heat capacity (J/kg K), k is the thermal conductivity (W/mK), and q_R is the radiative power flux = (W/m²). The terms on the right side of this equation represent the radiative heat transfer mechanism and heat generation, respectively (Metaxas, 1996; Knoerzer et al., 2008). The radiation heat transfer mechanism can be accounted by additional boundary conditions and Eq. (23) can be simplified to Eq. (24) (Knoerzer et al., 2008; Liu et al., 2013a);

$$\rho C_p \frac{\partial T}{\partial t} - \nabla \cdot (k \cdot \nabla T) = \Phi \tag{24}$$

where T is the temperature (K), ρCp is the specific heat per unit volume (J/m³K) and k is the thermal conductivity (W/mK) (Navarro and Burgos, 2017);

In cylindrical coordinates, for a one-dimensional case, Eq. (24) becomes

$$\frac{\rho C_p}{k}\frac{\partial T}{\partial t} = \frac{\partial^2 T}{\partial r^2} + \frac{1}{r}\frac{\partial T}{\partial r} + \frac{Q}{k} \qquad 0 < r < R \tag{25}$$

The initial and boundary conditions are;

$$T = T_0 \; at \; t = 0 \tag{26}$$

$$\frac{\partial t}{\partial r} = 0 \; at \; r = 0, t > 0 \tag{27}$$

$$-k\frac{\partial t}{\partial r} = h(T - T_a) \; at \; r = R, t > 0 \tag{28}$$

where h is the convective heat transfer coefficient (W/m²K), Ta is the ambient temperature and T_0 is the initial temperature of the food material (Navarro and Burgos, 2017).

The dimensionless forms with primes now omitted (Eqs. (29)–(32)) can be written by rescaling $r' = \frac{r}{R}, t' = \frac{kt}{(\rho C_p R^2)}, T' = \frac{k(T - T_0)}{(Q_0 R^2)}, \; Q' = \frac{Q}{Q_0}$, where Q_0 is the initial volumetric power dissipated (Navarro and Burgos, 2017).

$$\frac{\partial T}{\partial t} = \frac{\partial^2 T}{\partial t^2} + \frac{1}{r}\frac{\partial T}{\partial r} + Q \; for \; 0 < r < 1 \tag{29}$$

$$T = 0 \; at \; t = 0 \tag{30}$$

$$\frac{\partial t}{\partial r} = 0 \; at \; r = 0, \, t > 0 \tag{31}$$

$$-\frac{\partial T}{\partial r} = \frac{hR}{k}T + \frac{h}{Q_0 R} \; (T_0 - T_a) \; at \; r = 1, \, t = 0 \tag{32}$$

3. Studies on Mathematical Modelling of Microwave Heating in Solid Foods

The modelling of microwave heating is complicated because the nonlinear behavior needs to be solved by simultaneous electromagnetic and heat transfer mechanisms (Zhang and Datta, 2000; Zhu et al., 2007; Geedipalli et al., 2007; Knoerzer et al., 2008; Pitchai et al., 2012). In recent years, the partial differential equations governing them can be solved by using simulation software packages (Knoerzer et al., 2008).

Several efforts have been conducted to simulate microwave heating from using simple analytical approaches (Watanabe et al., 1978) to computational approaches. Hossan et al. (2010) presented a closed-form analytic solution predicting the temperature distribution by solving an unsteady energy equation and the electric field distribution by using Poynting theorem. They discussed that uniform heating depends on the frequency of the incident electromagnetic wave, geometric parameters and dielectric properties of the food material. Bhattacharya and Basak (2017) carried out a detailed theoretical analysis to investigate the effect of the shape of the sample on the microwave heating dynamics of food materials. They reported that the presence of the corner can have a pronounced effect on the heating dynamics for thick samples.

Simulation techniques based on Finite Element Method (FEM) and Finite Difference (FDM) have been commonly used for solving the microwave heating of food materials (Yakovlev, 2006; Pitchai et al., 2012; Navarro and Burgos, 2017). Knoerzer et al. (2008) developed a simulation approach to predict temperature distributions of arbitrarily shaped products in three dimensions, by using a user-friendly interface coupling two commercial software packages. The electromagnetic part of the model was solved with the finite difference time domain method (FDTD) while the heat transfer part was solved by using the finite element method.

Pitchai et al. (2012) solved a coupled electromagnetic and heat transfer problem by using finite-difference time-domain based commercial software. They compared the predicted temperature distribution on three planes in a cylindrical model food (1% gellan gel) with the corresponding thermal images for 30s heating period and experimental temperature measurements of twelve locations by using fiber optic sensors. Predicted temperature profiles agreed well with experimental data.

In another recent study, Navarro and Burgos (2017) solved Maxwell's equations, and the resulting electric field distribution was incorporated as a source term in the heat transfer equation. The model included the temperature dependence of the dielectric properties. The predicted results agreed well with the experimental temperature data in the literature.

Several assumptions have been performed to simplify the problem and reduce the computational time (Pitchai et al., 2012). In general, the problem is simplified by using Lambert's law, instead of modeling Maxwell's equation. The heat generation was modelled by taking into account the exponential decrease of the microwave power relative to the penetration depth (Yang and Gunasekaran, 2004; Pitchai et al., 2012; Navarro and Burgos, 2017). Ayappa et al. (1991) discussed that the heat generation rate for small size foodstuff can be analyzed well by using Poynting theorem while Lambert law was more accurate for bulky and thick samples. Similarly, Hossan et al. (2010) discussed that the Lambert's law approach was valid only for semi-infinite samples, and cannot be used for small food items. Yang and Gunasekaran (2001) reported that the Poynting theorem was much more accurate than Lambert's law, especially in the boundary region. In addition, Oliveira and Franca (2002) concluded that the sample size over which the Lambert's law can be applied is bigger for cylinders compared to slabs. However, they did not consider the influence of the sample geometry in the formulation of Lambert's law (Romano et al., 2005).

On the other hand, Lin et al. (1995) modified the power source term by considering the decreasing control volume. Zhou et al. (1995) reported Lambert' law could have described the power concentration along the central axis of the cylindrical sample. Similarly, simplified numerical results of Lambert's law were comparable with the experimental methods (Liu et al., 2005). Romano et al. (2005) discussed that the concentrated absorbing power effect was predominant in cylinders of smaller radii, given an attenuation factor. As the sample radius increased, the absorbed power decreased in the center and increased at the surface. Radial temperature profiles coincided with the temperature profiles along *z* when the radius of the cylinder was higher than the critical radius, defined by Oliveira and Franca (2002).

The local hot spot is less significant during pulsed microwave applications, especially when longer intermittent power off times are employed (Yang and Gunasekaran, 2001). Gunasekaran and Yang (2007) discussed similarly that the pulsed microwave heating provided more uniform heating than continuous mode at the same average output power.

The simpler way to achieve a relatively uniform heating is to use mode stirrers or rotating sample (Liu et al., 2013a). Chatterjee et al. (2007) heated containerized liquid by microwave in a rotating cylindrical cavity. They simulated the microwave heating by Lambert's law, and discussed that the simulation did not agree with experimental data if the radius of the container was too small since the focusing effect of microwave might have been happening. However, Liu et al. (2013) successfully developed a three-dimensional computer model based on FEM, considering the continuous turntable rotation during microwave heating.

In summary, there are numerous studies on the modelling of microwave heating in literature. Researchers have tried to predict temperature distribution within the food during microwave heating by the employment of different modelling approaches and different modelling methods. Some of the different modelling approaches could improve the prediction of surface temperature while some of them improve center temperature prediction. However, the improvement of temperature predictions for every location within the food by a single model is yet to be achieved.

In the case study section of this chapter, a general mathematical model, developed for microwave heating of solid foods and the calculation of absorbed power for heat generation term, was broadly explained. The model equations were numerically solved by using finite difference and finite element approaches. The accuracy of the different numerical models used (finite difference and finite element) were assessed by comparing with literature microwave heating data for cylindrical foods.

4. Case Study

4.1 Development of the mathematical model

The unsteady state heat conduction equation for solids with varying thermo-physical properties for a finite cylinder can be modified for microwave heating by the addition of a microwave heat generation term.

$$\rho Cp(T)\frac{\partial T}{\partial t} = \frac{1}{r}\frac{\partial}{\partial r}\left(rk(T)\frac{\partial T}{\partial r}\right) + \frac{\partial}{\partial z}\left(K(T)\frac{\partial T}{\partial z}\right) + \Phi \tag{33}$$

Equation (33) has been solved for simultaneous heat and mass transfers in a cupcake baked in a convection oven (Sakin et al., 2007). The finite difference numerical model (FDM) developed has been discussed in detail in this reference and will not be repeated here. This model has been extended to account for microwave heat generation (Icier and Ilicali, 2012). Only heat transfer has been considered. The microwave term used in the finite difference numerical model described in Icier and Ilicali (2012) has been modified in this work. In addition, this model was also solved with the finite element method (FEM) by modifying the method described in Lin et al. (1995) and Romano et al. (2005). The grid system used in the computations is shown in Fig. (1).

Microwave power density (absorbed microwave power per volume) has been formulated as follows by Romano et al. (2005);

$$\Phi = \frac{2\alpha RP_o''}{r}\left[e^{-2\alpha(R-r)} + e^{-2\alpha(R+r)}\right] + 2\alpha P_o''\left[e^{-2\alpha\left(\frac{L}{2}-z\right)} + e^{-2\alpha\left(\frac{L}{2}+z\right)}\right] \tag{34}$$

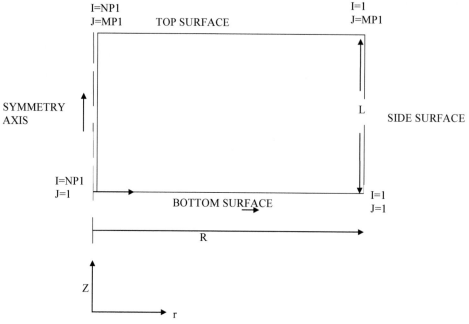

Figure 1: Schematic diagram of the grid system used.

where $P''_o = \dfrac{P_o}{A_{TOT}}$, assuming a uniform flux distribution, microwave power flux at a surface. A_{TOT} is the total surface area of the cylinder. Romano et al. (2005) used the same equation for the propagation of microwave energy in their finite element computations. Equation (34) has been modified for FDM and FEM simulations in this work considering the grid system given in Fig. 1, and obtained as Eq. (35);

$$\Phi = \frac{2\alpha R P''_o}{r} \left[e^{-2\alpha(R-r)} + e^{-2\alpha(R+r)} \right] + 2\alpha P''_o \left[e^{-2\alpha(L-z)} + e^{-2\alpha(z)} \right] \tag{35}$$

For the grid system shown in Fig. (1) excluding the radial symmetry axis, the finite difference form of Eq. (34) for volumetric heat generation may be obtained as

$$\Phi = \frac{2\alpha N P''_o}{(N+1-I)} \left[e^{-2\alpha(I-1)\Delta r} + e^{-2\alpha(2N-(I-1)\Delta r)} \right] + 2\alpha P''_o \left[e^{-2\alpha(J-1)\Delta z} + e^{-2\alpha(M-(J-1))\Delta z} \right] \tag{36}$$

where $R = N\Delta r$ and $L = M\Delta z$.

For the radial symmetry axis, the radial distance r is zero and power density is infinite. Therefore, microwave power density for the radial symmetry axis was evaluated at the volume-half radius between nodes NP1 and N.

For radial symmetry axis, the finite difference equation for volumetric heat generation was as follows;

$$\Phi = \frac{2\alpha N P''_o}{0.707} \left[2e^{-2\alpha R} \right] + 2\alpha P''_o \left[e^{-2\alpha(J-1)\Delta z} + e^{-2\alpha(L-(J-1)\Delta z)} \right] \tag{37}$$

A solution of the system of equations shown in Eqs. (36) and (37) as such is tedious. For a 20x20 grid system, 400 simultaneous equations have to be solved at one-time step. The implicit alternating direction method (Carnahan et al., 1969) overcomes this difficulty by solving the systems of equations implicitly in one direction for a time step of $\Delta t/2$ and using these computed temperatures as input for the next time step of $\Delta t/2$, which is solved implicitly this time in the alternating direction, thereby reducing the number of equations to be solved for a time step to 40.

Microwave energy absorbed was added as a source term to Eq. (33). Nominally, a 20x20-grid system was used for the solution of Eq. (33). Small adjustments to the number of nodes in the radial direction were made to cope with the experimental temperature measurement locations. The implicit alternating direction method was used. The time step was taken as 1 second. The system of equations obtained was solved by a FORTRAN 95 computer program. Before using the prepared computer program in microwave calculations, the predictions of the model were compared with the predictions of analytical solutions for finite cylinder (Cengel, 2003) undergoing unsteady state conduction heating. Excellent correlation was obtained between analytical and numerical results.

For FEM simulation, the COMSOL 5.2 (COMSOL Inc., USA) computer program was utilized. Governing heat transfer equations given above were solved by using radial symmetry approach in FEM analysis. Physical properties of the material were defined to the program as built-in material. Afterward, the geometry of material was generated according to experimental data. For the outer surface of the body, heat flux was defined according to data given in Table 1 and Table 3. Heat generation term (Eq. (35)) was defined for the entire generated geometry by utilizing a heat transfer module. Finally, the meshing process was performed, and 3646 elements with 0.95 element quality was obtained.

4.2 Evaluation of surface microwave power flux

Accurate estimation of the surface microwave power flux is essential to predict the temperature profiles during microwave heating. Lin et al. (1995) used a calorimetric method to determine the power absorbed. This technique consists of the calculation of the absorbed power by different water volumes submitted to heating, under the same operating conditions (position, power and container size) as used in the experiments.

Campañone and Zaritzky (2005) used a similar procedure. The surface power was taken to be equal to the power absorbed and the surface power absorbed was expressed as a function of the weight of the sample. Equations proposed by Lin et al. (1995) and Campañone and Zaritzky (2005) are shown in Eqs. (38) and (39), respectively, as examples to this approach for cylindrical samples.

$$P_0 = 453.2 + 59.8 \ln(weight) \qquad 25 \text{ g} \leq \text{weight} \leq 2500 \text{ g} \tag{38}$$

$$P_0 = \frac{1246.5 \, Weight}{1.45 \, Weight + 79.53} \qquad 50 \text{ g} \leq \text{weight} \leq 400 \text{ g} \tag{39}$$

However, the powers calculated by the calorimetric model are the average absorbed powers rather than the surface powers. Therefore, a simple equation relating absorbed power to surface power will be useful in modelling studies.

Padua (1993) proposed a method for evaluating the absorbed power to surface power for an infinite cylinder. The microwave power density was integrated over the volume of the object and surface power was evaluated from experimental absorbed microwave power data. This approach may be extended to finite cylinders as follows:

$$P_{abs} = \int_V \Phi dV \tag{40}$$

Equation (40) may be integrated independently for the radial and axial directions as

$$\Phi = \Phi_r + \Phi_z \tag{41}$$

where Φ_r and Φ_z are the power densities for the radial and axial directions.

$$\Phi_r = \frac{2\alpha R P_o''}{r} \left[e^{-2\alpha(R-r)} + e^{-2\alpha(R+r)} \right] \tag{42}$$

$$\Phi_z = 2\alpha P_o'' \left[e^{-2\alpha\left(\frac{L}{2}-z\right)} + e^{-2\alpha\left(\frac{L}{2}+z\right)} \right] \tag{43}$$

For the radial direction, the absorbed power may be calculated by taking into account heat microwave power flux at the surface;

$$P_{abs,\,r} = 4a\pi LRP''_o \int_0^R \frac{\left[e^{-2a(R-r)} + e^{-2a(R+r)}\right]}{r}\, rdr \tag{44}$$

After integration

$$P_{abs,\,r} = 2\pi LRP''_o (1 - e^{-4aR}) \tag{45}$$

Similarly, for the z direction,

$$P_{abs,\,z} = 2\pi R^2 P''_o (1 - e^{-2aL}) \tag{46}$$

Therefore, total absorbed power will be

$$P_{abs,\,total} = 2\pi LRP''_o (1 - e^{-4aR}) + 2\pi R^2 P''_o (1 - e^{-2aL}) \tag{47}$$

The total microwave power absorbed may be experimentally determined by the calorimetric method (Lin et al., 1995) and the surface power may be estimated from Eq. (47). If the geometry is an infinite cylinder, only the first term in Eq. (47) will exist. Similarly, for an infinite slab, only the second term will exist.

It may be easily shown that, if $aR \geq 1.15$ for an infinite cylinder, $P_{abs,r}/P_{0,r} \geq 0.99$. Similarly, for an infinite slab, if $aL \geq 2.3$, then $P_{abs,z}/P_{0,z} \geq 0.99$.

4.3 Experimental data

The experimental data used for the verification of the numerical models (FDM and FEM) were obtained from the literature. Only reproducible experimental data were used.

Case study 1: Predictions for 2% agar gel finite cylinders

Gunasekaran and Yang (2007) used continuous and pulsed microwave heating and reported temperature profiles after 1, 2 or 3 minutes. The pulsing ratio was defined as (Yang and Gunasekaran, 2001);

$$PR = \frac{t_{on} + t_{off}}{t_{on}} \tag{48}$$

Table 1 shows the experimental test conditions (Set ID G1-G6) for microwave heating of 2% agar gel finite cylinders (Gunesakaran and Yang, 2007). The input thermo-physical data for the numerical model are shown in Table 2. Surface powers calculated from Eq. (47) for the experimental conditions given in Table (1) were within 1% of the corresponding absorbed powers. Therefore, for these experimental conditions, surface powers were assumed to be equal to the absorbed powers given in Table (1).

Table 1: Experimental conditions (Set ID G1-G6) for microwave heating of 2% agar gel finite cylinders [Data from Gunasekaran and Yang (2007)].

Set ID	Pulse Ratio (PR)	R (m)	L (m)	T_i (°C)	T_a (°C)	h (W/m²K)	P_{abs} (W)
G1	1	0.04	0.07	4	25*	40	234
G2	2	0.04	0.07	4	25*	40	334
G3	1	0.035	0.07	4	25*	40	225
G4	2	0.035	0.07	4	25*	40	331
G5	1	0.04	0.07	4	25*	40	225
G6	1.47	0.04	0.07	4	25*	40	331

* assumed

Table 2: Thermo-physical input data to the numerical models (FDM and FEM) for finite cylinders of 2% agar gel.

k (W/mK)	ρ (kg/m³)	Cp (J/kgK)	α (m⁻¹)	Source
0.6	1070	4184	38	Yang and Gunesakaran (2001)

Case study 2: Predictions for 2% agar gel infinite cylinders

The experimental data of Padua (1993) were used to test the applicability of the present model to infinite cylinders. Padua (1993) measured the temperature profiles of 2% agar gel cylinders containing 0%, 40% and 60% sucrose in a 1500 W microwave oven operating at 2450 MHz. To ensure one-dimensional energy transfer, the bottom and the top of the cylinders were shielded with aluminium foil. An average temperature rise of known volumes of water and sucrose solutions were measured in the same experimental set-up under identical conditions with the gels.

The present numerical model for finite cylinders can also be used to compute the temperature distributions in microwave heating of infinite slab or infinite cylinder-shaped foods by proper modifications in the model. For an infinite cylinder, Eq. (36) may be modified as follows noting that the height of the infinite cylinder is much greater than the diameter:

$$\Phi = \frac{2\alpha NP_o''}{(N+1-I)} \left[e^{-2\alpha(I-1)\Delta r} + e^{-2\alpha(2N-(I-1)\Delta r)} \right] \tag{49}$$

Similarly, for the axis of the infinite cylinder, power density Φ becomes

$$\Phi = \frac{4\alpha NP_o''}{0.707} \left[e^{-2\alpha R} \right] \tag{50}$$

The area to be used in the calculation of surface microwave power flux may be calculated as $2\pi RH$, where H is the height of the cylinder through which microwave power enters the infinite cylinder.

Table 3: Experimental test conditions (Set ID P1-P4) for 2% infinite cylinder agar gels containing different percentages of sucrose [Data from Padua (1993)].

Set ID	Sucrose, %	R (m)	H (m)	T_i (°C)	T_a (°C)	h (W/m²K)	P_{abs} (W)
P1	0[a]	0.03[a]	0.086[a]	21[a]	25[c]	40[b]	298.4[a]
P2	40[a]	0.03[a]	0.086[a]	21[a]	25[c]	40[b]	471.4[a]
P3	60[a]	0.03[a]	0.086[a]	21[a]	25[c]	40[b]	477.3[a]
P4	0[a]	0.035[a]	0.086[a]	21[a]	25[c]	40[b]	369.3[a]

[a] Padua (1993)
[b] Yang and Gunesakaran (2001)
[c] assumed

Table 4: Thermo-physical input data to the numerical models (FDM and FEM) for infinite cylinders of 2% agar gel containing different percentages of sucrose.

Sucrose, %	k (W/mK)	ρ (kg/m³)	Cp (J/kgK)	α (m⁻¹)
0[a]	0.6[c]	1070[c]	4184[a]	38[a]
40[a]	0.51[b]	1176[b]	3091[b]	67[a]
60[a]	0.43[b]	1286[b]	2544[b]	78[a]
0[a]	0.6[c]	1070[c]	4184[a]	38[a]

[a] Padua (1993)
[b] Calculated from the composition
[c] Yang and Gunesakaran (2001)

Table 3 shows the experimental test conditions (Set ID P1-P4) for microwave heating of 2% agar gel infinite cylinders containing different percentages of sucrose (Padua, 1993). The input thermo-physical data for the infinite cylinder numerical model are shown in Table 4.

The surface power P_0 was taken to be equal to the power absorbed measured by the calorimetric method.

4.4 Comparison of numerical results with experimental data

The compatibility of numerical models (FDM and FEM) developed in the present study has been made with experimental temperature data from the literature for microwave heating of food analogs; case study 1 (Gunesakaran and Yang, 2007) and case study 2 (Padua, 1993).

Case study 1: Predictions for 2% agar gel finite cylinders

Figures 3–7 show the predictions of the present model for 2% agar gel finite cylinders for the experimental conditions in Table (1) and their comparison with the experimental data of Gunesakaran and Yang (2007). For FDM, all the temperatures presented were for the symmetry plane in the z direction (J = M/2 + 1), and a 20x20 mesh was used for this data set.

Error analysis for the comparisons of the experimental data (Gunesakaran and Yang, 2007) with the predictions of the present model was shown in Table 5. Error (%) was calculated using Eq. (51);

$$\text{Error, } \% = \frac{(T_{num} - T_{exp})}{T_{exp}} \times 100 \tag{51}$$

As may be observed from Table 5, the present model accurately predicted the surface temperatures for finite cylinders. However, the accuracy of the predictions of the present model became poorer for the interior nodes; the numerical predictions for the interior modes were on average 15% lower than the experimental temperatures. For the center temperatures, the mean percent error was +12.0%, indicating that the predicted temperature was higher than the experimental temperature. Δr, the distance between nodes in the radial direction used in the computations, was between 1.75 and 2 mm. The temperature gradient at the center is very steep and thermocouple positioning at the center may be a source of error in experimental temperatures.

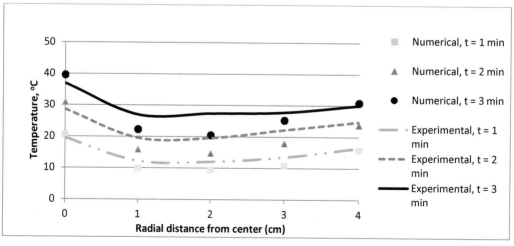

Figure 2: Comparison of temperature distributions in 2% agar gel cylinders (R = 0.04 m, L = 0.07 m) after 1, 2 and 3 minutes heating; Set ID G1; P_{abs} = 234 W, PR = 1 (Gunesakaran and Yang, 2007) with the predictions of the present numerical model (FDM) [Data from Gunasekaran and Yang (2007)].

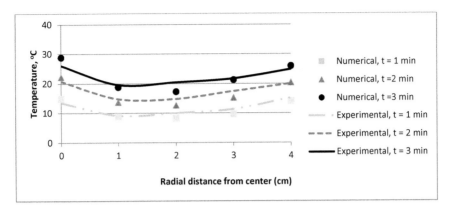

Figure 3: Comparison of temperature distributions in 2% agar gel cylinders (R = 0.04 m, L = 0.07 m) after 1, 2 and 3 minutes heating; Set ID G2; P_{abs} = 344W, PR = 2 (Gunesakaran and Yang, 2007) with the predictions of the present numerical model (FDM) [Data from Gunasekaran and Yang (2007)].

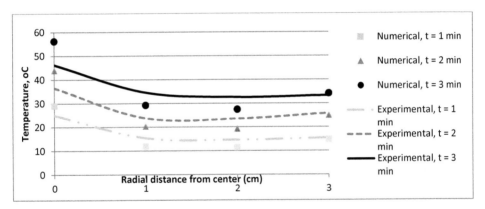

Figure 4: Comparison of temperature distributions in 2% agar gel cylinders (R = 0.035 m, L = 0.07 m) after 1, 2 and 3 minutes heating; Set ID G3; P_{abs} = 225 W, PR = 1 (Gunasekaran and Yang, 2007) with the predictions of the present numerical model (FDM) [Data from Gunasekaran and Yang (2007)].

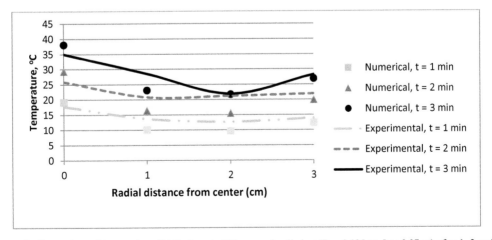

Figure 5: Comparison of temperature distributions in 2% agar gel cylinders (R = 0.035 m, L = 0.07 m) after 1, 2 and 3 minutes heating; Set ID G4; P_{abs} = 331 W, PR = 2 (Gunasekaran and Yang, 2007) with the predictions of the present numerical model (FDM) [Data from Gunasekaran and Yang (2007)].

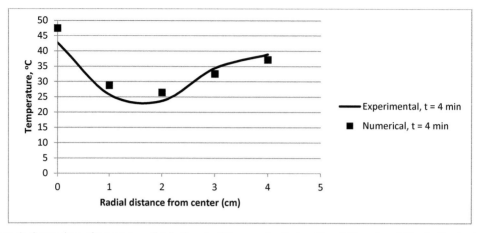

Figure 6: Comparison of temperature distributions in 2% agar gel cylinders (R = 0.04 m, L = 0.07 m) after 4 minutes heating; Set ID G5; P_{abs} = 225 W, PR = 1 (Gunesakaran and Yang, 2007) with the predictions of the present numerical model (FDM) [Data from Gunasekaran and Yang (2007)].

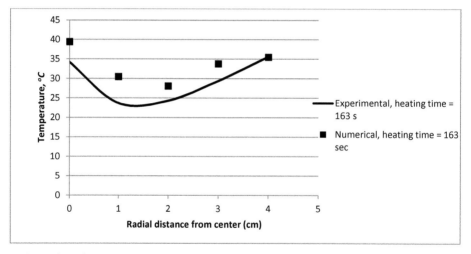

Figure 7: Comparison of temperature distributions in 2% agar gel cylinders (R = 0.04 m, L = 0.07 m) after 167 s heating and 73 s microwave power off; Set ID G6; P_{abs} = 331 W, PR = 1.47 (Gunesakaran and Yang, 2007) with the predictions of the present numerical model (FDM) [Data from Gunasekaran and Yang (2007)].

One other observation about the present finite difference model was the dependence of the predicted temperature profiles on the number of nodes used in the radial direction. The experimental temperature profile data of (Gunesakaran and Yang, 2007) used in model predictions were concave upward curves, maximum temperature being at the center, the second highest temperature being at the surface or near

Table 5: Error analysis for the comparison of the predictions of the present model (FDM) with the experimental data (Gunesakaran and Yang 2007) for different radial positions at the axial symmetry axis (J = M/2 + 1).

	Center	R = 2 cm	Surface
Mean error, %	11.4	−15.5	−3.0
Standard deviation, %	5.3	12.8	4.7
Minimum error, %	4.6	−26.4	−10.8
Maximum error, %	21.9	15.6	4.4

surface locations. Decreasing the number of nodes in the r direction would result in higher accuracy for the center temperatures. However, this will be at the expense of loss of accuracy for the surface or near-surface temperature predictions. What may be proposed to overcome this difficulty is to choose a mesh size that will minimise the error between the finite difference predictions and experimental data for one data set and then use the same number of nodes for other datasets from the same source. A similar approach has been used by Lin et al. (1995) in determining the element size for their finite element model.

Inaccurate predictions in the interior nodes may be attributed to inaccurate power density predictions for these nodes. Similarly, Pandit and Prasad (2003) conducted FEM analysis of microwave heating of rectangular and cylindrical shaped potato. They reported that predicted temperature values were lower than the experimental temperatures for both cylindrical and rectangular samples. They concluded that the reason for poor correlation with experimental data might be originated from error in calculated absorbed power values. Inaccurate electrical field density and dielectric loss factors would have led to inaccurate temperature predictions. Specification of the attenuation constant α is critical in temperature calculations. As discussed in the "Evaluation of surface power" section, if $\alpha R \geq 1.15$ for an infinite cylinder, $P_{abs,r}/P_{0,r} \geq$ 0.99. Similarly, for an infinite slab, if $\alpha L \geq 2.3$, then $P_{abs,z}/P_{0,z} \geq 0.99$. Larger α values mean shorter radial or axial distances, in which 99% of the surface power has been absorbed by the solid. In temperature predictions, larger input α values will increase the temperatures in the surface or near-surface locations, whereas smaller α values will increase the temperatures in central positions. Utilization of temperature dependent attenuation constants will increase the accuracy of the present model.

Temperature distributions and heat generation distributions of FEM predictions are given in Figures 8–10. The microwave heating pattern could be observed in the 3D display with high precision.

Also, distribution of heat generation in a 3D scale may assist in evaluating the temperature differences within the sample. For all heating conditions, temperature values at the center were higher than the surface ones, as expected. In addition, temperature increases at the edges of the samples became higher due to multidimensional (both from radial and lateral directions) heat generation. This phenomenon is explained by the well-known heating pattern of microwave heating due to wave penetration and wave structure in cylindrical samples (Hossan et al., 2010; Bhattacharya and Basak, 2017; Pitchai et al., 2015). The penetration of the wave is strongly related to the sample size and shape, dielectric properties and the microwave operation frequency. Depending on the sample size and shape, uniform temperature distribution could be observed with relatively small samples, while non-uniform temperature distribution is found in bigger samples (Bhattacharya and Basak, 2017; Pitchai et al., 2015).

As shown in Fig. 8, the final temperature of 2% agar gel increased as the diameter of the sample decreased, especially for three minutes continuous microwave heating. This situation could be explained by the increase in heat generation within the sample, especially at the center of the agar gel. In addition, this shows that temperature uniformity increased as the radius of the sample decreased.

In the case of pulsed microwave heating conditions (Fig. 9), similar trends to Fig. 8 were obtained. The final temperature of the sample reached higher degrees, and better temperature uniformity was obtained in the case of smaller radius. Pulse treatment increased the temperature uniformity throughout the sample more than continuous heating.

To evaluate the FDM and FEM predictions, chi-square and RMSE values were determined and given in Table 6.

In general, model predictions were improved or similar prediction results were obtained with FEM when compared to FDM predictions. Chi-Square error values for FDM predictions were much higher than those for FEM predictions, for some heating conditions, e.g., Set IDs G3 and G6. RMSE values for FEM and FDM predictions were in the range of 2.2–4.7°C and 1.4–6.3°C, respectively. These values showed that model predictions were in good agreement with experimental temperatures. Campañone et al. (2012) developed an FEM for the prediction of microwave heating patterns in various food samples. They investigated the effects of one side/both side microwave heating and pulse microwave process on temperature distribution, and their influences on FEM predictions. They reported that FEM simulated experimental temperature distributions satisfactorily, since simulation results were in good agreement with experimental results. In another study, Pandit and Prasad (2003) concluded that the reason for poor

(a)

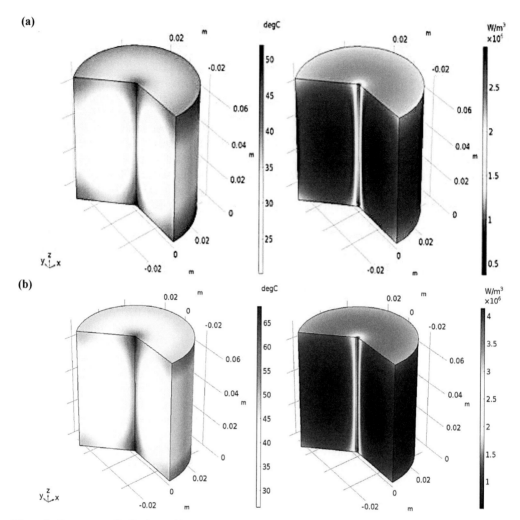

Figure 8: Temperature distribution and heat generation distribution of 2% agar gel with (a) diameter of 0.04 m (Set ID G1; Pabs = 234 W, PR = 1) (b) diameter of 0.035 m (Set ID G3; Pabs = 225 W, PR = 1) after 3 minutes heating (FEM predictions) [Data from Gunasekaran and Yang (2007)].

Color version at the end of the book

agreements of FEM predictions with experimental data might be due to errors in calculated absorbed power values.

Case study 2: Predictions for 2% agar gel infinite cylinders

Figures 11–14 show the predictions of the present model for 2% agar gel infinite cylinders containing different percentages of sucrose for the experimental conditions in Table 4, and their comparison with the experimental data of Padua (1993). For FDM, all the temperatures presented were for the symmetry plane in the z-direction (J = M/2 + 1). The numbers of nodes in the radial direction were adjusted slightly, considering the experimental points of measurement.

Error analysis for the comparisons of the experimental data (Padua, 1993) and predictions of the present model are shown in Table 7.

As may be observed from Table 6, the level of agreement between the numerical predictions (FDM) and experimental data was quite satisfactory for infinite cylinders for the surface, interior, and center positions.

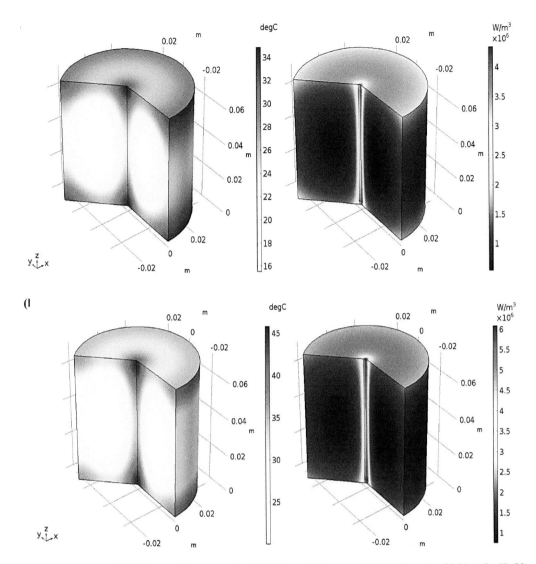

Figure 9: Temperature distribution and heat generation distribution of 2% agar gel with (a) diameter of 0.04 m (Set ID G2; $P_{abs} = 344$ W, PR = 2) (b) diameter of 0.035 m (Set ID G4; $P_{abs} = 331$ W, PR = 2) after 180 s pulsed microwave heating process (FEM predictions) [Data from Gunasekaran and Yang (2007)].

Color version at the end of the book

In other words, the present numerical method predicted the temperature profiles in infinite cylinders with higher accuracy than finite cylinders. Higher accuracy may be attributed to the fact that, for the infinite cylinder case, the top and the bottom of the cylindrical samples were shielded with aluminium foil in order to ensure one-dimensional microwave energy transfer. The electrical field surrounding the finite cylinder was less uniform than the infinite cylinder case, resulting in poorer predictions. FEM predictions for the infinite cylinder case shown in Figs. 15 and 16 showed that the power density at the surface was almost uniform. Uniform surface power generation resulted in accurate predictions by the FDM.

FEM predictions of temperature distributions and heat generation distributions for the 2% agar gel having different sucrose contents are given in Figs. 15 and 16. For the given experimental heating conditions, 3D heat generation distributions were obtained, as well as the temperature distributions. Heat generations were higher at the center of samples, as expected. A temperature gradient within the samples was predicted with good agreement. Due to the microwave heating pattern, center temperatures were

(a)

(b)

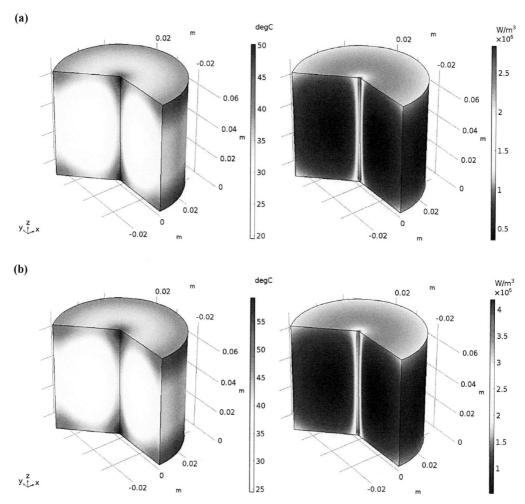

Figure 10: Temperature distribution and heat generation distribution of 2% agar gel with diameter of 0.04 m 4 minutes heating (a) continuous heating (Set ID G5; Pabs = 225 W, PR = 1) (b) pulsed heating (Set ID G6; P_{abs} = 331 W, PR = 1.47) (FEM predictions) [Data from Gunasekaran and Yang (2007)].

higher than interior temperatures. However, in Figs. 12 and 13, it may be observed that the temperatures of the center regions of 2% agar gel having 40% and 60% sucrose concentrations were lower than surface temperatures. This situation could be explained by increasing of attenuation coefficient due to the addition of sucrose. Sakai et al. (2004) modified the penetration characteristics of the microwave in agar gel with the addition of salt. They reported that the addition of NaCl shifted the hot point of the sample from the center to the surface for cylindrical shapes. Similarly, Bhattacharya and Basak (2017) reported the same dependency between attenuation constant and heat generation during microwave heating of turnip, pizza dough, ham and almond. They concluded that heat generation values were higher for those with lower penetration depth values (refers high attenuation constant). Similar results were reported by Navarro and Burgos (2017). They developed FEM simulation to predict temperature distributions given in Padua (1993) and Yang (2002) and discussed that the temperature profile focalized in the center similar to experimental results. They also stated that FEM predicted temperature values lower than experimental data, especially for inner parts of the sample.

Table 6: Statistical comparison of FDM and FEM predictions for heating of 2% agar gel after different heating times.

Set ID	Time (min)	FDM		FEM	
		Chi Square ($°C)^2$	RMSE (°C)	Chi Square ($°C)^2$	RMSE (°C)
G1	1	5.4	2.1	6.3	2.2
	2	14.9	3.5	14.7	3.5
	3	20.9	4.1	19.0	4.1
G2	1	2.3	1.4	6.4	2.3
	2	3.1	1.6	7.4	2.4
	3	4.0	2.0	10.1	2.8
G3	1	12.7	3.1	7.4	2.4
	2	28.7	4.6	13.7	3.2
	3	52.3	6.3	28.5	4.7
G4	1	8.8	2.6	9.4	2.7
	2	22.1	4.1	23.7	4.2
	3	13.5	3.2	15.4	3.4
G5	4	11.5	3.0	6.0	2.2
G6	4	27.0	5.0	7.7	2.5

Table 7: Error analysis for the comparison of the predictions of the present model (FDM) with experimental data (Padua, 1993) for different radial positions at the axial symmetry axis (J = M/2 + 1).

	Center	R = 2 cm	Surface
Mean error, %	3.7	−1.3	5.9
Standard deviation, %	6.4	6.0	6.9
Minimum error, %	−1.9	−7.3	−2.2
Maximum error, %	12.6	6.7	11.7

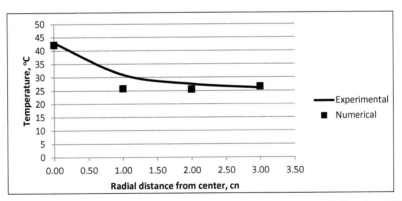

Figure 11: Comparison of temperature distributions in 2% agar gel infinite cylinders (R = 0.03 m, L = 0.086 m) after 15 s of heating; Set ID P1; P_{abs} = 298.4 W (Padua, 1993) with the predictions of the present numerical model [FDM; M = 20, N = 18] [Data from Padua (1993)].

For 2% agar gel, as the diameter decreased, the average temperature increase in the sample increased, since average heat generation in the thinner sample increased due to high penetration capability (Fig. 15). In addition, sucrose concentration affected the heat generation rate in samples (Fig. 16). Llave et al. (2016) investigated the effect of salt concentration (0%, 0.5%, 1% and 2% NaCl) on microwave

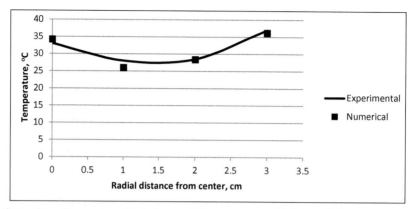

Figure 12: Comparison of temperature distributions in 40% sucrose agar gel infinite cylinders (R = 0.035 m, L = 0.086) after 15 s of heating; Set ID P2; P_{abs} = 471.4 W (Padua, 1993) with the predictions of the present numerical model [FDM; M = 20, N = 18] [Data from Padua (1993)].

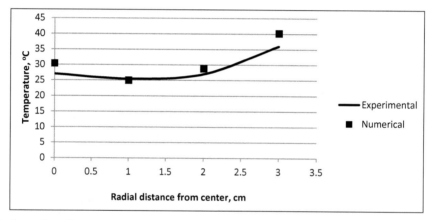

Figure 13: Comparison of temperature distributions in 60% sucrose agar gel infinite cylinders (R = 0.03 m, L = 0.086) after 15 s of heating; Set ID P3; P_{abs} = 477.3 W (Padua, 1993) with the predictions of the present numerical model [FDM; M = 20, N = 18] [Data from Padua (1993)].

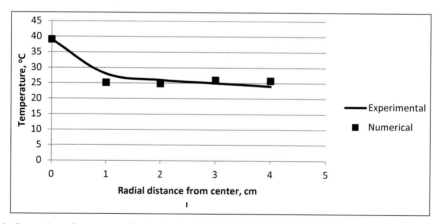

Figure 14: Comparison of temperature distributions in 2% agar gel infinite cylinders (R = 0.035 m, L = 0.086 m) after 15 s of heating; Set ID P4; P_{abs} = 369.3 W (Padua, 1993) with the predictions of the present numerical model [FDM; M = 20, N = 21] [Data from Padua (1993)].

(a)

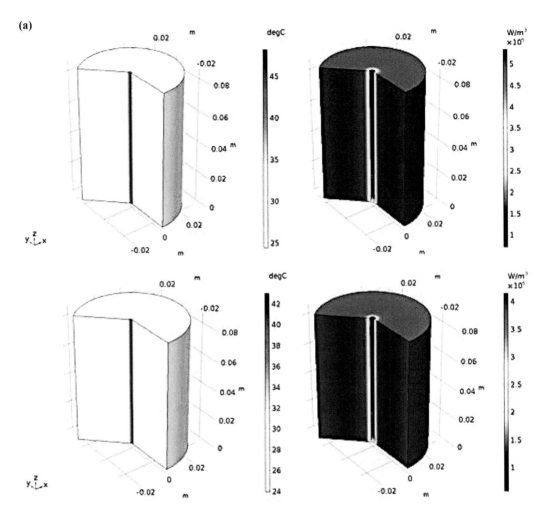

Figure 15: Temperature distribution and heat generation distribution of 2% agar gel with (a) diameter of 0.03 m (Set ID P1; P_{abs} = 298.4 W) (b) diameter of 0.035 m (Set ID P4; P_{abs} = 369.3 W) after 15 s of heating (FEM predictions) [Data from Padua (1993)].

Table 8: Statistical comparison of FDM and FEM predictions for heating of 2% agar gel with different sucrose contents after 15 s microwave heating.

Set ID	FDM		FEM	
	Chi Square $(°C)^2$	RMSE $(°C)$	Chi Square $(°C)^2$	RMSE $(°C)$
P1	13	3	21	4
P2	5	2	11	3
P3	22	4	6	2
P4	9	3	15	3

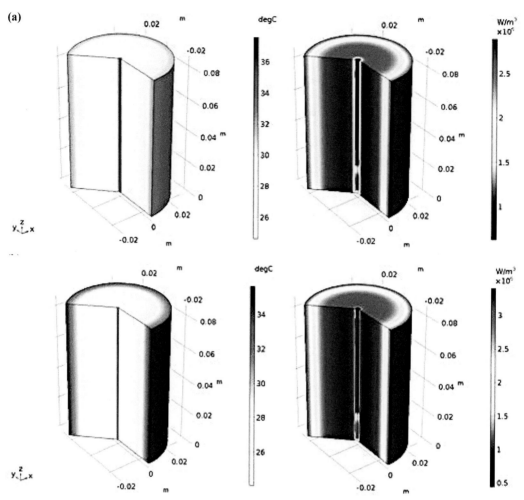

Figure 16: Temperature distribution and heat generation distribution of 2% agar gel with diameter of 0.03 m that contains (a) 40% sucrose (Set ID P3; P_{abs} = 477.3 W) (b) 60% sucrose (Set ID P4; P_{abs} = 369.3 W) after 15 s of heating (FEM predictions) [Data from Padua (1993)].

Color version at the end of the book

heating patterns of cylindrical tylose samples. They discussed that the addition of salt led to higher heat generation, thus, higher final temperature values. Similar results reported by Sakai et al. (2004).

Compatibility of FDM and FEM predictions to experimental data are given in Table 8. In general, FDM resulted in higher accuracy than FEM in infinite cylinder predictions. By comparing the results of FEM analyses for different cases, it can be concluded that the simulation of Case 2 resulted in more errors in temperature predictions than Case 1. This situation might be due to the differences in heat transfer solutions for finite and infinite geometries.

More extensive comparison with literature experimental data could not be performed since the experimental conditions under which the experimental data have been obtained and the parameters used in the model calculations were not fully reported.

Finite difference and finite element numerical models were developed in order to compute the temperature profiles in a finite cylinder during microwave heating. The resulting model was modified in order to account for microwave heating in an infinite cylinder. The numerical profiles were compared with experimental temperature profile data obtained from the literature. It was observed that the level of agreement between numerical and experimental temperature profiles was quite high for the surface, interior and center positions of the infinite cylinders. For the case of a finite cylinder, although the level of agreement between the experimental and predicted surface temperatures was high, the accuracy was poorer for the interior and center positions, indicating that the models must be refined in order to yield more accurate predictions in the interior and center nodes for finite cylinders.

5. Summary and Future Prospective

In recent years, wide-spreading software programs revealed their potential to provide solutions for practical applications in terms of system development by making available the simultaneous solution of the electromagnetic wave, heat and momentum modules. However, there is still a need for more research concerning the characterization of power distribution within the food. The characterization of the change of dielectric properties with system parameters would make the simulation of microwave heating more accurate and this will lead to more effective utilization of the results of these simulations. In contrast to other field of studies, foods have more complex structures, therefore, new approaches on consideration of structural, textural and chemical changes is an important future task. It is believed that further studies must be focused on the characterization of these changes during microwave heating.

Nomenclature

ATOT	the total surface area of the cylinder
$\vec{B}$	magnetic induction
Cp	specific heat capacity (J/kgK)
D	penetration depth (m)
$\vec{D}$	electric displacement
D_p	power penetration depth (m)
DR	distance between two nodes in the r direction, R/N
DZ	distance between two nodes in the z-direction, L/M
E	electrical field intensity (V/m)
$\vec{E}$	electrical field
E*	complex conjugate of electrical field intensity (V/m)
E_{rms}	root mean square of the electric field intensity (V/m)
$\vec{H}$	magnetic field
h	surface heat transfer coefficient (W/m²K)
I	the ith node
J	the Jth node
$\vec{J}$	current density
k	thermal conductivity (W/mK)
L	height of cylinder (m)
M	number of intervals in z direction
MP1	M+1, the top surface
N	number of intervals in r direction
NP1	N+1, the radial axis of symmetry
r	distance in a radial direction (m)
R	the radius of the cylinder (m)
P	power (W)

P_0	normal surface power (W)
$\vec{P}_0$	normal surface power flux (W/m^2)
PR	pulse ratio
q_R	radiative power flux density (W/m^2)
$\vec{q}$	Poynting vector
T	temperature (°C)
t	time (s)
V	volume (m^3)
z	distance in upward direction (m)

Subscripts

a	ambient
abs	absorbed
ij	any node
ini	initial
r, z	the r and z directions
eff	effective
num	numerical prediction
exp	experimental data

Greek symbols

α	attenuation factor (1/m)
Φ	volumetric heat generation (W/m^3)
F	frequency, Hz
ε^*	complex permittivity
ε'	dielectric constant
ε''	dielectric loss factor
ε_0	dielectric constant of free space (8.854 Farad/m)
μ'	relative permeability
μ_o	relative permeability of free space
λ_0	wave length (m)
σ	electric conductivity (S/m)
Δt	the time step (s)
ρ	density (kg/m^3)
ρ_v	electric charge density

References

Antonio, C. and Deam, R.T. 2005. Comparison of linear and non-linear sweep rate regimes in variable frequency microwave technique for uniform heating in materials processing. J. Mater Process. Technol. 169: 234–241.

Baıl, A. Le, Koutchma, T. and Ramaswamy, H. 2000. Modeling of temperature profiles under continuous tube-flow microwave and steam heating conditions. J. Food Process. Eng. 23: 1–24.

Barrınger, S.A., Davıs, E.A., Gordon, J., Ayappa, K.G. and Davıs, H.T. 1995. Microwave-heating temperature profiles for thin slabs compared to maxwell and lambert law predictions. J. Food Sci. 60: 1137–1142.

Basak, T. and Ayappa, K.G. 2002. Role of length scales on microwave thawing dynamics in 2D cylinders. Int. J. Heat Mass. Transf. 45: 4543–4559.

Baysal, T., İçier, F. and Baysal, A.H. 2011. Güncel Elektriksel Isıtma Yöntemleri. Sidas Yayınları, İzmir.

Bhattacharya, M. and Basak, T. 2017. A comprehensive analysis on the effect of shape on the microwave heating dynamics of food materials. Innov. Food Sci. Emerg. Technol. 39: 247–266.

Bows, J.R., Patrick, M.L., Janes, R. and Dibben, D.C. 1999. Microwave phase control heating. Int. J. Food Sci. Technol. 34: 295–304.

Buffler, C.R. 1993. Microwave Cooking and Processing: Engineering Fundamentals for the Food Scientist. AVI Book, New York.

Campañone, L.A. and Zaritzky, N.E. 2005. Mathematical analysis of microwave heating process. J. Food Eng. 69: 359–368.

Campañone, L.A., Paola, C.A. and Mascheroni, R.H. 2012. Modeling and simulation of microwave heating of foods under different process schedules. Food Bioprocess Technol. 5: 738–749.

Campañone, L.A., Bava, J.A. and Mascheroni, R.H. 2014. Modeling and process simulation of controlled microwave heating of foods by using of the resonance phenomenon. Appl. Therm. Eng. 73: 914–923.

Campañone, L.A. and Zaritzky, N.E. 2010. Mathematical modeling and simulation of microwave thawing of large solid foods under different operating conditions. Food Bioprocess Technol. 3: 813–825.

Carnahan, B., Luther, H.A. and Wilkes, J. 1969. Applied Numerical Methods. John Wiley & Sons. Inc., New York.

Cengel, Y. 2003. Heat Transfer a Practical Approach, 2nd ed. Mc Graw-Hill, New York.

Cha-um, W., Rattanadecho, P. and Pakdee, W. 2011. Experimental and numerical analysis of microwave heating of water and oil using a rectangular wave guide: Influence of sample sizes, positions, and microwave power. Food Bioprocess Technol. 4: 544–558.

Chatterjee, S., Basak, T. and Das, S.K. 2007. Microwave driven convection in a rotating cylindrical cavity: A numerical study. J. Food Eng. 79: 1269–1279.

Chiavaro, E., Barnaba, C., Vittadini, E., Rodriguez-Estrada, M.T., Cerretani, L. and Bendini, A. 2009. Microwave heating of different commercial categories of olive oil: Part II. Effect on thermal properties. Food Chem. 115: 1393–1400.

Curet, S., Rouaud, O. and Boillereaux, L. 2008. Microwave tempering and heating in a single-mode cavity: Numerical and experimental investigations. Chem. Eng. Process Process Intensif. 47: 1656–1665.

Datta, A.K. and Anantheswaran, R.C. 2001. Handbook of Microwave Technology for Food Applications. Marcel Dekker Inc., New York.

Datta, A.K. and Anantheswaran, R.C. 2005. Handbook of microwave technology for food applications (Datta, A.K. and Anantheswaran, R.C. (eds.). Marcel Dekker Inc, New York.

Decaureau, R.V. 1985. Microwaves in the Food Processing Industry. Academic Press Inc., Orlando.

Durka, T., Van Gerven, T. and Stankiewicz, A. 2009. Microwaves in heterogeneous gas-phase catalysis: Experimental and numerical approaches. Chem. Eng. Technol. 32: 1301–1312.

Farag, S., Sobhy, A., Akyel, C., Doucet, J. and Chaouki, J. 2012. Temperature profile prediction within selected materials heated by microwaves at 2.45 GHz. Appl. Therm. Eng. 36: 360–369.

Geedipalli, S.S.R., Rakesh, V. and Datta, A.K. 2007. Modeling the heating uniformity contributed by a rotating turntable in microwave ovens. J. Food Eng. 82: 359–368.

Gunasekaran, S. and Yang, H.-W. 2007. Effect of experimental parameters on temperature distribution during continuous and pulsed microwave heating. J. Food Eng. 78: 1452–1456.

Hossan, M.R., Byun, D. and Dutta, P. 2010. Analysis of microwave heating for cylindrical shaped objects. Int. J. Heat Mass. Transf. 53: 5129–5138.

İçier, F. and Baysal, T. 2004. Dielectric properties of food materials—1: Factors affecting and industrial uses. Crit. Rev. Food Sci. Nutr. 44: 465–471.

Icier, F. and Ilicali, C. 2012. A two dimensional finite difference model for predicting the temperature profiles during microwave heating of foods having finite cylinder geometry. Manas Univ. Fen Bilim. Derg.

James, C., Purnell, G. and James, S.J. 2015. A review of novel and innovative food freezing technologies. Food Bioprocess Technol. 8: 1616–1634.

Knoerzer, K., Regier, M. and Schubert, H. 2008. A computational model for calculating temperature distributions in microwave food applications. Innov. Food Sci. Emerg. Technol. 9: 374–384.

Knoerzer, K., Regier, M., Hardy, E.H., Schuchmann, H.P. and Schubert, H. 2009. Simultaneous microwave heating and three-dimensional MRI temperature mapping. Innov. Food Sci. Emerg. Technol. 10: 537–544.

Li, Z., Raghavan, G.S.V. and Orsat, V. 2010. Temperature and power control in microwave drying. J. Food Eng. 97: 478–483.

Lin, Y.E., Anantheswaran, R.C. and Puri, V.M. 1995. Finite element analysis of microwave heating of solid foods. J. Food Eng. 25: 85–112.

Liu, C.M., Wang, Q.Z. and Sakai, N. 2005. Power and temperature distribution during microwave thawing, simulated by using Maxwell's equations and Lambert's law. Int. J. Food Sci. Technol. 40: 9–21.

Liu, D., Sun, D.-W. and Zeng, X.-A. 2013b. Recent advances in wavelength selection techniques for hyperspectral image processing in the food industry. Food Bioprocess Technol. 7: 307–323.

Liu, S., Fukuoka, M. and Sakai, N. 2013a. A finite element model for simulating temperature distributions in rotating food during microwave heating. J. Food Eng. 115: 49–62.

Liu, S., Yu, X., Fukuoka, M. and Sakai, N. 2014. Modeling of fish boiling under microwave irradiation. J. Food Eng. 140: 9–18.

Llave, Y., Mori, K., Kambayashi, D., Fukuoka, M. and Sakai, N. 2016. Dielectric properties and model food application of tylose water pastes during microwave thawing and heating. J. Food Eng. 178: 20–30.

Mao, W., Watanabe, M. and Sakai, N. 2005. Analysis of temperature distributions in Kamaboko during microwave heating. J. Food Eng. 71: 187–192.

Metaxas, A.C. and Meredith, R.J. 1983. Industrial Microwave Heating. Polu Pelegrinus Ltd., London.

Metaxas, A.C. 1996. Foundations of electroheat. A unified approach. Fuel Energy Abstr. 37: 193.

Mutyala, S., Fairbridge, C., Paré, J.R.J., Bélanger, J.M.R., Ng, S. and Hawkins, R. 2010. Microwave applications to oil sands and petroleum: A review. Fuel Process Technol. 91: 127–135.

Navarro, M.C. and Burgos, J. 2017. A spectral method for numerical modeling of radial microwave heating in cylindrical samples with temperature dependent dielectric properties. Appl. Math Model. 43: 268–278.

Oliveira, M.E.C. and Franca, A.S. 2002. Microwave heating of foodstuffs. J. Food Eng. 53: 347–359.

Padua, G.W. 1993. Microwave heating of agar gels containing sucrose. J. Food Sci. 58: 1426–1428.

Pitchai, K., Birla, S.L., Subbiah, J., Jones, D. and Thippareddi, H. 2012. Coupled electromagnetic and heat transfer model for microwave heating in domestic ovens. J. Food Eng. 112: 100–111.

Romano, V.R., Marra, F. and Tammaro, U. 2005. Modelling of microwave heating of foodstuff: Study on the influence of sample dimensions with a FEM approach. J. Food Eng. 71: 233–241.

Roussy, G. and Pearce, J.A. 1995. Foundations and Industrial Applications of Microwave and Radio Frequency Fields: Physical and Chemical Processes. John Wiley & Sons Inc.

Sakai, N., Wang, C., Toba, S. and Watanabe, M. 2004. An analysis of temperature distributions in microwave heating of foods with non-uniform dielectric properties. J. Chemical Engineering Japan 37: 858–862.

Sakin, M., Kaymak-Ertekin, F. and Ilicali, C. 2007. Simultaneous heat and mass transfer simulation applied to convective oven cup cake baking. J. Food Eng. 83: 463–474.

Schubert, H. and Regier, M. 2005. The Microwave Processing of Foods. Woodhead Publishing Limited, Cambridge, England.

Taher, B.J. and Farid, M.M. 2001. Cyclic microwave thawing of frozen meat: Experimental and theoretical investigation. Chem. Eng. Process 40: 379–389.

Tang, S.-Y., Xia, Z.-N., Fu, Y.-J. and Gou, Q. 2008. Advances and applications of microwave spectroscopy. Chinese J Anal Chem 36: 1145–1151.

Turabi, E., Sumnu, G. and Sahin, S. 2008. Optimization of baking of rice cakes in infrared–microwave combination oven by response surface methodology. Food Bioprocess Technol. 1: 64–73.

Walker, C.E. and Li, A. 1993. Impingement oven technology. Part III: Combining impingement with microwave. Am. Inst. Bak. Technol. Bull. 15: 1–6.

Wäppling-Raaholt, B., Risman, P.O. and Ohlsson, T. 2006. Microwave heating of ready meals – FDTD simulation tools for improving the heating uniformity. pp. 243–255. *In*: Willert-Porada, E. (ed.). Advances in Microwave and Radio Frequency Processing. Springer Verlag.

Watanabe, M., Suzuki, M. and Ohkawa, S. 1978. Analysis of power density distribution in microwave ovens. J. Microw. Power 13: 173–181.

Wray, D. and Ramaswamy, H.S. 2015. Novel concepts in microwave drying of foods. Dry Technol. 33: 769–783.

Yakovlev, V.V. 2006. Examination of contemporary electromagnetic software capable of modeling problems of microwave heating. pp. 178–190. *In*: Willert-Porada, E. (ed.). Advances in Microwave and Radio Frequency Processing. Springer Verlag.

Yang, H. 2002. Analysis of Temperature Redistribution in Model Food during Pulsed Microwave Heating. University of Wisconsin-Madison.

Yang, H.W. and Gunasekaran, S. 2001. Temperature profiles in a cylindrical model food during pulsed microwave heating. J. Food Sci. 66: 998–1004.

Yang, H.W. and Gunasekaran, S. 2004. Comparison of temperature distribution in model food cylinders based on Maxwell's equations and Lambert's law during pulsed microwave heating. J. Food Eng. 64: 445–453.

Zhang, H. and Datta, A.K. 2000. Coupled electromagnetic and termal modeling of microwave oven heating of foods. J. Microw Power Electromagn Energy 35: 71–85.

Zhong, X., Dolan, K.D. and Almenar, E. 2015. Effect of steamable bag microwaving versus traditional cooking methods on nutritional preservation and physical properties of frozen vegetables: A case study on broccoli (Brassica oleracea). Innov. Food Sci. Emerg. Technol. 31: 116–122.

Zhou, L., Puri, V.M., Anantheswaran, R.C. and Yeh, G. 1995. Finite element modeling of heat and mass transfer in food materials during microwave heating—Model development and validation. J. Food Eng. 25: 509–529.

Zhu, J., Kuznetsov, A.V. and Sandeep, K.P. 2007. Mathematical modeling of continuous flow microwave heating of liquids (effects of dielectric properties and design parameters). Int. J. Therm. Sci. 46: 328–341.

CHAPTER **24**

Microwave Drying of Food Materials Modelled by the Reaction Engineering Approach (REA)– Framework

Aditya Putranto[1] and *Xiao Dong Chen*[2]*

1. Introduction

Microwave drying is commonly implemented for drying of food materials (Maskan et al., 2001; Cui et al., 2005; Alibas et al., 2007; Kowalski and Pakowski, 2010). For hot air drying, heat is supplied by drying air whose temperature is higher than the sample temperature. However, for microwave drying, heat is generated internally through the movement of polar molecules induced by electromagnetism. This makes the product temperature higher than the ambient temperature. In line with this, microwave drying results in more uniform moisture content and temperature than convective drying. Therefore, microwave play an important role in preserving bioactive content, colour and texture during drying. In addition, due to faster heating in comparison to hot air drying, microwave drying yields higher energy efficiency.

For development and optimisation of microwave drying schemes, mathematical modelling is important. An accurate model can help predict the drying kinetics as a basis for determining the key variables in microwave drying. Among the published models, the diffusion-based models give sound physical understanding. The diffusion-models usually result in a good agreement towards experimental data (Ahmad et al., 2001; Dincov et al., 2004; Mara et al., 2010). Nevertheless, this model requires a number of experiments in order to generate the drying parameters.

The reaction engineering approach (REA) is a heat mass transfer approach which sees drying as a zero order reaction with activation energy and first order reaction without activation energy (Chen and Putranto, 2013). The REA has been developed and is currently presented in its lumped and spatial format, which is labelled as lumped reaction engineering approach (L-REA) and spatial reaction engineering approach (S-REA), respectively. In L-REA, the REA is employed to model the global drying rate and project the average moisture content. In S-REA, the REA is used to model the local evaporation rate

[1] School of Chemistry and Chemical Engineering, Queen's University Belfast, Belfast, United Kingdom.
[2] School of Chemical and Environmental Engineering, College of Chemistry, Chemical Engineering and Materials Science, Soochow University, Suzhou, P.R. China.
* Corresponding author: xdchen@mail.suda.edu.cn

inside porous materials being dried. Both L-REA and S-REA have been proven to accurately model challenging drying cases, including the ones under time-varying conditions (Putranto et al., 2011a,b; Putranto and Chen, 2013a–c). It has also been applied to heat and mass transfer processes other than drying, such as baking, roasting and self-heating (Putranto and Chen, 2012, 2015a,b, 2017).

Due to the applicability of the REA, the REA may be a good alternative for describing microwave drying. However, the REA may have challenges as its equilibrium activation energy is defined according to drying air temperature. This seems to be not applicable for microwave drying, since for microwave drying, under equilibrium conditions, the product temperature will exceed the drying air temperature.

In this chapter, the development of the REA framework to cope with this aspect is discussed. The L-REA and S-REA are then used to model the microwave drying of food materials. The accuracy of the REA framework is assessed by validating towards experimental data. In this chapter, first, the principles of modelling using L-REA and S-REA are reviewed, followed by the application for microwave drying of food materials.

2. Mathematical Modelling of Microwave Drying Using the Lumped Reaction Engineering Approach (L-REA)

The basic principle for modelling using the lumped reaction engineering approach (L-REA) has been summarized in Chen and Putranto (2013). For modelling using the REA, initially, the relative activation energy needs to be generated. The procedures for generating the parameters are outlined below:

a) Based on the experimental data of moisture content, surface area and temperature of materials being dried, generate the activation energy (ΔE_v).
b) Evaluate the equilibrium activation energy ($\Delta E_{v,b}$). For microwave drying, the equilibrium activation energy can be expressed as

$$\Delta E_{v,b} = -RT_f \ln(RH) \tag{1}$$

where T_f is the final product temperature (K) and the relative humidity at the corresponding humidity and final product temperature.
c) Divide the activation energy by the equilibrium activation energy to yield the relative activation energy ($\Delta E_v/\Delta E_{v,b}$).
d) Express the relative activation energy as a function of moisture content difference ($X–X_b$) by using any algebraic equations.

For yielding the temporal profiles of moisture content and temperature during microwave drying, the mass balance is solved simultaneously with the heat balance. By using the REA, the mass balance is expressed as:

$$m_s \frac{d\overline{X}}{dt} = -h_m A \left[\exp(\frac{-\Delta E_v}{RT_s}) \rho_{v,sat}(T_s) - \rho_{v,b} \right] \tag{2}$$

where m_s is the dried mass of materials (kg), $\overline{X}$ is the average moisture content (kg water kg dry solids^{-1}), t is the time (s), h_m is the mass transfer coefficient (m s^{-1}), A is the surface area of materials being dried (m^2), T_s is the surface temperature (K), ΔE_v is the activation energy (J mol^{-1} K^{-1}), $\rho_{v,sat}$ is the saturated water vapor concentration (kg m^{-3}) and $\rho_{v,b}$ is the water vapor concentration in drying (kg m^{-3}). The mass balance implements the relative activation energy generated in the previous step.

The heat balance during microwave drying can be expressed as:

$$\frac{d\left[m_s(1+\overline{X})C_p T \right]}{dt} \approx hA(T - T_b) + m_s \frac{dX}{dt} \Delta H_v + Q_{microwave} \tag{3}$$

where C_p is the heat capacity of the sample (J.kg^{-1}.K^{-1}), T is the temperature of the sample (K), h is the heat transfer coefficient (W.m^{-2}.K^{-1}), ΔH_v is vaporization heat of water (J.kg^{-1}), T_b is the ambient temperature (K) and $Q_{microwave}$ is the received microwave power level (W).

3. Mathematical Modelling of Microwave Drying Using the Spatial Reaction Engineering Approach (S-REA)

The detailed description of the model has been presented in Chen and Putranto (2013). For modeling using the S-REA, the REA is coupled with a set of equations of conservation of heat mass transfer.

The mass balance of liquid water can be represented as:

$$\frac{\partial (C_s X)}{\partial t} = \frac{1}{r^2} \frac{\partial}{\partial r} (D_w r^2 \frac{\partial (C_s X)}{\partial r}) - \dot{I} \tag{4}$$

where t is the time (s), r is the sample radius (m), X is the concentration of liquid water (kg H$_2$O kg dry solids^{-1}), D_w is the capillary diffusivity (m^2 s^{-1}), C_s is the solids concentration (kg dry solids m^{-3}), $\dot{I}$ is the local evaporation or condensation rate (kgm^{-3}s^{-1}) and $\dot{I} > 0$ when evaporation occurs locally.

The mass balance of water vapor can be written as:

$$\frac{\partial C_v}{\partial t} = \frac{1}{r^2} \frac{\partial}{\partial r} (D_v r^2 \frac{\partial C_v}{\partial r}) + \dot{I} \tag{5}$$

where C_v is the concentration of water vapor (kg H$_2$O m^{-3}) and D_v is the water vapor diffusivity (m^2 s^{-1}).

In addition, the heat balance can be represented as:

$$\rho C_p \frac{\partial T}{\partial t} = \frac{1}{r^2} \frac{\partial}{\partial r} (k r^2 \frac{\partial T}{\partial r}) - \dot{I} \Delta H_v \tag{6}$$

where T is the sample temperature (K). ρ is the sample density (kg m^{-3}), C_p is the sample specific heat capacity (J kg^{-1} K^{-1}), ΔH_v is the water vaporization enthalpy (J kg^{-1}) and k is the sample thermal conductivity (W m^{-1} K^{-1}).

4. Application of the L-REA and S-REA to Model Microwave Drying of Green Peas

The L-REA is applied to model the microwave drying of green peas whose experimental details were presented earlier (Souraki and Mowla, 2008). By using the procedures highlighted in Section 2, the relative activation energy of green peas can be expressed as:

$$\frac{\Delta E_v}{\Delta E_{v,b}} = 0.197(\overline{X} - X_b)^3 - 0.64\,(\overline{X} - X_b)^2 - 0.003(\overline{X} - X_b) + 0.981 \tag{7}$$

Combining Eqs. (2), (3) and (7) results in the temporal profiles of moisture content and temperature during microwave drying.

Similarly, the S-REA is attempted to describe the microwave drying of green peas. For the modelling, Eq. (6) is used to describe the 'local' relative activation energy by replacing the average moisture content with the local moisture content. Coupling Eqs. (4) to (7) gives the spatial profiles of moisture content and temperature during microwave drying.

5. Results of Modelling of Microwave Drying of Green Peas Using the L-REA

Figure 1 describes the temporal profiles of average moisture content and temperature during the drying at microwave power of 0.7 W/g. Compared to drying at microwave power of 0.25 W/g, a higher drying rate is exhibited. This indicates that the microwave heating occurs internally in order to supply a larger amount of heat for evaporation. Here, at the end of microwave drying, the final temperature is around 340 K. Understanding of the temperature profiles is important when estimating whether the biochemical changes would occur so that fine-tuning of drying variables to manufacture products with the desirable quality can be made. When compared to Fig. 2, it appears that drying at 1.3 W/g gives lower profiles of moisture content and higher temperature.

The L-REA is able to yield reasonable profiles of both moisture content and temperature during microwave drying. The capability of the L-REA is most likely because of the accuracy of activation energy. It seems that the effects of internal heating on the drying rate are captured well by the relative activation energy. The combination of the equilibrium activation energy and the relative activation energy yields flexible relationships which accurately suggests the structural changes during microwave drying.

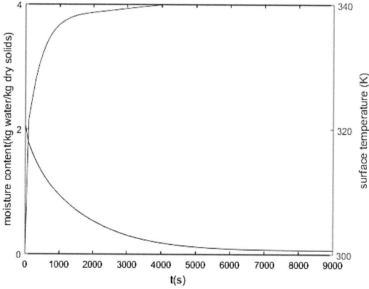

Figure 1: The temporal profiles of moisture content and temperature during microwave drying of green peas at power level of 0.7 W/g, modelled using the L-REA.

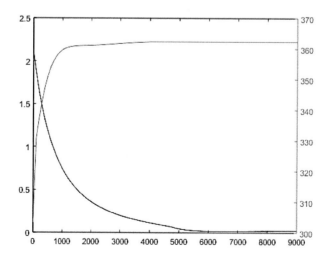

Figure 2: The temporal profiles of moisture content and temperature during microwave drying of green peas at power level of 1.3 W/g, modelled using the L-REA.

6. Results of Modelling of Microwave Drying of Green Peas Using the S-REA

Figures 3 to 7 show the results of modelling the microwave drying using the spatial reaction engineering approach (S-REA). A good agreement between the predicted and experimental moisture content data of microwave drying of green peas at microwave power level of 1.3 W/g is shown in Fig. 3. The results of modelling match well with the experimental data (R^2 of 0.992). The S-REA accurately describes the temporal profiles of moisture content during microwave drying at power level of 1.3 W/g. In addition, the profiles of temperature are modelled well by the S-REA, as described in Fig. 4. At the end of drying period, no noticeable change in moisture content is observed. In parallel with this, the temperature reaches a plateau at 360 K.

Figure 5 indicates the spatial profiles of moisture content during microwave drying at power level of 1.3 W/g. The highest moisture content is attained at the core of the samples, which indicates that the moisture migrates outwards during microwave drying. It is reasonable that the lowest one is located at the edge of the samples since the concentration of water vapor in the drying medium is lower than the surface water vapor concentration. The gradient is relatively large at the beginning, but this reduces

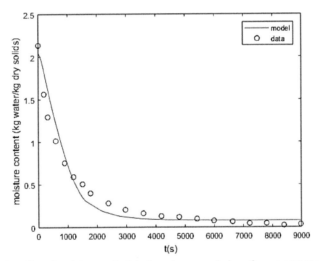

Figure 3: The temporal profiles of moisture content during microwave drying of green peas at power level of 1.3 W/g, modelled using the S-REA.

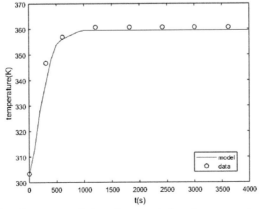

Figure 4: The temporal profiles of temperature during microwave drying of green peas at power level of 1.3 W/g, modelled using the S-REA.

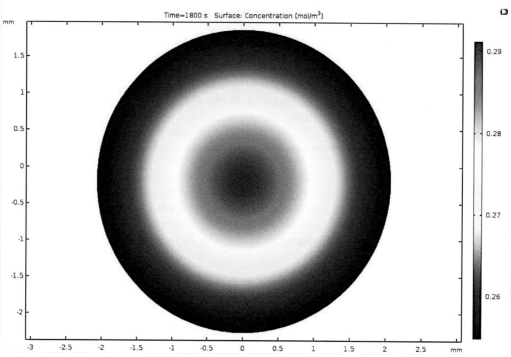

Figure 5: The spatial profiles of temperature during microwave drying of green peas at power level of 1.3 W/g, modelled using the S-REA.

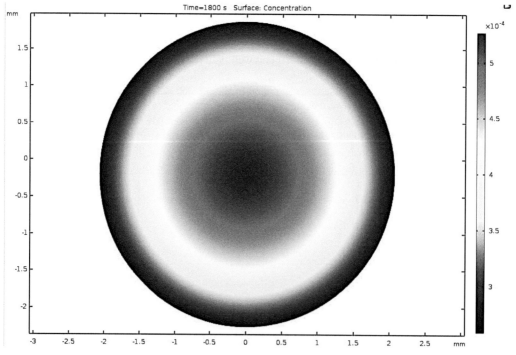

Figure 6: The spatial profiles of concentration of water vapor during microwave drying of green peas at power level of 1.3 W/g, modelled using the S-REA.

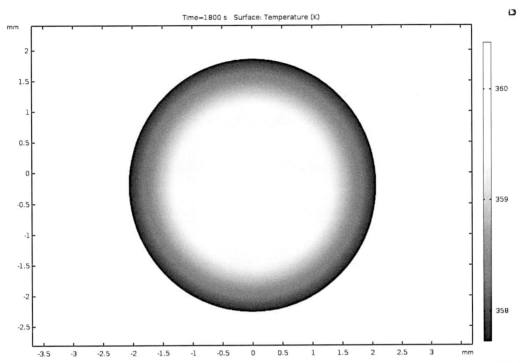

Figure 7: The spatial profiles of temperature during microwave drying of green peas at power level of 1.3 W/g, modelled using the S-REA.

as drying progresses. At the end of drying, the gradient becomes smaller as the equilibrium condition is approached. The spatial profiles of moisture content are similar to those of convective drying, but the gradient inside the samples may not be similar due to the difference in distribution of temperature. The spatial profiles of moisture content correspond well with those of concentration of water vapor and temperature described below.

The spatial profiles of water vapor are depicted in Fig. 6. The REA is used to link the equation of conservation of liquid water and water vapor, as shown in Eqs. (3) and (4). To the best of our knowledge, the REA-based model is the only model which has been able to generate such profiles of concentration of water vapor. The profiles at the core of the samples are higher than those at the outer part of the samples. This could be because of the higher local evaporation rate at the core as a result of higher moisture content. This may be further pronounced by the higher temperature at the core of the samples. The relatively high moisture content at the inner part may actually decrease the porosity of the samples which may retard the local evaporation rate. Nevertheless, this effect may be less pronounced than those resulted by the higher moisture content and temperature at the core.

Figure 7 shows the spatial profiles of temperature modelled using the S-REA. The temperature at the inner part of the samples is higher than that at the outer part due to the internal heating. The microwave generates the heat which is used for evaporating the water inside the samples and penetrated outwards by conduction. This is unlike convective drying where the drying air temperature is higher than the sample temperature. In this case, the gradient of temperature inside the samples is relatively low, which is most likely due to the small sample. The spatial profiles here are in agreement with those of concentration of water vapor highlighted above. Towards the end of drying, the gradient reduces to approach the equilibrium condition.

It has been demonstrated here that the S-REA accurately models the profiles of moisture content, concentration of water vapor and temperature during microwave drying. The S-REA is applicable not only to model the convective and intermittent drying highlighted previously (Putranto and Chen, 2013a–c; 2015a,b) but also the microwave drying. This applicability is largely because of the applicability of the REA to serve as the local evaporation/condensation rate. The 'local' relative activation energy, implemented by adopting Eq. (6), seems to be valid to represent the characteristics of materials under microwave drying. Here, basically, the relative activation energy is extrapolated to model the evaporation at micro-scale. The accuracy of the S-REA here indicates that the 'local' activation energy successfully integrates the equilibrium activation energy with the 'local' relative activation energy in order to represent the microstructural behaviors of the materials as influenced by the microwave power, humidity, temperature and other external conditions.

Based on the S-REA predictions, better estimations of the local phenomena, underlying the quality changes, can be made. In addition, the S-REA can be used in process design of microwave drying by embedding it into multi-scale simulation, in which two-way coupling between drying air and materials being dried is undertaken. Through the simulation, the flow-field of drying air (humidity, temperature, velocity and temperature) is predicted along with the internal profiles of the samples. The simulation can also serve as a reliable tool for fine-tuning microwave drying schemes.

Conclusions

In this chapter, both the L-REA and S-REA are employed to model the microwave drying of food materials. The L-REA successfully models the microwave drying of food materials. The accuracy is due to the applicability of the relative activation energy to describe the materials characteristics along with the applicability of the newly defined equilibrium activation energy. The S-REA also accurately describes the spatial profiles during microwave drying. The REA can describe the local evaporation/condensation rate inside porous food materials. Through the S-REA, more comprehensive understanding of interactions of structures, transport processes and operating conditions inside food materials can be made. Both the L-REA and S-REA are ready for adoption in industrial settings for designing new processes and optimising current processes of microwave drying systems.

References

Ahmad, S.S., Morgan., M.T. and Okos, M.R. 2001. Effect of microwave drying on the drying, checking and mechanical strength of baked biscuits. Journal of Food Engineering 50: 63–75.

Alibas, I. 2007. Microwave, air and combined microwave–air-drying parameters of pumpkin slices. LWT 40: 1445–1451.

Chen, X.D. and Putranto, A. 2013. Modeling Drying Processes: A Reaction Engineering Approach, Cambridge: Cambridge University Press.

Cui, Z.W., Xu, S.Y., Sun, D.W. and Chen, W. 2005. Temperature changes during microwave-vacuum drying of sliced carrots. Drying Technology 23: 1957–1974.

Dincov, D.D., Parrot, K.A. and Pericleous, K.A. 2004. Heat and mass transfer in two-phase porous materials under intensive microwave heating. Journal of Food Engineering 65: 403–412.

Kowalski, S.J. and Pawlowski, A. 2010. Drying of wet materials in intermittent conditions, Drying Technology 28, 636–643.

Marra, F., De Bonis, M.V. and Ruocco, G. 2010. Combined microwaves and convection heating: A conjugate approach. Journal of Food Engineering 97: 31–39.

Maskan, M. 2001. Drying, shrinkage and rehydration characteristics of kiwifruits during hot air and microwave drying. Journal of Food Engineering 48: 177–182.

Putranto, A. and Chen, X.D. 2013a. Spatial reaction engineering approach (S-REA) as a multiphase drying approach to model heat treatment of wood under constant heating rate. Industrial and Engineering Chemistry Research 52: 6242–6252.

Putranto, A. and Chen, X.D. 2013b. Spatial reaction engineering approach (S-REA) as an alternative for non-equilibrium multiphase mass transfer model for drying of food and biological materials. AIChE Journal 59: 55–67.

Putranto, A. and Chen, X.D. 2013c. Multiphase modeling of intermittent drying using the spatial reaction engineering approach (S-REA). Chemical Engineering and Processing: Process Intensification70, 169–183.

Putranto, A. and Chen, X.D. 2014. Examining the suitability of the reaction engineering approach (REA) to modeling local evaporation/condensation rates of materials with various thicknesses. Drying Technology 32: 208–221.

Putranto, A. and Chen, X.D. 2015a. Spatial reaction engineering approach (S-REA): An effective approach to model drying, baking and water vapor sorption process. Chemical Engineering Research and Design 101: 135–145

Putranto, A. and Chen, X.D. 2015b. Bread baking and its color kinetics modeled by the spatial reaction engineering approach (S-REA). Food Research International 71: 58–67.

Putranto, A. and Chen, X.D. 2017. A new model to predict diffusive self-heating during composting incorporating the reaction engineering approach (REA) framework. Bioresource Technology 232: 211–221.

Putranto, A. and Chen, X.D. 2012. Roasting of barley and coffee modeled using the lumped-reaction engineering approach (L-REA). Drying Technology 30: 475–483.

Putranto, A., Chen, X.D., Devahastin, S., Xiao, Z. and Webley, P.A. 2011b. Application of the reaction engineering approach (REA) to model intermittent drying under time-varying humidity and temperature. Chemical Engineering Science 66: 2149–2156.

Putranto, A., Chen, X.D., Xiao, Z. and Webley, P.A. 2011a. Modeling of high-temperature treatment of wood by using the reaction engineering approach (REA). Bioresource Technology 102: 6214–6220.

Souraki, B.A. and Mowla, D. 2008. Simulation of drying behaviour of a small spherical foodstuff in a microwave assisted fluidized bed of inert particles. Food Research International 41: 255–265.

CHAPTER 25

Modelling of Heat Transfer During Deep Fat Frying of Food

*KK Dash, * Maanas Sharma and MA Bareen*

1. Introduction

Frying is a unit operation, which is mainly used to alter the eating quality of food. A secondary consideration is the preservative effect that results from thermal destruction of micro-organisms and enzymes and a reduction in water activity at the surface of the food. The shelf life of fried foods is mostly determined by the moisture content after frying. Deep frying is a simultaneous heat and mass transfer process which involves submerging the food in the hot edible oil at a temperature (150–200°C) above the boiling point of the water contained in the food (Hubbard and Farkas, 1999; Weisshaar, 2014). Heat is transferred from the frying medium to the surface of the food by convection and from the surface to the inside of the food product by conduction. Alternatively, the moisture is evaporated and transported from the interior to the surface of food product by diffusion, which then migrates from the surface through the frying medium (Blumenthal, 1991). The thermal and physical properties that affect heat and mass transfer during frying are thermal diffusivity, heat transfer coefficient, specific heat, density, moisture diffusivity and effective mass transfer coefficient of the food and the oil. The different process parameters involved in frying are the geometry of the food, the temperature of the oil and time of frying (Moreira et al., 1995). All these thermo-physical parameters and process parameters were analysed, quantified and optimised during the frying of different food products, such as sweet potato, pumpkin and taro (Ahromrit and Nema, 2010), gulab jamun (Eljeeva et al., 2013), chhena jhilli (Mondal and Dash, 2017), alexia potatoes (Koerten et al., 2017) and cassava root slices (Oyedeji et al., 2016).

In the frying process, when food is placed in the hot oil the surface temperature rises rapidly and water is vaporized as steam. The surface then begins to dry out in a similar way to baking and roasting. The plane of evaporation moves inside the food, and a crust is formed. The surface temperature of the food then rises to that of the hot oil, and the internal temperature rises more slowly towards 100°C. The rate of heat transfer is controlled by the temperature difference between the oil and the food and by the surface heat transfer coefficient. The rate of heat penetration into the food is controlled by the

Department of Food Engineering and Technology, Tezpur University, Tezpur, Assam, 784018, India.
Emails: kshirod@tezu.ernet.in, manass@tezu.ernet.in, bareenmd@gmail.com

thermal conductivity of the food. The thermal conductivity of a food depends on its porosity, structure and chemical constituents. Deep frying causes instant cooking, browning on the surface of food, texture and flavor enhancement (Rimac-Brncic et al., 2004; Pedreschi et al., 2006). The heat transfer coefficient during frying is affected by bubble flow direction, velocity, bubble frequency, and the magnitude of oil agitation. Bubbles that form at the top surface escape upward immediately, providing greater oil agitation. Bubbles leaving the lateral surfaces move upward along the lateral surface, promoting a different type of agitation. Bubbles that leave the bottom surface coalesce before flowing against the product surface to the edges, and then move to the top oil surface along the lateral surfaces (Baik and Mittal, 2002). The minimal values of convective heat transfer coefficient were observed after 90 and 120s of frying, which depended on formulation of potato dough (Budzaki et al., 2005). Specific heat is mainly dependent on the moisture content of food, frying time and temperature (Farid and Kizilel, 2009). The moisture content in foods has a large effect on the specific heat (c_p) due to the relatively high specific heat of water and heat of sorption.

The time taken for food to be completely fried depends on (i) the type of food, (ii) the temperature of the oil, (iii) the method of frying (shallow or deep-fat frying), and (iv) the thickness of the food. The temperature used for frying is determined mostly by economic considerations and the requirements of the product. At high temperatures (180–200°C), processing times are reduced and production rates are, therefore, increased. However, high temperatures also cause accelerated deterioration of the oil and formation of free fatty acids, which alter the viscosity, flavor, and color of the oil and promote foaming. This increases the frequency with which oil must be changed and, hence, increases costs.

The different sections of this book chapter present types of frying, heat transfer in frying, effect of heat on fried foods, *thermal properties involved in frying and* dimensionless numbers used in deep fat frying. This chapter also focuses on characterising transient heat transfer during the frying in slab-, cylinder- and sphere-shaped food products at different values of heat transfer Biot number.

2. Types of Frying

There are four different types of commercial frying techniques.

i. Pan frying: Pan frying is the method of cooking in a very minimal amount of oil (just enough to lubricate the pan). In pan frying, because of partial coverage, the food has to be flipped at least once to get both sides cooked. Usually every 15–30 seconds.

ii. Shallow frying: Shallow frying method is suited to foods with large surface-area-to-volume ratio. Heat is transferred to the food mostly by conduction from the hot surface of the pan through a thin layer of oil. Here, the food is only partly submerged, about halfway up to the side of the food to be cooked, and it must be flipped in between. It is usually used to prepare cuts of fish, meat and eggs.

iii. Stir-frying: Stir-frying is a technique in which ingredients are fried in a small amount of very hot oil while being stirred in a bowl shaped frying pan. In stir-frying, heat is transferred to the frying pan, then to the food in it.

iv. Deep frying: Deep frying is a cooking process that involves submerging a food in extremely hot oil in the presence of air at a high temperature of 150 to 200°C. Because of the high temperature, the food gets cooked extremely quickly.

3. Heat Transfer in Frying

Deep fat frying is culinary technique where oil or fat is mainly chosen as the heat transfer medium. In this process, the food and oil come in direct contact with each other at a temperature above the boiling point of water. The surface frying shows the crust formation by increasing heat transfer coefficient. It is mostly subjected to convection, conduction, and evaporation. It also includes physiochemical alteration. The heat transfer of oil to food material during frying shows the evaporation of bound water from inside capillaries of food and intracellular structure toward the crust. During the frying process, in the first step, water escapes from the crust and, in the second step, the water in the core of the product migrates to the crust (Abtahi et al., 2016; Isleroglu and Kaymak-Ertekin, 2016; Kurek et al., 2017).

In the frying process, there are two general ways that heat can be transferred from one material to another, i.e., conduction and convection.

i. Conduction is heat transfer due to contact of molecules. Thermal energy, which can be thought of as the vibration of molecules in place, is transferred directly from one material to another in contact with it.

ii. Convection is heat transfer due to the bulk movement of molecules. Molecules move, changing places, not just vibrating in place, and take their heat with them.

The main purpose of frying is the development of characteristic colours, flavors, and aromas in the crust of fried foods. These eating qualities are developed by a combination of Maillard reactions and compounds absorbed from the oil. The main factors that control the changes to colour and flavour in a given food are, therefore: (i) pre-treatments, such as blanching or partial drying (ii) the type of oil used for frying, (iii) the interfacial tension between the oil and the product, (iv) the temperature and time of frying, (v) the size, moisture content and surface characteristics of the food.

The process of deep frying can be divided into four stages:

i. Initial heating
ii. Surface boiling
iii. Crust formation
iv. Bubble endpoint

3.1 Initial heating

In this first phase, no water evaporation takes place as the fry surface is heated to the boiling temperature of water. Heat is transported from the oil to the fry surface through free convection, while conduction takes place from the fry surface to the interior.

3.2 Surface boiling

This phase starts as soon as the fry surface reaches the boiling temperature of the water. Water starts evaporating while the fry surface temperature remains at the boiling temperature of the water. Tiny exploding bubbles are found to be sizzling at the surface of the food. Hence, the hot oil surrounding the food causes water inside the food to evaporate. The movement caused by the bubbling circulates the currents of the frying oil, which increases the rate of heat transfer and cooks the food faster.

3.3 Crust formation

As all the surface water is evaporated, crust formation begins and forms an additional barrier for heat transfer and vapour expulsion. The external heat transfer coefficient increases due to forced convection as a function of the evaporation rate.

3.4 Bubble endpoint

This is the last stage of deep frying, in which very few bubbles appear on the surface of the fried food. At this stage, water from inside the food is no longer evaporating, either because all the water from inside the food is gone, or heat transfer from the crust to the core has reduced to the point where it becomes improbable that the water will evaporate. At this point, the fried product should be removed from the oil, or else the oil will begin to seep into the fried product and make it soggy since there are no more water vapor bubbles to counteract the diffusion of oil inwards.

4. Thermal Properties Involved in Frying

4.1 Specific heat

Heat capacity of a system can be defined as the quantum of heat required to increase the temperature of the system by unit. Molar heat capacities C_p and C_v can be defined as the heat required per mole of the substance to increase the temperature by unity under constant pressure and constant volume, respectively. Hence, $C_p = m\,c_p$ and $C_v = m\,c_v$. For an ideal gas, the relation between the molar heat capacities can be given as: $C_p - C_v = R$ (Universal gas constant).

Specific heat is a property of the material. For a given material, specific heat is defined as the quantum of heat required to increase the temperature by unity, per unit mass of the material.

The two important types of specific heat capacities are:

i. specific heat capacity at constant volume (c_v)

ii. specific heat capacity at constant pressure (c_p)

Specific heat at constant volume is the change in internal energy per unit mass of the substance, per unit temperature rise. This relates to the heat absorbed by the system for unit temperature rise when the volume of the system remains unchanged. Specific heat at constant pressure is the change in internal energy per unit mass of the substance, per unit temperature rise. This heat capacity relates to the heat absorbed by the system for unit temperature rise, when the system pressure remains constant.

When heat energy is added to a substance, the temperature will change by a certain amount. The relationship between heat energy and temperature and the specific heat is presented in Eq. (1).

$$Q = mc\Delta T \tag{1}$$

Heat energy = (Mass of substance) (Specific heat) (Change in temperature)

Q = heat energy (Joules, J)

m = mass of a substance (kg)

c = specific heat (units J/kg·K)

Δ = symbol meaning "the change in"

ΔT = change in temperature (Kelvins, K)

The specific heat of food can be predicted from knowledge of the composition and the specific heat of each component. The specific heat of foodstuffs using the mass fraction of its constituents (water, protein, fat, carbohydrate and ash) can be estimated with Eq. (2) (Heldman, 1975).

$$c_p = 1.424M_C + 1.549M_P + 1.672M_f + 0.837M_a + 4.18M_w \tag{2}$$

where M_c = mass fraction of *carbohydrate*, M_P = mass fraction of *protein*, M_f = mass fraction of fat, M_a = mass fraction of ash, M_w = mass fraction of water.

The general model for prediction of specific heat of food containing n components can be estimated as per Eq. (3).

$$c_p = \sum (c_{pi}M_i) \tag{3}$$

where M_i = mass fraction of component i, c_{pi} = specific heat of component i (J/kg K).

The analysis of specific heat data for many liquid foods with different compositions and over a temperature range of 20–100°C (Choi and Okos, 1986) are presented in Eqs. (4)–(10).

The data presented by Choi and Okos (1986) are based on an extensive study and analysis of specific heat data for many liquid foods with different compositions and generally over a temperature range of 20–100°C.

$$c_{protein} = 1.9842 + 1.4733 \times 10^{-3}T - 4.8008 \times 10^{-6}T^2 \tag{4}$$

$$c_{carbohydrate} = 1.54884 + 1.9625 \times 10^{-3}T - 5.9399 \times 10^{-6}T^2 \tag{5}$$

$$c_{fat} = 1.9842 + 1.4733 \times 10^{-3}T - 4.8008 \times 10^{-6}T^2 \tag{6}$$

$$c_{ash} = 1.0926 + 1.8896 \times 10^{-3}T - 3.6817 \times 10^{-6}T^2 \tag{7}$$

$$c_{water} \,(> 0°C) = 4.0817 - 5.3062 \times 10^{-3}T + 9.9516 \times 10^{-4}T^2 \tag{8}$$

$$c_{water} \,(< 0°C) = 4.1762 - 9.0864 \times 10^{-5}T + 5.4731 \times 10^{-6}T^2 \tag{9}$$

$$c_{ice} = 2.0623 + 6.0769 \times 10^{-3}T \tag{10}$$

4.2 Density

Density expresses the proportional relationship between mass and volume of the substance. Density is defined as mass per unit volume; mathematically,

$$\rho = \frac{M}{V} \tag{11}$$

For any pure substance or homogenous mixture, density is an intensive quantity, which means that its value is the same no matter how much of the substance is present.

The general model for density prediction of food material was proposed by Choi and Okos (1986) and involves the use of product composition (M_i) for protein, fat, carbohydrate, ash and water and density (moisture-free) for each component (ρ_i).

$$\rho = \frac{1}{\Sigma \left(\dfrac{M_i}{\rho_i} \right)} \tag{12}$$

To predict the bulk density of a high-moisture food from its composition, the densities of different food components, i.e., protein, fat, carbohydrate, ash and water, were expressed as a function of temperature, as shown in Eqs. (13)–(18). These models are similar to the general model for density prediction, as proposed by Choi and Okos (1986). Equation (12) involves the use of product composition (M_i) for protein, fat, carbohydrate, ash and water and density (moisture-free) for each component (ρ_i). The density data to be used as inputs to the proposed model were published by Choi and Okos (1986).

The proposed model predicts the bulk density of a high-moisture food from typical composition information and density temperature relationships, as presented in Eqs. (13)–(18).

$$\rho_{protein} = 1.3299 \times 10^3 - 5.184 \times 10^{-1}T \tag{13}$$

$$\rho_{carbohydrate} = 1.59919 \times 10^3 - 3.1046 \times 10^{-1}T \tag{14}$$

$$\rho_{fat} = 9.2559 \times 10^2 - 4.1757 \times 10^{-1}T \tag{15}$$

$$\rho_{ash} = 2.4238 \times 10^3 - 2.8063 \times 10^{-1}T \tag{16}$$

$$\rho_{water} = 9.9718 \times 10^2 + 3.1439 \times 10^{-3}T - 3.7574 \times 10^{-3}T^2 \tag{17}$$

$$\rho_{ice} = 9.1689 \times 10^2 - 1.3071 \times 10^{-1}T \tag{18}$$

4.3 Thermal conductivity

Thermal conductivity is a material property that describes the ability to conduct heat. This also describes how readily a material will give or take heat through conduction. Thermal conductivity can be defined as the quantity of heat transmitted through a unit thickness of a material in a direction normal to a surface of unit area due to a unit temperature gradient under steady state conditions. Those materials with high thermal conductivity will transfer heat rapidly, either by receiving heat from a hotter material or by giving heat to a colder material. On the contrary, materials with low thermal conductivity act as thermal insulators, preventing the transfer of heat.

Based on generalised experimental data, an empirical relation for estimating thermal conductivity of food can be presented as per Eq. (19).

$$k = k_c M_c + k_p M_p + k_f M_f + k_a M_a + k_w M_w \tag{19}$$

where M_c, M_p, M_f, M_a and M_w are the mass concentrations of components with thermal conductivities k_c, k_p, k_f, k_a and k_w W/m°C, respectively.

Temperature dependence of thermal conductivities of pure water, carbohydrate (CHO), protein, ash, fat and ice at different temperatures can be empirically expressed according to Choi and Okos (1986) as follows.

$$k_{water} = 5.7109 \times 10^{-1} + 1.76265 \times 10^{-3} T - 6.7036 \times 0^{-6} T^2 \tag{20}$$

$$k_{CHO} = 2.0141 \times 10^{-1} + 1.3874 \times 10^{-3} T - 4.3312 \times 10^{-6} T^2 \tag{21}$$

$$k_{protein} = 1.7881 \times 10^{-1} + 1.1958 \times 10^{-3} T - 2.7178 \times 10^{-3} T \tag{22}$$

$$k_{ash} = 3.2962 \times 10^{-1} + 1.4011 \times 10^{-3} T - 2.9069 \times 10^{-6} T^2 \tag{23}$$

$$k_{fat} = 1.8071 \times 10^{-1} - 2.7604 \times 10^{-3} T - 1.7749 \times 10^{-7} T^2 \tag{24}$$

$$k_{ice} = 2.2196 - 6.2489 \times 10^{-3} T + 1.0154 \times 10^{-4} T^2 \tag{25}$$

In the above equations, k is thermal conductivity in W/m°C; T is temperature in °C and T varies between 0 and 90°C.

4.4 Dimensionless numbers used in deep frying

The heat transfer coefficient is a critical processing parameter that plays a vital role in frying. The heat transfer coefficient is a measure of how fast heat transfer occurs between a surface and a fluid. For transient heat transfer analysis, the concepts of the Reynolds number (Re), Prandtl number (Pr), Grashof number (Gr), and Nusselt number (Nu), which are dimensionless terms used to determine the overall convection heat transfer coefficient, h, need to be introduced.

Reynold's number

The Reynold's number 'Re' is the ratio of Inertial forces to the Viscous forces and is used in forced convection. It is primarily used to analyze different flow regimes, i.e., laminar, turbulent or transient flow. When viscous forces are dominant (i.e., low value of Re) it is a laminar flow and when inertial forces are dominant (i.e., high value of Re) it is a Turbulent flow. The Reynold's number is calculated by:

$$Re = \frac{\rho u x}{\mu} \tag{26}$$

where x is the position along the interface in the direction of fluid flow, u is the velocity, ρ is the density, and µ is the viscosity of the fluid.

Fluid flow will be laminar, if Re < 2300

Fluid flow will be transient, if Reynolds number is in between 2300 and 4000

Fluid flow will be Turbulent, if Re > 4000

Prandtl number

The Prandtl Number is a dimensionless number approximating the ratio of momentum diffusivity to thermal diffusivity. Prandtl number is a characteristic of the fluid only. Prandtl Number is also the ratio of velocity boundary layer to thermal boundary layer. Small value of Pr implies that rate of thermal diffusion (heat) is more than the rate of momentum diffusion (velocity) and also the thickness of thermal boundary layer is much larger than the velocity boundary layer.

The Prandtl number is calculated by:

$$Pr = \frac{\upsilon}{\alpha} = \frac{c_p \mu}{k} \tag{27}$$

where for the fluid, υ is the kinematic viscosity, α is the thermal diffusivity, c_p is the specific heat, and k is the heat transfer coefficient.

Prandtl number is solely dependent upon the fluid properties. For air at room temperature, Pr is 0.71, and Prandtl number of water at 17°C is 7.56. Liquids, in general, have high Prandtl numbers. The values of Pr for different materials are: For gases, Pr = 0.7 to 1.0; For water, Pr = 1 to 10; For liquid metals, Pr = 0.001 to 0.03; For oils, Pr = 50 to 2000.

Grashof number

Grashof number is the ratio of buoyancy force to viscous force in natural convection. In general, Reynolds number is used in forced convection of fluid flow, whereas Grashof number is used in natural convection. The Grashof number is calculated by:

$$Gr = \frac{g\beta (T_w - T_\infty)x^3}{\upsilon^3} \tag{28}$$

where,

g is the gravitational constant

β is the coefficient of thermal expansion (equal to approximately $1/T$, for ideal gases)

T_w is the temperature at the interfacial wall between the fluids

T_∞ is the bulk temperature

L is the vertical length

D is the diameter

υ is the kinematic viscosity.

At higher Grashof numbers, the boundary layer is turbulent; at lower Grashof numbers (in the range $10^3 < Gr_L < 10^6$), the boundary layer is laminar. The transition to turbulent flow occurs in the range $10^8 < Gr_L < 10^9$ for natural convection from vertical flat plates.

Nusselt number

The Nusselt number is the ratio of convective to conductive heat transfer coefficient across the boundary layer. It can also be viewed as conduction resistance to convection resistance of the material.

Free convection: Nu = f (Ra, Pr)

Forced Convection: Nu = f (Re, Pr)

The Nusselt number is calculated by:

$$Nu = \frac{hx}{k} \tag{29}$$

where h is the convective heat transfer coefficient of the fluid, L is the characteristic length, k is the thermal conductivity of the fluid.

Low Nu => conduction is more => Laminar flow

High Nu => convection is more => Turbulent flow.

The typical procedure is to calculate Pr, Re, and Gr and characterize the geometry and flow conditions about the interface of the two materials, then select the proper function (Nu = f (Pr, Re, Gr)) for the Nusselt number.

Fourier number

Fourier number represents the dimensionless time. It may be interpreted as the ratio of current time to time to reach steady-state. The Fourier number is a dimensionless number that characterizes heat conduction. It is the ratio of diffusive/conductive transport rate by the quantity storage rate and arises from non-dimensionalization of the heat equation. Fourier number can be obtained by multiplying the dimensional time by the thermal diffusivity and dividing by the square of the characteristic length:

$$F_o = \frac{\alpha t}{L^2} \tag{30}$$

For transient mass transfer by diffusion, there is an analogous mass Fourier Number to the thermal Fourier number.

$$F_o = \frac{Dt}{L^2} \tag{31}$$

Biot number

The Biot number (Bi) is a dimensionless quantity used in unsteady state (transient) heat transfer conditions. It represents the ratio of heat transfer resistance inside the body to heat transfer resistance at the surface of the body. It also can be represented as the ratio of internal thermal resistance to external thermal resistance. The Biot number is dimensionless, and it is represented as the ratio

$$Bi = \frac{hL}{k} \tag{32}$$

Biot number shows the variation of temperature inside the body with respect to time.

The difference between the Biot number and the Nusselt number is that the denominator of the Nusselt number involves the thermal conductivity of the fluid at the solid-fluid convective interface; the denominator of the Biot number involves the thermal conductivity of the solid at the solid-fluid convective interface.

The Nusselt Number is used to characterize the heat flux from a solid surface to a fluid. In that case, the thermal conductivity is for the fluid. The Biot number is used the characterize the heat transfer resistance "inside" a solid body. In that case, k is the thermal conductivity of the solid body, and h is the heat transfer coefficient that describes the heat transferred from the "surface of the solid body" to the surrounding fluid. The Biot Number can be thought of as the ratio of internal diffusion resistance to external convection resistance. Note that 1/h is the external convection resistance and L/k is the internal diffusion resistance.

Whenever the Biot number is small, i.e., Bi < 0.1

- Heat transfer resistance inside the body is very low.
- Inside the body, conduction takes place faster compared to convection at the surface.
- No temperature gradient inside the body (uniformity in temperature).
- Under above conditions, a transient problem can be treated by the "lumped thermal capacity" approach. The lumped capacity assumption implies that the object for analysis is considered to have a single mass-averaged temperature.

5. Transient Diffusion

Transient processes, in which the concentration at a given point varies with time, are referred to as unsteady state processes or time-dependent processes. This variation in concentration is associated with a variation in the mass flux.

These generally fall into two categories:

i. the process which is in an unsteady state only during its initial startup, and

ii. the process which is in a batch operation throughout its operation.

In unsteady state processes, there are three variables: Concentration, time and position. Therefore, the diffusion process must be described by partial rather than ordinary differential equations.

Although the differential equations for unsteady state diffusion are easy to establish, most solutions to these equations have been limited to situations involving simple geometries and boundary conditions, and a constant diffusion coefficient.

Many solutions are for one-directional mass transfer as defined by Fick's second law of diffusion:

$$\frac{\partial C_A}{\partial t} = D_{AB} \frac{\partial^2 C_A}{\partial z^2} \tag{33}$$

This partial differential equation describes a physical situation in which there is no bulk–motion contribution, and there is no chemical reaction. This situation is encountered when the diffusion takes place in solids, in stationary liquids, or in a system having equimolar counter diffusion. Due to the extremely slow rate of diffusion within liquids, the bulk motion contribution of flux equation (i.e., $y_A \sum N_i$) approaches the value of zero for dilute solutions; accordingly, this system also satisfies Fick's second law of diffusion.

The solution to Fick's second law usually has one of the two standard forms. It may appear in the form of a trigonometric series which converges for large values of time, or it may involve a series of error functions or related integrals which are most suitable for numerical evaluation at small values of time. These solutions are commonly obtained by using the mathematical techniques of separation of variables or Laplace transforms. To formulate the mathematical model, the following assumptions were made

1. One-dimensional unsteady-state heat and mass transfer in the radial direction was considered.

2. Initial temperature and moisture content distribution in foods was uniform.

3. The product was considered to be homogeneous, isotropic and spherical with constant dimensions throughout frying.

4. Product shrinkage during frying is negligible and crust was assumed to be negligible and have same properties as the whole sample.

5. There was no internal heat generation in the product.

6. Moisture movement is by diffusion and moisture diffusivity encompasses other mechanisms, including convection.

7. A microscopically uniform porous medium was formed after frying and most oil diffuses into the product only during the cooling period.

6. Modelling of Heat Transfer

The heat transfer coefficient was calculated from the time-temperature data using lumped capacitance method (Holman et al., 1997) and one-dimensional transient heat conduction equation (Kopelman and Pflug, 1968) approaches. Transient heat transfer is important because of the large number of heating and cooling problems that occur industrially. In food processing operations, it is necessary to predict cooling and heating rates of various geometries of food products in order to predict the time required to reach the desired temperature. In food processing, such as in the canning industry, perishable canned foods are heated by immersion in steam baths or chilled by immersion in cold water. Similarly, in the frying process, the processed products are immersed in hot oil for frying. In the simplified case, consider a solid which has a very high thermal conductivity or very low internal conductive resistance, compared to the external surface resistance, where the convection occurs from the external fluid to the surface of the solid.

Since the internal resistance is very small, the temperature within the solid is essentially uniform at any given time.

6.1 Lumped thermal capacity model

The lumped capacitance method assumes that the temperature inside the object is uniform at any time instance during the transient heat transfer process and there is negligible internal temperature gradient inside the system. The inside thermal resistance is very small compared to the outside surface resistance. The temperature is only a function of time.

Suppose the food material was initially at a uniform temperature, T_i. At time t = 0, the product was immersed in an oil bath at temperature T_∞. The temperature of the product will approach T_∞ with a rate that depends on the surface heat transfer coefficient "h" at the solid-liquid interface. Making a heat balance on the solid object for small time interval of time, dt, the heat transfer from the bath to the object, must equal the change in internal energy of the object.

$$hA(T_\infty - T)dt = C_p\rho VdT \tag{34}$$

where,

A = surface area of the object, m^2

T = average temperature of the object at time t, s

ρ = density of the object, kg/m^3

V = volume, m^3

Rearrange the equation and integrating between the limits of $T = T_0$ when t = 0 and $T = T_t$ at t = t

$$\int_{T=T_o}^{T=Tt} \frac{dT}{T_\infty - T} = \frac{hA}{C_p\rho V}dT \tag{35}$$

$$\frac{T_\infty - T_t}{T_\infty - T_i} = exp\left(\frac{-hA}{\rho C_p V}t\right) = exp\left(\frac{-hA}{mC_p}t\right) \tag{36}$$

where,

T_i = Initial temperature of the product (K)

T_∞ = Temperature of the frying medium (K)

T_t = Temperature of the product at any time (K)

A = Surface area of the product (m^2)

C_p = Specific heat of the product (J/kg K)

m = Mass of the product (kg)

h = Convective heat transfer coefficient (W/m^2K)

t = Frying time (s).

Transient heat conduction in different geometries when Bi > 0.1

In transient heat transfer, the temperature in a body keeps on changing with time. Transient analysis is important for calculations of heating and cooling processes in food processing. In the frying process, heat is conducted through the food as it is immersed in oil. The internal temperature of the food, therefore, is usually increasing during frying. Figure 1 shows the schematic diagram for transient heat transfer in plane wall or slab, cylinder and sphere.

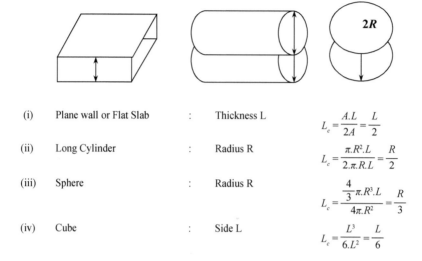

(i)	Plane wall or Flat Slab	:	Thickness L	$L_c = \dfrac{A.L}{2A} = \dfrac{L}{2}$
(ii)	Long Cylinder	:	Radius R	$L_c = \dfrac{\pi.R^2.L}{2.\pi.R.L} = \dfrac{R}{2}$
(iii)	Sphere	:	Radius R	$L_c = \dfrac{\frac{4}{3}\pi.R^3.L}{4\pi.R^2} = \dfrac{R}{3}$
(iv)	Cube	:	Side L	$L_c = \dfrac{L^3}{6.L^2} = \dfrac{L}{6}$

Figure 1: One-dimensional transient conduction in plane wall/slab, cylinder and sphere.

6.2 Case studies on measurement of convective heat transfer coefficient by lumped capacitance method

Case 1: Heat transfer in meatball

Spherical meatballs of 4 cm diameter, initially at a uniform temperature of 5°C, are immersed in hot oil at 160°C. The convection heat transfer coefficient was 195 W/m²°C, Specific heat was 2.7 kJ/kg°C, thermal conductivity was 15 W/m°C, and density was 1100 kg/m³. Determine time required by the meatball to reach 110°C.

Step I

Given
Spherical meat balls of diameter D = 4 cm = 0.04 m;
Temperature of the oil T_∞ = 160°C;
Initial temperature of the meat ball T_0 = 5°C;
Temperature of the meat ball at time t sec is T_t = 110°C;
Heat transfer coefficient h = 195 W/m²°C;
Density of the meat ball ρ = 1100 kg/m³;
Specific heat C_p = 2.7 kJ/kg°C = 2700 J/kg°C;
Thermal conductivity k = 15 W/m°C;

Step II

Characteristic length $L_c = \dfrac{V \, (volume)}{A_s \, (surface \; area)} = \dfrac{\frac{4}{3}\pi R^3}{4\pi R^2} = \dfrac{R}{3} = \dfrac{D}{2 \times 3} = \dfrac{0.04}{6} = 0.0067$ m;

Step III

Biot number $B_i = \dfrac{hL_c}{k} = \dfrac{195 \times 0.0067}{15} = 0.087$

As $B_i < 0.1$, the internal thermal resistance for conduction heat flow can be neglected and Lumped capacitance method can be applied.

Step IV

The time versus temperature relation is given by

$$\frac{T_\infty - T_t}{T_\infty - T_0} = exp\left[-\frac{hA_s}{\rho C_p V}t\right] = exp\left[-\frac{h}{\rho C_p L_c}t\right]$$

$$\frac{160 - 110}{160 - 5} = exp\left[-\frac{195}{1100 \times 2700 \times 0.0067}t\right]$$

$$0.323 = exp[-0.0099t]$$

$$t = \frac{1.1301}{0.0099} = 114.1515\,s = 1.90\ minutes$$

Case 2: Heat transfer in cylinder-shaped cut fish fillets

Cylinder-shaped cut fish fillets of 2 cm diameter and 8 cm length were initially at a uniform temperature of 20°C. They were exposed to hot oil at 165°C with a heat transfer coefficient of 70.6 W/m²°C on the surface. If the average thermal conductivity of the product is 5 W/m°C and thermal diffusivity 7.92×10^{-6} m²/s, find out the time required to achieve 90% of the oil temperature.

Step I

Given
Cylinder-shaped fish fillets of diameter D = 2 cm = 0.02 m and length L = 8 cm;
Temperature of the oil T_∞ = 165°C;
Initial temperature of the fish fillets T_0 = 20°C;
Temperature of the fish fillets at time t sec is $T_t = \frac{165 \times 90}{100}$°C = 148.5°C;
Heat transfer coefficient h = 70.6 W/m²°C;
Thermal conductivity k = 5 W/m°C;
Thermal diffusivity $\alpha = 7.92 \times 10^{-6}$ m²/s;

Step II

Characteristic length $L_c = \dfrac{V\ (volume)}{A_s\ (surface\ area)} = \dfrac{\pi R^2 L}{2\pi RL} = \dfrac{R}{2} = \dfrac{D}{2 \times 2} = \dfrac{0.02}{4} = 0.005$ m;

Step III

Biot number $B_i = \dfrac{hL_c}{k} = \dfrac{70.6 \times 0.005}{5} = 0.0706$;

As $B_i < 0.1$, the internal thermal resistance for conduction heat flow can be neglected and Lumped capacitance method can be applied.

Step IV

The time versus temperature relation is given by

$$\frac{T_\infty - T_t}{T_\infty - T_0} = exp\left[-\frac{hA_s}{\rho C_p V}t\right] = exp\left[-\frac{h}{\left(\dfrac{k}{\alpha}\right)L_c}t\right]$$

$$\frac{165-148.5}{165-20}=exp\left[-\frac{70.6}{\left(\dfrac{5}{7.92\times10^{-6}}\right)\times0.005}t\right]$$

$$0.114=exp[-0.0223t]$$

$$t=\frac{2.173}{0.0223}=97.16\,s=1.62\;minutes$$

Case 3: Heat transfer in vegetable slice

Consider a vegetable slice of 6 mm thickness and an average thermal conductivity k value of 5.7 W/m°C in the direction normal to the surface. The slice, initially at a temperature of 22°C, was fried in oil at a temperature of 150°C with a convective heat transfer coefficient of 31 W/m²°C. Density of the vegetable slice is ρ = 950 kg/m³ and the specific heat is 2.25 kJ/kg°C. Is it possible to use the lumped capacitance method to model transient heat transfer in the vegetable slice? Find the temperature of the slice after three minutes?

Step I

Given

Vegetable slice of thickness 2L = 6 mm = 0.006 m;

Temperature of the oil T_∞ = 150°C;

Initial temperature of the vegetable slice T_0 = 22°C;

Time t = 3 minutes = 180 sec;

Heat transfer coefficient h = 31 W/m²°C;

Density of the vegetable slice ρ = 950 kg/m³;

Specific heat C_p = 2.25 kJ/kg°C = 2250 J/kg°C;

Thermal conductivity k = 5.7 W/m°C

T_t = ?

Step II

Characteristic length $L_c=\dfrac{V\,(volume)}{A_s\,(surface\ area)}=\dfrac{A\,L}{2\,A}=\dfrac{L}{2}=0.003$ m;

Step III

Biot number $B_i=\dfrac{hL_c}{k}=\dfrac{31\times0.003}{5.7}=0.016$

As B_i < 0.1, the internal thermal resistance for conduction heat flow can be neglected and Lumped capacitance method can be applied

Fourier number $F_o=\dfrac{\alpha t}{L_c^2}=\dfrac{k\,t}{\rho\,C_p\,L_c^2}=\dfrac{5.7\times180}{950\times2250\times(0.003^2)}=53.33$

Step IV

The time versus temperature relation is given by

$$\frac{T_\infty-T_t}{T_\infty-T_0}=exp\left[-\frac{hA_s}{\rho C_p V}t\right]=exp\left[-B_iF_0\right]$$

$$\frac{150-T_t}{150-22} = exp\left[-0.016 \times 53.33\right] = 0.426$$

$$150 - T_t = 54.527$$

$$T_t = 95.473°C$$

Case 4: Heat transfer in spherical sweet potato ball

For heat transfer purposes, a spherical sweet potato ball of 40 mm diameter was dropped into hot oil at 180°C. The temperature of the ball was measured by two thermocouples, one located in the center and the other near the surface. The specific heat of the product was 2.9 kJ/kg°C and density was 950 kg/m³. Both the thermocouples registered the same temperature at a given instance. In one test, the initial temperature of the ball was 10°C and after 1.31 minutes the temperature was recorded as 124°C. Calculate the heat transfer coefficient for this case.

Step I

Given
Diameter of sweet potato ball D = 40 mm = 0.04 m;
Temperature of the oil T_∞ = 180°C;
Initial temperature of the sweet potato ball T_0 = 10°C;
Temperature of the sweet potato ball at time 1.31 minutes is T_t = 124°C;
Density of the meat ball ρ = 950 kg/m³;
Specific heat C_p = 2.9 kJ/kg°C = 2900 J/kg°C;

Step II

Characteristic length $L_c = \dfrac{V \, (volume)}{A_s \, (surface\ area)} = \dfrac{\frac{4}{3}\pi R^3}{4\pi R^2} = \dfrac{R}{3} = \dfrac{D}{2 \times 3} = \dfrac{0.04}{6} = 0.0067$ m;

Step III

The time versus temperature relation was given by

$$\frac{T_\infty - T_t}{T_\infty - T_0} = exp\left[-\frac{hA_s}{\rho C_p V}t\right] = exp\left[-\frac{h}{\rho C_p L_c}t\right]$$

$$\frac{180-124}{180-10} = exp\left[-\frac{h}{950 \times 2900 \times 0.0067}(1.31 \times 60)\right]$$

$$0.329 = exp[-0.004h]$$

$$h = \frac{1.1301}{0.004}$$

$$h = 260.78 \ \text{W/m}^2°C$$

Case 5: Deep frying of mashed potato balls

The mashed potato ball initially at a temperature of 18°C is suddenly immersed in hot oil at 155°C. The thermal properties of mashed potato ball are:
k = 1.16 W/m°C,
c_p = 2896 J/kgK
Mass of each ball m = 16 gm

Time, s	0	10	20	30	40	50	60	70
Temperature, °C	22	29.54	37.08	44.62	52.16	59.7	67.24	74.78

Time, s	80	90	100	110	120	130	140
Temperature, °C	82.32	88.92	95.52	102.12	108.72	115.32	121.92

Step I

Given
Spherical potato balls of diameter D = 7 cm = 0.07 m;
Temperature of the oil T_∞ = 155°C;
Initial temperature of the potato ball T_0 = 18°C;
Specific heat C_p = 2896 J/kg °C;
Thermal conductivity k = 1.16 W/m °C
Mass of potato ball m = 16 g = 0.016 kg

Step II

Characteristic length $L_c = \dfrac{V\,(volume)}{A_s\,(surface\ area)} = \dfrac{\frac{4}{3}\pi R^3}{4\pi R^2} = \dfrac{R}{3} = \dfrac{D}{2 \times 3} = \dfrac{0.07}{6} = 0.0116$ m;

Step III

Density $\rho = \dfrac{m\,(Mass)}{V\,(volume)} = \dfrac{m}{\frac{4}{3}\pi R^3} = \dfrac{0.016}{\frac{4}{3} \times 3.14 \times 0.035^2}$ 81.98 kg/m³

Step IV

The variation of Temperature ratio (TR) with respect to time is plotted in order to obtain the slope (Fig. 2)
Y = –0.0089x; R² = 0.9681

$$\ln(TR) = \ln\left(\frac{T_\infty - T_t}{T_\infty - T_o}\right) = -\frac{h}{\rho C_P L_c}t$$

$$Slope = -\frac{h}{\rho C_P L_c} = -0.0089$$

$$\frac{h}{81.98 \times 2896 \times 0.012} = 0.0089$$

h = 25.36 W/m²°C

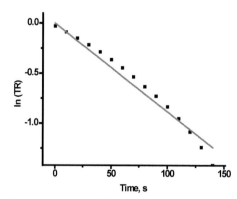

Figure 2: Temperature ratio and time plot for frying of mashed potato balls.

Case 6: Deep frying of Rice kernel

A rice grain of diameter 2 mm and length 6 mm, with an initial temperature of 18°C, is fried in oil at a temperature of 155°C. Find the heat transfer coefficient of rice grain and calculate the density, specific heat and thermal conductivity of the rice grain using empirical formulas. The nutritional composition of rice is given in the table

Moisture	Protein	Fat	Carbohydrate	Ash
12%	6.7%	0.4%	80.4%	0.5%

Time, s	0	10	20	30	40	50	60	70	80	90	100	110	120	130	140
Temperature, °C	18	27	36	45	54	62	70	78	86	94	101	108	115	120	125

Step I

Given

Diameter of rice grain D = 2 mm = 0.00 2 m and length is 6 mm;

Temperature of the oil T_∞ = 155°C;

Initial temperature of the potato ball T_0 = 18°C;

Composition of nutrient in rice:

Moisture = 12%

Protein = 6.7%

Fat = 0.4%

Carbohydrate = 80.4%

Ash = 0.5%

Step II

Thermal conductivity (k), specific heat (C_p) and density (ρ) of rice grain calculated empirically from Eqs. 1, 2 and 3 respectively

$k \quad = \Sigma k_i M_i = 0.251$ W/m°C

$C_p = \Sigma C_{pi} M_i = 1948$ J/kg°C

$\rho \quad = \Sigma \rho_i M_i = 1504.41$

Step III

Characteristic length $L_c = \dfrac{V\,(volume)}{A_s(surface\ area)} = \dfrac{\pi R^2 L}{2\pi R L} = \dfrac{R}{2} = \dfrac{D}{2 \times 2} = \dfrac{0.002}{4} = 0.0005$ m;

Step IV

The variation of Temperature ratio (TR) with respect to time is plotted in order to obtain the slope (Fig. 3). Y = −0.0097x; R^2 = 0.9694

$$\ln(TR) = \ln\left(\frac{T_\infty - T_t}{T_\infty - T_0}\right) = -\frac{h}{\rho C_p L_c} t$$

$$Slope = -\frac{h}{\rho C_p L_c} = -0.0097$$

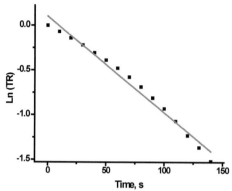

Figure 3: Temperature ratio and time plot for deep fat frying of rice kernel.

$$\frac{h}{154.41 \times 1948 \times 0.0005} = 0.0097$$

$$h = 14.21 \text{ W/m}^2\text{°C}$$

7. Transient Heat Conduction in Different Geometries when Bi > 0.1

In transient heat transfer, the temperature in a body keeps on changing with time. Transient analysis is important for calculations of heating and cooling processes in food processing. In the frying process, heat is conducted through the food as it is immersed in oil. The internal temperature of the food, therefore, is usually increasing during frying. Figure 1 shows the schematic diagram for transient heat transfer in plane wall or slab, cylinder and sphere.

Assumptions

i. The heat and mass transfer parameters and physical properties of the products are constant.
ii. The products are homogeneous and isotropic.
iii. The interacting effects between heat transfer and moisture transfer are negligible.
iv. The initial temperatures and moisture contents of the products are uniform, and the medium temperature is constant.
v. The effects of the internal heat generation and oil-uptake on the temperature distribution and moisture distribution are negligible.

Basic Solution Techniques

To solve the unsteady state heat transfer equations in stationary solids, the following dimensionless parameters are required.

i. $\theta_{slab} = \dfrac{T_\infty - T(x,t)}{T_\infty - T_o}$ Dimensionless Temperature

ii. $X = \dfrac{x}{L}$ Dimensionless distance from center

iii. $F_o = \dfrac{\alpha t}{L^2}$ Dimensionless time

iv. $Bi = \dfrac{hL}{k}$ Dimensionless heat transfer coefficient

In this case, in order to evaluate the Biot number for a plane wall, half of the thickness is taken as the characteristic length and for long cylinder and sphere, the radius is considered as the characteristic length.

7.1 Transient heat transfer equation in plane wall or slab

Consider a plane wall of thickness 2L, and there is thermal symmetry about the mid plane passing through x = 0, thus, the temperature distribution must be symmetrical about the mid plane. In order to characterize this transient behavior, the governing unsteady equations for stationary solids are presented in Eq. (37). Rectangular coordinates:

$$\frac{\partial T}{\partial t} = \alpha \left[\frac{\partial^2 T}{\partial x^2} + \frac{\partial^2 T}{\partial x^2} + \frac{\partial^2 T}{\partial x^2} \right] + \frac{A}{\rho c_p} \tag{37}$$

where $\alpha = \dfrac{k}{\rho c_p}$ is the thermal diffusivity of the food material,

ρ = density of the food material,

k = thermal conductivity of the food material,

c_p = specific heat of food material,

$\dfrac{A}{\rho c_p} = q_{gen}$ = heat generated per unit volume.

In most of the food processing operation, the heat generation q_{gen} is equal to zero.

The height and the width of the wall are large, relative to its thickness, and thus, heat conduction in the wall can be approximated to be one-dimensional. Hence, considering spatial variation of temperature only in x-direction, i.e., T = T(x, t) temperature only depends on x and t, the Eq. (37) reduces to the following equation.

$$\frac{\partial T}{\partial t} = \alpha \frac{\partial^2 T}{\partial x^2}$$

At time t = 0, the wall is immersed in a fluid at temperature T_∞ and is subjected to convection heat transfer from both sides with a convection coefficient of h.

Initial condition

$T(x, t = 0) = T_0 = constant$

Boundary conditions:

$k \dfrac{\partial T}{\partial x} \bigg|_{x=0} = 0$

$T(x = L/2, t) = T_\infty = constant$

$k \dfrac{\partial T}{\partial x} \bigg|_{x=L/2} = h(T_\infty - T(x = \frac{L}{2}, t))$

T_0 = Initial temperature of the plane wall at t = 0,

T_∞ = Temperature of the plane wall at t = ∞,

T_t = Temperature of the plane wall at t = t,

For an infinite slab, Infinite heat-transfer coefficient: Bi > 40

With uniform initial temperature and constant surface temperature, negligible surface resistance

$$\theta_{slab} = \frac{T_\infty - T(x,t)}{T_\infty - T_o} = \sum_{m=1}^{\infty} \frac{2(-1)^{m+1}}{\lambda_m} \cos\left(\lambda_m \frac{x}{L} \right) exp(-\lambda_m^2 F_o)$$

$$\lambda_m = (2m-1)\frac{\pi}{2} \tag{38}$$

$$F_o = \frac{\alpha t}{\delta^2} = \frac{\alpha t}{L^2}$$

Here, δ is characteristic dimension for an infinite slab.

For an infinite slab, Finite heat-transfer coefficient: 1 < Bi < 40

With uniform initial temperature and both surface and internal resistances are significant, i.e., $1 < Bi < 40$

$$\theta_{slab} == \frac{T_\infty - T(x,t)}{T_\infty - T_0} = \sum_{m=1}^{\infty} \frac{2\sin(\beta_n)}{\beta_n + \sin(\beta_n)\cos(\beta_n)} \cos\left(\beta_n \frac{x}{L}\right) \exp(-\beta_n^2 F_o) \qquad (39)$$

Where is β_n the nth positive root of $\beta \tan \beta = Bi_{Slab}$

$$Bi_{Slab} = \frac{h\delta}{k} = \frac{hL}{k}$$

The series solutions, Eq. (39), converge rapidly to the first term beyond a F_o value of 0.2 and the resulting equation is presented in the following equation.

$$\theta_{slab} = \frac{T_\infty - T(x,t)}{T_\infty - T_0} = \frac{2\sin(\beta_1)}{\beta_1 + \sin(\beta_1)\cos(\beta_1)} \cos\left(\beta_1 \frac{x}{L}\right) \exp(-\beta_1^2 F_o) \qquad (40)$$

$$\theta_{slab} = \frac{T_\infty - T(x,t)}{T_\infty - T_0} = A_1 \cos\left(\beta_1 \frac{x}{L}\right) \exp(-\beta_1^2 F_o) \qquad (41)$$

At the center of the plane wall x = 0, cos 0 = 1

$$\theta_{slab,x=0} = \frac{T_\infty - T}{T_\infty - T_0} = A_1 \exp(-\beta_1^2 F_o) \qquad (42)$$

7.2 Transient heat transfer equation in an infinite cylinder

The basic governing partial differential equation in cylindrical coordinates to describe the temperature distribution in an infinite solid cylinder of finite radius R but infinite length is presented in Eq. (43)

Cylindrical coordinates:

$$\frac{\partial T}{\partial t} = \alpha \left[\frac{1}{r} \frac{\partial}{\partial x}\left(r \frac{\partial T}{\partial r}\right) + \frac{1}{r^2} \frac{\partial^2 T}{\partial \varnothing^2} + \frac{\partial^2 T}{\partial z^2} \right] + \frac{A}{\rho C_p} \qquad (43)$$

$$\frac{1}{\alpha} \frac{\partial T}{\partial t} = \frac{1}{r} \frac{\partial}{\partial r}\left(r \frac{\partial T}{\partial r}\right) \qquad (44)$$

$$\frac{1}{\alpha} \frac{\partial T}{\partial t} = \frac{\partial^2 T}{\partial r^2} + \frac{1}{r} \frac{\partial T}{\partial r} \qquad (45)$$

Initial Condition

$T(r, t = 0) = T_0$ = constant

Boundary conditions:

$$k \frac{\partial T}{\partial t}\bigg|_{r=0} = 0$$

$T(r = R, t) = T_\infty$ = constant

$$k \frac{\partial T}{\partial t}\bigg|_{r=R} = h(T_\infty - T(r = R, t))$$

For an infinite cylinder, Infinite heat-transfer coefficient, if Bi > 40

$$\theta_{infinite\,cylinder} = \frac{T_\infty - T(r,t)}{T_\infty - T_0} = \sum_{n=1}^{\infty} \frac{2}{\beta_n J_1(\beta_n)} J_0\left(\beta_n\left(\frac{r}{R}\right)\right) exp(-\beta_n^2 Fo_c) \qquad (46)$$

where β_n is the nth positive root of $J_0(\beta_n) = 0$

Considering the first term of the series:

$$\theta_{infinite\,cylinder} = \frac{T_\infty - T(r,t)}{T_\infty - T_0} = \frac{2}{\beta_1 J_1(\beta_1)} J_0\left(\beta_1\left(\frac{r}{R}\right)\right) exp(\beta_1^2 Fo_c) \qquad (47)$$

$$\theta_{infinite\,cylinder} = A_1 J_0\left(\beta_1\left(\frac{r}{R}\right)\right) exp(-\beta_1^2 Fo_c) \qquad (48)$$

At center of the cylinder $J_0(0) = 1$

Hence, $\theta_{infinite\,cylinder} = \dfrac{T_\infty - T(r,t)}{T_\infty - T_0} = \dfrac{2}{\beta_1 J_1(\beta_1)} exp(-\beta_1^2 Fo_c) = A_1 exp(-\beta_1^2 Fo_c) \qquad (49)$

For an infinite cylinder, Finite heat-transfer coefficient: 0.1 < Bi < 40

$$\theta_{infinite\,cylinder} = \frac{T_\infty - T(r,t)}{T_\infty - T_0} = \sum_{n=1}^{\infty} \frac{2Bi}{(\beta_n^2 + \beta i^2)} \frac{J_0\left(\beta_n\left(\frac{r}{R}\right)\right)}{J_0(\beta_n)} exp(-\beta_n^2 Fo_c) \qquad (50)$$

$$Bi = \frac{hR}{k}$$

β_n is the positive root of $\beta_n J_1(\beta_n) = Bi\, J_0(\beta_n)$ and, hence, $Bi = \dfrac{\beta_n J_1(\beta_n)}{J_0(\beta_n)}$

The series solutions, Eq. (50), converge rapidly to the first term beyond a Fo value of 0.2 and the resulting equation is presented in the following equation.

$$\theta_{infinite\,cylinder} = \frac{T_\infty - T(r,t)}{T_\infty - T_0} = \frac{2Bi}{(\beta_1^2 + \beta i^2)} \frac{J_0\left(\beta_1\left(\frac{r}{R}\right)\right)}{J_0(\beta_1)} exp(-\beta_1^2 Fo) \qquad (51)$$

At center

$$\theta_{infinite\,cylinder} = \frac{T_\infty - T(r,t)}{T_\infty - T_0} = \frac{2Bi}{(\beta_1^2 + \beta i^2)} \frac{1}{J_0(\beta_n)} exp(-\beta_1^2 Fo) \qquad (52)$$

Table 1: The first five roots of the Bessel function.

m	$J_0(x)$	$J_1(x)$	$J_2(x)$	$J_3(x)$	$J_4(x)$	$J_5(x)$
1	2.4048	3.8317	5.1356	6.3802	7.5883	8.7715
2	5.5201	7.0156	8.4172	9.761	11.0647	12.3386
3	8.6537	10.1735	11.6198	13.0152	14.3725	15.7002
4	11.7915	13.3237	14.796	16.2235	17.616	18.9801
5	14.9309	16.4706	17.9598	19.4094	20.8269	22.2178

Bessel function for cylindrical geometry

For frying of food having cylindrical geometry, the linear second-order equation representing the temperature distribution with time and distance involves Bessel's equation. The analytical solution of this equation requires the use of Bessel functions designated $J_n(x)$. The first few roots of the Bessel function are given in the following table for small nonnegative integer values of n and k.

In the above table

The m-th roots of $J_n(x) = 0$.

The 1st root of $J_0(x) = 0$ at x value of 2.404

The 2nd root of $J_0(x) = 0$ at x value of 5.520

The 1st root of $J_1(x) = 0$ at x value of 3.831

The 2nd root of $J_1(x) = 0$ at x value of 7.015

Bessel function of zero order and first order:

For zero order, the first solution of Bessel's equation is given by Eq. (53).

$$J_o(x) = \sum_{r=0}^{\infty} \frac{(-1)^r \left(\frac{x}{2}\right)^{2r}}{r!\,\Gamma(r+1)} = 1 - \frac{x^2}{2^2} + \frac{x^4}{2^2 4^2} - \frac{x^6}{2^2.4^2.6^2} + \ldots \tag{53}$$

$$J_1(x) = \sum_{r=0}^{\infty} \frac{(-1)^r \left(\frac{x}{2}\right)^{2r+1}}{r!\,(r+1)!} = \frac{x}{2} - \frac{x^3}{2^3 1! 2!} + \frac{x^5}{2^5 2! 3!} - \ldots \tag{54}$$

In Excel, the syntax of the BESSELJ function is BESSELJ(x, n).

The value of 'x' in this argument is the value wherein the Bessel function is evaluated. The 'n' in the function is the order of the Bessel function.

Transcendental function

If $f(x)$ is a transcendental function, then the equation $f(x) = 0$ is called a transcendental equation.

7.3 Transient heat transfer equation in sphere

Spherical coordinates:

$$\frac{\partial T}{\partial t} = \alpha \left[\frac{1}{r^2} \frac{\partial}{\partial x}\left(r^2 \frac{\partial T}{\partial r}\right) + \frac{1}{r^2 \sin\theta} \frac{\partial}{\partial x}\left(\sin\theta \frac{\partial T}{\partial \theta}\right) + \frac{1}{r^2 \sin^2\theta} \frac{\partial^2 T}{\partial \varnothing^2} + \frac{\partial^2 T}{\partial z^2} \right] + \frac{A}{\rho C_p} \tag{55}$$

The one-dimensional, unsteady state, temperature distribution for a solid sphere of radius R can be presented by the following differential equation.

$$\frac{1}{\alpha} \frac{\partial T}{\partial t} = \frac{1}{r} \frac{\partial}{\partial r}\left(r \frac{\partial T}{\partial r}\right) \tag{56}$$

$$\frac{1}{\alpha} \frac{\partial T}{\partial t} = \frac{\partial^2 T}{\partial r^2} + \frac{1}{r} \frac{\partial T}{\partial r} \tag{57}$$

Initial Condition

$T(r, t = 0) = T0 = $ constant

Boundary conditions

$$k\frac{\partial T}{\partial t}\bigg|_{r=0} = 0$$

$$T(r = R, t) = T_\infty = \text{constant}$$

$$k\frac{\partial T}{\partial t}\bigg|_{r=R} = h(T_\infty - T(r = R, t))$$

For sphere, Infinite heat-transfer coefficient: if Bi > 40

$$\theta_{sphere} = \frac{T_\infty - T(r,t)}{T_\infty - T_0} = \frac{2R}{\pi r}\sum_{n=1}^{\infty}\frac{(-1)^{n+1}}{n}\sin\left(\frac{n\pi r}{R}\right)\exp(-n^2\pi^2 Fo_s) \qquad (58)$$

At the center of the sphere, the limit of sin(x)/x is also 1

$$\theta_{sphere} = \frac{T_\infty - T(r,t)}{T_\infty - T_0} = \frac{2R}{\pi r}\sum_{n=1}^{\infty}\frac{(-1)^{n+1}}{n}\exp(-n^2\pi^2 Fo_s) \qquad (59)$$

For sphere, Finite heat-transfer coefficient: 0.1 < Bi < 40

Finite Internal and Surface Resistance to Heat Transfer

$$\theta_{sphere} = \frac{T_\infty - T(r,t)}{T_\infty - T_0} = \sum_{n=1}^{\infty}\frac{\sin(\beta_n) - \beta_n\cos(\delta\beta_n)}{\beta_n - \sin(\beta_n)\cos(\beta_n)}\frac{\sin\left(\beta_n\frac{r}{R}\right)}{\beta_n\frac{r}{R}}\exp(-\beta_n^2 Fo_s) \qquad (60)$$

$$\theta_{sphere} = \frac{T_\infty - T(r,t)}{T_\infty - T_0} = \sum_{n=1}^{\infty}A_n\frac{\sin\left(\beta_n\frac{r}{R}\right)}{\beta_n\frac{r}{R}}\exp(\overline{n}\beta_n^2 Fo_s) \qquad (61)$$

where β_n is the nth positive root of $\beta_n\cot(\beta_n) = 1 - Bi$

The series solutions, Eq. (61), converge rapidly to the first term beyond a Fo value of 0.2, and the resulting equation is presented in the following equation.

$$\theta_{sphere} = \frac{T_\infty - T(r,t)}{T_\infty - T_0} = \frac{\sin(\beta_1) - \beta_1\cos(\delta\beta_1)}{\beta_1 - \sin(\beta_1)\cos(\beta_1)}\frac{\sin\left(\beta_1\frac{r}{R}\right)}{\beta_1\frac{r}{R}}\exp(-\beta_1^2 Fo_s) \qquad (62)$$

$$\theta_{sphere} = \frac{T_\infty - T(r,t)}{T_\infty - T_0} = A_1\frac{\sin\left(\beta_1\frac{r}{R}\right)}{\beta_1\frac{r}{R}}\exp(-\beta_1^2 Fo_s) \qquad (63)$$

At the center of the sphere, the limit of sin(x)/x is also 1

$$\theta_{sphere} = \frac{T_\infty - T(r,t)}{T_\infty - T_0} = A_1\exp(-\beta_1^2 Fo_s) \qquad (64)$$

For Finite heat-transfer coefficient; 1 < Bi < 40

Equations (40), (51) and (62) can be generalized for the first root situation and written as:

$$\theta = A \exp(-B\,Fo) \tag{65}$$

where A and B are shaped dependent functions of Biot number based on the first root of characteristic functions and the expression for A is presented in Table 2 below.

Table 2: Mathematical expression for "A" for infinite slab, infinite cylinder and sphere.

	A	A at center	Transcendental equation
Infinite Slab	$\dfrac{2\sin(\beta_1)}{\beta_1 + \sin(\beta_1)\cos(\beta_1)}\cos\left(\beta_1\dfrac{x}{L}\right)$	$\dfrac{2\sin(\beta_1)}{\beta_1 + \sin(\beta_1)\cos(\beta_1)}$	$\beta\tan\beta = Bi_p$
Infinite Cylinder	$\dfrac{2Bi}{(\beta_1^2 + Bi^2)}\dfrac{J_0(\beta_1(\frac{r}{R}))}{J_0(\beta_1)}$	$\dfrac{2Bi}{(\beta_1^2 + Bi^2)}\dfrac{1}{J_0(\beta_1)}$	$\dfrac{\beta J_1(\beta)}{J_0(\beta)} = Bi_c$
Sphere	$\dfrac{\sin(\beta_1) - \beta_1\cos(\delta\beta_1)}{\beta_1 - \sin(\beta_1)\cos(\beta_1)}\dfrac{\sin\left(\beta_1\frac{r}{R}\right)}{\beta_1\frac{r}{R}}$	$\dfrac{\sin(\beta_1) - \beta_1\cos(\delta\beta_1)}{\beta_1 - \sin(\beta_1)\cos(\beta_1)}$	$1 - \beta\cot(\beta) = 2Bi_s$

References

Abtahi, M.S., Hosseini, H., Fadavi, A., Mirzaei, H. and Rahbari, M. 2016. The optimization of the deep-fat frying process of coated zucchini pieces by response surface methodology. Journal of Culinary Science and Technology 14: 176–189.

Ahromrit, A. and Nema, P.K. 2010. Heat and mass transfer in deep-frying of pumpkin, sweet potato and taro. Journal of Food Science and Technology 47: 632–637.

Alvis, A., Vélez, C., Rada-Mendoza, M., Villamiel, M. and Villada, H.S. 2009. Heat transfer coefficient during deep-fat frying. Food Control 20: 321–325.

Baik, O.D. and Mittal, G.S. 2002. Heat transfer coefficients during deep-fat frying of a tofu disc. Transactions of the ASAE 45: 14–93.

Baumann, B. and Escher, F. 1995. Mass and heat transfer during deep-fat frying of potato slices—I. Rate of drying and oil uptake. LWT-Food Science and Technology 28: 395–403.

Blumenthal, M.M. and Stier, R.F. 1991. Optimization of deep-fat frying operations. Trends in Food Science & Technology 2: 144–148.

Budžaki, S. and Šeruga, B. 2005. Determination of convective heat transfer coefficient during frying of potato dough. Journal of Food Engineering 66: 307–314.

Choi, Y. 1986. Effects of temperature and composition on the thermal properties of food. Food Engineering and Process. 1: 93–101.

Debnath, S., Rastogi, N.K., Krishna, A.G. and Lokesh, B.R. 2012. Effect of frying cycles on physical, chemical and heat transfer quality of rice bran oil during deep-fat frying of poori: An Indian traditional fried food. Food and Bioproducts Processing 90: 249–256.

Dincer, I. 1996. Modelling for heat and mass transfer parameters in deep-frying of products. Heat and Mass Transfer 32: 109–113.

Eljeeva, M., Franklin, E., Pushpadass, H.A., Menon, R.R., Rao, K.J. and Nath, B.S. 2013. Modeling the heat and mass transfer during frying of gulab jamun. Journal of Food Processing and Preservation 1–9.

Farid, M. and Kizilel, R. 2009. A new approach to the analysis of heat and mass transfer in drying and frying of food products. Chemical Engineering and Processing: Process Intensification 48: 217–223.

Farinu, A. and Baik, O.D. 2005. Deep fat frying of foods—Transport Phenomena. Food Reviews International 21: 389–410.

Farinu, A. and Baik, O.D. 2007. Heat transfer coefficients during deep fat frying of sweet potato: Effects of product size and oil temperature. Food Research International 40: 989–994.

Heldman, D.R. and Gorby, D.P. 1975. Prediction of thermal conductivity in frozen food. ASAE Trans 18: 156.

Holman, S.J., Robinson, R.A., Beardsley, D., Stewart, S.F.C., Klein, L. and Stevens, R.A. 1997. Hyperbaric dye solution distribution characteristics after pencil-point needle injection in a spinal cord model. Anesthesiology: The Journal of the American Society of Anesthesiologists 86: 966–973.

Hubbard, L.J. and Farkas, B.E. 1999. A method for determining the convective heat transfer coefficient during immersion frying. Journal of Food Process Engineering 22: 201–214.

Kita, A., Lisińska, G. and Gołubowska, G. 2007. The effects of oils and frying temperatures on the texture and fat content of potato crisps. Food Chemistry 102: 1–5.

Kopelman, I.J. and Pflug, I.J. 1968. The relationship of the surface, mass average and geometric center temperatures in transient conduction heat flow. Food Technology 22: 799–804.

Kurek, M., Ščetar, M. and Galić, K. 2017. Edible coatings minimize fat uptake in deep fat fried products: A review. Food Hydrocolloids 71: 225–235.

Mondal, I.H. and Dash, K.K. 2017. Textural, color kinetics and heat and mass transfer modeling during deep fat frying of chhena jhili. 1–13.

Moreira, R., Palau, J. and Sun, X. 1995. Simultaneous heat and mass transfer during the deep fat frying of tortilla chips. Journal of Food Process Engineering 18: 307–320.

Moreira, R., Castell-Perez, M.E., Barrufet, M.A. and Yanniotis, S. 2004. Deep fat frying. 210–236. Heat Transfer in Food Processing: Recent Developments and Applications. WIT Press, Southampton, Boston.

Oyedeji, A.B., Sobukola, O.P., Henshaw, F.O., Adegunwa, M.O., Sanni, L.O. and Tomlins, K.I. 2016. Kinetics of mass transfer during deep fat frying of yellow fleshed cassava root slices. Heat and Mass Transfer/Waerme-Und Stoffuebertragung 52: 1061–1070.

Pedreschi, F., Kaack, K. and Granby, K. 2006. Acrylamide content and color development in fried potato strips. Food Research International 39: 40–46.

Rimac-Brnčić, S., Lelas, V., Rade, D. and Šimundić, B. 2004. Decreasing of oil absorption in potato strips during deep fat frying. Journal of Food Engineering 64: 237–241.

Sahin, S., Sastry, S.K. and Bayindirli, L. 1999. Heat transfer during frying of potato slices. LWT-Food Science and Technology 32: 19–24.

Torres-Gonzalez, J.D., Alvis-Bermudez, A., Gallo-Garcia, L.A., Acevedo-Correa, D., Castellanos-Galeano, F. and Bouchon-Aguirre, P. 2018. Effect of deep fat frying on the mass transfer and color changes of arepa con huevo. Indian Journal of Science and Technology 11: 6.

van Koerten, K.N., Somsen, D., Boom, R.M. and Schutyser, M.A.I. 2017. Modelling water evaporation during frying with an evaporation dependent heat transfer coefficient. Journal of Food Engineering 197: 60–67.

Vitrac, O., Trystram, G. and Raoult-Wack, A.L. 2000. Deep-fat frying of food: Heat and mass transfer, transformations and reactions inside the frying material. European Journal of Lipid Science and Technology 102: 529–538.

Weisshaar, R. 2014. Quality control of used deep-frying oils. European Journal of Lipid Science and Technology 116: 716–722.

Index

Color Section

Chapter 3: Fig. 7, p. 33

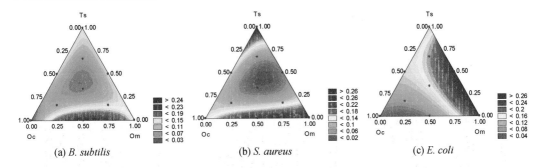

(a) *B. subtilis* (b) *S. aureus* (c) *E. coli*

Chapter 5: Fig. 9, p. 61

Chapter 7: Fig. 11, p. 93

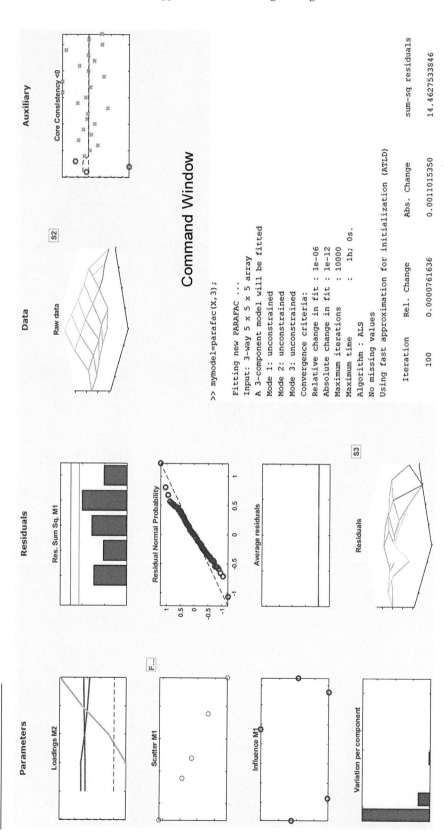

Chapter 8: Fig. 4, p. 106

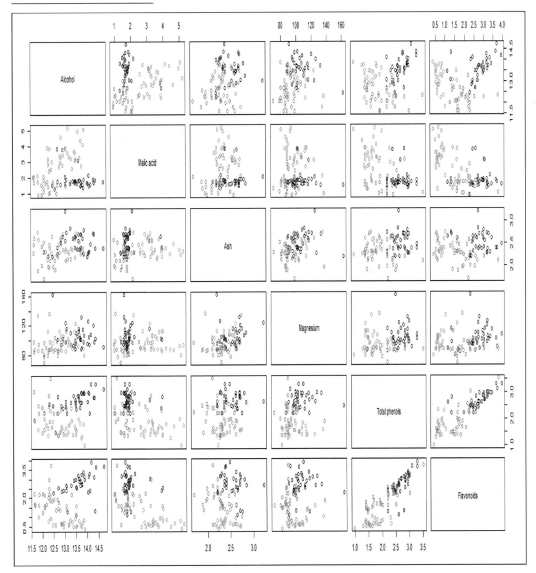

Chapter 10: Fig. 2, p. 133

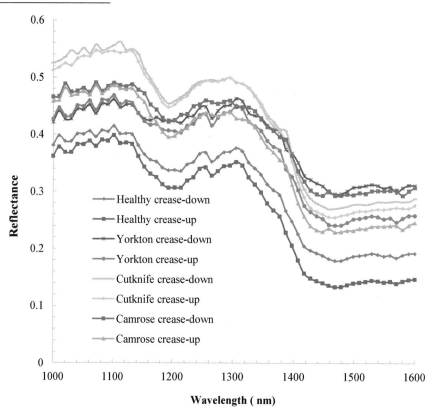

Chapter 10: Fig. 3, p. 135

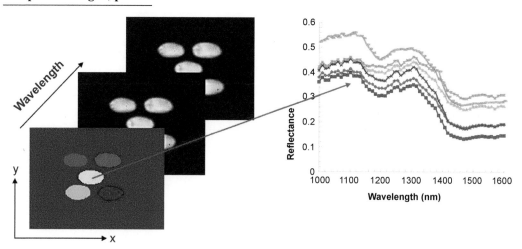

Chapter 10: Fig. 4, p. 136

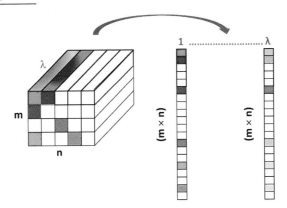

Chapter 10: Fig. 5, p. 137

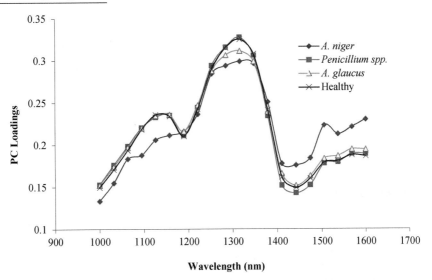

Chapter 10: Fig. 6, p. 137

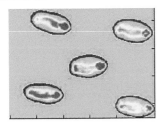
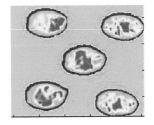
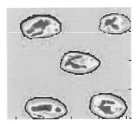

Healthy kernels   Sprouted kernels   Midge damaged kernels

Chapter 13: Fig. 3, p. 191

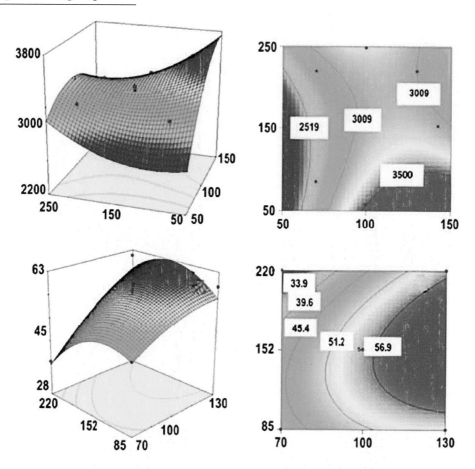

Chapter 14: Fig. 2, p. 209

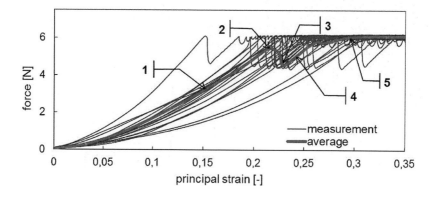

Chapter 16: Fig. 3, p. 254

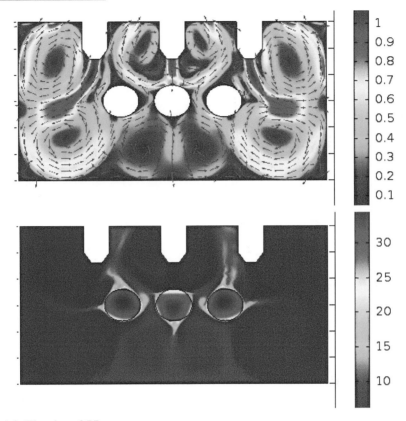

Chapter 16: Fig. 4, p. 255

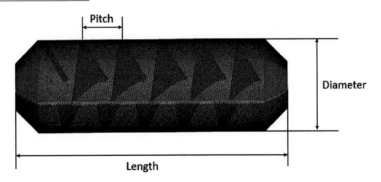

Chapter 23: Fig. 8, p. 378

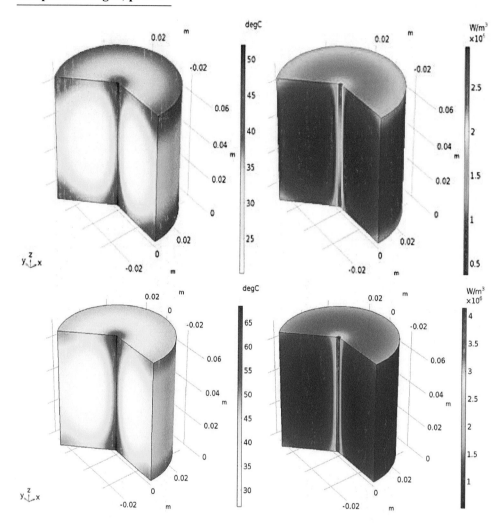

Chapter 23: Fig. 9, p. 379

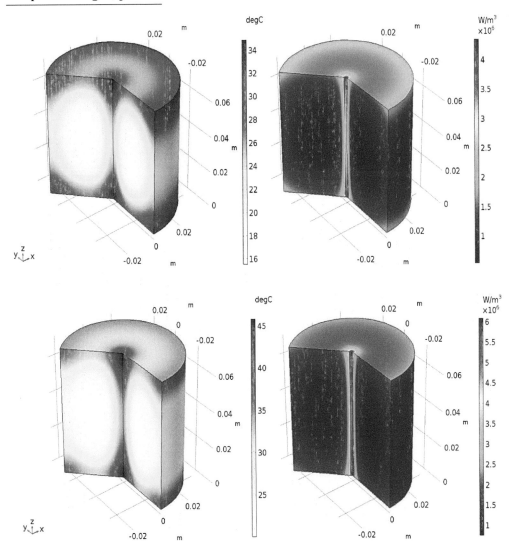

Chapter 23: Fig. 16, p. 384

(a)

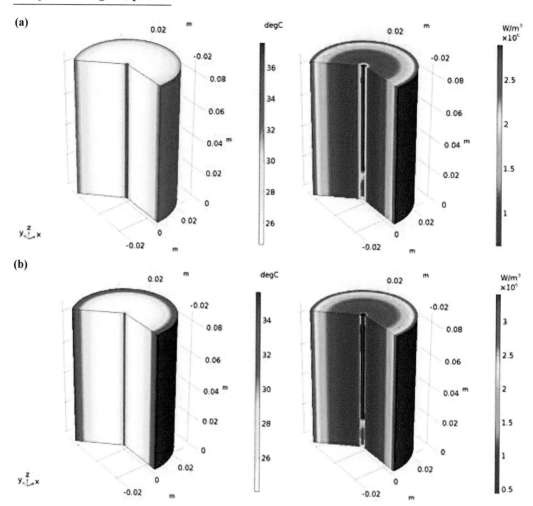

(b)